AF782560

From Ion Channels to Cell-to-Cell Conversations

Series of the Centro de Estudios Científicos de Santiago

Series Editor: Claudio Teitelboim
Centro de Estudios Científicos de Santiago
Santiago, Chile
and Facultad de Ciencias
Universidad de Chile
Santiago, Chile

FROM ION CHANNELS TO CELL-TO-CELL CONVERSATIONS
Edited by Ramon Latorre and Juan Carlos Sáez

IONIC CHANNELS IN CELLS AND MODEL SYSTEMS
Edited by Ramon Latorre

PHYSICAL PROPERTIES OF BIOLOGICAL MEMBRANES AND THEIR FUNCTIONAL IMPLICATIONS
Edited by Cecilia Hidalgo

PRINCIPLES OF STRING THEORY
Lars Brink and Marc Henneaux

QUANTUM MECHANICS OF FUNDAMENTAL SYSTEMS 1
Edited by Claudio Teitelboim

QUANTUM MECHANICS OF FUNDAMENTAL SYSTEMS 2
Edited by Claudio Teitelboim and Jorge Zanelli

QUANTUM MECHANICS OF FUNDAMENTAL SYSTEMS 3
Edited by Claudio Teitelboim and Jorge Zanelli

TRANSDUCTION IN BIOLOGICAL SYSTEMS
Edited by Cecilia Hidalgo, Juan Bacigalupo, Enrique Jaimovich, and Julio Vergara

From Ion Channels to Cell-to-Cell Conversations

Edited by

Ramón Latorre

Centro de Estudios Científicos de Santiago
and Facultad de Ciencias
Universidad de Chile
Santiago, Chile, and
University of California at Los Angeles
Los Angeles, California

and

Juan Carlos Sáez

Universidad Católica
Santiago, Chile, and
Albert Einstein College of Medicine
Bronx, New York

PLENUM PRESS • NEW YORK AND LONDON

Library of Congress Cataloging-in-Publication Data

From ion channels to cell-to-cell conversations / edited by Ramon Latorre and Juan Carlos Sáez.
p. cm. -- (Series of the Centro de Estudios Cientificos de Santiago)
Includes bibliographical references and index.
ISBN 0-306-45605-2
1. Ion channels--Congresses. 2. Gap junctions (Cell biology)--Congresses. I. Latorre, Ramon. II. Sáez, Juan Carlos. III. Series.
QH603.I54F76 1997
571.6'4--dc21 97-13654
CIP

ISBN 0-306-45605-2

© 1997 Plenum Press, New York
A Division of Plenum Publishing Corporation
233 Spring Street, New York, N. Y. 10013

http://www.plenum.com

All rights reserved

10 9 8 7 6 5 4 3 2 1

No part of this book may be reproduced, stored in a retrieval system, or transmitted in any form or by any means, electronic, mechanical, photocopying, microfilming, recording, or otherwise, without written permission from the Publisher

Printed in the United States of America

Contributors

Julio Aguilera, Departamento de Ciencias Fisiológicas, Facultad de Ciencias Biológicas, Pontificia Universidad Católica de Chile, Alameda 340, Santiago, Chile

Richard Aldrich, Department of Molecular and Cellular Physiology, Howard Hughes Medical Institute, Stanford, CA 94305

Juan Bacigalupo, Universidad de Chile, Facultad de Ciencias, Departamento de Biología, Las Palmeras # 3425, Santiago, Chile

P. Baldelli, Department of Neurosciences, University di Torino, Torino, Italy

Thaddeus A. Bargiello, Department of Neuroscience, Albert Einstein College of Medicine, Bronx, NY 10461

Francisco Barrantes, Instituto de Investigaciones Bioquímicas, Universidad Nacional del Sur, Conicet, Camino de La Carrindanga 7 C.C. 857-800, Bahia Blanca, Argentina

B. Beltrán, Instituto de Biotecnologia, Departamento de Genética y Fisiología Molecular, Universidad Nacional Autónoma de México, Apartado 510-3, Cuernavaca, Morelos 62271, Mexico

Michael Bennett, Department of Neuroscience, Albert Einstein College of Medicine, Bronx, NY 10461

Dale Benos, Department of Physiology and Biophysics, The University of Alabama at Birmingham, Birmingham, AL 35294-0005

B. K. Berdiev, Department of Physiology and Biophysics, The University of Alabama at Birmingham, Birmingham, AL 35294-0005

Viviana Berthoud, Department of Pediatrics, Washington University School of Medicine, St. Louis, MO 63110

Eric C. Beyer, Department of Pediatrics, Washington University School of Medicine, St. Louis, MO 63110

Francisco Bezanilla, Department of Physiology and Anesthesiology, UCLA School of Medicine, Los Angeles, CA 90024-1751

Lutz Birnbaumer, Department of Anesthesiology, UCLA School of Medicine, Los Angeles, CA 90024-1778

Mauricio P. Boric, Departamento de Ciencias Fisiológicas, Facultad de Ciencias Biológicas, Pontificia Universidad Católica de Chile, Alameda 340, Santiago, Chile

A. L. Bradford, Department of Physiology and Biophysics, The University of Alabama at Birmingham, Birmingham, AL 35294-0005

J. K. Bubier, Department of Physiology and Biophysics, The University of Alabama at Birmingham, Birmingham, AL 35294-0005

Mateo Budinich, Departamento de Ciencias Fisiológicas, Facultad de Ciencias Biológicas, Pontificia Universidad Católica de Chile, Alameda 340, Santiago, Chile

Angela D. Burgstahler, Liver Study, Yale University School of Medicine, New Haven, CT 06510

Grischa Chandy, Department of Physiology and Biophysics, University of California, Irvine, Irvine, CA 92717

L. Pablo Cid, Centro de Estudios Científicos de Santiago, Avenida Presidente Errázuriz # 3132, Las Condes, Santiago, Chile

Gabriel Cota, Centro de Investigacion y de Estudios Avanzados del IPN, Departamento de Fisiologia Biofisica and Neurosciences, Mexico D.F. C.P. 07000, Mexico

Andrew J. Czernik, Laboratory of Molecular and Cellular Neuroscience, The Rockefeller University, New York, NY 10021-6399

Alberto Darszon, Instituto de Biotecnologia, Departamento de Genética y Fisiología Molecular, Universidad Nacional Autónoma de México, Cuernavaca, Morelos 62271, Mexico

Ricardo Delgado, Centro de Estudios Cientificos de Santiago and Universidad de Chile, Avda Presidente Errázuriz # 3132, Las Condes, Santiago, Chile

Lain Diaz, Centro de Estudios Cientificos de Santiago and Universidad de Chile, Avda Presidente Errázuriz # 3132, Las Condes, Santiago, Chile

A. L. Eguiguren, Centro de Estudios Científicos de Santiago, Avda Presidente Errázuriz # 3132, Las Condes, Santiago, Chile

Eliseo Eugenin, Departamento de Ciencias Fisiologicas, Facultad de Ciencias Biologicas, Pontificia Universidad Católica de Chile, Alameda 340, Santiago, Chile

Michael B. Fallon, Liver Study, Yale University, School of Medicine, New Haven, CT 06510

Glenn I. Fishman, Department of Medicine, Division of Cardiology, Albert Einstein College of Medicine, Bronx, NY 10461

Robert J. French, Department of Medical Physiology, University of Calgary, 3330 Hospital Dr. NW Calgary, Alberta T2N 4N1, Canada

C. M. Fuller, Department of Physiology and Biophysics, The University of Alabama at Birmingham, Birmingham, AL 35294-0005

Yang Gao, Department of Neuroscience, Albert Einstein College of Medicine, Bronx, NY 10461

Sarah S. Garber, Department of Physiology, Allegheny University of the Health Sciences, Philadelphia, PA 19129

Carlos González, Departamento de Fisiología, Universidad Austral de Chile, Valdivia, Chile

James E. Hall, Department of Physiology and Biophysics, University of California, Irvine, Irvine, CA 92717

Marcel Hermoso, Departamento de Ciencias Fisiológicas, Facultad de Ciencias Biológicas, Pontificia Universidad Católica de Chile, Alameda 340, Santiago, Chile

Elliot Hertzberg, Department of Neuroscience, Albert Einstein College of Medicine, Bronx, NY 10461

Patricia Hidalgo, Department of Biological Chemistry and Molecular Pharmacology, Harvard Medical School, Boston, MA 02115

Richard Horn, Department of Physiology, Thomas Jefferson University Medical College, Philadelphia, PA 19107-6799

R. Hurst, Department of Anesthesiology, UCLA School of Medicine, Los Angeles, CA 90024-1778

I. I. Ismailov, Department of Physiology and Biophysics, The University of Alabama at Birmingham, Birmingham, AL 35294-0005

Patricio Jara, Departamento de Ciencias Fisiológicas, Facultad de Ciencias Biológicas, Pontificia Universidad Catolica de Chile, Alameda 340, Santiago, Chile

O. Jorquera, Centro de Estudios Cientificos de Santiago y Universidad de Chile, Avda Presidente Errázuriz # 3132, Las Condes, Santiago, Chile

Nicola Konig, MRC Laboratory of Molecular Biology, Hills Road, CB2 2QH Cambridge, England

Pedro Labarca, Centro de Estudios Cientificos de Santiago and Universidad de Chile, Avda Presidente Errázuriz # 3132, Las Condes, Santiago, Chile

Ramón Latorre, Centro de Estudios Cientificos de Santiago and University of Chile, Avda President Errázuriz # 3132, Las Condes, Santiago, Chile

I. Levitan, Department of Physiology, Allegheny University of the Health Sciences, Philadelphia, PA 19129

A. Liévano, Instituto de Biotecnologia, Departamento de Genética y Fisiologia Molecular, Universidad Nacional Autónoma de México, Cuernavaca, Morelos 62271, Mexico

Donald D. F. Loo, Department of Biology, UCLA School of Medicine, Los Angeles, CA 90024

Roderick MacKinnon, Department of Neurobiology, Harvard Medical School, Boston, MA 02115

R. Madrid, Universidad de Chile, Facultad de Ciencias, Departamento de Biologia, Las Palmeras # 3425, Santiago, Chile

C. Maureira, Centro de Estudios Cientificos de Santiago, Universidad de Chile, Avenida Presidente Errázuriz # 3132, Las Condes, Santiago, Chile

Albert Mennone, Liver Study, Yale University School of Medicine, New Haven, CT 06510

U. Meza, Centro de Investigación y de Estudios Avanzados del IPN, Departamento de Fisiologia, Biofisica and Neurosciences, Cinvestav, México D.F. C.P. 07000, México

Alan Miller, Department of Molecular and Cellular Physiology, Howard Hughes Medical Institute, Stanford, CA 94305

Christopher Miller, Department of Biochemistry–HHMI, Brandeis University, Waltham, MA 02254

Bernardo Morales, Universidad de Chile, Facultad de Ciencias, Departamento de Biologia, Las Palmeras # 3425, Santiago, Chile

Angus Nairn, Laboratory of Molecular and Cellular Neuroscience, The Rockefeller University, New York, NY 10021-6399

David Naranjo, Department of Biochemistry–HHMI, Brandeis University, Waltham, MA 02254

Michael H. Nathanson, The Section of Digestive Diseases, Department of Internal Medicine, Yale University, New Haven, CT 06520-8019

Bruce J. Nicholson, Center for Advanced Molecular Biology and Immunology, Buffalo, NY 14260-1300

F. Noceti, Department of Anesthesiology, UCLA School of Medicine, Center for the Health Sciences, Los Angeles, CA 90024-1778

Andrés Oberhauser, Centro de Estudios Cientificos de Santiago and Universidad de Chile, Avenida Presidente Errázuriz # 3132, Las Condes, Santiago, Chile

Peter O'Day, Institute of Neuroscience, University of Oregon, Eugene, OR 97403-1254

R. Olcese, Department of Anesthesiology, UCLA School of Medicine, Center for the Health Sciences, Los Angeles, CA 90024-1778

Diane Papazian, Department of Physiology and Anesthesiology, UCLA School of Medicine, Los Angeles, CA 90024-1751

Alex Peinado, Department of Neuroscience, Albert Einstein College of Medicine, Bronx, NY 10461

Alejandro Roth, Departamento de Ciencias Fisiológicas, Pontificia Universidad Católica de Chile, Alameda 340, Santiago, Chile

Renato Rozental, Department of Neuroscience, Albert Einstein College of Medicine, Bronx, NY 10461

Juan Carlos Sáez, Departamento de Ciencias Fisiológicas, Pontificia Universidad Católica de Chile, Alameda # 340, Santiago, Chile

Claudia G. Sáez, Departamento de Ciencias Fisiológicas, Pontificia Universidad Católica de Chile, Alameda # 340, Santiago, Chile

Y. Sandoval, Instituto de Biotecnologia, Departamento de Genética y Fisiología Molecular, Universidad Nacional Autónoma de México, Cuernavaca, Morelos 62271, Mexico

C. Santis, Instituto de Biotecnologia, Departamento de Genética y Fisiología Molecular, Universidad Nacional Autónoma de México, Cuernavaca, Morelos 62271, Mexico

Stephan F. Schlosser, Liver Study, Yale University School of Medicine, New Haven, CT 06510

Francisco Sepúlveda, Centro de Estudios Científicos de Santiago, Avda Presidente Errázuriz # 3132, Las Condes, Santiago, Chile

S. A. Sheoh, Department of Physiology and Anesthesiology, UCLA School of Medicine, Los Angeles, CA 90024-1751

D. Sigg, Department of Physiology and Anesthesiology, UCLA School of Medicine, Los Angeles, CA 90024-1751

David Spray, Department of Neuroscience, Albert Einstein College of Medicine, Bronx, NY 10461

Enrico Stefani, Department of Anesthesiology, UCLA School of Medicine, Los Angeles, CA 90024-1778

Andrés Stutzin, Centro de Estudios Científicos de Santiago, Avda Presidente Errázuriz # 3132, Las Condes, Santiago, Chile

Ligia Toro, Department of Anesthesiology, UCLA School of Medicine, Los Angeles, CA 90024-1778

E. Brady Trexler, Department of Neuroscience, Albert Einstein College of Medicine, Bronx, NY 10461

Gonzalo Ugarte, Universidad de Chile, Facultad de Ciencias, Departamento de Biología, Las Palmeras # 3425, Santiago, Chile

Cecilia Vergara, Universidad de Chile, Facultad de Ciencias, Departamento de Biología, Las Palmeras # 3425, Santiago, Chile

Vytas K. Versalis, Department of Neuroscience, Albert Einstein College of Medicine, Bronx, NY 10461

Manuel Villalón, Departamento de Ciencias Fisiológicas, Pontíficia Universidad Católica de Chile, Alameda # 340, Santiago, Chile

Monique Vink, Department of Neuroscience, Albert Einstein College of Medicine, Bronx, NY 10461

Martin Wallner, Department of Anesthesiology, UCLA School of Medicine, Los Angeles, CA 90024

C.A. Warren, Department of Molecular and Cellular Physiology, Howard Hughes Medical Institute, Stanford, CA 94305

Eileen M. Westphale, Department of Pediatrics, Washington University School of Medicine, St. Louis, MO 63110

O. Zapata, Instituto de Biotecnologia, Departamento de Genética y Fisiología Molecular, Universidad Nacional Autónoma de Mexico, Cuernavaca, Morelos 62271, Mexico

Guido Zampighi, Department of Anatomy and Cell Biology, UCLA School of Medicine, Los Angeles, CA 90024

Preface

Ion channels allow us to see nature in all its magnificence, to hear a Bach suite, to smell the aroma of grandmother's cooking, and, in this regard, they put us in contact with the external world. These ion channels are protein molecules located in the cell membrane. In complex organisms, cells need to communicate in order to know about their metabolic status and to act in a coordinate manner. The latter is also accomplished by a class of ion channels able to pierce the lipid bilayer membranes of two adjacent cells. These intercellular channels are the functional subunits of gap junctions. Accordingly, the book is divided in two parts: the first part is dedicated to ion channels that look to the external world, and the second part is dedicated to gap junctions found at cell interfaces. This book is based on a series of symposia for a meeting on ion channels and gap junctions held in Santiago, Chile, on November 28-30, 1995. The book should be useful to graduate students taking the first steps in this field as well as a reference for the aficionado. The aim of the meeting was mainly to show the impact of various modern techniques, including cell biology, molecular biology, biophysics, and molecular-genetics techniques in the study of these ubiquitous intrinsic membrane proteins. Molecular-genetics techniques paved the road to the manipulation of the channel-forming molecules. It also allows us to control the composition of subunits expressed and we can modify the primary structure by means of point mutations or deletions. On the other hand, electrophysiological methods, and in particular, the patch clamp technique, allow the functional consequences of these molecular manipulations to be measured with an exquisite sensitivity. The identification, isolation, and functional expression of the *Shaker* Kchannel was the beginning of a new era in our understanding of a class of channels, the voltage-dependent K^+ channels. Thus, in the first part, we learn about the molecular origin of voltage dependence in ion channels, how K^+ channels are assembled, and how the *Shaker* K^+ channel, normally an outward rectifier, can be converted into an inward rectifier by means of a few point mutations performed in the voltage sensor. We also learn here about the structure of calcium-activated K^+ channels of large conductance, the nature of their voltage dependence, and the functional characteristics and the molecular nature of their interaction with auxilliary subunits. Electrostatic interactions between the voltage sensor of sodium channels with charged toxins that plug the sodium channel pore and how an auxiliary subunit modulates the coupling between charge movement and pore opening in calcium channels are also discussed.

Once open, ion channels are able to let millions of ions per second go through them and in the case of K^+ channels they do so with a high degree of ion selectivity. The combined molecular-genetic/electrophysiological approach has also shaped our view of how a protein molecule may fold up in a membrane to form a K^+-selective pore. Therefore, the first part of the book also deals with the pore structure and its gross architecture as determined using blocking ions as structural probes. Ion channels may be considered as the guardians of the cellular electrical homeostasis and, therefore, there must be many of them and all of them finely tuned. The modulation of ion channels by second messengers, hormones, the lipid environment, and the functional role of ion channels in cell volumen regulation and signal transduction are clearly illustrated in various sections of this book

In the second part of the book, we learn about the abrupt progress in the understanding of gap junction channels gained during the last decade. As for non-junctional ion channels, the use of diverse modern techniques has allowed to elucidate many features of the intercellular channels. In this part of the book, evidences supporting the role of gap junction channels during the normal tissue function, as well as in the pathogenesis of genetic diseases are discussed. Biochemical and molecular biology approaches have made possible to recognize highly homologous proteins, termed connexins, that are coded by a gen family . We note here that the cloning of the major intrinsic protein (MIP) of the ocular lens, initially thought to be a gap junction protein, has clarified that MIP might rather be a water permeable channel. Moreover, it has been elucidated that while most connexins form functional intercellular channels with characteristic unitary conductances few of them are negative dominants. Measurements of the hemichannel unitary conductance, as well as the generation of mutated connexins has started to reveal the principles that govern the channel structure and function.

In several tissues, such as lens, heart and liver, gap junctions are abundantly expressed. Evidence presented in several chapters supports that the intercellular kiss between cells of these and other tissues is a highly regulated phenomenon. Regulation may occur at various cellular levels, including transcriptional rates, post-transcriptional mechanisms (mRNA stability), protein modification (phosphorylation) and presumably turnover of the protein. Moreover, migratory cells, such as white blood cells do not express connexins under resting conditions but upon activation by insults such as infections or tissue damage, they transiently express gap junction channels. It is accepted that in electrical excitable tissues gap junctions control the intercellular propagation of electrical signals, but the functional role of gap junctions expressed by white blood cells remains speculative.

Considering the importance of these cells in various specific and non-specific immune responses, it is predictable that the latter issue will be an active field of research. The intercellular propagation of Ca^{2+} waves through gap junction-dependent or -independent pathways are presented and clearly offers a mechanism to further understand how cells gossip. Ca^{2+} waves can be triggered by various extracellular signals, including neurotransmitter, such as glutamate and ATP. The last chapter of this book is devoted to the the process of exocytosis. Exocytosis is viewed through high temporal resolution techniques that allow to study directly the activation pathways that mediate vesicle fusion and the release of the chemical messengers.

In summary, this book gives a current opinion on diverse issues of molecular events that occur in different channels and specialized membrane domains during cell interactions with the external world as well as between cells of the same or different organs.

We would like to thank Mrs. Susana Claro for putting so much effort and care in the organization of the course and to Ana Maria Navia and Paulina Rodriguez for excellent secretarial assistance. We are also thankful to all institutions that made this event financially viable, including Fondo Nacional de Desarrollo y Tecnologia (FONDECYT), Facultad de Ciencias Biológicas of the Pontificial Catholic University of Chile, Fundación Chilena para Biologia Celular, Direccion de Investigación de la Universidad Católica de Chile, Facultad de Ciencias of the University of Chile, and Fondo Argentino de Cooperación Horizontal FO-AR. The Centro de Estudios Científicos de Santiago is supported by a Group of Chilean companies (Empresas CMPC, CGE, CODELCO, COPEC, Minera Escondida, NOPVAGAS, BUSINESS DESIGN ASS., XEROX Chile). R.L. is a recipient of a Cátedra Presidencial from the Chilean Government.

Ramón Latorre
Juan Carlos Sáez

Santiago, Chile

Contents

I. Ion Channels in Contact with the External World

Chapter 3

Assembly of Shaker K-Channels from a Random Mixture of Subunits Carrying Different Mutations

David Naranjo

Chapter 4

New Insights into the Topography, Gating and Modulation of MaxiK Channels

Ligia Toro, Martin Wallner, and Prataap Neera

Chapter 5

Shifts of Macroscopic Current Activation in Partially Blocked Sodium Channels. Interaction between the Voltage Sensor and a μ-Conotoxin

Robert French and Richard Horn

Chapter 8

Barium as a Probe of the Molecular Architecture of the Pore of K^+ Channels

Ramon Latorre, Raymond Hurst, Felipe Diaz, Ligia Toro, and Enrico Stefani

Chapter 9

Possible Participation of a cAMP Regulated K^+ Channel from the Sea Urchin Sperm in the Speract Response

Pedro Labarca, Celia Santi, Otilia Zapata, Carmen Beltrán, Arturo Liévano, Yolanda Sandoval, and Alberto Darszon

Chapter 10

Gating and Selectivity Properties of a cAMP-Activated K^+ Selective Channel from* Drosophila *Larval Muscle

Pedro Labarca, Ricardo Delgado, Orlando Jorquera, and Carlos Maureira

Chapter 11
Regulation of Ca^{2+} and Na Channels in GH3 Cells by Epidermal Growth Factor
Gabriel Cota, Ulises Meza, and Eduardo Monjaraz

Chapter 12
Physical State of the Nicotinic Acetylcholine Receptor Membrane and Modulation of the Receptor Channel by the Lipid Environment
Francisco Barrantes

Chapter 13
Amiloride-Sensitive Na^+ Channels and Human Hypertension
Dale J. Benos, Bakhram K. Berdiev, Ann Lynn Bradford, James K. Bubien, Catherine M. Fuller, and Iskander I. Ismailov

Chapter 14

Anion Channels Involved in Volume Regulation: A Common Pathway for Chloride and Organic Osmolyte Permeation?

Andres Stutzin, Ana Luisa Eguiguren, L. Pablo Cid, and Francisco V. Sepulveda

Chapter 15

Volume-Regulated Anion Channels and Cytoskeletal Interaction

Irena Levitan and Sarah S. Garber

Chapter 16

Inhibitory Responses to Odorants in Vertebrate Olfactory Neurons

Juan Bacigalupo, Pedro Labarca, Bernardo Morales, Gonzalo Ugarte, Rodolfo Madrid

Chapter 19
Studies of Voltage Gating Gap Junctions and Hemichannels Formed by Connexin Proteins
Vytas K. Versalis, E. Brady Trexler, Thaddeus A. Bargiello, and Michael V. L. Bennett

Chapter 20
Gap Junctions between Leukocyte and Endothelium: Expression of Connexin43 in Adherent or Activated Cells
Mauricio P. Boric, Alejandro Roth, Patricio Jara, and Juan C. Sáez

Chapter 21
Regulation of Gap Junctions in Rat Liver during Acute and Chronic CCl_4-Induced Liver Injury
Claudia G. Sáez, Eliseo Eugenin, Elliot L. Hertzberg, and Juan C. Sáez

Chapter 25

A Possible Physiological Role of the Major Intrinsic Protein of the Ocular Lens

Grischa Chandy, Guido Zampighi, and James E. Hall

Chapter 26

Regulation of Smooth Muscle Activity and Gap Junctions by Sexual Hormones in the Rat Oviduct

Manuel Villalón, Marcela Hermoso, Mateo Budinich, Julio Aguilera, and Juan C. Sáez

Chapter 27

Intercellular Calcium Signaling in Liver

Michael H. Nathanson, Michael B. Fallon, Angela D. Burgstahler, Albert Mennone, Stephan F. Schlosser, Carlos Gonzalez, and Juan C. Sáez

Part I

Ion Channels in Contact with the External World

THE MOLECULAR ORIGIN OF VOLTAGE DEPENDENCE IN IONIC CHANNELS

F. Bezanilla, S-A. Seoh , D. Sigg and D. M. Papazian

Depts. of Physiology and Anesthesiology
UCLA School of Medicine
Los Angeles, CA 90095
U.S.A.

INTRODUCTION

Excitable membranes play a fundamental role in the generation, transmission and transduction of information in the nervous system, muscle and other tissues. The generation of an action potential, a fast transient change in the membrane potential of the cell, is produced by the increase in permeability of ion-selective channels. Hodgkin and Huxley (1952) described the changes in permeability of sodium and potassium specific pathways that occur in response to changes in membrane potential. The classical sodium channel is an intrinsic membrane protein that conducts preferentially Na ions but it does so only at depolarized membrane potentials and for short periods of time. In the generation of the action potential, a small decrease in the membrane potential opens a few Na channels allowing Na to rush into the cell. This produces a larger depolarization which in turn opens more Na channels, developing the rising phase of the impulse. This rising phase of the action potential is slowed down and reversed by the late opening of the potassium channels that produce outward current, repolarizing the membrane to its initial state. Thus, changes in membrane conductances are driven by changes in membrane potential and this is the fundamental property of voltage dependent channels.

From Ion Channels to Cell-to-Cell Conversations
Edited by Latorre and Sáez, Plenum Press, New York, 1997

In their original description Hodgkin and Huxley (1952) separated the voltage dependent mechanism and the conduction pathway and proposed that this voltage dependent process changes the probability that the conduction pathway is open or closed to ion flow. It is remarkable that this basic formulation has proven to be correct through many experiments done on macroscopic currents, single channel recordings and gating currents and even by the inferences obtained from the structure of the channels as deduced from sequence analysis. In addition, they recognized the steep voltage dependence of the conductances and proposed that this voltage dependence is the result of orientation of dipoles or charges within the membrane which gate the conducting pathway open or closed in response to changes of the membrane electric field. With this proposal, Hodgkin and Huxley predicted the existence of gating currents, corresponding to the movement of the charges or dipoles and also predicted the presence of charges or dipoles in the channel structure.

When the sodium channel of the electric eel was cloned (Noda et al, 1984) the authors noted a peculiar sequence in the fourth putative transmembrane segment of all four homologous domains of the channel. The sequence consists of four to seven basic residues, arginines and lysines, separated by two hydrophobic residues. This sequence was proposed as the prime candidate to fulfill the Hodgkin and Huxley prediction of charged groups in the membrane that may move in response to changes in membrane potential and confer the voltage dependence on the channel (Catterall, 1986; Guy and Seetharamulu, 1986). This idea was reinforced by the finding that voltage dependent potassium channels showed the same peculiar sequence in the fourth transmembrane segment of its subunit structure (Tempel et al, 1987). The same sequence was found in cloned voltage dependent calcium channels (Tanabe et al, 1987). These findings prompted many investigators to test the hypothesis that the S4 segment was the voltage sensor of voltage-dependent channels by changing the basic residues using site-directed mutagenesis (Stuhmer et al, 1989; Papazian et al 1991; Liman et al 1991; Logothetis et al 1992). Although the results were consistent with the idea that the S4 segment was involved in voltage sensing they did not demonstrate that the S4 segment was actually the voltage sensor. The main difficulty in answering whether the S4 is part of the sensor resides in the limitations of interpreting results of ionic currents, the approach that was used in the mutagenesis experiments just mentioned.

The voltage dependence of the conductance reflects the sensitivity of the voltage sensor in a complex way. The relative position of the conductance-voltage (g-V) curve on the voltage axis may be affected by the charges involving the sensor but other changes in the molecule, totally unrelated to the voltage sensor, may also shift the voltage dependence. Therefore shifts are poor indicators of possible modifications of the voltage sensor. A parameter that better reflects the charges involved in voltage sensing is the steepness of the g-V relationship.

In general, the higher the voltage dependence, the steeper is the relation between conductance and membrane potential but, in a general way, it is not possible to extract the correct number of charges involved in the sensing process by just studying the voltage dependence of the conductance in the normal voltage range of activation. The study of the gating charge movement reflects more closely the properties of the voltage sensor (Perozo et al, 1994) but only a combination of conductance and gating current studies give a more accurate description of the voltage dependent process (Seoh et al, 1996).

In what follows we will introduce the concept of the voltage sensor, its behaviour with changes in voltage, and ways to determine the charges that are involved in voltage dependent activation of ion channels.

THE VOLTAGE SENSOR

A fundamental property of voltage dependent channels is that the probability of the channel of being open is a function of the membrane potential. In sodium and potassium channels this probability function is sigmoid and approaches zero at negative membrane potentials and attains its maximum value at depolarized membrane potentials.

The basic question is, how is it possible to change the open probability with changes in membrane potential? It is clear that to achieve this, an electric charge or an electric (or magnetic) dipole must be involved in sensing the changes in membrane potential and their reörientation results in a change in the probability that the conducting path will be open or closed. We call this charge or dipole system, together with its surrounding electric field, the *voltage sensor*. We emphasize that the sensor includes the charges plus the field where they move because, as we will see below, both contribute to the voltage dependence and they are not separable in electrical studies.

Depending on the value of the membrane potential, the sensor will present different positions or conformations and these positions are, in principle, detectable from the external circuit as changes in the polarization of the dielectric. It follows that at extreme potentials the polarization will show a maximum and a minimum describing a sigmoid curve which has been described as the gating charge *vs* voltage (Q-V) curve. The name of gating charge comes from the assumption that all this extra charge reorientation is involved in gating the channel open or close. It is conceivable, however , that some of the charge that reorients is not directly connected to the gating process and that portion should not be called gating charge.

The detection of the gating charge is usually done by stepping the membrane potential from an initial value to a final value and recording a current that has the charactersitics of a capacitive current. The time integral of this *gating current*, as it is called, constitutes the gating charge. To record the gating current, it must be separated from ionic currents and the capacitive currents needed to charge the membrane linear capacitor. This can be achieved by blocking the ionic currents and by subtracting the linear components based on their linear property. (Armstrong and Bezanilla, 1973). With the large density of channels achieved with expression of cloned channels it is even possible to detect the gating currents with minimum contamination of the linear capacitive component (Stefani et al, 1994). In addition, in the *Shaker* B potassium channel it is possible to eliminate ionic conduction altogether by a specific mutation in the pore region that renders the channel non-conductive but preserves the gating process (Perozo et al, 1993).

THE ACTIVATION CHARGE

Hodgkin and Huxley (1952) estimated the minimum charge that must move in the field to account for the voltage dependence of the conductance assuming that the voltage sensor had only two stable states and that the proportion of charges in these states was governed by a Boltzmann distribution in which the energy contained a term directly proportional to the voltage of the membrane. In this way they computed that the voltage dependence of the sodium channel could be explained if 4 elementary charges moved across the entire electric field. Almers (1978) extended this argument to a linear sequential model with a series of closed states ending in a single open state. By assuming that at very negative potentials the closed states were favored and at positive potentials the open state was favored he obtained the following limiting relation

$$\lim_{V \to -\infty} kT \frac{\mathrm{d} \ln (P_0)}{\mathrm{d} V} = q_{\max} \tag{1}$$

where V is the membrane potential, P_o is the open probability, q_{max} is the total charge per channel needed for channel opening and k and T are the Boltzmann constant and absolute temperature respectively. This expression, if it could be evaluated from experimental data, would give the total amount of charge involved in the activation of one channel; consequently it is a direct measurement of the voltage dependence of the channel. Many authors have used equation (1) to estimate the gating charge but it is easy to show that to obtain the actual value of q_{max} requires going to vanishingly small values of P_o (eg. Bezanilla and Stefani, 1994) which is very difficult experimentally. In addition, the derivation of equation (1) was restricted to a linear sequential model with one open state at the end of the sequence.

Recently Sigg and Bezanilla (1996) have generalized the derivation to a general model with an arbitrary number of open and closed states with any type of connections between them (Fig. 1). In this derivation the coordinate is the

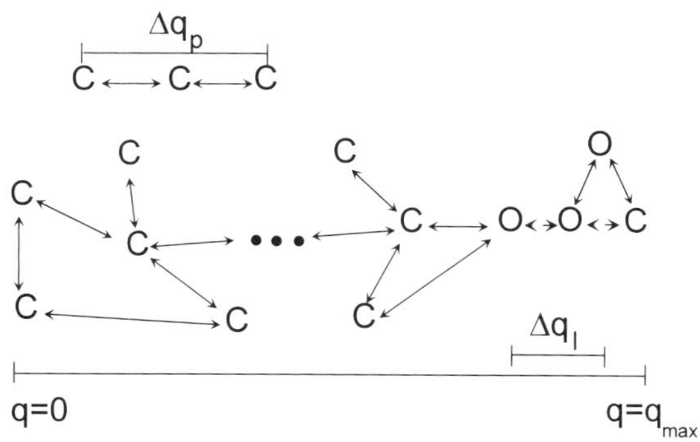

Figure 1. General model of activation. Reaction coordinate is the charge q. The charge Δq_p has no connection with the activation charge. Δq_l is the charge moving between open states.

charge q and the states are defined with subindex i. The basic assumption is that the free energy F_i in each state i is a linear function of the applied potential and it is given by the following equation

$$F_i = G_i + q_i \delta_i V \qquad \textbf{(2)}$$

where G_i is the energy of state i in the absence of an electric field, V is the membrane potential and δ_i is the fraction of the field where q_i moves. It is important to notice that q_i and δ_i appear always as product meaning that we can only determine the charge times the fraction of the field where it moves and not the absolute charge. As we have no direct measurement of δ_i, we will equate it to one to simplify the notation which will mean that from now on a value of q will always be understood as the product of $q\delta$. The probability p_i of being in a state i is computed from the Boltzmann distribution as

$$p_i = \frac{\exp \frac{-F_i}{kT}}{\sum_i \exp \frac{-F_i}{kT}} \qquad \textbf{(3)}$$

The channel may have several open states with fractional conductance f_i . Defining the *activation charge* $\langle q_a \rangle$ as

$$\langle q_a \rangle = kT \frac{d \ln P_o}{dV} \qquad \textbf{(4)}$$

and the Q-V curve as $<q>$, the result of the derivation (Sigg and Bezanilla, 1996) can be written as

$$\langle q_a \rangle + \langle q \rangle = q_{\max} - \frac{\sum_i \theta_i f_i \exp\frac{-F_i}{kT}}{\sum_i f_i \exp\frac{-F_i}{kT}} \quad (5)$$

where $\theta_i = q_{max} - q_i$ corresponding to the charge displacement with reference to the positive end of the activation sequence. The second term on the right is the charge distribution moving between open states or *latent charge* which we may call $<q_l>$. We may rewrite equation (5) as follows

$$\langle q_a \rangle + \langle q \rangle + \langle q_c \rangle = q_{\max} \quad (6)$$

Notice that if there is no charge movement between open states (Fig. 2) we obtain

$$kT\frac{d\ln P_o}{dV} + <q> = q_{\max} \quad (7)$$

and it reduces to Almers expression (equation (1)) when $<q>= 0$ which will occur at very negative potentials. An example of the relation between the $<q_a>$, $<q>$ and P_o is shown in Fig. 2 where it is clear that $<q_a>$ is equal to the q_{max} minus the Q-V curve. Equation (7) can be used to estimate the value of q_{max} if the Q-V curve and the conductance are known in a range of voltage without the need to measure the slope at vanishingly small values of probabilities. In addition, eq. (7) is general and it does not assume a linear sequential model but it requires that there is no charge movement between open states. If there is charge movement between open states, equation (6) must be used but it becomes inpractical to determine the value of q_{max}.

In summary, it is then possible to estimate the total charge per channel involved in the activation q_{max}, by measuring the activation charge $<q_a>$ with the slope of the logarithm of the open probability (or simply using the conductance) as a function of voltage together with the Q-V curve obtained from gating current measurements. q_{max} is the fundamental quantity that determines the voltage dependence of the channel and its value may be correlated to the presence of putative residues that are suspected to be involved in the voltage sensor.

CHARGE PER CHANNEL

In the previous paragraph we have indicated a way to determine the charge per channel directly connected to the opening of the channel (q_{max} when $<q_l>= 0$). Another way to determine the charge per channel is to measure the total gating charge in a patch of membrane and divide that amount by the number of channels present in the same patch, a method referred to as Q/N. The number of channels can be determined by fluctuation analysis of the ionic currents (Sigworth, 1980) or by counting the number of channels in the membrane with toxin binding (Levinson and Meves, 1975; Aggarwal and MacKinnon, 1996). It should be noted that for this method to be accurate, the gating current measurement and the estimation of the number of channels must be applied to the same segment of membrane (Schoppa et al, 1992; Seoh et al, 1996; Aggarwal and MacKinnon, 1996). However, this method does not necessarily give the same charge per channel as estimated by the activation charge method described in the previous paragraph. This is because charge movement that occurs in

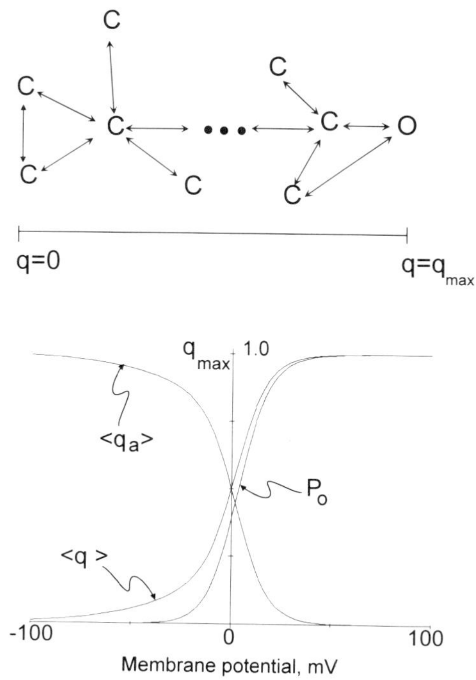

Figure 2. *Top*: Activation model that conforms to equation (7) with no charge moving between open states. *Bottom*: Plots of the activation charge, Q-V curve and open probability as a function of membrane potential.

parallel with the activation charge (in Fig. 1 it corresponds to the *peripheral charge* Δq_p) will be detected by the Q/N method, although it would be invisible to the activation charge method. On the other hand, if both methods give the same charge per channel, it shows that all the charge movement is connected to the activation of the channel.

THE GATING CHARGE OF *SHAKER* K^+ CHANNEL

Schoppa el al (1992) applied the Q/N method (determining the number of channels with fluctuation analysis) to the *Shaker* B potassium channel and found a total of 12 to 13 electronic charges moving across the whole electric field (e_o). This number has been confirmed by Seoh et al (1996) who, using the same method, found a value of 12.9 e_o and by Aggarwal and MacKinnon (1996), who determined the number of channels using toxin binding and found a value of 13.6 e_o.

The limiting slope method has been applied by Schoppa et al (1992) and by Zagotta el al (1994a) and both groups obtained a value of about 10 e_o. Given that the determination of the limiting slope was done at potentials at which there is still charge movement, those estimates must be considered as a lower limit. Seoh et al (1996) applied equation (7) to determine the value of q_{max} and found a value of 12.6 e_o. This value agrees with the value found with the Q/N method indicating that all the non-linear charge movement in *Shaker* B channel is energetically coupled to channel opening.

THE RESIDUES THAT MAKE UP THE VOLTAGE SENSOR

Although alpha helices are electric dipoles that may orient in the field, it would be difficult to account for the large charge per channel determined for the *Shaker* K channel if they were the only dipolar structures that can orient in the field. On the other hand the voltage sensor may be constituted by charged amino acids reorienting in the electric field. In the sequence of *Shaker* B channel, the basic and acid residues of putative transmembrane segments are distributed as two acidics in S2 , one acidic in S3 and 7 basic residues in S4 (see Fig. 3). Our approach was to test whether any or all of these residues are part of the voltage sensor by measuring the charge per channel in mutants of the inactivation-removed *Shaker* B channel (ShB-IR) where each residue was neutralized one at a time. The measurement of charge per channel was performed with the Q/N method and by the use of the slope of the logarithm of the conductance together with the Q-V curve using equation (7). A full account of this work has been published by Seoh et al (1996).

One of the consequences of neutralizing the first outermost charge in S4 (R362) was that the mutant showed evidence for more than one open state. This

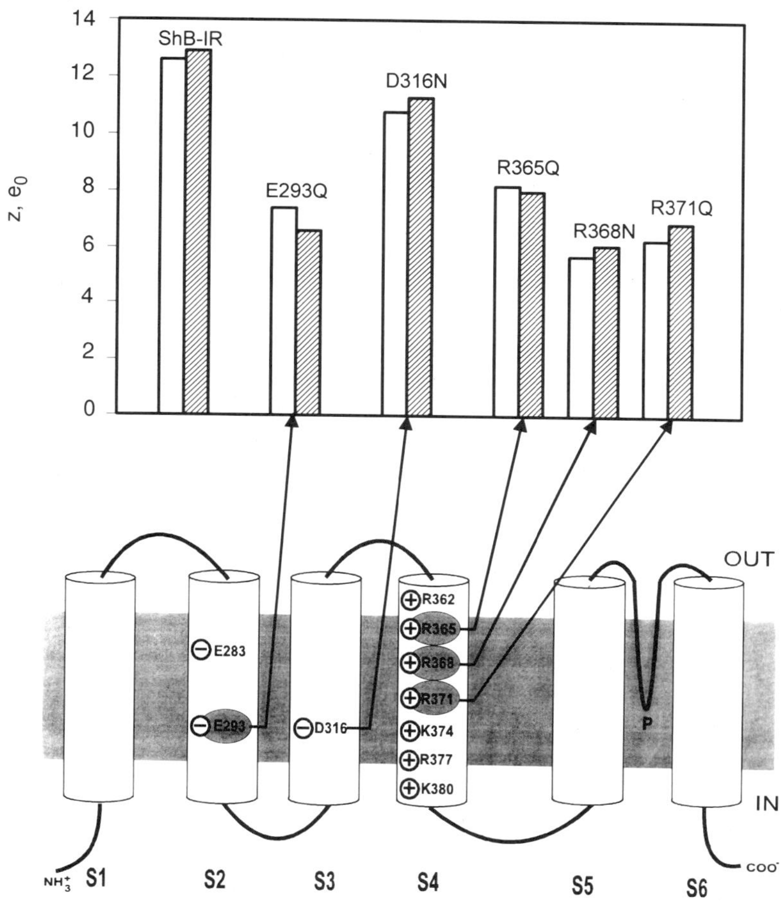

Figure 3. Charges that contribute to the voltage sensor in *Shaker* B channel. *Bottom:* schematic representation of the transmembrane segments of the *Shaker* channel subunit showing the acidic and basic residues. Shaded residues contribute significantly to the gating charge. *Top*: the charge per channel of wild type (ShB-IR) and neutralization mutants as indicated. The open bars were estimated with the activation charge and the closed bars were estimated with the Q/N method.

made the estimation of the charge per channel unreliable, consequently we did not determine the contribution of this residue to the gating charge. The sixth residue (R377) could not be neutralized without loss of function of the channel, therefore its contribution to the gating charge was not determined.

When the outermost acidic residue in S2 (E283) was neutralized, the charge per channel was not affected. However, the neutralization of the innermost acidic (E293) showed a drastic decrease in the charge per channel, as determined with both methods. This is shown in the upper panel of Fig. 3 along with the charge per channel of the control channel (ShB-IR) and the mutants that showed some change in the charge per channel. For each case, the left bar (open) represents the charge per channel of the neutralized mutant as determined with the help of eq. (7) and the right bar (hatched) represents the charge per channel determined using the Q/N method.

When the acidic residue D316 in S3 was neutralized the charge per channel was decreased by about 2 e_0 which was consistent but not statistically significant.

Residues R365, R368 and R371 in S4 showed a drastic reduction of the charge per channel when any of them was neutralized. The next residue, K374, could not be neutralized to a glutamine without loss of function unless a negative residue, either E293 or D316 was simultaneously neutralized (Papazian et al, 1995). The results showed that K374Q + E293Q showed the same charge per channel as E293Q alone. Also, K374Q + D316N showed the same charge per channel than D316N alone indicating that K374 does not contribute significantly to the charge per channel.

The diagram of Fig. 3 summarizes the above results by showing the important charges in shaded ovals and indicating the *reduction* of the charge per channel when that charge is neutralized. The gating charge is a contribution of one acidic residue in S2 and three basic residues in S4. It is interesting to note that these basic residues of S4 are in the outermost part of the segment.

The reduction of the charge upon neutralization as assessed by the Q/N method (shaded bars) or the activation charge method (open bars) is the same. This implies that all the charge that moves is coupled energetically to channel opening not only in the wild type ShB-IR but also in mutants in which some of the charges have been neutralized. In addition, in all the neutralization mutants the maximum open probability was not very different than the ShB-IR channel. These results indicate that the movement of the charge is tightly coupled to the gating process and the neutralization mutations have not disrupted this coupling significantly.

MOVEMENT OF THE CHARGES AND CHANGES IN THE FIELD

The results of Fig. 3 identifies residues that contribute to the gating currents but it does not allow the assignment of each particular charge to a particular fraction of the gating charge. If each charge moved in the membrane electric field without affecting the movement of the other charges and without affecting the field itself, it would be expected that the neutralization of one charge would produce at the most a decrease of 4 e_0, because four identical subunits form the functional *Shaker* B channel (MacKinnon, 1991). In contrast, Fig. 3 shows that the neutralization of the four residues that reduce charge movement produce a decrease larger than 4 e_0 indicating that the neutralization not only reduces the moving charges but also has other indirect effects. This result is not surprising because the measurement of charge per channel resolves the product of the charge times the fraction of the field and a neutralization may contribute to a reduction of the charge simultaneously with a change in the field seen by the rest of the charges.

To illustrate how the neutralization of a particular charge could induce a change in the local field, Figure 4 presents an hypothetical mechanism that includes the movement of the S4 segment across the plane of the membrane. The negative charge in position 293 is assumed to induce a crevice that effectively would extend the internal medium into the transmembrane region of the protein making the voltage drop across a very narrow region when the channel is in its resting or most closed position (Fig 4, top left). When the membrane is depolarized, positive charges are pushed outwardly and the negative charge moves inwardly translocating a total of 3.5 e_0 per subunit or 14 e_0 per channel. If position 293 is neutralized (Fig. 4, middle panel), the crevice is obliterated and the voltage drop would occur over a larger extent of the protein. Thus, some of the charges that originally moved across the whole field now span only a quarter or half of the field giving a reduction to about 9 charges, as opposed to 12 that would be expected if the contribution of 293 had been 0.5 charge per subunit. Another example is given whereby position 368 is neutralized (Fig. 4, bottom). The assumption here is that the neutral position also obliterates part of the crevice, extending the region of the voltage drop and producing a final effect of reducing the charge by 7 e_0 instead of 4 as would be expected if the neutralization only had the effect of reducing the charge.

The mechanism illustrated in Fig. 4 is highly speculative and its details will certainly prove to be incorrect but it serves the purpose of showing that neutralization mutations can change the profile of the electric field as well as reducing the number of movable charges. Some features shown in Fig. 4 are consistent with results of experiments using modifications of residues using cysteine mutations. Larsson et al (1996) have shown that some of the charged residues when replaced by cysteine may be reactive in the open position but not in the closed position. Yang et al (1996) have shown in the muscle sodium channel that the exposure of some the residues of the S4 segment of the fourth domain change from the inside to the outside depending on the state of the

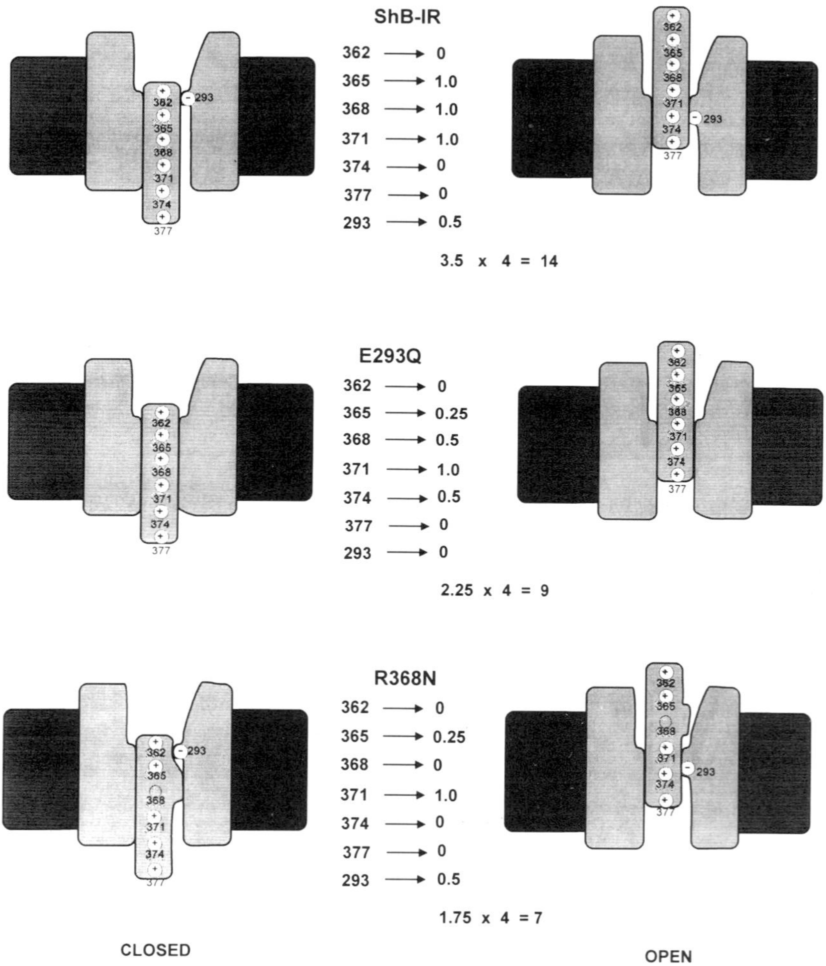

Figure 4. Speculation on the effects of charge neutralization on the shape of the local field sensed by the other gating charges. Left panels are the closed states and right panels the open states. Three cases are illustrated. *Top panel*: the ShB-IR channel, *middle panel:* neutralization of E293 and *bottom panel:* neutralization of R368.

channel. These results imply that the cysteine residues (that originally were in the positions of the charged residues) may be in solvent accesible crevices where the cysteine reagents can reach them for chemical modification. At the same time, it reaffirms the concept that some of the charges may in fact completely cross the membrane field that is focused in a narrow region of the protein. On the other hand, some of the simplifications shown in Fig. 4 do not attempt to explain the electrostatic interactions between D316 and E293 with K374 that were clearly demonstrated by Papazian et al (1995).

Another study on S4 charge neutralization done by Aggarwal and MacKinnon (1996) counting the channels with an specific toxin and measuring the total non linear charge in the same oocyte, shows qualitatively similar results to the study of Seoh et al (1996). The main differences are that they found a contribution of positions 362 and 374 to the gating charge and that the charge reduction rarely exceeded 4 e_0. Although there is no obvious explanation for these differences, the values of Q/N measured by Seoh et al (1996) (Fig. 3) were confirmed by the activation charge approach which measures exclusively the charges energetically connected to channel activation.

THE VOLTAGE-DEPENDENT STEPS

A sudden change in membrane potential induce movements of the gating charges which in turn may modify the profile of the electric field producing a displacement current that can be recorded as gating currents under voltage clamp conditions. Although we now can assign some of the residues in the channel sequence to the function of this voltage sensor we still do not know how the movement of each of these residues produce the gating current. The Q-V curve is a complicated function of voltage that can be approximated by at least two Boltzmann distributions (Bezanilla et al, 1994). The kinetics of the charge movement and channel opening indicate that there must be many more states and a minimum of eight states have been proposed to account for the experimental results (Bezanilla et al, 1994) while another model proposed up to 16 states (Zagotta et al, 1994b). The functional relationship between Q and V is drastically altered in some of the neutralization mutants (Seoh et al, 1996) as it would be expected by the elimination of part of the gating charge, and presumably, by the elimination of some of the voltage dependent steps. Unfortunately it is not straightforward to correlate the changes in the voltage dependence of the Q-V relation and the neutralization of a praticular charge. For example, the neutralization of R368 by a glutamine produces a clear separation of the two basic components of the gating charge with a drastic reduction of the second gating component predicting a decrease of the charge per channel (Perozo et al, 1994). The neutralization of R368 by an aspargine does not separate the two components as clearly but the charge per channel is reduced drastically (Seoh et al, 1996). On the other hand, the neutralization of R371 by a glutamine decreases the charge per channel although the Q-V relation is hardly changed with respect to the wild-type. The overall steepness of the Q-V curve increases with the total charge but it decreases as the number of steps is increased. Therefore the neutralization of R371 must somehow compensate the decrease in steepness due to charge reduction by an increase in steepness due to a reduction in the number of steps traversed by the gating charge. These results suggest that a detailed correlation of the kinetics and steady-state properties of the gating currents studied in the neutralization mutants may yield the contribution of each of the charged residues to the steps traversed by the total charge.

In the multiple series of steps leading to channel opening, there is at least one that carries about 2.4 e_0, as was determined by fluctuation analysis of the gating currents (Sigg et al, 1994). This is the most voltage dependent step. It is very near the open state, but it occurs before the opening step. This transition may be the result of a cooperative action between channel subunits and a major challenge is to correlate this large step with the movement of the residues that have been identified as part of the voltage sensor.

AKNOWLEDGEMENT

This work was supported by USPHS grant GM30376.

REFERENCES

Aggarwal, S.K. and MacKinnon, R. (1996). Contribution of the S4 segment to gating charge in the *Shaker* K+ channel. *Neuron* 16:1169-1177.

Almers, W. (1978). Gating currents and charge movements in excitable membranes. *Rev. Physiol. Biochem. Pharmacol.* 82, 96-190.

Armstrong, C. M. and Bezanilla, F. (1973). Currents related to movement of the gating particles of the sodium channels. *Nature* 242, 459-461.

Bezanilla, F., Perozo, E., and Stefani, E. (1994). Gating of *Shaker* K^+ channels: II. The components of gating currents and a model of channel activation. *Biophys. J.* 66, 1011-1021.

Catterall, W. A. (1986). Molecular properties of voltage-sensitive sodium channels. *Annu. Rev. Biochem.* 55, 953-985.

Guy, H. R., and Seetharamulu, P. (1986). Molecular model of the action potential sodium channel. *Proc. Natl. Acad. Sci. USA* 83, 508-512.

Hodgkin, A. L. and Huxley, A. F. (1952). A quantitative description of membrane current and its application to conduction and excitation in nerve. *J. Physiol.* 117, 500-544.

Larsson, H.P., Baker, O.S., Dhillon, D.S., Isacoff, E.Y. (1996). Transmembrane movement of the *Shaker* K+ channel S4. *Neuron.* 16, 387-397.

Levinson, S.R. and H. Meves. (1975). The binding of tritiated tetrodotoxin to squid giant axons. *Philo. Trans. R. Soc. Lond. B Biol. Sci.* 270:249-352.

Liman, E. R., Hess, P., Weaver, F., and Koren, G. (1991). Voltage-sensing residues in the S4 region of a mammalian K^+ channel. *Nature* 353, 752-756.

Logothetis, D. E., Movahedi, S., Satler, C., Lindpaintner, K. and Nadal-Ginard, B. (1992). Incremental reductions of positive charge within the S4 region of a voltage-gated K+ channel result in corresponding decreases in gating charge. *Neuron* 8, 531-540.

MacKinnon, R. (1991). Determination of the subunit stochiometry of a voltage-activated potassium channel. *Nature* 350, 232-235.

Noda, M., Shimizu, S., Tanabe, T., Takai, T., Kayano, T., Ikeda, T., Takahashi, H., Nakayama, H., Kanaoka, Y., Minamino, N.; et al. (1984). Primary structure of Electrophorus electricus sodium channel deduced from cDNA sequence. *Nature* 312, 121-127.

Papazian, D. M., Shao, X. M., Seoh, S. -A., Mock, A. F., Huang, Y., and Wainstock, H. (1995). Electrostatic Interactions of S4 Voltage Sensor in *Shaker* K^+ Channel. *Neuron* 14, 1293-1301.

Papazian, D. M., Timpe, L. C., Jan, Y. N., and Jan, L. Y. (1991). Alteration of voltage-dependence of *Shaker* potassium channel by mutations in the S4 sequence. *Nature* 349, 305-310.

Perozo, E., MacKinnon, R., Bezanilla, F., and Stefani, E. (1993). Gating currents from a nonconducting mutant reveal open-closed conformations in *Shaker* K^+ channels. *Neuron* 11, 353-358.

Perozo, E., Santacruz-Toloza, L., Stefani, E., and Bezanilla, F. and Papazian, D. M. (1994). S4 mutations alter gating currents of *Shaker* K channels. *Biophys. J.* 66, 345-354.

Schoppa, N. E., McCormack, K., Tanouye, M. A., and Sigworth, F. J. (1992). The size of gating charge in wild-type and mutant *Shaker* potassium channels. *Science* 255, 1712-1715.

Seoh, S-A., Sigg, D., Papazian, D.M. and F. Bezanilla. (1996). Voltage-sensing residues in the S2 and S4 segments of the *Shaker* K^+ channel. *Neuron* 16:1159-1167.

Sigg, D., Stefani, E., and Bezanilla, F. (1994). Gating current noise produced by elementary transitions in *Shaker* potassium channels. *Science* 264, 578-582.

Sigg, D. and Bezanilla, F. (1996). Total charge movement per channel: The relation between gating charge displacement and the voltage sensitivity of activation. (Submitted for publication)

Sigworth, F. J. (1980). The variance of sodium current fluctuations at the node of Ranvier. *J. Physiol.* 307, 97-129.

Stefani, E. and Bezanilla, F. (1994). Voltage-dependent gating of ionic channels. *Annu. Rev. Biophys. Biomol. Struct.* 23:819-846.

Stefani, E., Toro, L., Perozo, E., and Bezanilla, F. (1994). Gating of *Shaker* K^+ channels: I. Ionic and gating currents. *Biophys. J.* 66, 996-1010.

Stuhmer, W., Conti, F., Suzuki, H., Wang, X. D., Noda, M., Yahagi, M., Kubo, H., and Numa, S. (1989). Structural parts involved in activation and inactivation of the sodium channel. *Nature* 339, 597-603.

Tanabe, T., Takeshima, H., Mikami, A., Flockerzi, V., Takahashi, H., Kangawa, K, Kojima, M., Matsuo, H., Hirosi, T. and Numa, S. (1987). Primary structure of the receptor for calcium channel blockers from skeletal muscle. *Nature* 328: 313-318.

Tempel, B.L., Papazian, D.M., Schwarz, T.L, Jan, Y.N and Jan, L.Y. (1987). Sequence of a probable potassium channel component encoded at *Shaker* locus of *Drosophila. Science* 237:770-775.

Yang, N., George, A.L., and Horn, R. (1996). Molecular basis of charge movement in voltage-gated sodium channels. *Neuron.* 16, 113-122.

Zagotta, W.N., Hoshi, T., Dittman, J. and Aldrich, R.W. (1994a) *Shaker* potassium channel gating II: transitions in the activation pathway. *J. Gen. Physiol.* 103:279-319.

Zagotta, W.N., Hoshi, T. and Aldrich, R.W. (1994b) *Shaker* poatssium channel gating III: evaluation of kinetic models for activation. *J. Gen. Physiol.* 103:321-362.

INWARD RECTIFICATION BY AN INACTIVATION GATING MECHANISM

A.G. Miller, C.A. Warren, R.W. Aldrich

Dept. of Mol. and Cellular Physiology, Stanford University and Howard Hughes Medical Institute, Stanford, Ca. 94305

INTRODUCTION

Potassium channels are found in every organ and virtually in every cell, and appear to play divergent physiological roles (Hille, 1992). In the last 9 years, the identification of a number of potassium channel amino acid sequences suggests only a small number of classes with different putative membrane spanning topologies. Presently, there appear to be two major classes of K^+ channels. One class, whose members include the first cloned potassium channel (*Shaker*) contains voltage gated channels with at least six putative hydrophobic transmembrane segments (Chandy and Gutman, 1995), including a positively charged segment and a highly conserved signature sequence (MacKinnon and Yellen, 1990, Hartmann, et al., 1991, Yool and Schwarz, 1991) in between the fifth and sixth transmembrane segment. The other class consists of smaller channels with only two putative transmembrane segments and the conserved signature sequence and whose members are mainly inward rectifiers whose rectification is a result of block of an internal particle (Ficker, et al., 1994, Lopatin, et al., 1994, Nichols, et al., 1994).

This chapter focuses on the first class of K^+ channels (those gated by voltage) and presents a framework for understanding the functional

From Ion Channels to Cell-to-Cell Conversations
Edited by Latorre and Sáez, Plenum Press, New York, 1997

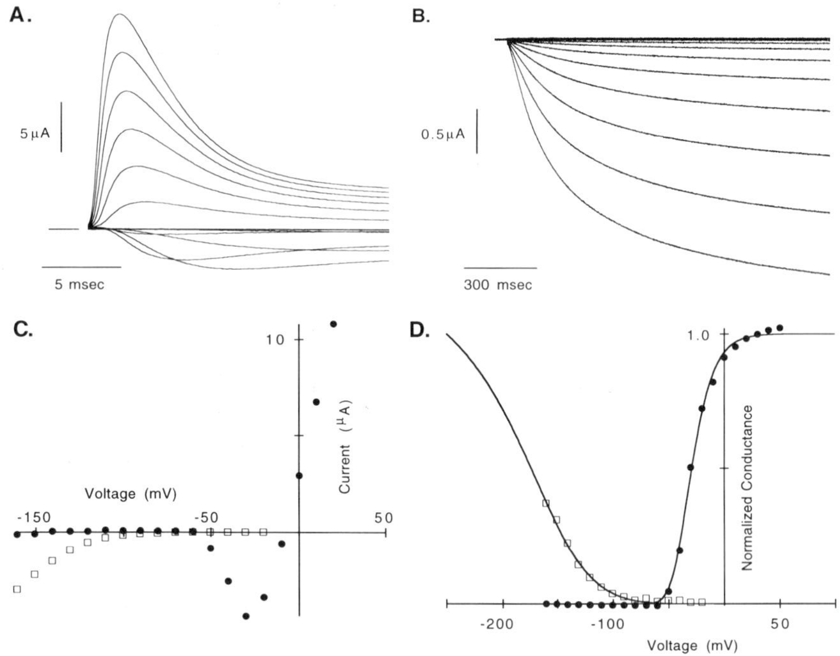

Figure 1: A. A family of current traces for WT *Shaker*. The oocyte was held at -80 mV and stepped from -160 mV to +50 mV in increments of 10 mV for 20 msec. B. Family of current traces for WT KAT1. The oocyte was held at -80 mV and stepped from -160 mV to -20 mV in increments of 10 mV for 1.35 sec. C. Peak I-V plots for *Shaker* (●) and for KAT1 (□). The points were obtained from the data in 1A and 1B. D. Normalized conductance voltage plots for *Shaker* (●) and for KAT1 (□). For *Shaker* the conductance is measured as the peak current divided by the driving force, which is approximately equal to the command voltage since $E_{rev} \approx 0$. For KAT1 the conductance is measured as the amplitude of the tail current at +40 mV. The points were obtained from the data in 1A and 1B.

diversity found within this class, despite the overall sequence similarity. Figure 1 shows a comparison of the sequence topology and the kinetic and steady state properties of two members of this family with contrasting functional properties: the fast inactivating *Shaker* channel, which is an outward rectifier, and a channel cloned from the Arabidopsis plant (KAT1), which is an inward rectifier and whose rectification is not due to relief of block by magnesium (Hoshi, 1995). The positively charged S4 sequence has been implicated in the voltage sensing of *Shaker* family channels (Liman, et al., 1991, Papazian, et al., 1991, Papazian, et al., 1995, Larsson, et al., 1996, Mannuzzu, et al., 1996), and sodium channels (Stuhmer, et al., 1989, Yang and Horn, 1995, Yang, et al., 1996). Since both KAT1 and *Shaker* contain a similar S4 sequence this raises an interesting question: how could two channels with a similar voltage sensor open with voltage of opposite sign? In this chapter we provide one answer to this question, by describing a single kinetic scheme which can account for the behavior of both outward rectifiers and inward rectifiers. We then show experimental manipulations of the *Shaker* channel to convert it into an inward rectifier with similar properties to KAT1, thus providing a model framework from which functional diversity can be achieved given very similar channel structure (Miller and Aldrich, 1996).

INWARD RECTIFICATION BY AN INACTIVATION GATING MECHANISM

Figure 2a shows a simple kinetic scheme which can describe the gating of the fast inactivating *Shaker* channel (Hoshi, et al., 1990). At negative voltages *Shaker* channels are in the closed state. A depolarizing pulse opens the channels, which then inactivate, as the N-terminal region binds to the inner mouth of the channel and blocks the pore (Armstrong and Bezanilla, 1977, Hoshi, et al., 1990, Zagotta, et al., 1990). If it were possible to shift *Shaker* gating to more negative voltages, a larger fraction of channels would be in the inactivated state at these negative voltages. It might be possible to shift *Shaker* gating far enough in the hyperpolarizing direction, so that all of the channels would be inactivated at -80 mV. In this case, a hyperpolarizing pulse from -80 mV would result in recovery from inactivation through the open state, (ie: unbinding of the ball from the inner mouth of the channel (Demo and Yellen, 1991, Ruppersberg, et al., 1991, Gomez-Lagunas and Armstrong, 1994), and then deactivation back to the closed state. Conversely, a depolarizing pulse will result in further inactivation and thus little or no current response since the channels are predominantly in the inactivated state. Such a channel with

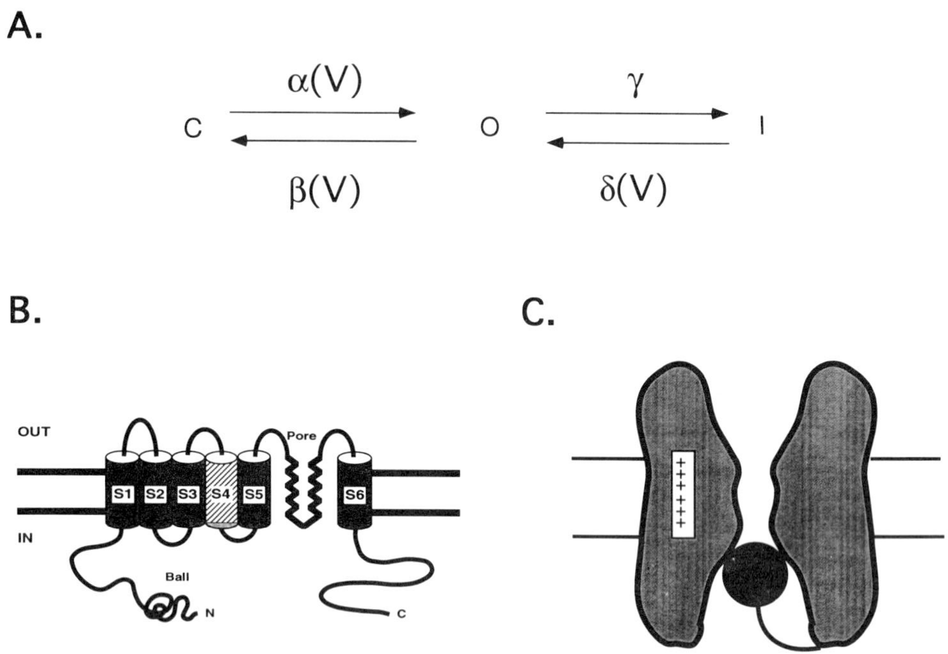

Figure 2. A. 3 state scheme used to describe the conversion of an outward rectifier into and inward rectifier. C (closed state), O (open state), I (inactivated state). B. Schematic showing the 6 putative transmembrane segments found in both *Shaker* and KAT1, including the positively charge S4 segment (hatched segment). C. Cartoon illustrating the mechanism of N-type inactivation in which the N-terminal domain (depicted as a ball) blocks the internal mouth of the channel (Hoshi, et al., 1990, Zagotta, et al., 1990).

relatively large current upon hyperpolarization and relatively little current upon depolarization will be an inward rectifier.

CONVERSION OF AN OUTWARD RECTIFIER INTO AN INWARD RECTIFIER

Three point mutants in the S4 region have been shown to shift *Shaker* activation to more negative voltages (Liman, et al., 1991, Lopez, et al., 1991, Tytgat and Hess, 1992). Figure 3 shows that combinations of these three mutations progressively shift the conductance versus voltage curve to more hyperpolarized voltages. For the most negatively shifted mutant, the triple mutant, a large fraction of the channels will be in the inactivated state at -80 mV. Figure 4 shows a series of current traces in response to voltage steps from -160 mV to

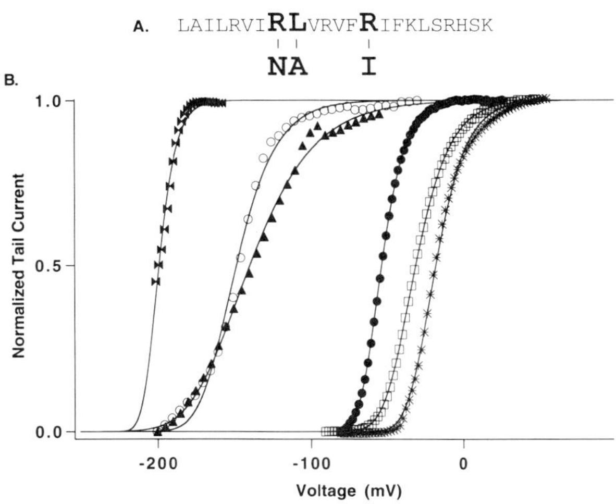

Figure 3. A. *Shaker* S4 sequence showing the position of the three amino acid replacements. B. Representative normalized conductance-voltage plots for ShB Δ4-46 (∗), R365N (□), R371I (●), R365N:R371I (O), L366A:R371I(▲), R365N:L366A:R371I(▶◀) in the *Shaker* background. The constructs lack the N-terminal inactivation domain, a 40 amino acid segment from position 4 to 46, so that the steady state conductance-voltage relationship can be measured without contamination by inactivation. Thus, ShB Δ4-46 denotes the wild-type ShB channel without the N-terminal inactivation domain. Similar shifts are seen in peak conductance-voltage plots with the intact N-terminal domain (data not shown). Isochronal measurements were made approximately 0.5-1 msec after the step to the tail voltage. The tail current voltage was as follows: ShB Δ4-46 (-80 mV), R365N (-80 mV), R371I (-120 mV), R365N:R371I (-140 mV), L366A:R371I (-140 mV), R365N:L366A:R371I (-100 mV). The data were fit with a Boltzman function raised to the fourth power.

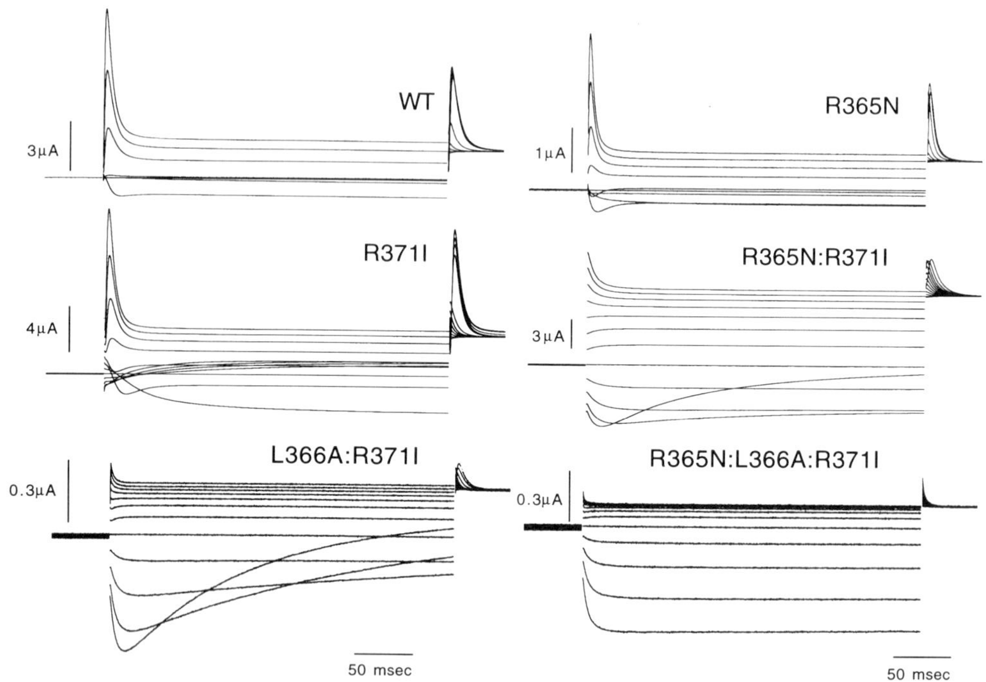

Figure 4. The 6 constructs are the same as in Figure 3, except that they all contain an intact N-terminal domain. From top to bottom: ShB, R365N, R371I, R365N:R371I, L366A:R371I, R365N:L366A:R371I. For all constructs except R365N:L366A:R371I, the oocyte was held at -80 mV, stepped through a range of voltages from -160 mV to +60 mV in increments of 20 mV for 300 msec, and then stepped to +40 mV for 100 msec. For R365N:L366A:R371I the most positive voltage is +50 mV instead of 60 mV, all else is the same as for the other constructs. n values: ShB (6), R365N (7), R371I (6), R365N:R371I (4), L366A:R371I (4), R365N:L366A:R371I (9).

+60 mV for the 5 mutants and the wild type channel and illustrates the gradual conversion from an outward rectifier to an inward rectifier as *Shaker* activation is shifted to more negative voltages. The bottom right set of traces shows the triple mutant inward rectifier with characteristics similar to KAT1. What is usually considered activation for an inward rectifier is for the triple mutant recovery from inactivation and what is usually considered deactivation is inactivation. The two single mutants and the two double mutants show characteristics intermediate to the wild type outward rectifier and the triple mutant inward rectifier. This is because there is a gradual increase in the fraction of inactivated channels at -80 mV as *Shaker* activation is shifted in the negative direction. As the *Shaker* channel behaves more like an inward rectifier the rising phase upon hyperpolarization becomes more pronounced and the subsequent decay is reduced. This is due to a gradual decrease in the rate of deactivation (O → C) relative to recovery from inactivation (I → O).

This conversion of an outward rectifier into an inward rectifier requires a minimum of three states. We eliminated the inactivated state by truncating the N-terminal region for each of the constructs (Hoshi, et al., 1990, Zagotta, et al., 1990). Figure 5 shows current families for all constructs with the N-terminal domain deleted. In contrast to the conversion of an outward rectifier to an inward rectifier shown in Figure 4, Figure 5 shows channels that become open over a larger voltage range as *Shaker* activation is shifted negative. The triple mutant without an N-terminal domain (Figure 5, bottom right) shows little time dependent current over the voltage range shown. This implies that the activation of the triple mutant inward rectifier (Figure 4, bottom right) is due to the inactivation gate of the *Shaker* channel.

Figure 5 summarizes the rectification properties of the six constructs with an intact N-terminal domain and shows a gradual increase in inward rectification as *Shaker* activation is shifted to more hyperpolarized potentials. As the fraction of channels in the inactivated state at -80 mV increases, the current response to a hyperpolarized potential will increase. In addition the current response to a depolarized potential should decrease.

Figure 6 shows a prepulse inactivation curve for the triple mutant together with an *activation* curve for KAT1 and shows that the voltage range of activation is similar for both channels. This illustrates that it is possible to shift *Shaker* activation far enough to the left to generate an inward rectifier with a similar range of *activation* to KAT1. However, the absolute shift along the voltage axis necessary to convert an outward rectifier into an inward rectifier depends on the steepness of the conductance-voltage curve of the outward rectifier and the effects

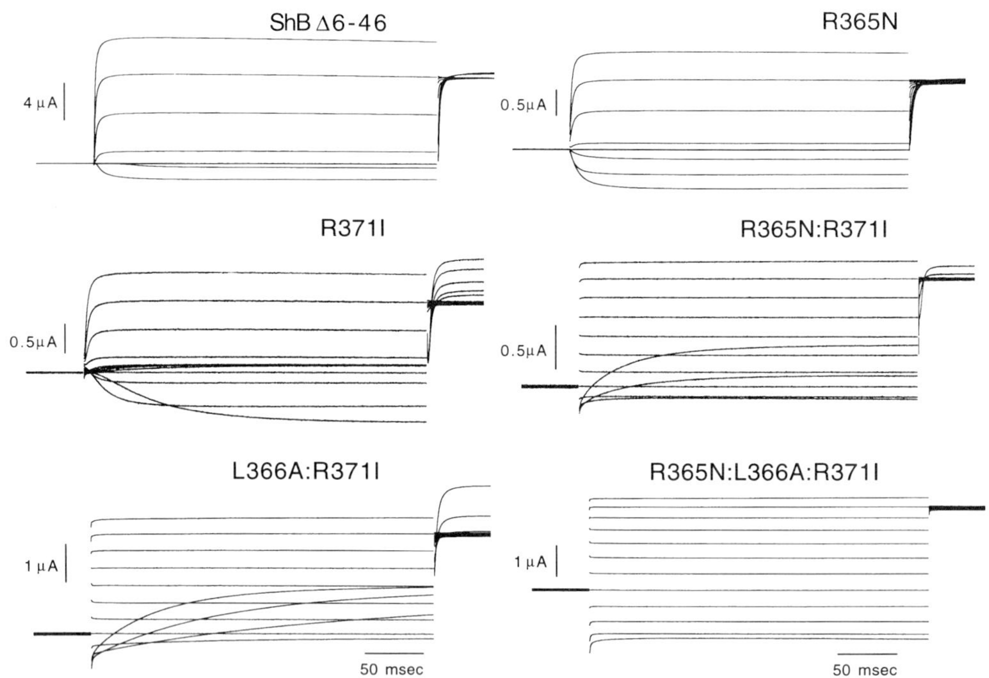

Figure 5. The 6 constructs are the same as in figure 4, except that a 40 amino acid segment from amino acid 6 to 46 has been deleted to remove inactivation (Hoshi, et al., 1990). From top to bottom: ShBΔ4-46, R365N, R371I, R365N:R371I, L366A:R371I, R365N:L366A:R371I. For all constructs the oocyte was held at -80 mV, stepped through a range of voltages from -160 mV to +60 mV in increments of 20 mV for 300 msec, and then stepped to +40 mV for 100 msec, n values: ShBΔ4-46 (3), R365N (3), R371I (3), R365N:R371I (5), L366A:R371I (4), R365N:L366A:R371I (6).

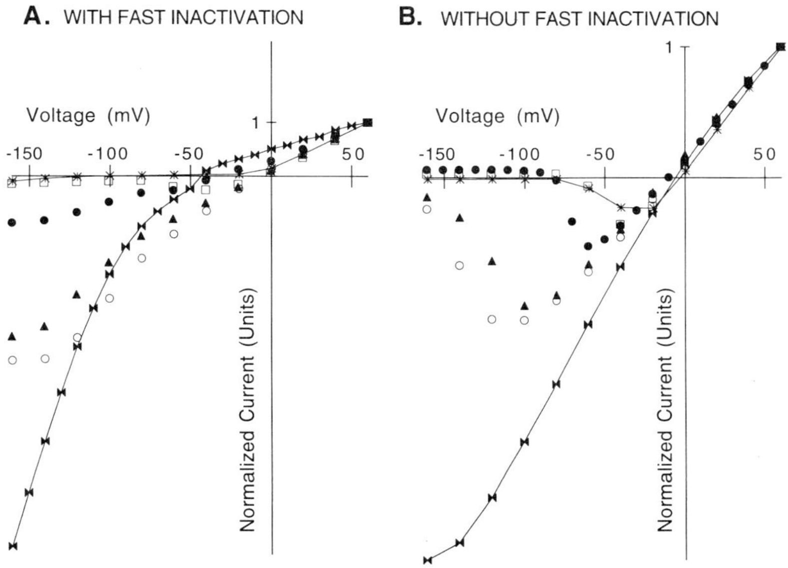

Figure 6. Representative normalized peak current-voltage for all constructs, with (A) and without (B) an N-terminal domain. The current was normalized to the value at +60 mV. For clarity the wild type data (ShB) and that from the triple mutant are shown with lines connecting the points. Leak current is not subtracted. n values and symbols: (*) ShB (6) , (□) R365N (7), (●) R371I (6), (O) R365N:R371I (4), (▲) L366A:R371I (4), (▶◀) R365N:L366A:R371I (9).

of the shift-mutants on inactivation. Thus it could be possible to convert *Shaker* into an inward rectifier with a mutant that shifts activation to a greater or lesser extent than the triple mutant described here.

CONCLUSION

We have shown that it is possible to convert the *Shaker* outward rectifier into an inward rectifier with only three amino acid substitutions. In the kinetic scheme that describes this conversion, there are two gates, defined relative to the normal *Shaker* gating behavior: an activation gate (C ↔ O), and an inactivation gate (O ↔ I). In the normal *Shaker* channel the activation gate opens the channel (after depolarization), whereas in the triple mutant *Shaker* channel the inactivation gate opens the channel (after hyperpolarization). In this scenario the gating charges, which could be amino acid charges in the

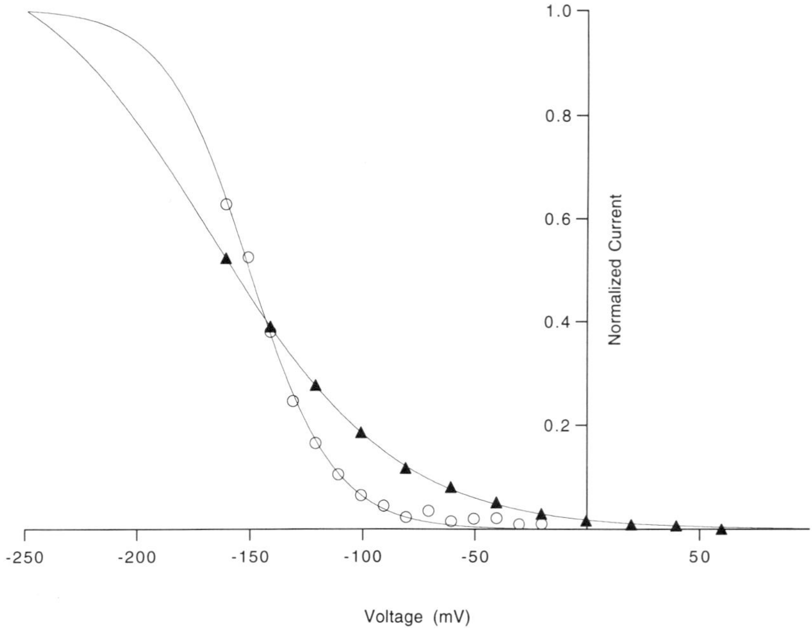

Figure 7. Normalized conductance-voltage plot of WT KAT1(O) and the *Shaker* triple mutant R365N:L366A:R371I (▲). Plots were obtained as normalized tail current versus voltage. The tail voltage was +40 mV for KAT1 and +40 mV for R365N:L366A:R371I. For the *Shaker* triple mutant this would be equivalent to a prepulse inactivation curve, and thus measures the relative number of inactivated channels at a given voltage.

S4 sequence, move in the same direction for both the outward rectifier and the inward rectifier, since opening of each channel involves a different gate. The voltage dependence of opening of the inward rectifier is not due to the same charge movement as the voltage dependence of the outward rectifier. The voltage dependence of the triple mutant is presumably due to external potassium knockoff of the N-terminal ball at a site within the electric field (Demo and Yellen, 1991).

The three state scheme used to describe the conversion of an outward rectifier into an inward rectifier is not restricted to specific molecular gating mechanisms. In particular, outward rectifiers with slow C-type inactivation (Hoshi, et al., 1991)could behave like inward rectifiers if activation were shifted far enough in the negative direction. The inward rectification behavior of the HERG channel, a member of the six transmembrane segment class of K^+ channels (Warmke and

Ganetzky, 1994, Sanguinetti, et al., 1995, Trudeau, et al., 1995) has recently been shown to be due to recovery from an inactivation mechanism with properties more similar to C-type inactivation than to N-type inactivation (Smith, et al., 1996).

Subtle changes in amino acids can dramatically change the relative rates of the different gating mechanisms in a channel and can have profound functional consequences. In this way channels with similar structures can be adjusted to have very different functional properties and thus possibly very different physiological roles. We have shown how at least two different functions can be generated. It remains to be seen whether other types of functions can be generated and if such channels are actually present in cells.

EXPERIMENTAL PROCEDURES

Electrophysiology: Data in figures 1 and 6 were leak subtracted. For *Shaker* a hyperpolarizing pulse was used, for KAT1 a depolarizing pulse was used. Data in figures 3,4 and 5 were collected without leak subtraction since the hyperpolarized shifted mutants are open over a large voltage range. Approximately 0.8-1 msec at the beginning of each step was blanked. The cut-open oocyte clamp (Dagan, Minneapolis, MN, Taglialatela, et al., 1994)was used for all experiments. The oocyte was permeabilized with 0.3% saponin; the external solution contained 110 mM KOH, 2 mM $MgCl_2$, 5 mM Hepes, pH to 7.1 with Methanesulfonic acid; the internal solution contained 110 mM KOH, 2 mM $MgCl_2$, 0.1 mM $CaCl_2$, 11 mM EGTA, 10 mM Hepes, pH to 7.2 with Methanesulfonic acid. The microelectrodes used had resistances less than 1 M-W, and were filled with 3 M KCl. The agar bridges contained 1 M Na-Methanesulfonic acid, with platinum-iridium wires placed in the agar. All recordings were done at approximately 20 ^oC.
Molecular Biology: Five sets of two oligos with the appropriate mutations were synthesized (Millipore, Bedford, MA) and inserted into the ShBΔ4-46 background to generate the five constructs lacking the N-terminal domain. These 5 mutants were then subcloned into the ShB gene to generate the five mutants with an intact N-terminal inactivation domain.

REFERENCES

Armstrong, C. M., and F. Bezanilla. 1977. Inactivation of the sodium channel. II. Gating current experiments. *Journal of General Physiology.* 70:567-590.

Chandy, K. G., and G. A. Gutman. 1995. Voltage-gated potassium channel genes. *In* Ligand-and voltage-gated ion channels. CRC Press, Inc., Ann Arbor.

Demo, S. D., and G. Yellen. 1991. The Inactivation gate of the *Shaker* K^+ channel behaves like an open-channel blocker. *Neuron.* 7:743-753.

Ficker, E., M. Taglialatela, B. A. Wible, C. M. Henley, and A. M. Brown. 1994. Spermine and spermidine as gating molecules for inward rectifier K^+ channels. *Science.* 266:1068-1072.

Gomez-Lagunas, F., and C. M. Armstrong. 1994. The relation between ion permeation and recovery from inactivation of *ShakerB* K^+ channels. *Biophysical Journal.* 67:1806-1815.

Hartmann, H. A., G. E. Kirsch, J. A. Drewe, M. Taglialatela, R. H. Joho, and A. M. Brown. 1991. Exchange of conduction pathways between two related channels. *Science.* 251:942-944.

Hille, B. 1992. Ionic channels of excitable membranes. Sinauer Associates, Inc., Sunderland, MA.

Hoshi, T. 1995. Regulation of voltage dependence of the KAT1 channel by intracellular factors. *J. Gen. Physiol.* 105:309-328.

Hoshi, T., W. N. Zagotta, and R. W. Aldrich. 1990. Biophysical and molecular mechanisms of *Shaker* potassium channel inactivation. *Science.* 250:533-538.

Hoshi, T., W. N. Zagotta, and R. W. Aldrich. 1991. Two types of inactivation in *Shaker* K^+ channels: Effects of alterations in the carboxy-terminal region. *Neuron.* 7:547-556.

Larsson, P. H., O. S. Baker, D. S. Dhillon, and E. Y. Isacoff. 1996. Transmembrane movement of the *Shaker* K^+ channel S4. *Neuron.* 16:387-397.

Liman, E. R., P. Hess, F. Weaver, and G. Koren. 1991. Voltage-sensing residues in the S4 region of a mammalian K+ channel. *Nature.* 353:752-756.

Lopatin, A. N., E. N. Makhina, and C. G. Nichols. 1994. Potassium channel block by cytoplasmic polyamines as the mechanism of intrinsic rectification. *Nature.* 372:366-369.

Lopez, G. A., Y. N. Jan, and L. Y. Jan. 1991. Hydrophobic substitution mutations in the S4 sequence alter voltage-dependent gating in *Shaker* K^+ channels. *Neuron.* 7:327-336.
MacKinnon, R., and G. Yellen. 1990. Mutations affecting TEA blockade and ion permeation in voltage-activated K+ channels. *Science.* 250:276-279.
Mannuzzu, L. M., M. M. Moronne, and E. Y. Isacoff. 1996. Direct physical measure of conformation rearrangement underlying potassium channel gating. *Science.* 271:213-216.
Miller, A. G., and R. W. Aldrich. 1996. Conversion of a delayed rectifier K+ channel to a voltage-gated inward rectifier K+ channel by three amino acid substitutions. *Neuron.* 16:853-858.
Nichols, C. G., K. Ho, and S. Hebert. 1994. Mg2+ dependent inward rectification of ROMK1 potassium channels expressed in *Xenopus* oocytes. *J. Physiol.* 476.3:399-409.
Papazian, D. M., X. M. Shao, S. Seoh, A. F. Mock, Y. Huang, and D. H. Wainstock. 1995. Electrostatic interactions of S4 voltage sensor in *Shaker* K^+ channel. *Neuron.* 14:1293-1301.
Papazian, D. M., L. C. Timpe, Y. N. Jan, and L. Y. Jan. 1991. Alteration of voltage-dependence of *Shaker* potassium channel by mutations in the S4 sequence. *Nature.* 349:305-310.
Ruppersberg, J. P., R. Frank, O. Pongs, and M. Stocker. 1991. Cloned neuronal Ik(A) channels reopen during recovery from inactivation. *Nature.* 353:657-660.
Sanguinetti, M. C., C. Jiang, M. E. Curran, and M. T. Keating. 1995. A mechanistic link between an inherited and an acquired cardiac arrhythmia: HERG encodes the I_{Kr} potassium channel. *Cell.* 81:299-307.
Smith, P.L.,T. Baukrowitz, and G. Yellen. 1996. The inward rectification mechanism of the HERG cardiac potassium channel. *Nature.* 379:833-836.
Stuhmer, W., F. Conti, H. Suzuki, X. Wang, M. Noda, N. Yahagi, H. Kubo, and S. Numa. 1989. Structural parts involved in activation and inactivation of the sodium channel. *Nature.* 339:597-603.
Taglialatela, M., B. A. Wible, R. Caporaso, and A. M. Brown. 1994. Specification of pore properties by the carboxyl terminus of inwardly rectifying K^+ channels. *Science.* 264:844-847.
Trudeau, M. C., J. W. Warmke, B. Ganetzky, and G. A. Robertson. 1995. Herg, a human inward rectifier in the voltage-gated potassium channel family. *Science.* 269:92-95.
Tytgat, J., and P. Hess. 1992. Evidence for cooperative interactions in potassium channel gating. *Nature.* 359:420-423.
Warmke, J. W., and B. Ganetzky. 1994. A family of potassium channel genes related to EAG in *Drosophila* and mammals. *Proc. Natl. Acad. Sci. USA.* 91:3438-3442.

Yang, N., A. L. George, and R. Horn. 1996. Molecular basis of charge movement in voltage-gated sodium channels. *Neuron.* 16:113-122.

Yang, N., and R. Horn. 1995. Evidence for voltage-dependent S4 movement in sodium channels. *Neuron.* 15:213-218.

Yool, A. J., and T. L. Schwarz. 1991. Alteration of ionic selectivity of a K^+ channel by mutation of the H5 region. *Nature.* 349:700-704.

Zagotta, W. N., T. Hoshi, and R. W. Aldrich. 1990. Restoration of inactivation in mutants of *Shaker* potassium channels by a peptide derived from ShB. *Science.* 250:568-571.

ASSEMBLY OF *SHAKER* K-CHANNELS FROM RANDOM MIXTURE OF SUBUNITS CARRYING DIFFERENT MUTATIONS

David Naranjo

Graduate Department of Biochemistry
Howard Hughes Medical Institute
Brandeis University
Waltham, MA 02254
USA.

Homo-multimeric membrane proteins are functional entities formed by the stable association of identical subunits that recognize each other to form an homogeneous population. With the advent of recombinant DNA technology, the manipulation of protein phenotypes combined with their heterologous expression has given rise to a whole new set of phenomena: heterogeneous populations with intermediate or combined phenotypes. They appear when different variants of the same protein are expressed in the same cell at the same time. The study of the statistical composition of these heterogeneous protein populations may lead to a better understanding of the mechanisms of synthesis, subunit association, and assembly of membrane proteins.

RANDOM ASSOCIATION OF PROTEIN SUBUNITS

When two *almost* identical and independent types of subunit are randomly mixed, in addition to the two almost identical homo-multimeric channels, a plethora of hetero-multimeric channels carrying different subunit composition can be formed. The number and proportions of heteromeric populations depend on the *n*-meric nature of the channel protein assembled and can be predicted according to the

From Ion Channels to Cell-to-Cell Conversations
Edited by Latorre and Sáez, Plenum Press, New York, 1997

binomial distribution. If two subunit types, X and Y, are independently mixed in the membrane to assemble an n-meric channel protein, $n+1$ populations containing 0,1,..., n X-subunits are formed. Then, the fraction of channels containing i X-subunits (ϕ_i) is described by:

$$\phi_i = \frac{n!}{i!(n-i)!}\theta^i(1-\theta)^{n-i} \tag{1}$$

where θ is the molar fraction of the X-subunit. In most of the multimeric channels known, subunits are assembled with radial symmetry, leaving the pore at the axis. For example: the tetrameric voltage activated K channels (MacKinnon, 1991), the pentameric acetylcholine receptor (Raftery et al., 1980) and the hexameric hemi-channel of the gap-junction (Unwin and Zampighi, 1980). For these geometrical arrangements, only for dimers and trimers the number of populations created is predicted by the binomial distribution. For oligomers having $n \geq 4$, positional isomers are created because some permutations of the i X-subunits generate non-equivalent spatial arrangements (Fig. 1). For example, a tetramer containing two X-subunits can form two types of positional isomer; both X-subunits can be either diagonally or adjacently positioned. Because positional isomers increase the number of heteromeric populations beyond the 'n+1' expected solely from the binomial distribution, they may potentially complicate the interpretation of combinatorial based experiments. As a study case, I will use the tetrameric *Shaker* K-channels to illustrate goods and evils of the binomial analysis to the multimeric proteins.

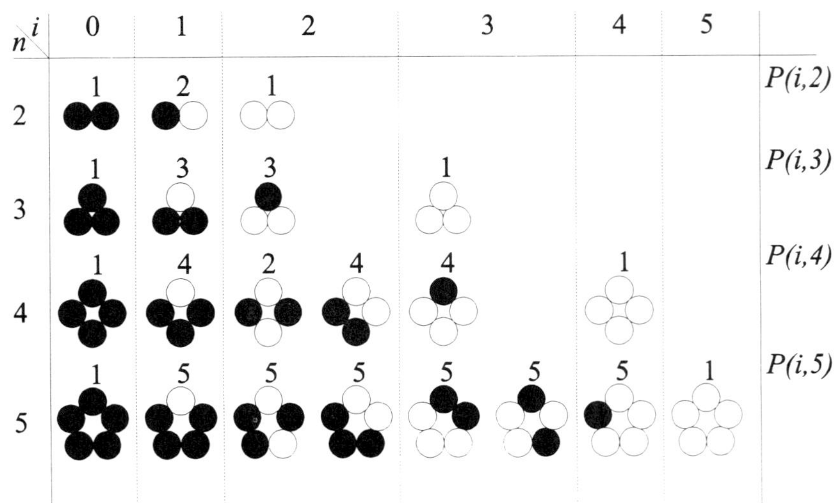

Figure 1. Populations of symmetric n-mers produced by the mixture of two subunits types. The number on top of each species, $P(i,n)$, is the possible number of geometrically equivalent permutation for i subunits in a n-mer

RANDOM ASSEMBLY OF HETEROMULTIMERIC *SHAKER* K-CHANNELS

The first assumption is that mutant and wild-type K channel subunits are randomly mixed in the plasma membrane to form tetrameric channels (a circular argument because the tetrameric nature of K-channels was shown by assuming random subunit mixing (MacKinnon, 1991; see below). This assumption may not always be legitimate, in particular if the mutual recognition domain is altered in the mutant subunit (Li et al., 1992).

If X and Y are two types of almost identical *Shaker* subunits, their combinatorial mixture generates six different populations of this tetrameric channel, including both positional-isomeric ϕ_2 forms. For a given value of θ, the composition of each population ϕ_i can be predicted from Eq. 1, and figure 1 as follows:

$$\phi_0 = (1-\theta)^4 \quad (2)$$

$$\phi_1 = 4\theta(1-\theta)^3 \quad (3)$$

$$\phi_{2a} = 4\theta^2(1-\theta)^2 \quad \text{equal subunits adjacent} \quad (4a)$$

$$\phi_{2d} = 2\theta^2(1-\theta)^2 \quad \text{equal subunits diagonally opposed} \quad (4b)$$

$$\phi_3 = 4\theta^2(1-\theta) \quad (5)$$

$$\phi_4 = \theta^4 \quad (6)$$

If the conductance and kinetics of each kind of tetramer are not distinguishable from the others (a verifiable assumption), equations 2-6 represent the contribution of each population to the total current measured from the expressed channels. Experimentally, a useful measure is to calculate the fraction of channels containing at least *i* X-subunits, $\Phi_i = \phi_i + \phi_{i+1} + ... + \phi_n$. Calculations of $\phi_i(\theta)$ and $\Phi_i(\theta)$ are shown on figure 2.

BALL-TAGGING

As an additional indicator of subunit composition of the channel population, the mutant subunit is often 'tagged' with other easily recognizable phenotype that does not alter the random nature of the subunit mixing. In *Shaker*, a useful strategy to verify the random composition of the mixture is to tag one of the subunit types with the N-type inactivation domain (MacKinnon et al., 1993; Gross and MacKinnon, 1996). Upon depolarization, N-inactivating *Shaker* K-channels

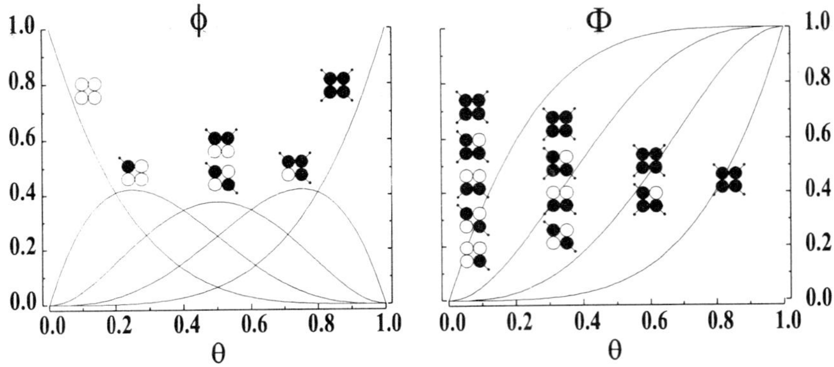

Figure 2. Predicted population composition of radially symmetric tetramers as function of the molar fraction of the filled-type of subunits. Curves where calculated from Eqs. 2-6, however, both two positional isomers for i=2 are added together. Filled phenotype is also represented schematically as ball-tagged. **A** plots the expected fraction of each population and **B** plots the fraction of tetramers containing *at least i* filled subunits.

open and then inactivate by the movement of the positively charged domain located near the N-terminus to physically occlude the intracellular mouth of the conduction pore. This process has been christened *the ball and chain* inactivation mechanism (Armstrong and Bezanilla, 1977; Armstrong, 1981, Hoshi et al., 1990). Thus, at positive membrane potentials a negligible fraction of channels are closed and the predominant equilibrium occurs between the open and inactivated sates:

$$....C \rightarrow O \leftrightarrow I \qquad \text{I}$$

In this simple scheme, because each N-inactivating *Shaker* subunit contributes a single *ball* domain to the tetramer, the inactivation rate increases linearly with the number of N-type inactivating subunits carried. Additionally, the rate of recovery is found to be independent of the number of inactivating subunits, meaning that channel inactivation is produced by one *ball* only (MacKinnon et al., 1993; Gomez-Lagunas and Armstrong, 1995). For each channel population that carries i ball-domains, it is possible to decompose the equilibrium of the scheme I:

$$O \underset{\beta}{\overset{i\alpha}{\rightleftarrows}} I \qquad \text{II}$$

where α is the microscopic inactivation rate for the channel population carrying a single ball-inactivating domain, and β is the rate of recovery from the inactivated state. Thus, for a given value of θ, the macroscopic inactivation rate can be predicted from the binomial composition of the mixture. The time course of *the inactivating fraction* of the current follows a multi-exponential relaxation, $r(t)$, produced by the kinetic contribution of each population of channels containing a variable number i of tagged subunits:

$$\mathrm{r}(t) = \sum_{i=1}^{4} w_i \,\mathrm{e}^{-t/\tau_i} \tag{7}$$

where $\tau_i = 1/(i\alpha+\beta)$ is the inactivation time constant of a population of channels containing i ball-tagged subunits and w_i is the fractional contribution of each inactivating population to the macroscopic relaxation. i.e.

$$w_i = \frac{\phi_i}{\sum_{i=1}^{4} \phi_i} \tag{8}$$

Because $x_0 = \int_0^\infty e^{-x/x_0} dx$, a macroscopic *ensemble* time constant $<\tau>$ can be defined as follows:

$$<\tau> \equiv \int_0^\infty \mathrm{r}(t)dt = \sum_{i=1}^{4} w_i \tau_i \tag{9}$$

This equation does not take into account the steady state current due to channels coming in and out of the inactivated state, however, it can predict the relaxation of the currents due to multiple binomial populations (Fig 3A). In practice, it is difficult to measure the ensemble time constant by integration of the relaxation; the capacitative transient and the activation kinetics often interfere with the determination of the inactivation kinetics at times near the beginning of the depolarization pulse. The time constant of a single exponential fit to the relaxation, although not mathematically rigorous, is a good approximation to $<\tau>$ as is shown in figure 3A. In this figure, the predicted multi-exponential relaxations and single exponential decays with time equal to $<\tau>$ are compared. The single-exponential approximation is also good for the experimental current decay. Figure 3B, shown are single-exponential fits to the measured relaxation in response to a voltage positive enough to maximize the activation and inactivation kinetics. The expected value for $<\tau>$ and the measured single exponential time constants are in good agreement for various values of θ (Fig. 3C).

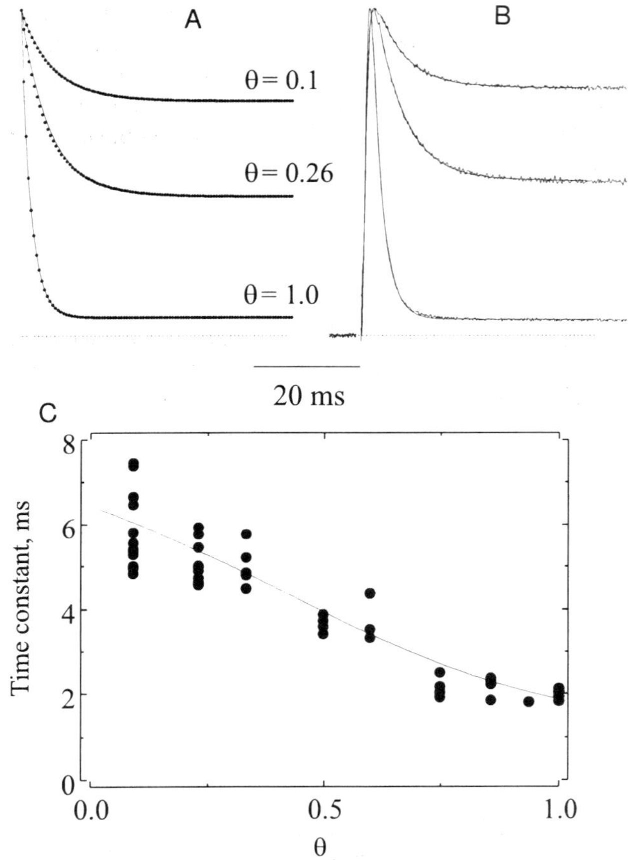

Figure 3. Kinetics of macroscopic inactivation as function of the molar fraction of the ball-tagged subunits. **A** plots the expected relaxation kinetics for K-channels at three different values of θ. Black dots are the predicted current-relaxation from the randomly distributed channels and was calculated from the equation:

$$r(t)=\sum_{i=0}^{4}\phi_i\left[\frac{\beta}{i\alpha+\beta}+\left(1-\frac{\beta}{i\alpha+\beta}\right)\cdot\exp\left\langle\frac{-t}{i\alpha+\beta}\right\rangle\right]$$

where α=0.125 ms^{-1} and β=0.03 ms^{-1} (MacKinnon et al., 1993). The continuous line traces are single-exponential relaxations with time constants equal to the weighted average of the all inactivating time constants ($<\tau>$ in Eq. 9). These curves have been scaled up to match the steady state value of the predicted relaxation. **B** shows experimentally measured relaxation currents in response to a +40 mV pulse from -90 mV holding potential. Values for θ are the same as in A. Superimposed on the records are single exponential fits to the current relaxations. Records shown are leak and capacitance subtracted. **C** shows a comparison of the time constants, measured as in B, and $<\tau>$ calculated from Eq. 9 as function of θ. Figure modified from Naranjo and Miller, 1996.

COUNTING SUBUNITS AND THE PROBLEM OF ESTIMATING θ

Heterologous expression, either in *Xenopus leavis* oocytes or in transfected cell lines, is the tool of choice for designing combinatorial based experiments. A central assumption in the previous discussion is that we know θ, the subunit molar fraction. In fact θ is not an easy variable to measure with confidence. For a rigorous determination of θ, it is not only necessary to measure the amount of each subunit type in the oocyte membrane, but also, to test for random subunit association. Several experimental approaches have been applied to obtain control of this variable, all them aiming to control the subunit composition by setting the molar ratio of the mixture of the cDNA or cRNA introduced into the expressing cell type. In the following paragraphs, I will introduce two case-studies of the manipulations of cRNA mixtures injected into *Xenopus* oocytes.

Wild type and mutant cRNAs express equally well: The tetrameric nature of *Shaker* K-channels.

This assumption was made by MacKinnon (1991) for the determination of the tetrameric nature of the *Shaker* K-channel. In this case θ should correspond to the measured ratio of radioactively labeled RNA's that are injected into the oocyte. To recapitulate this elegant strategy: The n-meric nature of the K-channel is unknown and is under quest. Wild type cRNA is co-injected with an excess of cRNA coding for a point-mutant having 250-fold lower affinity for a peptide toxin specific for K-channels. Under these mixing conditions, most of the channels are expected to be homomeric toxin-insensitive, with a few containing a single wild type subunit and fewer two or more. For example: for θ=0.1, ϕ_1=0.24 and ϕ_2=0.024 for a trimeric protein; ϕ_1=0.29 and ϕ_2=0.048 for a tetrameric; and ϕ_1=0.33 and ϕ_2=0.073 for a pentameric one. Thus for θ=0.1 most of the heteromeric channels carrying toxin-sensitive subunits contain only one of them. Then, the fractional K-current, f_{mix} in the presence of toxin [T] is approximated by:

$$f_{mix} \approx \phi_0 \frac{K_0}{K_0 + [\mathrm{T}]} + \phi_1 \frac{K_1}{K_1 + [\mathrm{T}]} + \phi_2 \frac{K_2}{K_2 + [\mathrm{T}]} \qquad (10)$$

where K_0, K_1 and K_2 correspond to the toxin's dissociation constants for the channel species containing 0, 1 and 2 toxin-sensitive subunits respectively. By increasing toxin concentration to [T] >>K_1 and K_2, channel populations carrying one or two high-affinity subunits are blocked, thus, the current is carried by the existing fraction of the homomeric low-affinity population. Then:

$$f_{\text{mix}} \approx (1-\theta)^n f_{\text{mut}},$$

where $f_{mut}=K_0/(K_0+[T])$.

Thus:

$$n \approx -\log\left(\frac{f_{mix}}{f_{mut}}\right)\log(1-\theta) \tag{11}$$

In separate experiments: with [T]=400 nM, $f_{mix}=0.49$ measured for oocytes injected with mixed cRNA, and $f_{mut}=0.73$ for those injected with the mutant cRNA only. Thus, solving Eq. 11 a value $n=3.8$ is obtained. In a similar manner, for $\theta=0.2$, f_{mut} measured 0.32, then $n= 3.7$. Thus, MacKinnon concluded that *Shaker* K-channels are tetramers. As [T] gets bigger and θ gets smaller, the approximation

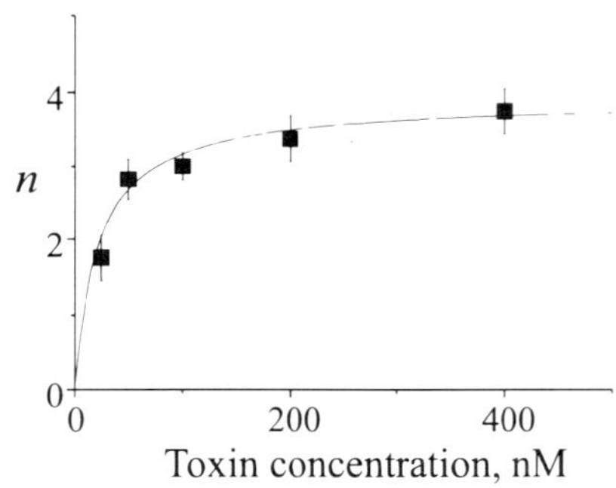

Figure 4. Determination of the subunit stoichiometry of the K-channels. Experimental values for n where calculated according to Eq. 11. The continuous line was drawn according to Eqs. 10 and 11, with $\theta=0.1$, $K_0=1020$ nM; $K_1=K_2=19$ nM and [T]=400 nM. Figure modified from MacKinnon, 1991.

made in Eq.10 gets more precise, thus n converges to 4 (Figure 4). Of course, this conclusion is critically dependent on the assumption that both cRNAs express equally well. For example: if one of the cRNAs expresses with two-fold higher efficiency than the other, application of Eq. 11 to the same results can yield to the conclusion that the channel is either an octamer or a dimer.

Wild type and mutant cRNA do not express equally well: Counting interaction points between *Shaker* K-channels and Charybdotoxin.

Charybdotoxin is a peptide toxin that binds with high specificity to the external vestibule of K-channels. This toxin, whose structure does not reveal apparent structural symmetry, can interact in four energy-equivalent orientations with the four-fold symmetric K-channel. Thus, in a given binding orientation, the toxin may not interact identically with four equivalent residues of each subunit. Studying the mechanism by which a single point mutation, *Shaker*-T449F, changes the affinity 20 thousand-fold for a charybdotoxin variant (M29I), Naranjo and Miller (1996) injected mixtures of wild-type and T449F cRNAs and asked: What is the minimum number of high-affinity subunits necessary to make a heterotetrameric channel bind toxin as well as the high-affinity homotetrameric channel? To answer this question, the strategy involving cRNA mixing with ball-tagging of the high-affinity phenotype was chosen because it allows visualization of the toxin binding effect as a reduction of the inactivating component of the current (MacKinnon et al., 1993, Lu and Miller, 1995, Gross and MacKinnon, 1996). To compute θ, cRNA expression efficiency of the inactivating high-affinity and non-inactivating low-affinity variants were estimated by measuring the current amplitude expressed after separate injections of known amount of each cRNA. For similar amounts of injected cRNA, the expressed currents differed by ~30 fold. In mixtures where θ=0.1, most of the inactivating component of the current is carried by channels having a single high-affinity subunit ϕ_1. As expected, the inactivation time constant was ~3.5-fold slower than that of channels containing four inactivating subunits (MacKinnon et al., 1993; Gomez-Lagunas and Armstrong, 1995; see also Fig. 3). Application of 10 nM toxin, a concentration that is 200-fold higher and 100-fold lower than the dissociation constants of the high-affinity and low-affinity homotetramers, respectively, produces an almost complete inhibition of the inactivating fraction of the current (mostly ϕ_1) without a detectable effect on ϕ_0, the non-inactivating population (Fig. 5A). However, as evidenced by the 20-fold faster unbinding rate (Fig. 5B upper trace), the binding to ϕ_1 has much lower affinity than the binding to the high-affinity homotetramer. At increasing values of θ, in addition to an intensified inhibition, the dissociation kinetics also show an increasing amplitude of the slow unbinding rate characteristic of the high-affinity homotetramer. At θ>0.7, the fast component is undetectable. Figure 5C plots the amplitude of the two unbinding kinetic components as a function of θ. The fast component amplitude is compared with the predicted curve for ϕ_1, and the slow component is compared with the prediction for Φ_2, the fraction of channels carrying *at least* two high-affinity subunits (compare with Fig.2). Although the data for the slow unbinding component do not match perfectly with the Φ_2 curve, these results suggest that only two high-affinity subunits are necessary to confer high-affinity binding to the tetrameric K-channel. Thus, they suggest that two 449 residues in the channel's vestibule are close enough to interact with a single residue 29 in the toxin. Moreover, because of the four-fold symmetry of the K-channel, the four 449 residues should be very close to each other (see also Heginbotham and MacKinnon, 1992).

Because *Shaker* macroscopic currents show inactivation before the activation is complete, the peak-current measurements underestimate the actual number of expressed ball-tagged channels. This underestimation, that at +40 mV can be

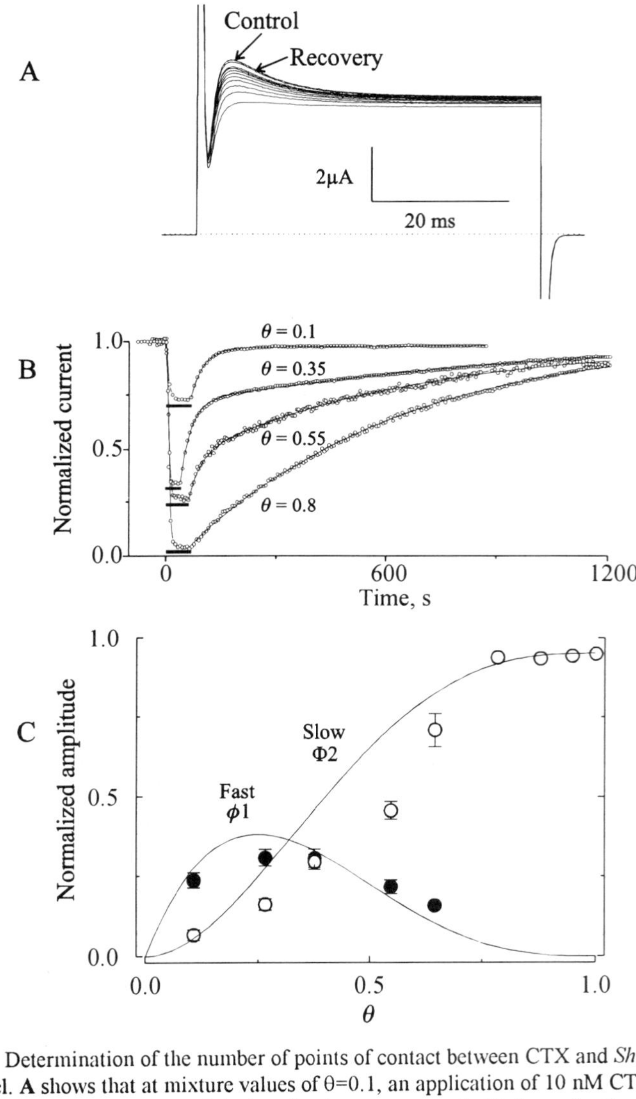

Figure 5. Determination of the number of points of contact between CTX and *Shaker* K-channel. **A** shows that at mixture values of θ=0.1, an application of 10 nM CTX-M29I produces a selective inhibition of the inactivating currents only. **B** shows the time-course of toxin inhibition of peak currents measured at various values of θ (horizontal bars). Double-exponential fits to the toxin release are superimposed on to the current measurements. On average, fast and slow time constants were 30 and 650 s. **C** shows a comparison of the amplitude of both unbinding components and the expected amplitude for ϕ_1 and Φ_2. Figure modified from Naranjo and Miller, 1996.

between 10-30% for the inactivating homotetramer (Gomez-Lagunas and Armstrong, 1995), affects the computation of θ. It also underestimate the current amplitude in mixtures. By assuming that the fraction of undetected current increases linearly with θ, it is possibly to compensate partially for deviations from Φ_2 (not shown).

RANDOM MIXTURE?

Random subunit mixing is so far a very robust phenomenon and, because of that, combinatorial analysis has been widely used as a powerful tool to address a number of inquires: i).- Mutants that do not express currents as homotetramers can be studied in subunit mixtures (Lu and Miller, 1995). ii).-Questions about functional stoichiometry as those presented above (MacKinnon, 1991; MacKinnon et al., 1993; Gomez-Lagunas and Armstrong, 1995). iii).- Questions about the cell biology of ion channels as is to address if heterologous expressed channels mix with natively expressed population (Panyi and Deutsch, 1996). However, little is known about the intermediate steps between the insertion of genetic material and the assembly and expression of K-channels. Binomially distributed populations indicate that channels are assembled from a common pool of available subunits probably in the endoplasmic reticulum. However, given the complexity of the translation and trafficking processes, random mixture is a remarkable and somewhat unexpected result. Translation may be occurring in polyribosomal structures, therefore locating identical subunits in close proximity to each other in the membrane of the rough endoplasmic reticulum. Additionally, random association may be difficult by the high degree of compartmentalization of the reticulum and the Golgi.

In a simplistic approximation to this problem, at least three processes can be distinguished: synthesis, incorporation to a common pool, and assembly. Assembly is found to be irreversible (Panyi and Deutsch, 1996). Thus, association of locally synthesized identical subunits (non-random association) will result if the rate of synthesis is faster than the rate of incorporation to the common pool. Typically, 24 hours after the injection of 0.25 ng of *Shaker*-cRNA, 1-10 μA of K^+ current at +40 mV can be measured. Hypothetically, $2x10^8$ copies of the *Shaker*-cRNA are translated into $4\text{-}40x10^6$ copies of the channel subunit (unitary current of 1 pA and open probability near unity). Thus, on average only 2-20% of all cRNA copies ever make a functional K-channel in a ~20-hour period. Even if 90% of the synthesized channels never get to the oocyte plasma membrane, this apparent low copy number suggest a very slow rate of synthesis for the *Shaker* subunits. Although extremely simplistic, these calculations may suggest that random association in the oocyte membrane originates from a slow rate of synthesis of channel subunits relative to the rate of insertion into the common pool to be assembled.

ACKNOWLEDGEMENTS

Thanks to Chris Miller and lab crew for critical reading and usually very insightful discussions. Supported by NIH-GM-31768 and Howard Hughes Medical Institute to C. Miller.

REFERENCES

Armstrong C. M. 1981. Sodium channels and gating currents. *Physiol. Rev.* 61:644-683.

Armstrong C. M. and Bezanilla F. M. 1977. Inactivation of the sodium channel. II. Gating currents experiments *J. Gen. Physiol.* 70:567-590.

Gomez-Lagunas F. and Armstrong C. M. 1995. Inactivation in *Shaker* B K-channels: a test for the number of inactivating particles on each channel. *Biophys. J.* 68:89-95.

Gross A. and MacKinnon R. 1996. Agitoxin footprinting the *Shaker* potassium channel pore. *Neuron.* 16:399-406.

Heginbotham L. and MacKinnon R. 1992. The aromatic binding site for tetraethylammonium ion on potassium channels. *Neuron.* 8:483-491.

Hoshi T., Zagotta W. N. and Aldrich R. W. 1990. Biophysical and molecular mechanisms of *Shaker* potassium channel inactivation. *Science* 250:533-538.

Li, M. Jan, Y.N. and Jan. L.Y. 1992. Specification of subunit assembly by the hydrophilic amino-terminal domain of the *Shaker* potassium channel. *Science* 257:1225-1230.

Lu, Q. and Miller. C. 1995. Silver as a probe for pore forming residues in a potassium channel. *Science*.268:304-307.

MacKinnon, R. 1991. Determination of the subunit stoichiometry of a voltage-activated potassium channel. *Nature* 350:232-235.

MacKinnon, R. Aldrich, R.W. and Lee, A. 1993. Functional stoichiometry of *Shaker* potassium channel inactivation. *Science*.262:757-759.

Naranjo D. and Miller C. 1996. A strongly interacting pair of residues on the contact surface of charybdotoxin and a *Shaker* K-channel. *Neuron.* 16:123-130.

Panyi G. and Deutsch C. 1996. Assembly and suppression of endogenous KV1.3 channels in human T cells. *J. Gen. Physiol.* 107:409-420

Raftery, M.A., Hunkapiller M.W., Stader C.D and Hood L.E. 1980. Acetylcholine receptor: Complex of homologous subunits *Science* 208:1454-1457.

Unwin, P.T.N. and Zampighi G. 1980. Structure of the junction between communicating cells. *Nature*. 283:545-549.

NEW INSIGHTS INTO THE TOPOGRAPHY, GATING, AND MODULATION OF MAXIK CHANNELS

L. Toro, M. Wallner, and P. Meera

Department of Anesthesiology
UCLA, CHS
Los Angeles, CA 90095-1778

INTRODUCTION

Large conductance voltage and calcium sensitive K channels (MaxiK, K_{Ca}, or BK channels) are stimulated upon membrane depolarization and/or by a raise in intracellular calcium. Mammalian MaxiK channels are distinguished from other calcium-sensitive potassium channels, by their large conductance and sensitivities to toxins. For example, they are insensitive to apamin (which inhibits small conductance calcium-activated K channels) and sensitive to charybdotoxin and iberiotoxin (for review *see* Garcia et al., 1991). MaxiK channels are present in almost every tissue with an important exception, the heart myocytes. Their abundance in the plasma membrane varies from tissue to tissue; the most notable cell type for their high expression are smooth muscles. MaxiK channels are modulators of neuronal excitability in brain (Crest and Gola, 1993; Robitaille et al., 1993) and in smooth muscles they set the level of contractile activity (Anwer et al., 1993; Nelson et al., 1995). One of the mechanisms (perhaps the most important) by which MaxiK channels exert their function is by sensing the intracellular Ca^{2+} levels in their vicinity (for a comment *see* Fay, 1995). An increase in intracellular Ca^{2+}, via Ca^{2+} entry or Ca^{2+} release from intracellular stores, promotes their activation leading to a prompt hyperpolarization and decrease in neuronal excitability or to relaxation

From Ion Channels to Cell-to-Cell Conversations
Edited by Latorre and Sáez, Plenum Press, New York, 1997

in smooth muscles. Conversely, their inhibition and consequent membrane depolarization causes constriction of smooth muscles (Anwer et al., 1993; Nelson et al., 1995). In certain neurons MaxiK channels have been found colocalized with voltage dependent Ca channels (Gola and Crest, 1993; Robitaille et al., 1993; Wisgirda and Dryer, 1994) leading to an efficient coupling of these two channel activities. Their localization at the presynaptic compartment suggests a possible role in neural transmission (Knaus et al, 1996). In arterial smooth muscle, MaxiK channels effectively respond to spontaneous Ca sparks (Ca^{2+} transients originating from Ca release from ryanodine-sensitive internal stores) (Nelson et al., 1995) generating spontaneous transient outward currents (STOCS) (Benham and Bolton, 1986). These two mechanisms, Ca-sparks and the resulting STOCS, appear to regulate the degree of constriction of small arteries (Nelson et al., 1995). Yet, close association of membranes containing ryanodine receptors and plasma membrane MaxiK channels has not been demonstrated. In summary, its seems that once Ca increases locally (to possibly micromolar levels) in the vicinity of MaxiK channels, their activity increases leading to hyperpolarization of the cell membrane providing an efficient feedback and levering mechanism that regulates neuronal excitability and vascular tone.

MaxiK channels are multi-modulated. Besides Ca, their activity may be regulated by hormones, lipids and neurotransmitters. Various mechanisms have been described for these regulators, such as: direct action, release of Ca from intracellular stores, membrane-delimited G protein action, and phosphorylation (for review, *see* Toro and Stefani, 1993).

Molecular cloning (Atkinson et al., 1991; Adelman et al., 1992; Butler et al, 1993; Dworetzky et al., 1994; Tseng-Crank et al., 1994; Pallanck and Ganetzky, 1994; McCobb et al., 1995; Wallner et al., 1995), and biochemical purification of functional MaxiK channels revealed that the native channels may be conformed by at least two subunits, the pore-forming α-subunit and a stimulatory β-subunit (Garcia-Calvo et al., 1994; Knaus et al., 1994b,c; McManus et al., 1995; Wallner et al., 1995). For the MaxiK channels α-subunit from *Drosophila* and mammals, variability is generated by alternative splicing. Some of these splice variants show differences in their apparent Ca^{2+} sensitivity (Lagrutta et al., 1994; Tseng-Crank, et al., 1994). In the human cDNA, four sites for alternative splicing have been found (Tseng-Crank et al, 1994; Wallner et al., 1995; McCobb et al, 1995), providing a mechanism to explain observed differences in MaxiK channel activity (for review *see* Latorre et al., 1989) or sensitivities for phosphorylation found in native tissues (Reinhart et al., 1991; Pérez et al, 1994). In addition, as discussed later in this chapter, the level of expression or association of the β subunit may be another factor contributing to the diverse K_{Ca} channel Ca^{2+}-sensitivities found within and among tissues.

An interesting feature of the mammalian K_{Ca} channel cDNA clones are two GC-rich triplet repeats at the 5'-end, which upon translation would result in homopolymeric stretches of amino acids (shaded area in Fig. 1) (Wallner et al., 1995; McCobb et al., 1995). Enlargements of similar triplet repeats are associated with genetic diseases like the fragile X-syndrome and Huntington's disease (Caskey et al., 1992; for reviews *see* Ross et al., 1993; Ross, 1995). However, the human K_{Ca} channel gene has been localized to chromosome 10 at positions q22.3 (Tseng-Crank et al., 1994) and q23.1 (McCobb et al., 1995), where no cardiovascular, respiratory, gastrointestinal, neurological or muscular diseases have been mapped yet.

MOLECULAR DETERMINANTS OF MAXIK CHANNEL FUNCTION

Primary Sequence and Topography

Pore forming α-subunit. Molecular cloning of *Drosophila* (*dslo*) and mammalian MaxiK channels (*mslo* and *hslo*) cDNA's revealed that they share limited homology to voltage-dependent K^+(K_V) channels (Atkinson et al., 1991; Adelman et al., 1992; Butler et al, 1993; Dworetzky et al., 1994; Pallanck and Ganetzky, 1994; Tseng-Crank et al., 1994; Wallner et al., 1995; McCobb et al, 1995). Due to the homology in the S4 and pore regions, MaxiK channels are classified as members of the S4 superfamily of voltage dependent ion channels (Jan and Jan, 1992). However, MaxiK channels are unique in both N- and C-terminal regions (Figs. 1 and 2) (discussed later).

Initial analysis of sequence alignments and hydrophobicity plots lead to propose that MaxiK channels had S1-S6 regions, similar to K_V channels (Fig. 2A), but were distinguished by four additional hydrophobic regions (S7-S10) at the carboxyl terminus (e.g. Adelman et al., 1992; Butler et al., 1993) (Fig. 2B, old model). In analogy to K_V channels it was thought that MaxiK channel α-subunit had an intracellular N-terminus followed by six transmembrane regions (S1-S6) and a pore region lying between S5 and S6. The nature of the carboxyl terminus was uncertain (Fig. 2B, old model).

We have recently demonstrated that the topography of MaxiK channels is unique with respect to the S4 superfamily of ion channels, since they possess a seventh transmembrane region at the N-terminus (S0) that leads to an exoplasmic N-terminus (Fig. 2, new model). This additional S0 region is consistent with the alignment of critical residues involved in K channel gating and permeation (*see* Fig. 1). Preliminary evidence supports the view that S7-S10 regions are intracellular (our unpublished results) (Fig. 2, new model). This is consistent with the lack of glycosylation of mammalian MaxiK channels in

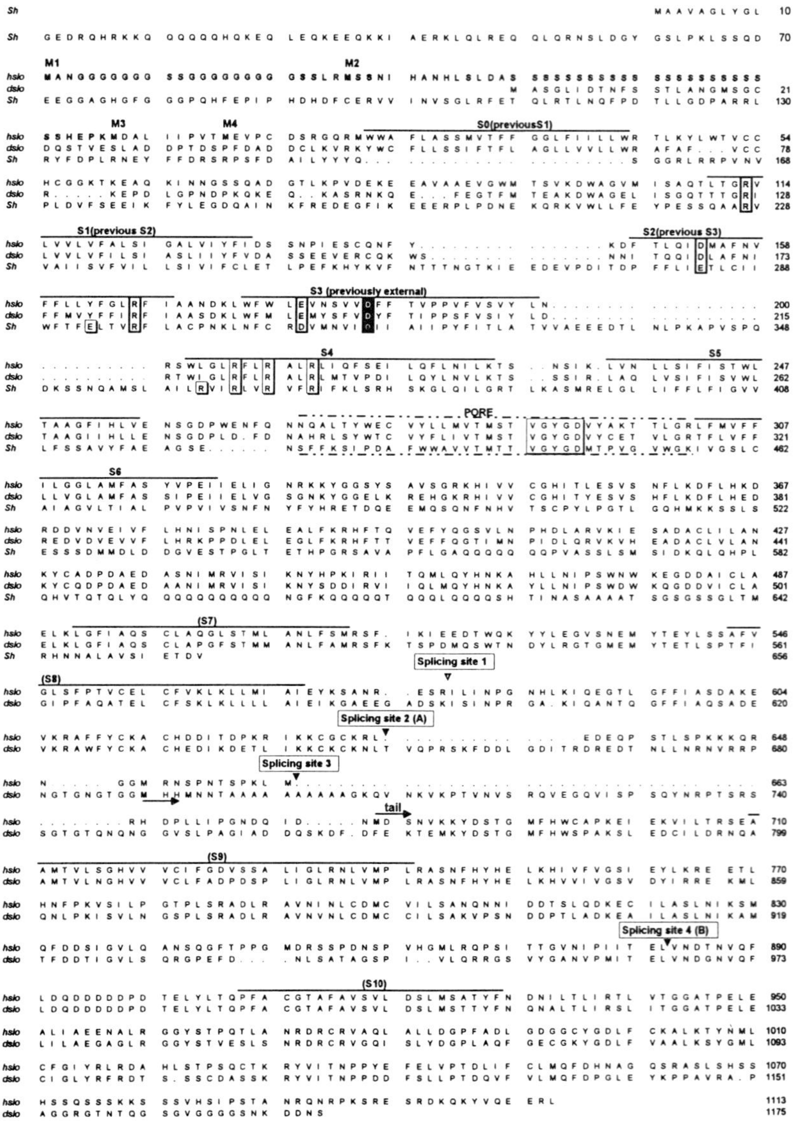

Figure 1. Sequence alignment of *hslo, dslo,* and *Shaker* K channels (GenBank Accession #s, U11058, JH0697; X06742). Arrow heads mark splicing sites in *hslo*. M1, M2, M3 and M4 correspond to putative sites for initiation of translation in *hslo*. Functional studies have used M as start codon (residue #1 in this alignment). Open boxes: aa involved in K channel voltage sensing (S2 and S4) and conduction (pore). Black box (S3), open box (S2): aa involved in electrostatic stabilization with S4. Gray boxes: conserved charged aa in S1-S3 domains. Horizontal arrows mark the beginning of clones expressing a carboxyl terminus region ("tail").

tissues (Knaus et al., 1995), despite the presence of several putative glycosylation sites in the long C-terminus.

The localization of S0 region as membrane spanning and of the N-terminus as exoplasmic, was determined by *in vitro* translation and glycosylation studies (Wallner et al., 1996). In vitro translation experiments in presence of microsomes demonstrated that S0 was localized in the membrane fraction. Only the *Drosophila* S0, but not *hslo* S0, could be glycosylated as expected from its unique N-glycosylation site. Additional evidence for an exoplasmic N-terminus was provided by signal peptide fusion experiments, where fusion constructs gave normal expression indicating that an exoplasmic N-terminus leads to perfect functional channels. This S0 region serves as a signal anchor sequence leading to an exoplasmic N-terminus, in absence of a cleavable signal sequence. Our finding that the N-terminus of MaxiK channels is *exoplasmic*, places this K channel as unique among other voltage gated K channels whose N-terminus is intracellular.

A sequence alignment of the human and the *Drosophila* MaxiK α-subunit with *Shaker* K channel is shown in Fig. 1. We have marked in parenthesis the previous assignment made by several authors (Atkinson, et al., 1991; Butler et al., 1993; Dworetzky et al., 1994; Tseng-Crank et al., 1994; McCobb et al., 1995), who proposed that the first hydrophobic segment corresponded to S1 (named S0 by us in Wallner et al., 1996), that the second hydrophobic segment was S2 (now S1), that the third hydrophobic region was S3 (now S2), and that the fourth hydrophobic region was external (now S3) (*see* also Fig. 2B). Note that our proposal is similar to the assignment suggested by Adelman, *et al.*, (1992) and Pallanck and Ganetzky for regions S1 to S3, but we now highlight the presence of S0 as an additional transmembrane region. Of particular significance in this sequence alignment are conserved charged amino acid residues that in other K_V channels participate in voltage-dependent gating (open boxes in S4 and S3) (Papazian et al., 1991; Planells-Cases et al., 1995; Seoh et al., 1996; Aggarwal and MacKinnon 1996), in K selective conduction (box in Pore region) (Heginbotham et al., 1994; Lipkind et al., 1995), and in stabilizing the voltage gating machinery by electrostatic interactions with S4 (black box in S3 and open box in S2) (Papazian et al., 1995). It would be interesting to test if other highly conserved charged residues in putative transmembrane segments S1-S3 (gray boxes) play structural or functional roles in MaxiK channels. Seoh et al. (1996) have shown that in Shaker K channel residues E283 (gray box in S2) and D316 (black box in S3) do not contribute to gating charge, and thus are not voltage sensing residues.

In this section we have summarized the evidence for a novel topography of MaxiK channels. This novel topography indicates that MaxiK channels differ from the S4 superfamily channels due to an additional transmembrane region at the N-terminus, which we have called S0.

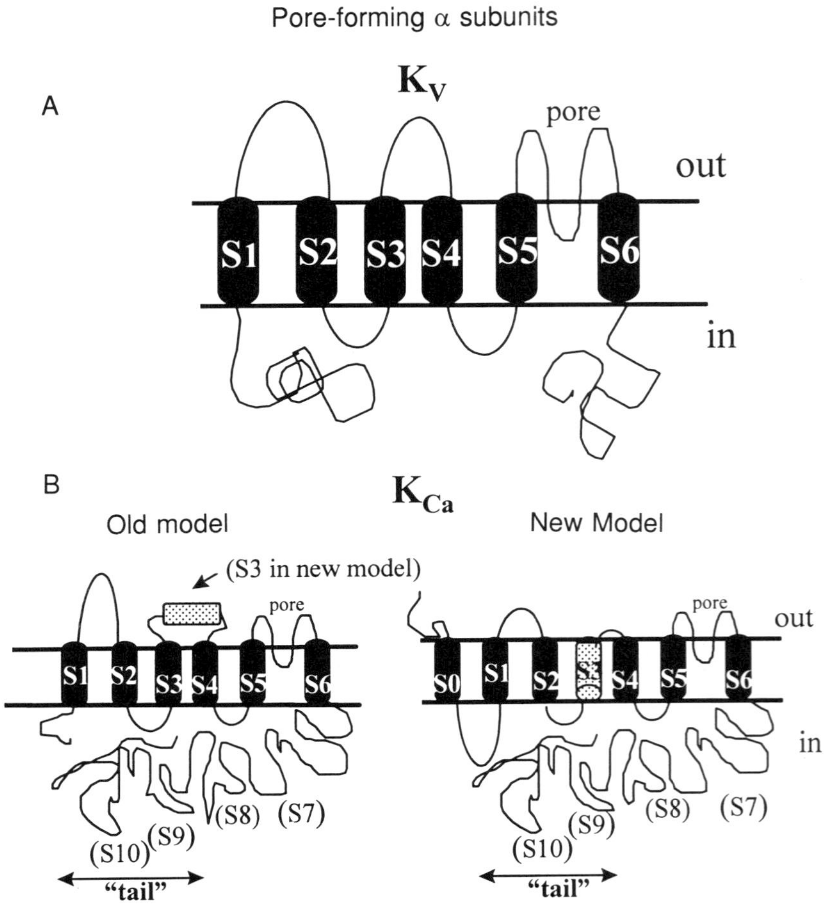

Figure 2. Topography models for K channel α-subunits of the S4 superfamily. **A**: Voltage dependent K channels (K_V). NH_2 and carboxyl termini are intracellular. Transmembrane regions S1 to S6. **B:** Old and new version for K_{Ca} channels. In "old model" an hydrophobic region (S3 in new model) was considered extracellular. (S7 to S10) are putative transmembrane regions (squared boxes). In "new model" the extracellular hydrophobic region is S3, previous S3 is S2, previous S2 is S1 and previous S1 is S0. The NH_2 terminus is exoplasmic and regions (S9-S10) are cytosolic.

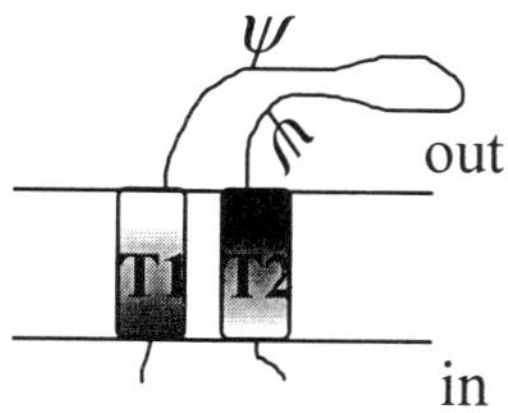

Figure 3. Topology of the human β-subunit of MaxiK channels. The extracellular loop of β subunit is glycosylated (Ψ).

Modulatory β-subunit. The β-subunit of MaxiK channels was first cloned from bovine smooth muscle (Knaus et al., 1994b). In contrast to β-subunits of K_V channels (which are cytoplasmic) (for review *see* Latorre et al., in press), the β-subunit of MaxiK channels is an integral membrane protein with two transmembrane domains (T1 and T2) (Fig. 3). Biochemical purification and treatment with N-glycanase showed that native β subunit from tracheal smooth muscle is heavily glycosylated (Garcia-Calvo et al., 1994). In Figure 3 we show the current view of the topography of the β subunit protein cloned from human myometrium (Meera et al., 1996; Wallner et al., in press), which is in agreement with the suggested topography by Knaus et al. (1994a,b) for the β subunit cloned from bovine smooth muscle.

Gating

The activity of MaxiK channel increases both upon depolarization and by increasing intracellular calcium (Latorre et al., 1989). Until recently, these two properties were indistinguishable, and the nature of the gating machinery was unclear. An unanswered question was: is the channel gated by Ca or by voltage?

Molecular cloning of MaxiK channels made possible their overexpression and consequently the analysis of channel gating under extreme conditions. One property that came to our attention was that MaxiK channels could open at Ca-concentrations that in native tissues made them undetectable because of the low number of channels present in the studied tissues. More important, was our discovery that MaxiK channels could open independently of the Ca concentration when intracellular Ca was kept at or below 100 nM (Meera et al., 1996) (Fig. 4). The only requirement for the channels to open was to provide enough electrical energy; large depolarized potentials were necessary to open the channels (half activation potentials of about 200 mV were necessary). We called this phase, which spans from 10 pM to 100 nM Ca^{2+}, as "Ca-

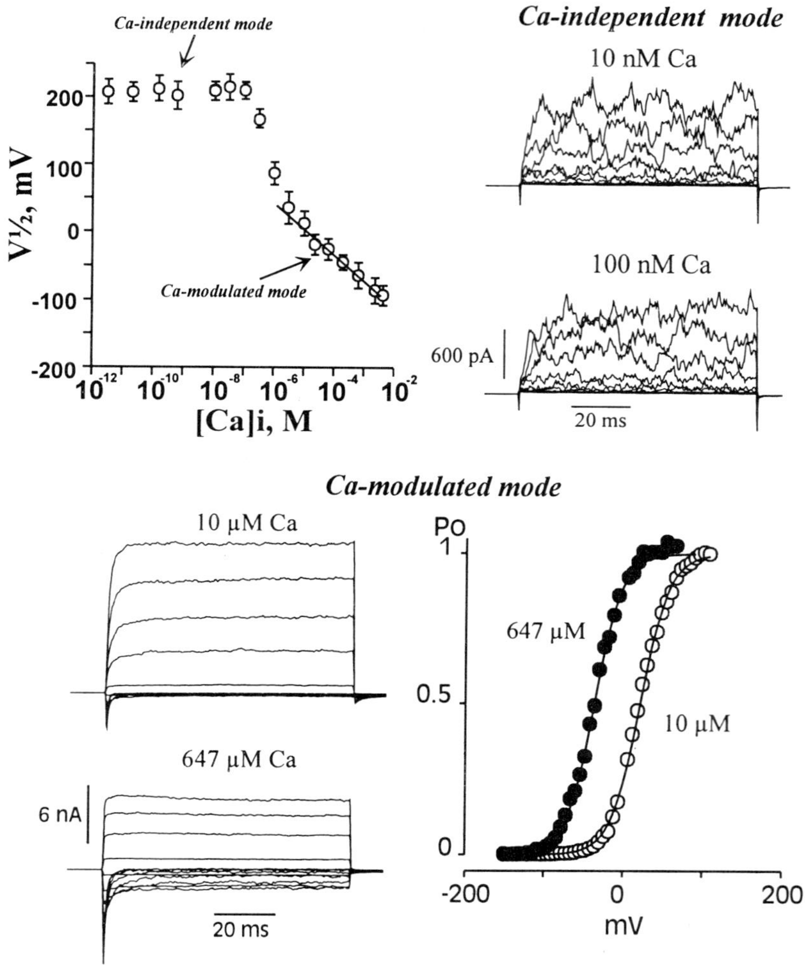

Figure 4. Ca-independent and Ca-modulated modes of gating in MaxiK channels (*hslo*). At low $[Ca^{2+}]$ ($\leq$ 100 nM) K currents flowing through *hslo* are the same and become independent of $[Ca^{2+}]i$. Micromolar Ca^{2+} or higher concentrations are needed to modulate MaxiK channels. Records are in isotonic K^+ from a holding potential of 0 mV. Half activation potential is 21 mV for 10 μM Ca^{2+} and -36 mV for 647 μM Ca^{2+}.

independent" mode. Between 100 nM and 1 μM Ca^{2+} there seems to be a huge transition between a conformation or mode that is Ca-independent, and a protein conformation which is susceptible of modulation by Ca ("Ca-modulated mode") (Meera et al., 1996). These experiments, together with the fact that MaxiK channels share homology with crucial residues involved in K_V channel gating (boxed residues in S4, Fig. 1), prompted us to seek for the presence of "gating currents", which are the result of the movement of the voltage sensor upon depolarization (Armstrong and Bezanilla, 1973; Bezanilla et al, 1991; for reviews *see* Bezanilla and Stefani, 1994; Sigworth, 1994). In agreement with our results of ionic currents, gating currents in MaxiK channels were generated independently of calcium when $[Ca^{2+}]i$ is lower than 100 nM. Voltage activation curves of charge movement (Q-V curves) were superimposed when measured in the cell-attached mode (about 100 nM Ca^{2+}) and in the presence of 5 nM Ca^{2+}. However, when Ca^{2+} is increased to the micromolar range, less voltage is necessary to move a given amount of charge. This indicates that the presence of micromolar Ca^{2+} induces a switch from a Ca-independent mode to a Ca-modulated mode. Furthermore, it indicates that voltage is sufficient to activate MaxiK channels (Stefani et al., in press).

The coupling between charge movement and pore opening of MaxiK channels differs substantially from the type of coupling seen in K_V channels. In Maxi K channels, small depolarizations induce fast transient currents (decay

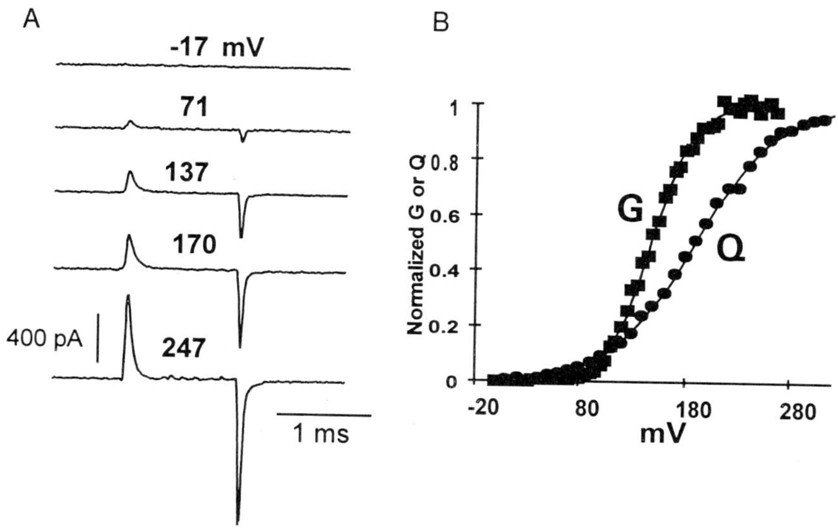

Figure 5. Gating currents in MaxiK channels. **A:** Gating currents from hslo recorded after blocking ionic currents with isotonic tetraethylammonium applied extracellularly. **B:** Normalized ionic (G) and gating (Q) activation curves in cell attached patches (nM Ca^{2+}). Most of the charge moves in open-open transitions.

time constant of about 70 μs) before channel opening. However, for larger depolarizations the decay of gating currents superimposes with the onset of the ionic currents. This behavior leads to Q-V curves that cross the conductance activation curves (G-V) near their feet, and are positive to them for most potentials (Fig. 5). This indicates that in MaxiK channels most of the charge moves during open transitions that are parallel to closed states (Stefani et al, in press), in contrast to K_V channels where charge moves in closed states that are sequentially related to the open state (Bezanilla et al., 1994; Bezanilla and Stefani 1994).

We have demonstrated that MaxiK channels are voltage gated, which goes along with the conserved charged residues within regions S3 and S4 thought to be critical for voltage dependent gating in K_V channels (structural in S3 and voltage sensing residues in S4) (Papazian et al., 1995; Plannells-Cases et al., 1995; Seoh et al., 1996; Aggarwal and MacKinnon, 1996) (*see* Fig. 1). Although S4 residues that in Shaker K channels contribute to the number of charges moved during the gating process (R365, R368 and R371) (Seoh et al., 1996) are conserved in MaxiK channels (R207, R210 and R213 in *hslo; see* Fig. 1), the number of charges per channel in MaxiK channels is less (4) (Stefani et al, submitted) than the number counted for Shaker K channels which is about 14 (Schoppa et al, 1992; Noceti et al, 1996; Seoh, et al., 1996; Aggarwal and

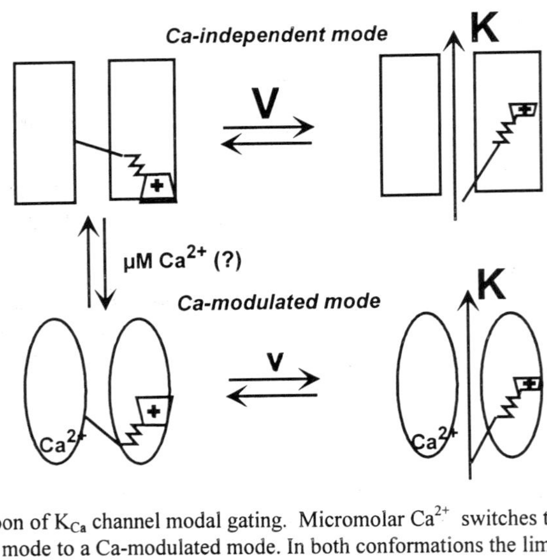

Figure 6. Cartoon of K_{Ca} channel modal gating. Micromolar Ca^{2+} switches the channel from a Ca-independent mode to a Ca-modulated mode. In both conformations the limiting charge (+) moved and current flow (K) are the same, but less voltage (V) is required when Ca^{2+} is present.

MacKinnon, 1996). This suggests that other residues in Shaker K channels may be contributing to the total number of charges moved during the gating process. In fact, it has been shown that residues E293 in S2 region (Seoh et al., 1996; open box in Fig. 1) and R362 in S4 region (Aggarwal and MacKinnon ,1996) also contribute to the total amount of charge in Shaker K channels.

The gating process of MaxiK channels seems to work in a modal way (Fig. 6). MaxiK channels switch from a Ca-independent mode to a Ca-modulated mode in the presence of micromolar calcium. In both modes, the channel moves the same amount of charge and may reach the maximum open probability, but in the Ca-independent mode much more electrical energy is needed for this to occur (Meera et al., 1996; Stefani et al., in press). MaxiK channel gating is not a simple process, single channel studies of the *Drosophila* homologue have lead Silberberg et al. (1996) to propose a modal way of gating also for the Ca-modulated mode.

Modulation other than intracellular Ca^{2+}

Intracellular and membrane delimited. MaxiK channels can be modulated (activated or inhibited) by intracellular metabolic pathways without a change in intracellular Ca^{2+}. In general, activation occurs when a process increases the voltage or apparent Ca^{2+} sensitivities of the channel; conversely, inhibition occurs when the voltage or apparent Ca^{2+} sensitivities of the channel decrease. In native tissues, much has been studied in relation to their regulation by protein kinases and phosphatases (for review *see* Toro and Stefani, 1993; Bielefeldt and Jackson, 1994; Reinhart and Levitan, 1995). Protein kinase A (cAMP-dependent) has dual effects on K_{Ca} channels expressed in the same tissue (Reinhart, et al., 1991; Perez and Toro, 1994). At the molecular level, this dual modulation could be due to the presence of different splice variants within the same tissues or due to differential coupling with β subunit as recently reported by Dworetzky et al., (1996). Protein kinase G (cGMP dependent) has been shown to activate this channel (Robertson et al., 1993; Alioua, et al., 1995); whereas protein kinase C (diacylglycerol and Ca^{2+} dependent) seems to inhibit K_{Ca} channels (Minami et al, 1993). Direct G-protein modulation has also been observed in smooth muscles coupled to β-adrenergic stimulation (Kume et al., 1992; Scornik et al, 1993). Direct phosphorylation of the cloned mammalian K_{Ca} channel has not yet been demonstrated. Recently, it was demonstrated that the timing of functional expression of MaxiK channels in developing chick cilliary ganglion neurons is determined by post-translational modifications induced by a target-derived factor(s) (Subramony et al., 1996).

β-subunit induced activation. Biochemical purification of a tightly bound ancillary subunit (β-subunit) followed by its molecular cloning (Knaus et al., 1994b,c) have permitted the elucidation of another mechanism of modulation of mammalian MaxiK channels. The β-subunit of the mammalian MaxiK channel (human and bovine homologues) increases the voltage-sensitivity of this channel in a dramatic way. The voltage activation curve is shifted to negative potentials by about 100 mV (McManus et al., 1995; McCobb et al., 1995; Wallner et al., 1995; Meera, et al., 1996) (Fig. 7, middle). However, the Drosophila MaxiK channel is unaffected by the mammalian β-subunit (Wallner et al., 1996) (Fig. 7, left). At present, it is not known if a Drosophila β-subunit exists.

The specific modulation of the mammalian MaxiK channels by the β-subunit does not occur at all Ca-concentrations tested. Our experiments have demonstrated that the stimulatory effect of the β-subunit is only observed when the MaxiK channels is in the Ca-modulated mode (in the presence of micromolar Ca^{2+}) (Meera et al., 1996). This requisite is another proof that micromolar Ca^{2+} changes the conformation of the MaxiK channel, in this case from an unresponsive interaction (activation kinetics change, but not steady-, state properties) to a mode where the β-subunit increases dramatically the open probability of the channel.

The β-subunit of the mammalian MaxiK channel modulates the activation pathway diminishing the voltage (or Ca^{2+}) necessary to activate the channel in its Ca-modulated mode. In Fig. 7 (middle) co-expression of *hslo* (α pore forming subunit) with its β-subunit caused a 100 mV shift of the voltage-activation curve to negative potentials. β-subunits of Kv channels, on the other

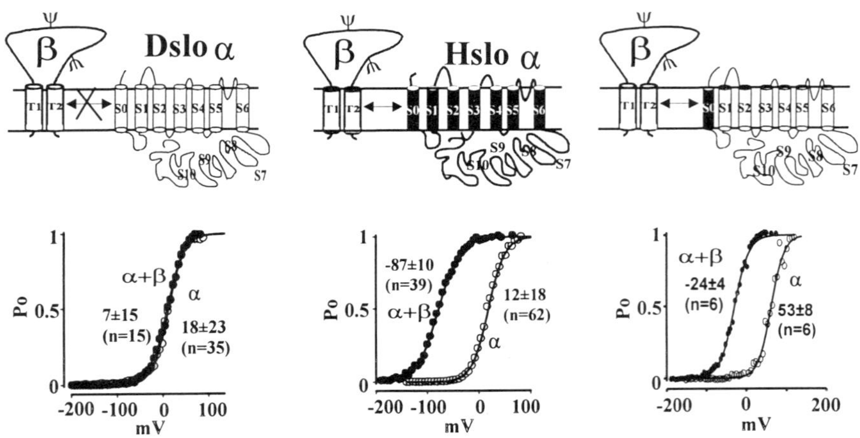

Figure 7. Molecular determinant for β-subunit modulation of *hslo*. Left panel: Dslo (light lines) is not modulated by the K_{Ca} β subunit. Po vs. voltage curves are the same when α or α+β are expressed. Middle panel: Hslo (dark lines) is positively regulated by its β subunit. Po vs. voltage curve is shifted by about 100 mV. Much less voltage is required to open the channels when β subunit is coexpressed. Right panel: Chimeric construct (S0 and N-terminus from Hslo; rest of protein from Dslo) that makes Dslo respond to β subunit modulation.

hand, are not only dissimilar in their topography (cytosolic vs. transmembrane) but also in their mechanism of modulation. β-subunits of K_V channels affect not the activation pathway, but the inactivation processes of K_V channels (for review *see* Latorre et al, in press). Up to date, only $K_V\beta2$ slightly affect channel ($K_V1.5$) activation producing a 10 mV shift of the voltage-activation curve (Heinemann et al., 1996).

Pharmacological properties of MaxiK channels are also modulated by its β subunit. For example, only in the presence of β-subunit are MaxiK channels opened by a dehydrosoyasaponin (McManus et al., 1995), and their sensitivity to iberiotoxin blockade diminishes (Dworetzky et al., 1996).

Molecular determinants in the pore forming α-subunit required for β-subunit modulation. To determine the region in the MaxiK channel required for β-subunit activation we made chimeras between the unresponsive Drosophila MaxiK channel (Fig. 7, left) and the responsive human MaxiK channel (*hslo*) (Fig. 7, center). Note that only when the β subunit is coexpressed with *hslo* the voltage activation curve (open probability, Po vs. voltage) is left shifted, indicating that less voltage is necessary to open the channel. A chimera where the N-terminus including S0 is from *hslo* makes the otherwise unresponsive *Drosophila* channel (*dslo*) to be modulated by the β-subunit (Fig. 7, right). This gain of function in *dslo* demonstrates that S0 is required for β-subunit modulation (Wallner et al., 1996).

CONCLUSIONS

In summary, the molecular cloning of two subunits of the MaxiK channel complex, the pore forming α-subunit and the modulatory β-subunit have lead to important changes in previous concepts of how MaxiK channels work. We now know that: **1)** *MaxiK channels are voltage-gated;* **2)** *MaxiK channels belong to the S4 superfamily, but are unique due to the presence of an additional seventh transmembrane domain (S0) at the N-terminus and four hydrophobic segments (S7-S10) at the C-terminus*; **3)** *they operate in a modal way;* ***4)*** *Ca-switches the channel from a Ca-independent mode to a Ca-modulated mode and to a β-subunit modulated mode; and* ***4)*** *S0 region is required for β-subunit modulation.*

ACKNOWLEDGMENTS

This work was supported by NIH grants to L.T. This work was done during the tenure of LT of an Established Investigatorship from the American Heart Association. We thank Dr. E. Stefani for his comments.

REFERENCES

Adelman, J.P., Shen, K.Z., Kavanaugh, M.P., Warren, R.A., Wu, Y.N., Lagrutta, A., Bond, C.T., and North, R.A., 1992, Calcium-activated potassium channels expressed from cloned complementary DNAs, *Neuron* 9:209.

Aggarwal, S.K., and MacKinnon, R., 1996, Contribution of the S4 segment to gating charge in the Shaker K^+ channel, *Neuron* 16:1169.

Alioua, A., Huggins, J.P., and Rousseau, E., 1995, PKG-Iα phosphorylates the α-subunit and upregulates reconstituted GK_{Ca} channels from tracheal smooth muscle, *Am. J. Physiol.* 268:L1057.

Anwer, K., Oberti, C., Pérez, G.J., Perez-Reyes, N., McDougall, J.K., Monga, M., Sanborn, B.M., Stefani, E., and Toro, L., 1993, Calcium-activated K^+ channels as modulators of human myometrial contractile activity, *Am. J. Physiol.* 265:C976.

Armstrong, C.M., and Bezanilla, F., 1973, Currents related to movement of the gating particles of the sodium channels, *Nature* 242:459.

Atkinson, N.S., Robertson, G.A., and Ganetzky, B., 1991, A component of calcium-activated potassium channels encoded by the *Drosophila slo* locus, *Science* 253:551.

Benham, C.D., and Bolton, T.B., 1986, Spontaneous transient outward currents in single visceral and vascular smooth muscle cells of the rabbit, *J. Physiol. (London)* 381:385.

Bezanilla, F., Perozo, E., Papazian, D.M., and Stefani, E., 1991, Molecular basis of gating charge immobilization in Shaker potassium channels, *Science* 254:679.

Bezanilla, F., Perozo, E., and Stefani, E., 1994, Gating of *Shaker* K^+ channels: II. The components of gating currents and a model of channel activation, *Biophys. J.* 66:1011.

Bezanilla, F., and Stefani, E., 1994, Voltage dependent gating of ionic channels, *Ann. Rev. Biophys. Biochem. Chem.* 23:819.

Bielefeldt, K., and Jackson, M.B., 1994, Phosphorylation and dephosphorylation modulate a Ca^{2+}-activated K^+ channel in rat peptidergic nerve terminals, *J. Physiol. (London)* 475:241.

Butler, A., Tsunoda, S., McCobb, D.P., Wei, A., and Salkoff, L., 1993, *mSlo*, a complex mouse gene encoding "maxi" calcium-activated potassium channels, *Science* 261:221.

Caskey, C.T., Pizzuti, A., Fu, Y.H., Fenwick, R.G.J., and Nelson, D.L., 1992, Triplet repeat mutations in human disease, *Science* 256:784.

Crest, M., and Gola, M., 1993, Large conductance Ca^{2+}-activated K^+ channels are involved in both spike shaping and firing regulation in *Helix* neurones, *J. Physiol. (London)* 465:265.

Dworetzky, S.I., Trojnacki, J.T., and Gribkoff, V.K., 1994, Cloning and expression of a large-conductance calcium-activated potassium channel, *Mol. Brain Res.* 27:189.

Dworetzky, S.I., Boissard, C.G., Lum-Ragan, J.T., McKay, M.C., Post-Munson, D.J., Trojnacki, J.T., Chang, C.P., and Gribkoff, V.K, 1996, Phenotypic alteration of a human BK (*hslo*) channel by *hsloβ* subunit coexpression: changes in blocker sensitivity, activation/relaxation and inactivation kinetics, and protein kinase A modulation, *J. Neurosc.* 16:4543.

Fay, F.S., 1995, Calcium sparks in vascular smooth muscle: relaxation regulators, *Science* 270:588.

Garcia, M.L., Galvez, A., Garcia-Calvo, M., King, V.F., Vazquez, J., and Kaczorowski, G.J., 1991, Use of toxins to study potassium channels, *J. Bioenerg. Biomembr.* 23:615.

Garcia-Calvo, M., Knaus, H.-G., McManus, O.B., Giangiacomo, K.M., Kaczorowski, G.J., Garcia, M.L., and Knaus, H.G., 1994, Purification and reconstitution of the high-conductance, calcium-activated potassium channel from tracheal smooth muscle, *J. Biol. Chem.* 269:676.

Gola, M., and Crest, M., 1993, Colocalization of active K_{Ca} channels and Ca^{2+} channels within Ca^{2+} domains in helix neurons, *Neuron* 10:689.

Heginbotham, L., Lu, Z., Abramson, T., and MacKinnon, R., 1994, Mutations in the K^+ channel signature sequence, *Biophys. J.* 66:1061.

Heinemann, S.H., Rettig, J., Graack, H.R., and Pongs, O., 1996. Functional characterization of K_V channel β-subunits from rat brain, *J. Physiol. (London)* 493:625.

Jan, L.Y., and Jan, Y.N., 1992, Tracing the roots of ion channels, *Cell* 69:715.

Knaus, H.G., Eberhart, A., Kaczorowski, G.J., and Garcia, M.L., 1994a, Covalent attachment of charybdotoxin to the β-subunit of the high conductance Ca^{2+}-activated K^{+} channel. Identification of the site of incorporation and implications for channel topology, *J. Biol. Chem.* 269:23336.

Knaus, H.G., Eberhart, A., Koch, R.O.A., Munujos, P., Schmalhofer, W.A., Warmke, J.W., Kaczorowski, G.J., and Garcia, M.L., 1995, Characterization of tissue-expressed α subunits of the high conductance Ca^{2+}-activated K^{+} channel, *J. Biol. Chem.* 270:22434.

Knaus, H.G., Folander, K., Garcia-Calvo, M., Garcia, M.L., Kaczorowski, G.J., Smith, M., and Swanson, R., 1994b, Primary sequence and immunological characterization of β-subunit of high conductance Ca^{2+}-activated K^{+} channel from smooth muscle, *J. Biol. Chem.* 269:17274.

Knaus, H.G., Garcia-Calvo, M., Kaczorowski, G.J., and Garcia, M.L., 1994c, Subunit composition of the high conductance calcium-activated potassium channel from smooth muscle, a representative of the *mSlo* and *slowpoke* family of potassium channels, *J. Biol. Chem.* 269:3921.

Knaus, H.G., Schwarzer, C., Koch, R.O., Eberhart, A., Kaczorowski, G.J., Glossmann, H., Wunder, F., Pongs, O., Garcia, M.L., and Sperk, G., 1996, Distribution of high-conductance Ca^{2+}-activated K^{+} channels in rat brain: targeting to axons and nerve terminals, *J. Neurosci.* 16:955.

Kume, H., Graziano, M.P., and Kotlikoff, M.I., 1992, Stimulatory and inhibitory regulation of calcium-activated potassium channels by guanine nucleotide-binding proteins, *Proc. Natl. Acad. Sci. USA* 89:11051.

Lagrutta, A., Shen, K.Z., North, R.A., and Adelman, J.P., 1994, Functional differences among alternatively spliced variants of *Slowpoke*, a *Drosophila* calcium-activate potassium channel, *J. Biol. Chem.* 269:20347.

Latorre, R., Stefani E., and Toro L., Balls, chains and potassium channels, *in*: "Calcium and Cellular Metabolism: Transport and Regulation," J. R. Sotelo and J. C. Benech, eds., Plenum Press, New York, In Press.

Latorre, R., Oberhauser, A., Labarca, P., and Alvarez, O., 1989, Varieties of calcium-activated potassium channels, *Ann. Rev. Physiol.* 51:385.

Lipkind, G.M., Hanck, D.A., Fozzard, H.A., 1995, A structural motif for the voltage-gated potassium channel pore, *Proc. Natl. Acad. Sci. USA* 92:9215.

McCobb, D.P., Fowler, N.L., Featherstone, T., Lingle, C.J., Saito, M., Krause, J.E., and Salkoff, L., 1995, A human calcium-activated potassium channel gene expressed in vascular smooth muscle, *Am. J. Physiol.* 269:H767.

McManus, O.B., Helms, L.M., Pallanck, L., Ganetzky, B., Swanson, R., and Leonard, R.J., 1995, Functional role of the beta subunit of high conductance calcium-activated potassium channels, *Neuron* 14:645.

Meera, P., Wallner, M., and Toro, L., 1996, A calcium switch for the functional coupling between α (*hslo*) and β subunits ($K_{V,Ca}\beta$) of maxi K channels, *FEBS Lett.* 382:84.

Minami, K., Fukuzawa, K., and Nakaya, Y., 1993, Protein kinase C inhibits the Ca^{2+}-activated K^+ channel of cultured porcine coronary artery smooth muscle cells, *Biochem. Biophys. Res. Commun.* 190:263.

Nelson, M.T., Cheng, M.R., Santana, L.F., Bonev, A.D., Knot, H.J., and Lederer, W.J., 1995, Relaxation of arterial smooth muscle by calcium sparks, *Science* 270:633.

Noceti, F., Baldelli, P., Wei, X., Qin, N., Toro, L., Birnbaumer, L., and Stefani, E., 1996, Effective gating charges per channel in voltage dependent K^+ and Ca^{2+} channels, *J. Gen. Physiol.* In press.

Pallanck, L., and Ganetzky, B., 1994, Cloning and characterization of human and mouse homologs of the *Drosophila* calcium-activated potassium channel gene, *slowpoke*, *Hum. Mol. Genet.* 3:1239.

Papazian, D.M., Shao, X.M., Seoh, S.-A., Mock, A.F., Huang, Y., and Wainstock, D.H., 1995, Electrostatic interactions of S4 voltage sensor in Shaker K^+ channel, *Neuron* 14:1293.

Papazian, D.M., Timpe, L.C., Jan, Y.N., and Jan, L.Y., 1991, Alteration of voltage-dependence of Shaker potassium channel by mutations in the S4 sequence, *Nature* 349:305.

Pérez, G., and Toro, L., 1994, Differential modulation of large conductance K_{Ca} channels by PKA in pregnant and nonpregnant myometrium, *Am. J. Physiol.* 266:C1459.

Planells-Cases, R., Ferrer-Montiel, A.V., Patten, C.D., and Montal, M., 1995, Mutation of conserved negatively charged residues in the S2 and S3 transmembrane segments of a mammalian K^+ channel selectively modulates channel gating, *Proc. Natl. Acad. Sci. USA* 92:9422.

Reinhart, P.H., Chung, S., Martin, B.L., Brautigan, D.L., and Levitan, I.B., 1991, Modulation of calcium-activated potassium channels from rat brain by protein kinase A and phosphatase 2A, *J. Neurosci.* 11:1627.

Reinhart, P.H., and Levitan, I.B., 1995, Kinase and phosphatase activities intimately associated with a reconstituted calcium-dependent potassium channel, *J. Neurosci.* 15:4572.

Robertson, B.E., Schubert, R., Hescheler, J., and Nelson, M.T., 1993, cGMP-dependent protein kinase activates Ca-activated K channels in cerebral artery smooth muscle cells, *Am. J. Physiol.* 265:C299.

Robitaille, R., Garcia, M.L., Kaczorowski, G.J., and Charlton, M.P., 1993, Functional colocalization of calcium and calcium-gated potassium channels in control of transmitter release, *Neuron* 11:645.

Ross, C.A., 1995, When more is less: pathogenesis of glutamine repeat neurodegenerative diseases, *Neuron* 15:493.

Ross, C.A., McInnis, M.G., Margolis, R.L., and Li, S.H., 1993, Genes with triplet repeats: candidate mediators of neuropsychiatric disorders, *Trends Neurosci.* 16:254.

Schoppa, N.E., McCormack, K., Tanouye, M.A., and Sigworth, F.J., 1992, The size of gating charge in wild-type and mutant *Shaker* potassium channels, *Science* 255:1712.

Scornik, F.S., Codina, J., Birnbaumer, L., and Toro, L., 1993, Modulation of coronary smooth muscle K_{Ca} channels by $G_s\alpha$ independent of phosphorylation by protein kinase A, *Am. J. Physiol.* 265:H1460.

Seoh, S.A., Sigg, D., Papazian, D.M., Bezanilla, F., 1996, Voltage-sensing residues in the S2 and S4 segments of the Shaker K^+ channel, *Neuron* 16:1159.

Sigworth, F.J., 1994, Voltage gating of ion channels, *Q. Rev. Biophys.* 27:1.

Silberberg, S.D., Lagrutta, A., Adelman, J.P., and Magleby, K.L., 1996, Wanderlust kinetics and variable Ca^{2+}-sensitivity of *Drosophila*, a large conductace Ca^{2+}-activated K^+ channel, expressed in oocytes, *Biophys. J.* 70:2640.

Stefani, E., Ottolia, M., Noceti, F., Olcese, R., Wallner, M., Latorre, R., and Toro, L. Voltage controlled gating in a large conductance Ca^{2+}-sensitive K^+ channel (*hslo*), *Proc. Natl. Acad. Sci. USA*. In press.

Subramony, P., Raucher, S., Dryer, L., and Dryer, S.E., 1996, Postranslational regulation of Ca^{2+}-activated K^+ currents by a target-derived factor in developing parasympathetic neurons, *Neuron* 17:115.

Toro, L., and Stefani, E., 1993, Modulation of maxi calcium-activated K channels. Role of ligands, phosphorylation and G-proteins, *in*: Handbook of Experimental Pharmacology. Vol. 108 "GTPases in Biology". B. Dickey and L. Birnbaumer, eds., Springer-Verlag, New York, pp. 561-579.

Tseng-Crank, J., Foster, C.D., Krause, J.D., Mertz, R., Godinot, N., DiChiara, T.J., and Reinhart, P.H., 1994, Cloning, expression, and distribution of functionally distinct Ca^{2+}-activated K^+ channel isoforms from human brain, *Neuron* 13:1315.

Wallner, M., Meera, P., Ottolia, M., Kaczorowski, G., Latorre, R., Garcia, M.L., Stefani, E., and Toro, L., 1995, Cloning, expression and modulation of a human maxi K_{Ca} channel isolated from myometrium, *Receptors and Channels* 3:185.

Wallner, M., Meera, P., and Toro, L., Determinant for β-subunit regulation in maxi K_{Ca} channels. An additional transmembrane region at the N-terminus, *Proc. Natl. Acad. Sci. USA* 93:14922-14926.

Wei, A., Solaro, C., Lingle, C., and Salkoff, L., 1994, Calcium sensitivity of BK-type K_{Ca} channels determined by a separable domain, *Neuron* 13:671.

Wisgirda, M.E., and Dryer S.E., 1994, Functional dependence of Ca^{2+}-activated K^+ current on L- and N-type Ca^{2+} channels: differences between chicken sympathetic and parasympathetic neurons suggest different regulatory mechanisms, *Proc. Natl. Acad. Sci. USA* 29:2858

SHIFTS OF MACROSCOPIC CURRENT ACTIVATION IN PARTIALLY BLOCKED SODIUM CHANNELS. INTERACTION BETWEEN THE VOLTAGE SENSOR AND A μ-CONOTOXIN

Robert J. French[1] and Richard Horn[2]

[1]Department of Physiology & Biophysics, University of Calgary
Calgary, Alberta, Canada, T2N 4N1
[2]Department of Physiology, Jefferson Medical College
Philadelphia, PA 19107, USA

INTRODUCTION

The μ-conotoxins (μCTXs) are peptide toxins from carnivorous marine snails of the genus Conus, that specifically block voltage-dependent sodium channels from adult skeletal muscle (Moczydlowski et al., 1986; Cruz et al., 1985). μ-Conotoxins GIIIA, GIIIB, and GIIIC, from Conus geographus, are highly charged, 22 amino-acid peptides with structures that are constrained by 3 intramolecular disulphide bonds (Figure 1). μCTX GIIIA has a nominal net charge of +6. Studies of point mutants of μCTX GIIIA reveal arginine 13 as the single most important residue for high affinity binding and block of the channel (French et al., 1996; Chahine et al., 1995; Becker et al., 1992; Sato et al., 1991), although several residues appear to participate in the toxin-channel interaction. In addition, our initial single channel studies revealed a striking property of the derivative R13Q, in which arginine 13 was replaced by glutamine: whereas the native toxin blocked single-channel current in an all-or-none manner, R13Q appeared to only partially occlude the conducting pore, reducing the single channel current to about 30% of its normal value. The gating machinery appears to continue operating in the face of this partial block of the channel (French et al., 1996; Becker et al., 1992), with complete closures, from the blocked state, occurring on a millisecond time scale. This

From Ion Channels to Cell-to-Cell Conversations
Edited by Latorre and Sáez, Plenum Press, New York, 1997

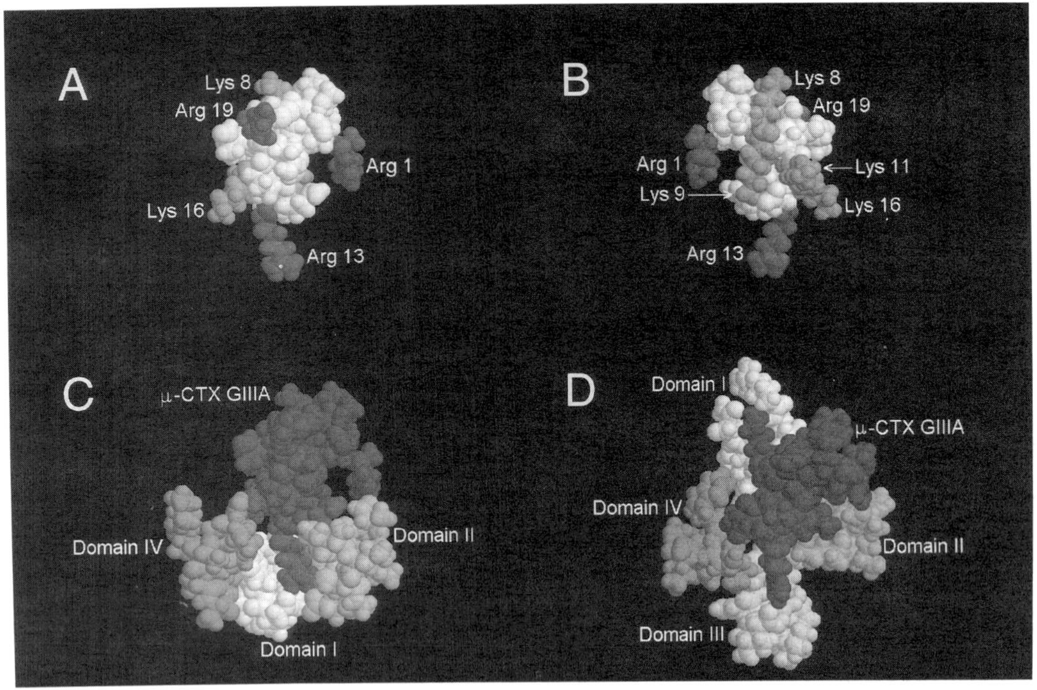

Figure 1. **A & B.** Solution structure of μ-CTX GIIIA (Lancelin et al., 1991) viewed from opposite sides. **C.** Working hypothesis of the docking of μ-CTX to the putative pore-forming (P) loops of the rSkM1 sodium channel, after Fozzard and collaborators (Dudley et al., 1995; Lipkind et al., 1994; Lipkind and Fozzard, 1994). The extracellular vestibule of the pore faces upwards, and the P loop of domain III has been removed to show the Arg-13 side chain extending into, but not completely filling, the narrowest part of the pore. **D.** View from the extracellular solution of μCTX bound to the pore.

partially blocking peptide thus offers a novel opportunity to study the response of the channel's voltage sensor, which controls gating, to an imposed, static electric field during periods when the peptide is bound to the channel.

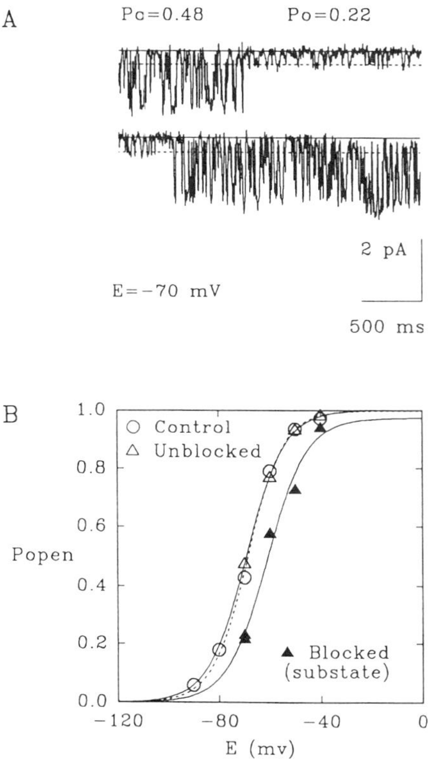

Figure 2. The sodium channel activation curve shifts to more positive voltages when the cationic peptide, R13Q, binds to the channel. **A.** A continuous recording from a single BTX-activated sodium channel taken in the presence of R13Q at a constant voltage (E = -70mV). The first and last sections of the trace show the voltage-dependent gating fluctuations for the unblocked channel at approximately the midpoint of the activation curve (from a longer segment, including the one displayed, for which P_o = 0.48). The middle section of the trace, continuing from the first to the second line, represents a single R13Q blocking event, with superposed channel gating fluctuations between the R13Q-blocked subconductance level (broken line) and the fully closed level (solid line). Note that the probability of the channel being 'open' drops dramatically while the R13Q is bound (P_o = 0.22). For display, the record was filtered using a digital Gaussian filter with a corner frequency of 75Hz (data sampled at 2000Hz and low-pass filtered, prior to sampling, with an 8-pole Bessel filter, -3dB at 1000Hz). **B.** Activation curves were constructed by plotting the open probability (P_{open}) before the addition of R13Q (Control), and, in the presence of R13Q, either in the unbound state (Unblocked) or in the peptide-bound state (Blocked). For the purposes of calculating P_o while R13Q is bound, the subconductance level is considered to be 'open'. Activation curves were fit using a Boltzmann function of the following form: $P_o = P_{omax}/(1+ \exp[z_aF(E_{mid}-E)/RT])$, where there are three adjustable parameters: P_{omax} is the maximal open probability; z_a is the apparent gating charge; E_{mid} is the half-activation voltage. The other symbols are as follows: E is the applied voltage; and F, R, and T are the Faraday, the gas constant, and the absolute temperature, respectively. In 6 experiments, the gating in the substate showed a significant positive shift in E_{mid}, (6.4 ± 2.7 mV, mean ± s.d.; $P<0.01$, t test), as compared to the value obtained from the unblocked state observed in the same trace. In the same experiments, no significant changes were observed in the maximal activation, P_{omax} (control: 0.96 ± 0.06; substate: 0.93 ± 0.08) or in the steepness of the activation curve reflected by the apparent gating charge, z_a (control: 4.09 ± 0.48; substate: 4.21 ± 0.60). Reproduced by permission (French et al., 1996).

The effect of peptide binding on gating is most directly observed in single channel recordings from non-inactivating, batrachotoxin(BTX)-treated channels. R13Q binds for an average duration of several seconds. Thus, many opening and closing fluctuations are normally observed within a single 'blocked' event while the peptide is bound to the channel, and at voltages on the steep part of the activation curve, there is an immediate, obvious decrease in the open probability following the decrease in amplitude that reflects peptide binding (Figure 2, and see (French et al., 1996)). The observed decrease in open probability, P_o, reflects a positive shift in the activation curve of about 6-7 mV in the depolarizing direction. Thus, when R13Q is bound, a larger depolarization is required for channel opening.

This phenomenon is similar to the depolarizing shift in Na channel activation induced by raising external calcium, or otherwise increasing external ionic strength (Cukierman, 1993; Cukierman et al., 1988; Hille et al., 1975; Frankenhaeuser and Hodgkin, 1957). The details of the mechanism, however, are likely to differ, in that Ca^{2+} appears to shift gating because of screening and/or binding interactions with charges on both the membrane lipid and on the channel protein (Cukierman, 1991; Cukierman et al., 1988), whereas μCTX and its derivatives probably bind with one-to-one stoichiometry (Moczydlowski et al., 1986; Cruz et al., 1985) to a discrete site at the mouth of the pore (French et al., 1996; Dudley et al., 1995). Furthermore, although gating shifts observed in bilayer experiments using BTX, and in cellular preparations, are generally similar, it must be remembered that BTX changes several specific aspects of ion conduction and gating of Na channels (Khodorov, 1985), and it was therefore desirable to determine whether the R13Q-induced shift reflects a general response of the channel's voltage sensor, or reflects a property peculiar to the behavior of the BTX-modified channel.

Thus, the bilayer experiments prompt a number of questions, including the following, which we explore in this chapter: 1) Does any substitution of R13 in μCTX GIIIA result in incomplete block of the sodium channel? 2) Is the lack of complete block by R13Q a property unique to BTX-modified channels? 3) Does the shift in activation gating produced by R13Q binding reflect a specific property of BTX-modified channels, or a fundamental property of the sodium channel voltage sensor? 4) Could the R13Q-induced activation shift result from a simple electrostatic inhibition of the movement of the voltage sensor in the transmembrane electric field?

Experimental Approach

In the work described in this chapter, we addressed the questions posed above using whole-cell, patch-clamp recording from cloned adult rat skeletal muscle Na channels (termed rSkM1, or μ1 (Trimmer et al., 1989)) expressed in a mammalian cell line, tsA-201, which is a transformed variant of the human embryonic kidney line, HEK-293 (Margolskee et al., 1993). Cells were raised off the bottom of the culture dish after attaining the whole-cell recording configuration, and solutions containing synthetic derivatives of μCTX were applied using a gravity-driven flow from a macropipet placed adjacent to the cell using a micromanipulator. Further details of the methods are provided in reference (French et al., 1996).

COMPLETE BLOCK BY μCTX DERIVATIVES REQUIRES RESIDUE 13 TO BE CATIONIC

In whole-cell recordings, the question of whether a channel blocker acts in an all-or-none manner must be addressed indirectly by determining the fraction of current remaining at saturating concentrations of the peptide, when all of the channels are bound. At a voltage at which the inward transient Na current is approximately maximal, application of high concentrations of peptides with different residues replacing arginine 13 (R13X, where X=glutamine, lysine, or ornithine), or with the adjacent substitution of glutamate for aspartate (D12E), produced dramatically different results (Figure 3). Application of R13Q left a residual current of about 30% of the control

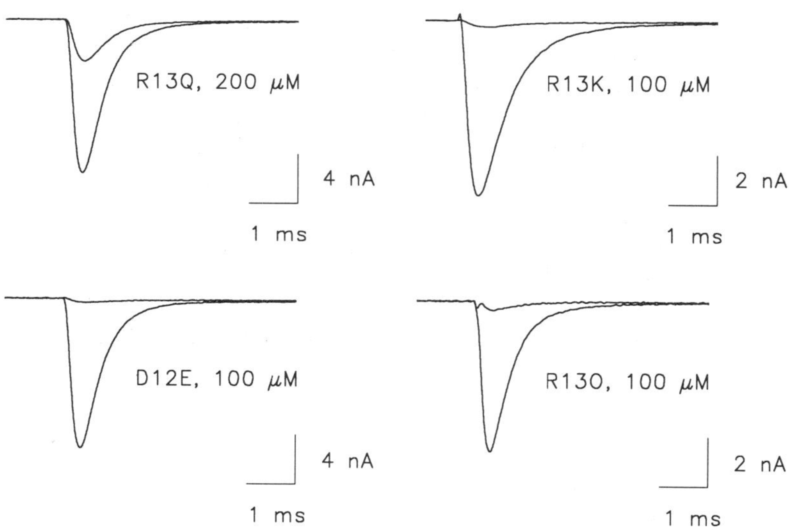

Figure 3. Whole-cell currents recorded in the absence and presence of near-saturating concentrations of various μ-CTX derivatives. In each case, the larger amplitude trace is the control. Of the μ-CTX derivatives tested, only R13Q allowed a substantial residual current to flow. The derivatives in which charge was conserved (R13K, R13O and D12E) caused almost complete block. Voltage step to -5mV, except -10mV for R13K; holding potential, -100mV.

amplitude. In contrast, similar concentrations of 3 different derivatives, produced by charge-conserving single-residue substitutions, at positions 12 or 13, led to almost complete block of the sodium channel currents.

Concentration-response curves for the three arginine-13 mutants (Figure 4) confirm the incomplete block by R13Q, showing an asymptotic limiting current amplitude of about 28% of the control. This is very close to the residual fraction of single-channel current observed during discrete, partial blocking events induced by R13Q interacting with BTX-modified channels (see Figure 2 and French et al., 1996; Becker et al., 1992). Much smaller residuals (<10% of control) were estimated from fits to dose-response data for the charge-conserving mutations, R13K and R13O. For each derivative, the apparent dissociation constant (see Table 1) indicates a significant reduction in binding affinity from that of the native μCTX (K_B=17-110nM, depending on conditions and preparation). Blocking parameters were determined by fitting the dose-response relations (Figure 4) to the following relation:

$$f_U = f_\infty + (1-f_\infty)/(1+[B]/K_B),$$

where f_U = the unblocked fraction of current, [B] = concentration of blocker, and the parameters adjusted during the fit are f_∞ = the unblocked fraction of current at saturating concentrations of blocker, and K_B = apparent dissociation constant for the blocker.

These results are consistent with the following hypotheses: 1) that a cationic side-chain at position 13 is essential for nearly-complete channel block, and 2) that both the positive charge and the conformation of the arginine side chain make important contributions to the high affinity binding of μCTX to the channel.

Although there is good qualitative agreement between the single-channel and whole-cell data for R13Q, the exact percentage of the residual current should be interpreted with caution. The residual currents observed in the presence of high concentrations of R13K and R13O are of an amplitude

Table 1. Blocking Parameters for arginine-13 substituted μ-CTX Derivatives

Toxin derivative	K_B (μM)	f_∞
R13Q	7.4	0.28
R13K	0.58	0.083
R13O	0.29	0.078

comparable to intrinsic inward transient currents sometimes seen in non-transfected HEK-293 cells. These residual currents appear to be insensitive to μCTX (Ukomadu et al., 1992). For this reason, the estimates of the residual unblocked current should be taken as an upper limit of the residual current flowing through rSkM1 channels. This issue can only be resolved unambiguously by single channel measurements, since it is not possible to record macroscopic currents from an individual cell before and after transfection. Although we have not yet done single-channel measurements using the expressed rSkM1 channnels, recent bilayer experiments using single BTX-modified channels do suggest a small (≈5%), residual single-channel current during discrete R13K-induced blocking events (K. Hui, I. Sierralta, and R. French, 1996, unpublished results).

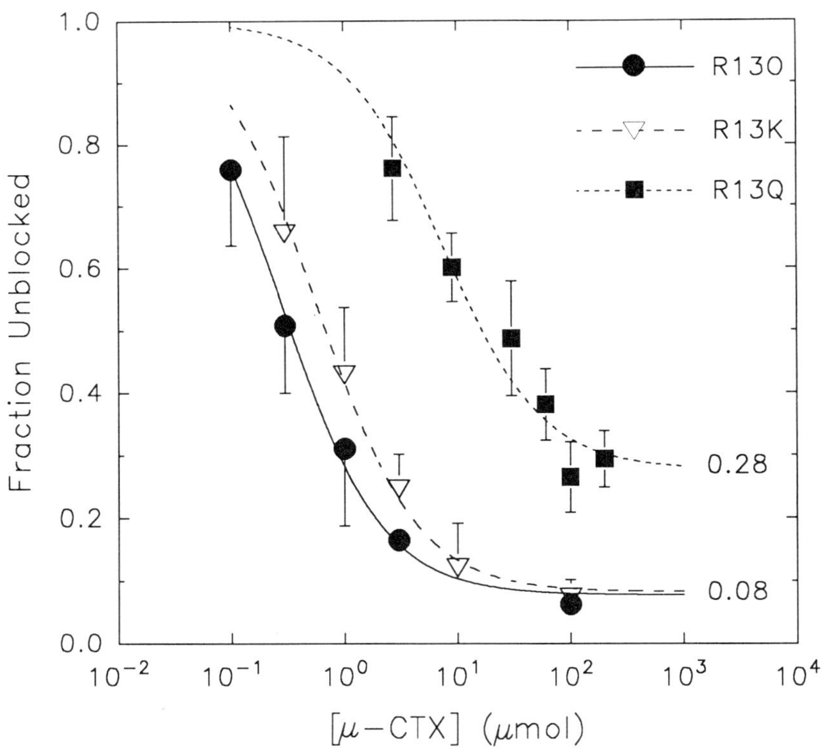

Figure 4. Dose-response Curves for μ-CTX derivatives formed by substituting either ornithine, lysine or glutamine for arginine-13. Note the substantial residual current at high concentrations of R13Q.

Overall, although the whole-cell data support the idea that block of single, normally inactivating rSkM1 channels by R13K and R13O may be an all-or-none event, the latest single channel recordings suggest that 'block' by R13K allows a small but significant residual current to flow when this derivative is bound. The physically smaller side chain of ornithine may allow a slightly larger residual current when R13O is bound; this possibility will be examined in single channel experiments. On the other hand, the bulkier arginine side chain would be expected to more completely occlude the channel. However, even with the R13 residue in place in the native μCTX, it may be sterically possible for a Na^+ ion to pass by the bound toxin (e.g. see model in Figure 1C). Thus block of single-channel current could rely on electrostatic repulsion, rather than complete steric occlusion of the pore.

RESIDUAL CURRENTS FROM CHANNELS BOUND BY R13Q SHOW SHIFTED ACTIVATION

The concentration-response curves for R13Q, and the single-channel data, suggest that R13Q binds to rSkM1 channels with 1:1 stoichiometry, and on binding, reduces the current to about 30% of that through an unbound channel. At concentrations above 100 μM R13Q, >92% of the channels would be bound by R13Q. In addition, given that even the mutated peptides dissociate slowly from the channels (dissociation rate constants ~0.1-1 sec^{-1}), the time course of whole cell currents should accurately reflect the kinetics of gating of the channels that were bound by R13Q at the beginning of the depolarizing voltage step. For this reason, the residual whole-cell currents allow us to examine the voltage-dependent gating of the peptide-bound, partially blocked channels.

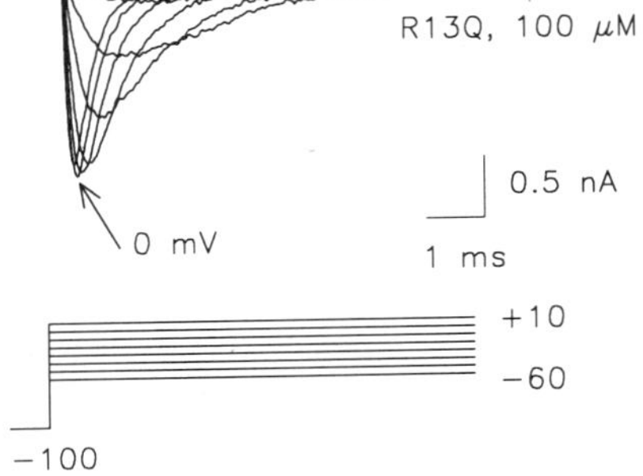

Figure 5. Families of inward sodium currents elicited by voltage steps to levels from -60 to +10 mV, at 10 mV intervals, from a holding potential of -100 mV. The maximal inward current occurred at -20 mV in the control case, and at 0 mV in the presence of R13Q, suggesting a positive shift in sodium channel activation when R13Q is bound to the channel. Note that the amplitude scale is expanded four-fold for the R13Q data to clearly display the reduced currents. Experiment A4202.

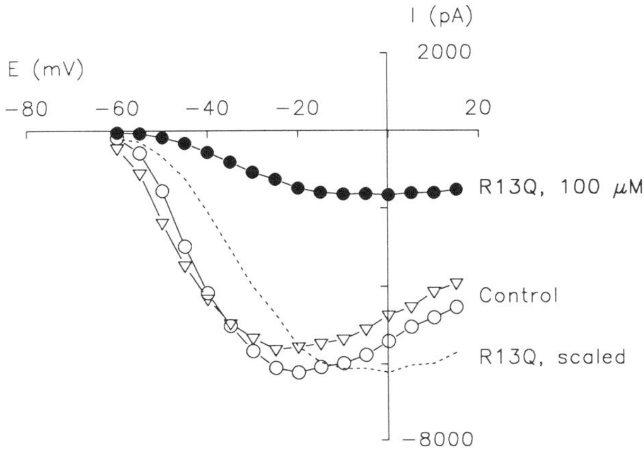

Figure 6. Peak inward current plotted against voltage for the experiment shown in Figure 5: open circles, control data before exposure to R13Q; closed circles, data collected while superfusing the cell with 100 μM R13Q; open triangles, after washout of R13Q. The dotted curve represents the R13Q data scaled to match the maximal inward current with the value observed before R13Q application. For 5 experiments in which 100 μM R13Q was applied, there was a reversible shift in the I-E relation of 7.6 ± 3.2 mV (mean ±s.d.), which is significantly greater than zero ($P<0.005$, t test). This agrees well with the shifts in the midpoint of the activation curves determined from the single channel data in Figure 2. To measure the shifts, we fit the peak I-E relationships with a cubic spline function, and the voltage, E_a, was determined at a point half-way to the minimum of the I-E curve, along the negative slope region. E_a is approximately equivalent to the parameter, E_{mid}, measured in the single channel measurements (see Figure 2B). The shift was expressed as the difference between E_a measured in the presence of R13Q, and the mean of the values from control I-E relations taken before and after R13Q application.

Examination of the records (Figure 5) shows immediately that attaining the maximal inward current amplitude requires a larger depolarization when the R13Q is bound than in its absence. The current-voltage relations (Figure 6) show that this requirement results from a shift in the negative slope region of the curves by about 6-7 mV in the positive direction. This observation closely parallels the shift in activation of single, BTX-activated channels described in the INTRODUCTION, suggesting that normally inactivating channels are also more reluctant to open when the R13Q is bound.

ACTIVATION SHIFTS REQUIRE SPECIFIC BINDING OF PEPTIDE

Before making any mechanistic interpretation, it is necessary to ensure, as far as possible, that the observed shifts in activation gating result from specific action of the R13Q, rather than a non-specific artifact of the experimental procedure. Two of the questions that we considered are: 1) Are micromolar concentrations of a non-blocking polycationic peptide sufficient to cause a similar shift in gating? 2) Could the shift arise artifactually from the reduction in current, due to incomplete compensation for the voltage drop across the resistance in series with the cell membrane?

The following experiments were done to address these issues. First, the potassium channel-blocking peptide, charybdotoxin, was applied at a concentration of 100 μM. In its general structure, charybdotoxin resembles the μ-conotoxins, in that it is a fairly small peptide (37 amino acids) whose structure is defined and stabilized by 3 disulphide bonds, with a net charge of ~+5 at physiological pH. It specifically blocks a number of voltage-dependent and calcium-dependent potassium channels with dissociation constants in the nanomolar range. Charybdotoxin (100 μM) had no significant effect on the rSkM1 sodium channels; we observed neither any block of currents, nor any change in the voltage-dependent activation of the whole-cell currents (Figure 7). This result excludes the possibility that the activation shift observed in the presence of R13Q arises from non-specific screening or binding of surface charges by the polyvalent peptide.

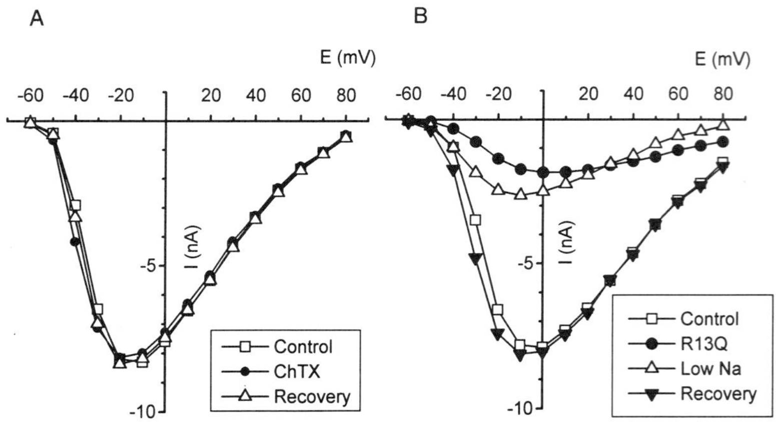

Figure 7. A. The polycationic peptide charybdotoxin (nominal net charge, +5), which does not bind to sodium channels, does not cause any significant shift in the I-E relations when present at 100 μM. **B.** A reduction of external [Na] to 10% of the control level reduces the currents by about the same fraction as 100 μM R13Q, but does not cause a depolarizing shift in the I-E relations.

Second, current magnitudes were reduced, at constant ionic strength, by reducing external sodium concentration. No significant shift in activation was observed. Series resistance errors were electronically compensated to a fixed percentage, usually 80%. If there were significant command voltage errors due to this incomplete compensation of the voltage drop of the series resistance, the larger currents in the control records would lead to larger absolute errors, and an apparent positive shift in the activation curve when measuring smaller currents (e.g. in the presence of R13Q) could result. A decrease in absolute current magnitude, resulting from the reduction of external sodium, by a fraction comparable to the block by saturating concentrations of R13Q, did not cause any systematic shift in current activation, ruling out the possibility that the activation shift in the presence of R13Q arises artifactually from a series resistance error.

WHICH TRANSITION IS AFFECTED?

The activation shift caused by R13Q is simplest to understand if we assume that gating is a 1-step voltage-dependent transition between a closed state and an open state. The free energy for this gating transition may be expressed as the sum of voltage-dependent and voltage-independent components. Since R13Q causes a depolarizing shift without a change in slope of the Po-E relationship, it must be affecting the voltage-independent free energy difference between these 2 states without affecting the number of charges that move during the transition. The free energy difference between the closed and open state in this 2-state model can be written as

$$\Delta G = \Delta G_0 - qE,$$

where ΔG_0 is the free energy difference at 0 mV, q is the charge moved across the transmembrane voltage during a closed-to-open transition, and E is the membrane potential. ΔG_0 is often expressed as a function of the midpoint of a P_o-E relationship:

$$\Delta G_0 = qE_{mid},$$

where E_{mid} is the membrane potential at which half the channels are open in the steady state. For this 2-state model the open probability is a function of a voltage-dependent equilibrium constant K(E):

$$P_o(E) = [1+K(E)]^{-1}$$

$$\text{where } K(E) = \exp[qE_{mid}-E)/kT]$$

$$\text{and } kT = 2.44 \text{ kJoule/mole.}$$

The depolarizing shift in the Po-E relationship caused by R13Q, ΔE_{mid},

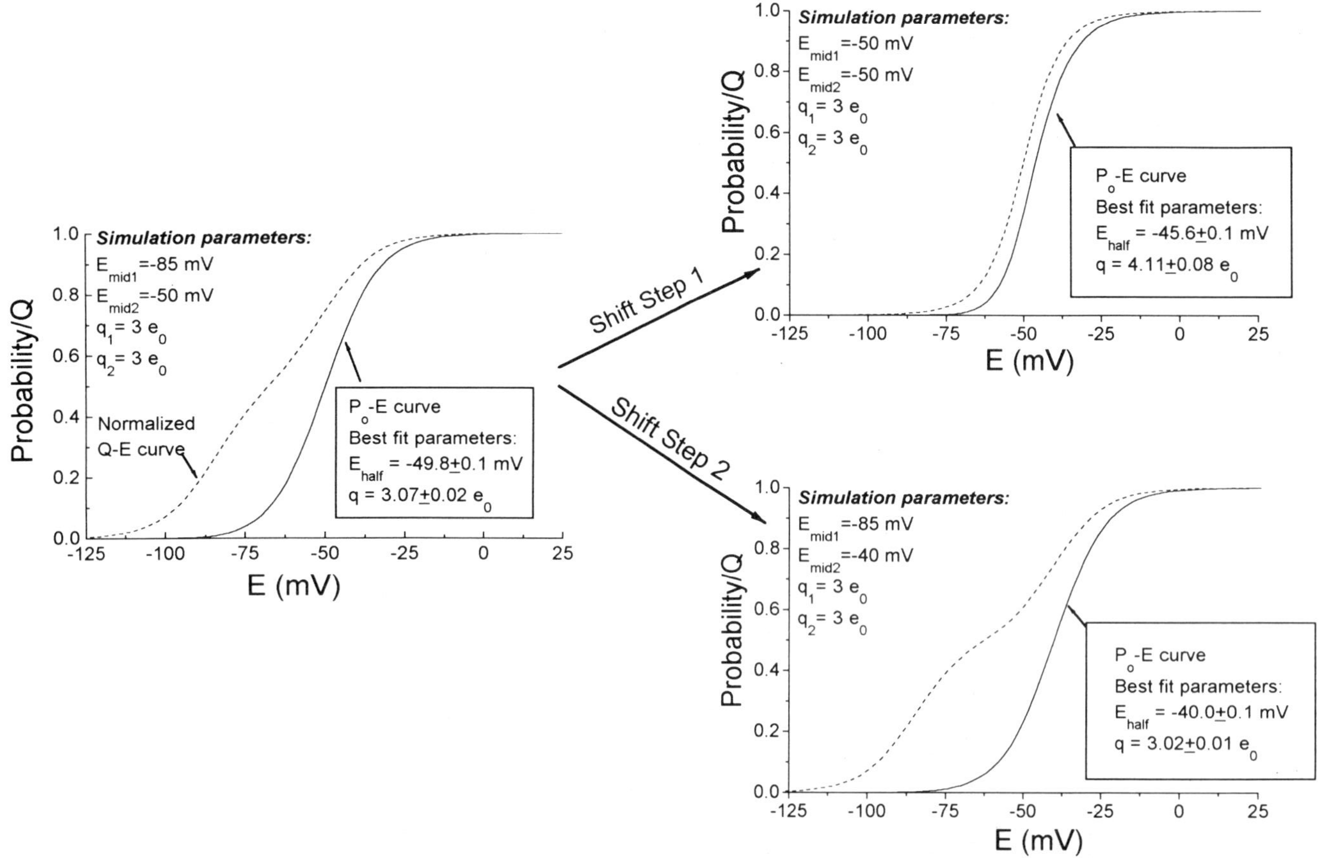

Figure 8. Simulations of steady-state charge movement (Q-E) and activation (Po-E) curves for the two-step, C⇔C⇔O model.

can be described formally as a change in the voltage-independent free energy difference between a closed and open state:

$$\Delta\Delta G_0 = q\Delta E_{mid}.$$

This extra work required to move the gating charges can be used to estimate the distance through which these charges move (see below).

The gating of sodium channels cannot, however, be reconciled with a 2-state kinetic scheme (e.g. French and Horn, 1983). Here we examine the consequences of this complication on $\Delta\Delta G_0$ estimated from gating shifts. Suppose, for example, that activation is modelled by a 3-state linear kinetic scheme having 2 closed and 1 open state:

$$C_0 \overset{K_1}{\longleftrightarrow} C_1 \overset{K_2}{\longleftrightarrow} O$$

where each of the gating transitions is represented by a separate voltage-dependent equilibrium constant $K_1(E)$ for the closed-closed transition, and $K_2(E)$ for the closed-open transition, each defined as above. For simplicity we assume that each transition in our simulations involves translocation of $q_1=q_2=3e_0$ elementary charges. The steady state properties of this gating scheme are determined by the midpoints of the 2 transitions, E_{mid1} and E_{mid2}, according to the following equations.

$$\text{Open probability,}\quad P_o=1/[1+K_2(1+K_1)]$$

$$\text{Probability of } C_1,\quad P_{C1}=K_2/[1+K_2(1+K_1)]$$

$$\text{Steady state charge movement,}\quad Q=3e_0\,P_{C1}+6e_0\,P_o$$

What is the meaning of an interaction of R13Q with such a channel, as opposed to the simpler 2-state channel? First, the number of gating charges estimated from a Boltzmann fit to a P_o-E relationship is usually an underestimate of the total gating charge in a multi-state model. Figure 8 shows 3 examples (solid lines) of such fits for different values of E_{mid1} and E_{mid2}. The total charge estimated from such fits is always between $3e_0$, the charge for a single transition, and $6e_0$, the total charge moved for both. If R13Q inhibits both transitions equally, *i.e.*, causes an equivalent depolarizing shift of ΔE in both E_{mid1} and E_{mid2}, this results in a shift of ΔE in the P_o-E curve without a change of shape (not shown).

However, there is also the possibility that R13Q, when bound, might preferentially inhibit either the first or the second transition in the 3-state model. For example, it may be closer to an S4 segment responsible for 1 of the 2 transitions. Alternatively, the S4 segments may move outward in 2 steps (Zagotta et al., 1994; Keynes, 1992), and the bound toxin could affect 1 of these transitions more than the other. The simulations in Figure 8 show that an effect on the first rather than the second transition is distinguishable, especially if gating currents can be measured (integrated Q-E curves are represented as dashed lines). For the specific example used here, a 35 mV change of E_{mid1} alone causes a 4.2 mV shift of the P_o-E curve, where a 10 mV change of E_{mid2} causes a 9.8 mV shift. Therefore, an alteration of the transition into the open state has a greater effect on the P_o-E curve than an equivalent alteration of the closed-closed transition. Furthermore, an effect on the second transition has very little effect on the slope of the P_o-E curve in this example, whereas there is a much greater effect on slope of an alteration in the first transition.

Activation gating of sodium channels is clearly a multistep process with an apparent voltage dependence that is much lower than the true limiting slope of P_o-E curves (Sigworth, 1993; Almers, 1978). Because R13Q causes a depolarizing shift with little change in the slope of the activation curve, these simulations suggest that it must be affecting a gating transition near the open state. In order to know if it is affecting all gating transitions equally, it would be desirable to obtain Q-E curves from gating currents (Figure 8, dashed lines), which add a further criterion for identification of the kinetic step that is affected.

The arguments in this section are strictly thermodynamic, i.e. they require only that binding of R13Q change the free energy difference between adjacent kinetic states of the channel. This could be brought about by a conformational change or an allosteric effect, within the channel protein, that selectively stabilizes the channel in a closed conformation when R13Q binds. Alternatively, it might result from a direct electrostatic effect of the bound R13Q (nominal net charge, +5) on channel gating. Below, we will explore the feasibility of a simple electrostatic mechanism.

DOES R13Q AFFECT INACTIVATION?

We have restricted our attention so far to activation gating. Because inactivation is also voltage dependent, it is reasonable to ask if it is also affected by bound R13Q. It is generally believed that the voltage dependence of inactivation is derived largely from that of activation (French and Horn, 1983). Therefore, if R13Q affects the voltage sensors underlying activation, it should also affect the voltage dependence of inactivation. We examined this possibility by measuring the effects of nearly saturating concentrations of R13Q ($\geq$100 μM) on the inactivation time constants (τ_h) during depolarizing pulses and on steady-state inactivation, reflected by h_∞.

During a depolarization the decay of rSkM1 current after the peak is typically well fit by a single exponential relaxation with a time constant τ_h.

This time constant is voltage dependent (Figure 9), but is not significantly affected by R13Q, in spite of the fact that R13Q caused depolarizing shifts of ~6 mV in the negative slope region of the peak I-E curves in the same 6 cells. We measured the effect of R13Q on steady-state inactivation in 3 cells. In 2 of these cells R13Q had no effect on E_{mid} of the h_∞ curves, whereas we observed a ~10 mV hyperpolarizing shift for the third cell (data not shown). Activation was shifted in the depolarizing direction in all 3 of these cells by R13Q. R13Q did have a small, but statistically insignificant, effect on the slopes of the h_∞ curves, a decrease of ~0.5 e_0 (P~0.07; ANOVA).

Clearly, more experiments need to be done. However, the consistent effects on activation, but not inactivation, suggest that R13Q has a greater influence on the former. This could be due to the fact that bound toxin is electrostatically closer to an activation voltage sensor (see below); alternatively, it may have a greater allosteric effect on the activation than on the inactivation process.

HYPOTHESIS: R13Q SHIFTS ACTIVATION BY ELECTROSTATIC INHIBITION S4 MOVEMENT

As noted in the INTRODUCTION, the activation shift induced by R13Q resembles that caused by a modest increase (a few mM) in external calcium. Such shifts are thought to result from screening or binding of negative surface charges on the membrane lipid or the channel protein (Hille et

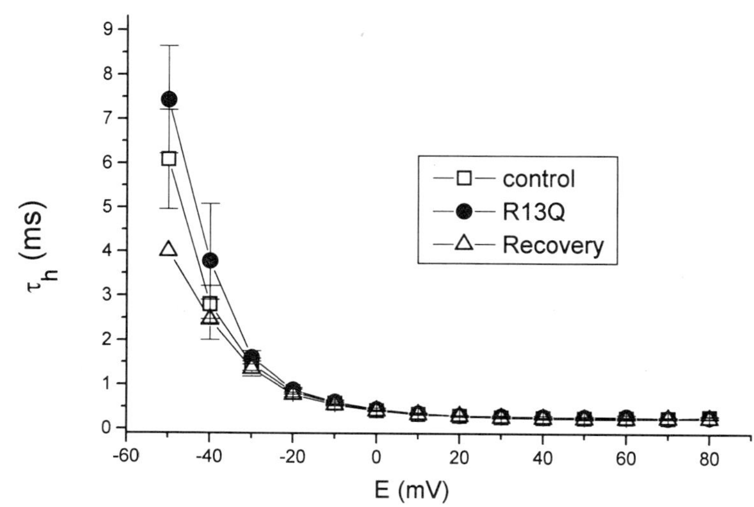

Figure 9. Inactivation time constants (τ_h) determined from the decay of currents during individual step depolarizations to the indicated voltage.

al., 1975; Frankenhaeuser and Hodgkin, 1957). This would change the local electric field sensed by the 'voltage sensor' that controls channel gating. If the phenomena are so similar, the question that arises is: what more might we learn about channel mechanism or structure from the R13Q data?

The significant point of difference is that the screening/binding action of calcium demonstrably acts through an extended distribution of charges on both channel protein and lipid (Cukierman et al., 1988). By contrast, the action of R13Q appears to be mediated by the discrete binding of a single charged peptide molecule to the channel. Moreover, data and models are emerging that begin to fix the location of R13Q binding within the primary sequence and the three-dimensional structure of the sodium channel protein. Over the past 10 years, evidence has been accumulating from mutagenesis studies that the guanidinium toxins, saxitoxin and tetrodotoxin, and the μ-conotoxins inhibit sodium current by binding to the pore-forming region of the channel and physically and/or electrostatically blocking current flow (Dudley et al., 1995; Heinemann et al., 1992; Terlau et al., 1991). We have presented arguments that the μCTX mutant, R13Q, despite its atypically incomplete blocking action and its lowered affinity, also acts by binding to the pore mouth and partially obstructing current flow (French et al., 1996). This raises the question of how fixed charges placed at a defined location perturb the local electric field, and thus alter the apparent voltage registered by the channel's voltage sensor. The geometrical arrangement of the bound R13Q with respect to the voltage sensor is obviously a crucial factor in determining the electrostatic effect of the toxin on gating. Here we pose only a question of feasibility: Could an electrostatic interaction between the R13Q, and the channel's voltage-sensing charges, separated by realistic molecular distances, account for the activation shift that we have observed?

QUESTION: HOW FAR DO S4 CHARGES MOVE?

In order to estimate distances, we equate an empirical estimate of the additional work performed to open the channel when R13Q is bound, $\Delta\Delta G_0$, with an estimate, from first principles, of the electrostatic work done in moving the S4 charges toward those of the bound R13Q. It is possible, for any arbitrary starting point, to calculate the distance that the voltage-sensing charge moves towards the R13Q. In this way a family of possible trajectories for charge movement is obtained. While keeping in mind that, because of the simplifications invoked, this is no more than an order of magnitude calculation, a striking point emerges: the calculated range of movement falls within the scope of realistic macromolecular distances. Had this result not arisen - for example, if estimated physical movement fell in subatomic or macroscopic ranges - the hypothesis of an electrostatic mechanism for genesis of the R13Q-induced shift in activation gating could have been immediately discarded.

Using the simple model depicted in the cartoon in Figure 10, we obtain the result that mobile S4 charges which start, in the closed state, a few tens of Ångstroms from the center of the peptide charge and move a few Ångstroms towards the bound peptide would energetically account for the shifts in gating (see Figure 10 and Table 2).

Approximations and Simplifications

For a more rigorous calculation, detailed knowledge of the structure of the R13Q-channel complex in both open and closed states would be required. Here we are concerned only with obtaining a qualitative, order-of-magnitude solution as one test of the feasibility of the following hypothesis: that the electrostatic inhibition of outward movement of voltage-sensing S4 charges (Stühmer et al., 1989) of the channel by the bound R13Q is sufficient to account for the induced shift in channel activation. Several workers have proposed that the outward motion of positive S4 charges in response to depolarization is an essential step in the channel opening (Yang et al., 1996; Yang and Horn, 1995; Guy and Seetharamulu, 1986; Catterall, 1984; Noda et al., 1986). In so doing when R13Q is bound, we would expect that the S4 charges would move closer to the positively charged peptide molecule. The presence of the peptide would thus require added electrical work to be done in order to open the channels, in the form of a larger depolarization for channel activation. As a first approximation, we consider a scenario which retains the essential elements of the problem. We envision a point charge in a membrane/protein phase, representing the S4 charges, moving toward a second point charge, representing the R13Q charges, positioned at the interface between the aqueous and the membrane/protein phases. More details of our calculations (French et al., 1996), and earlier results on which they are based (Stillinger, 1961; Mathias et al., 1992) have been presented elsewhere. Here, we focus on the underlying assumptions and limitations of the calculations and the interpretation of the results.

Probably the most critical consideration in evaluating the strength of the interaction between the bound peptide and voltage-sensing charges in the channel is that the charges on the peptide, partially exposed at the aqueous interface, are subject to screening by mobile counterions in the solution. Placement of an equivalent point exactly at the membrane/protein interface effectively assumes that half the charge is exposed to screening ions. If the peptide slipped intimately into a pocket in the channel protein so that less than half of its charged side chains were exposed to soluble counterions, it would cause a correspondingly stronger perturbation of the electric potential within the membrane/protein phase; the reverse would be true if only a relatively small area of the peptide made close contact with the protein and more were exposed to the aqueous solution. The charged residues of μCTX are distributed widely over its surface, and our initial survey of point mutants of μCTX showed that dissociation rate constants were changed by >4-fold when 5 different residues were individually mutated (mutations R1Q, K8Q, R13Q, K16Q, and Hyp17P, see (Becker et al., 1992)). Thus, the interaction surface appears to be fairly large, and the implicit assumption that roughly 50% of the charges are vulnerable to screening seems reasonable.

The broad distribution of charges over the surface of the peptide makes it reasonable to replace the peptide with an equivalent point charge, in that the electric field at the surface an isolated, uniformly charged sphere may be calculated using Gauss's Law by replacing the charge on the sphere by an equivalent point charge, placed at its center.

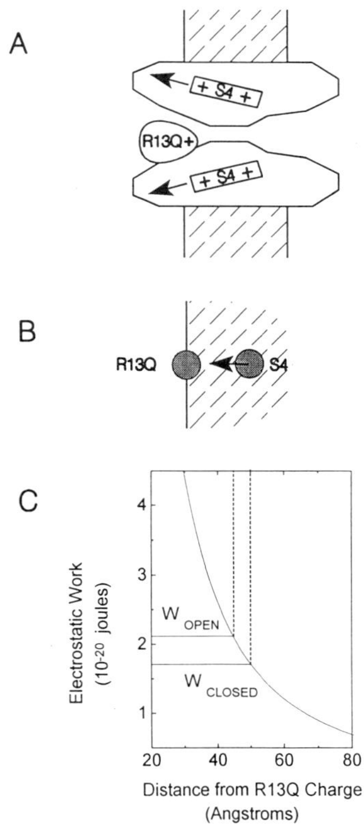

Figure 10. **A.** Cartoon showing the hypothetical relationship between the bound R13Q peptide and the charged, voltage-sensing S4 segments of the channel. Outward movement of the S4 segments during the opening transition would be electrostatically opposed by the bound cationic peptide, requiring larger depolarizations for channel opening when the peptide is bound. **B.** To simplify calculations, the net charge (+5) on R13Q was collapsed into a point charge at the membrane-solution interface, and the S4 charges (+23) were collapsed into an equivalent point charge on the 'pore' axis of symmetry - an axis extending from the R13Q perpendicularly into the membrane/protein dielectric. This allowed us to estimate the extra electrical work required, in the presence of R13Q, for S4 charge movement accompanying the transition from the closed to the open state (see description of calculations in French et al., 1996). **C.** For the simplified geometry of part B., the electrostatic work required to move the S4 charge from infinity towards the R13Q is plotted against the distance from the centre of charge of the R13Q. The vertical dotted lines mark the limits of one trajectory of S4 movement that accounts for the observed shifts in activation curves seen in Figs. 2 and 6. The distance that the S4 charge moves is completely determined by the starting (closed state) position, and the extra electrostatic work required to open the channel in the presence of bound R13Q, $W_s = W_{OPEN} - W_{CLOSED}$. Calculations were performed using Mathcad PLUS 5.0 (MathSoft, Cambridge, MA.). Figure reproduced by permission (French et al., 1996).

A fundamental problem in calculation of electrostatic potential in and around macromolecules is that the dielectric constant in the membrane/protein phase is both unknown, and presumably non-uniform. Although this is an important consideration for precise, detailed calculations, for our immediate goal of testing the feasibility of the hypothesis of an electrostatic interaction between the peptide and voltage-sensing charges, this issue is less serious than it might at first appear. The reason lies in a remarkable result derived by Mathias and coworkers (Mathias et al., 1992). They show that the dielectric coefficient that determines the potential profile around a charge located at the interface is effectively the average of the values within the membrane and the aqueous phases. Thus, even if the dielectric constant in the membrane phase is varied between 2 and 80, the potential at any point will change by a factor of no more than about 2.

Table 2. Estimated Distance Moved by S4 Charge (Ångstroms)

Starting distance from R13Q	S4 movement
40	3.0
50	5.5
60	8.5

A final approximation in our simple calculations is the assumption that the membrane effectively has infinite thickness. Again, this is unlikely to introduce major errors. The potential contribution from the R13Q charge decays steeply in distance, given that it is effectively the potential of a dipole formed by the R13Q and the local concentration of screening ions in the aqueous solution (Mathias et al., 1992).

FUTURE QUESTIONS

The calculations suggest that an electrostatic mechanism is sufficient to account for the gating shifts associated with R13Q binding to the channel, but they do not conclusively prove that this is the actual or complete mechanism. An obvious test of the electrostatic hypothesis is to determine whether the gating shifts depend on ionic strength, as would be expected if the R13Q charges are partially exposed to screening from counterions in the aqueous solution. Thinking more broadly, we note that any mechanism by which R13Q stabilizes the closed state with respect to the open state could cause the observed gating shift. Thus, a lack of dependence of the shift on ionic strength would suggest some other mechanism - perhaps a conformational change in the channel protein that is driven by R13Q binding and which favors channel closure.

Mutation of specific residues in the peptide, or the channel protein, may allow us to determine whether only a subset of charged residues contribute to the shift. Binding of μCTX to the channel pore is probably asymmetric (Chahine et al., 1995; Dudley et al., 1995), and thus it will be of interest to determine if R13Q interacts more strongly with the S4 charges of a particular domain. Recent evidence that charges outside the S4 segment may contribute to gating charge movement (Seoh et al., 1996) suggests that the simplistic view in which the S4 segments provide all of the voltage-sensing charge may have to be evaluated carefully.

A working hypothesis regarding block of sodium channels by μCTX is that R13 enters the pore most deeply and thus prevents conduction. It is clear, however, from the R13Q data that the charge of arginine is not required to induce a gating shift. The interaction of completely blocking toxin variants with the voltage sensor can be tested using gating current measurements. Effects on gating charge movement, as opposed to channel opening, may give further insight into the way in which the translocation of voltage-sensing charges is coupled to channel opening.

Finally, it is likely that probes analogous to the μ-conotoxins will allow similar studies in other voltage-dependent channels, given the recent observations of Mintz (1994).

ACKNOWLEDGMENTS. We thank Drs. S.C. Dudley, H.A. Fozzard, and G. Lipkind for providing the coordinates of their sodium channel pore model, and Aaron Sheldon for his excellent assistance with computer graphics. RJF was a Senior Scholar of the Alberta Heritage Foundation for Medical Research during the course of this work. Operating funds were provided by the MRC (Canada) and NIH grant AR41691. Figures 2,5,6, and 10 are reproduced from French et al. (1996) by permission of the copyright owner, Cell Press.

REFERENCES

Almers, W. 1978, Gating currents and charge movements in excitable membranes, *Rev. Physiol. Biochem. Pharmacol.* 82: 96-190.

Becker, S., Prusak-Sochaczewski, E., Zamponi, G., Beck-Sickinger, A.G., Gordon, R.D., and French, R.J. 1992, Action of derivatives of μ-conotoxin GIIIA on sodium channels. Single amino acid substitutions in the toxin separately affect association and dissociation rates, *Biochem.* 31(35): 8229-8238.

Catterall, W.A. 1984, The molecular basis of neuronal excitability, *Science* 223: 653-661.

Chahine, M., Chen, L.-Q., Fotouhi, N., Walsky, R., Fry, D., Santarelli, V., Horn, R., and Kallen, R.G. 1995, Characterizing the μ-conotoxin binding site on voltage-sensitive sodium channels with toxin analogs and channel mutations, *Receptors and Channels* 3: 161-174.

Cruz, L.J., Gray, W.R., Olivera, B.M., Zeikus, R.D., Kerr, L., Yoshikami, D., and Moczydlowski, E. 1985, *Conus geographus* toxins that discriminate between neuronal and muscle sodium channels, *J. Biol. Chem.* 260(16): 9280-9288.

Cukierman, S., Zinkand, W.C., French, R.J., and Krueger, B.K. 1988, Effects of membrane surface charge and calcium on gating of rat brain sodium channels in planar bilayers, *J. Gen. Physiol.* 92: 431-447.

Cukierman, S. 1991, Asymmetric electrostatic effects on the gating of rat brain sodium channels in planar lipid membranes, *Biophys. J.* 60: 845-855.

Cukierman, S. 1993, Divalent cations shift the voltage dependence of the open probabilty of Na channels even in very high ionic strength solutions, *Biophys. J.* A109 (Abstract)

Dudley, S.C., Todt, H., Lipkind, G., and Fozzard, H.A. 1995, A μ-conotoxin-insensitive Na^+ channel mutant: Implications for the structure of the outer vestibule. *Biophys. J.* 69: 1657-1665.

Frankenhaeuser, B. and Hodgkin, A.L. 1957, The action of calcium on the electrical properties of squid axons, *J. Physiol. (London)* 137: 218-244.

French, R.J., Prusak-Sochaczewski, E., Zamponi, G.W., Becker, S., Kularatna, A.S., and Horn, R. 1996, Interactions between a pore-blocking peptide and the voltage sensor of the sodium channel: an electrostatic approach to channel geometry, *Neuron* 16: 407-413.

French, R.J. and Horn, R. 1983, Sodium channel gating: models, mimics and modifiers, *Ann. Rev. Biophys. Bioeng.* 12: 319-356.

Guy, H.R. and Seetharamulu, P. 1986, Molecular model of the action potential sodium channel, *Proc. Natl. Acad. Sci. USA* 83: 508-512.

Heinemann, S.H., Terlau, H., and Imoto, K. 1992, Molecular basis for pharmacological differences between brain and cardiac sodium channels, *Pflügers Arch.* 422: 90-92.

Hille, B., Woodhull, A.M., and Shapiro, B.I. 1975, Negative surface charge near sodium channels of nerve: Divalent ions, monovalent ions, and pH, *Phil. Tran. R. Soc. Lond B.* 270: 301-318.

Keynes, R.D. 1992, A new look at the mechanism of activation and inactivation of voltage-gated ion channels, *Proc. R. Soc. Lond. B* 249: 107-112.

Khodorov, B.I. 1985, Batrachotoxin as a tool to study voltage-sensitive sodium channels of excitable membranes, *Prog. Biophys. Mol. Biol.* 45: 57-148.

Lancelin, J.-M., Kohda, D., Tate, S.-I., Yanagawa, Y., Abe, T., Satake, M., and Inagaki, F. 1991, Tertiary structure of conotoxin GIIIA in aqueous solution, *Biochem.* 30: 6908-6916.

Lipkind, G., Fozzard, H., and Dudley, S. 1994, A structural model of the μ-conotoxin binding site in the sodium channel, *Biophys. J.* 66(2): A281 (Abstract)

Lipkind, G.M. and Fozzard, H.A. 1994, A structural model of the tetrodotoxin and saxitoxin binding site of the Na^+ channel, *Biophys. J.* 66: 1-13.

Margolskee, R.F., McHendry-Rinde, B., and Horn, R. 1993, Panning transfected cells for electrophysiological studies, *BioTech.* 15(5): 906-911.

Mathias, R.T., Baldo, G.J., Manivannan, K., and McLaughlin, S. 1992, Discrete charges on biological membranes, *in* "Electrified Interfaces in Physics, Chemistry and Biology," R. Guidelli, ed., Kluwer Academic Publishers, Netherlands, 473-490.

Mintz, I.M. 1994, Block of Ca channels in rat central neurons by the spider toxin ω-Aga-IIIA, *J. Neurosci.* 14(5): 2844-2853.

Moczydlowski, E., Olivera, B.M., Gray, W.R., and Strichartz, G.R. 1986, Discrimination of muscle and neuronal Na-channel subtypes by binding competition between [^{3}H]saxitoxin and μ-conotoxins, *Proc. Natl. Acad. Sci. USA* 83: 5321-5325.

Noda, M., Ikeda, T., Kayano, T., Suzuki, H., Takeshima, H., Kurasaki, M., Takahashi, H., and Numa, S. 1986, Existence of distinct sodium channel messenger RNAs in rat brain, *Nature* 320: 188-191.

Sato, K., Ishida, Y., Wakamatsu, K., Kato, R., Honda, H., Nakamura, H., Ohya, M., Kohda, D., Inagaki, F., Lancelin, J., and Ohizumi, Y. 1991, Active site of μ-conotoxin GIIIA, a peptide blocker of muscle sodium channels, *J. Biol. Chem.* 266(26): 16989-16991.

Seoh, S.-A., Sigg, D., Papazian, D.M., and Bezanilla, F. 1996, Voltage-sensing residues in S2 and S4 segments of the *Shaker* K^+ channel, *Neuron* 16: 1159-1167.

Sigworth, F.J. 1993, Voltage gating of ion channels, *Q. Rev. Biophys.* 27: 1-40.

Stillinger, F.H.J. 1961, Interfacial solutions of the Poisson-Boltzmann equation, *J. Chem. Phys.* 35(5): 1584-1589.

Stühmer, W., Conti, F., Suzuki, H., Wang, X., Noda, M., Yahagi, N., Kubo, H., and Numa, S. 1989, Structural parts involved in activation and inactivation of the sodium channel, *Nature* 339: 597-603.

Terlau, H., Heinemann, S.H., Stühmer, W., Pusch, M., Conti, F., Imoto, K., and Numa, S. 1991, Mapping the site of block by tetrodotoxin and saxitoxin of sodium channel II, *FEBS Lett.* 293: 93-96.

Trimmer, J.S., Cooperman, S.S., Tomiko, S., Zhou, J., Crean, S.M., Boyle, M.B., Kallen, R.G., Sheng, Z., Barchi, R.L., Sigworth, F.J., Goodman, R.H., Agnew, W.S., and Mandel, G. 1989, Primary structure and functional expression of a mammalian skeletal muscle sodium channel, *Neuron* 3: 33-49.

Ukomadu, C., Zhou, J., Sigworth, F.J., and Agnew, W.S. 1992, μI Na^+ channels expressed transiently in human embryonic kidney cells: biochemical and biophysical properties, *Neuron* 8: 663-676.

Yang, N., George, A.L., Jr., and Horn, R. 1996, Molecular basis of charge movement in voltage-gated sodium channels, *Neuron* 16: 113-122.

Yang, N. and Horn, R. 1995, Evidence for voltage dependent S4 movement in sodium channels, *Neuron* 15: 213-218.

Zagotta, W.N., Hoshi, T., and Aldrich, R.W. 1994, *Shaker* potassium channel gating III: evaluation of kinetic models for activation, *J. Gen. Physiol.* 103: 321-362.

β SUBUNIT MODULATION OF THE COUPLING BETWEEN CHARGE MOVEMENT AND PORE OPENING IN CALCIUM CHANNELS

Riccardo Olcese[1], Francesca Noceti[1], Pietro Baldelli[6], Ning Qin[1], Lutz Birnbaumer[1,3,4,5] and Enrico Stefani[1,2,5]

[1]Department of Anesthesiology
[2]Department of Physiology
[3]Department of Biological Chemistry
[4]Molecular Biology Institute and [5]Brain Research Institute
UCLA School of Medicine, Los Angeles CA, 90095-1778.
[6]Current address: Dept. of Neuroscience, Università di Torino 10125, Italy.

INTRODUCTION

Like all voltage-dependent ion channels, Ca^{2+} channels react to changes in the potential across the plasma membrane with conformational changes leading to channel opening. A well defined region of the protein, particularly rich in charged residues (S4 transmembrane segment), is acting as voltage sensor whose movement during potential changes generates a fast transient current macroscopically detected as gating currents preceding the channel opening and therefore the ionic flux.

The pore forming α_1 subunit of Ca^{2+} channels encodes the voltage gating machinery (Perez-Reyes et al., 1989), while one or more auxiliary subunits (Catterall, 1991; Hofmann et al., 1994) can increase Ca^{2+} current amplitude by increasing the expression level or by facilitating channel opening. For example, the coexpression of the skeletal muscle type-S Ca^{2+} channel (α_{1S}) in murine fibroblasts with the skeletal β (β_{1a}) subunit, resulted in an increase in the

From Ion Channels to Cell-to-Cell Conversations
Edited by Latorre and Sáez, Plenum Press, New York, 1997

number of high affinity dihydropyridine binding sites, and in channels that activated faster than channels lacking β subunits (Lacerda et al., 1991). This Ca^{2+} current potentiation by the coexpression of β subunits seems to be a general phenomenon for Ca^{2+} channel α_1 subunits (Singer et al., 1991; Varadi et al., 1991; Hullin et al., 1992; Williams et al., 1992; Lory *et al.*, 1993; Olcese et al., 1994; De Ward and Campbell, 1995; Perez-Garcia et al., 1995; Kamp, Perez-Garcia and Marban, 1996). The potentiation of Ca^{2+} current by coexpression of the β subunit with the α_1 subunit may occur without an increase in the number of expressed channels in the *Xenopus* oocyte expression system. In fact, coexpression of the cardiac type-C α_{1C} (α_{1C}) with the type-2 β subunit (β_{2a}) showed that the presence of the β subunit stimulates Ca^{2+} channel opening by improving the coupling of the voltage sensor to the opening of the pore, causing a left shift of the conductance-voltage (G(V)) relationship without detectable changes in the size, voltage dependence and time course of the charge movement (Neely et al., 1993). Moreover, direct biochemical determinations of the Ca^{2+} channel α_{1C} protein showed that its expression level was increased by coexpression with the $\alpha_2\delta$ subunit, while it was not affected by coexpression of the β_{2a} subunit which greatly potentiated ionic currents (Shistik et al., 1995). On the other hand, similar experiments in *Xenopus* oocytes with a chimeric cardiac α_{1C} /skeletal α_{1S} Ca^{2+} channel co-expressed with the β_{2a} subunit, showed that the current potentiation was associated with an increase in the charge movement, suggesting that for this clone the increase in current was due to a larger number of expressed channels (Delbono et al., 1994). Similarly, in a mammalian expression system (embryonic kidney cell line), ionic currents and charge movement were enhanced by coexpression of the Ca^{2+} channel α_{1C} with the β_{1a} and β_3 subunits (Kamp et al. 1996; Josephson and Varadi, 1996).

Together these results suggest that the potentiation by coexpression of the β subunit may result from an increase of either channel opening or the number of inserted Ca^{2+} channels and that the relative contribution of these mechanisms will depend on the type of Ca^{2+} channel subunit and on the expression system. We have recently demonstrated that coexpression of β subunits with the neuronal α_{1E} in *Xenopus* oocytes shifted the conductance-voltage (G(V)) relationship to more negative potentials and that the effect of β subunits on channel activation was independent of the type or subtype of β subunit tested (Schneider et al., 1994; Olcese et al., 1994). This shift occurred without significant changes in the charge movement-voltage (Q(V)) curve or in the time course of gating current. The shift in the G(V) curve without changes in the Q(V) curve indicated that modulation of α_1 by the β subunit facilitated channel opening altering relatively uncharged transitions with undetectable contribution to the total charge movement.

EXPERIMENTAL RESULTS

We shall examine the mechanism of Ca^{2+} channel potentiation by coexpression of the β subunit with the neuronal Ca^{2+} channel α_{1E} subunit by simultaneously measuring charge movement (gating currents) and ionic currents. We shall also evaluate the role of regulatory β subunits in determining *z* and the limiting open probability (Po_{max}) (Neely et al., 1993).

The properties of ionic and gating currents of α_{1E} expressed in *Xenopus laevis* oocytes were investigated using the cut-open oocyte (Stefani et al.,1994) and macropatch techniques: ionic currents were measured in 10 mM external Ba^{2+}, gating currents were isolated in 2 mM external Co^{2+}. The data shown have been reported by Olcese et al., 1996 and Noceti et al.,1996

The Effect of β Subunit on the α_{1E} Activation Curve (G(V)).

G(V) curves reflect the voltage-dependence of ion channel activation. We expressed the neuronal α_{1E} with and without the β_{2a} subunit and we studied how the presence of the accessory β subunit modifies the sensitivity of the protein to the membrane potential.

Macroscopic conductance can be derived from macroscopic current during the test pulse by dividing it by the driving force or, alternatively, by measuring peak tail currents at a constant voltage following the various test potentials (Bean and Rios, 1989). The plot of peak tail current amplitude at a constant post-pulse potential, as a function of a test potential, yields the voltage dependence for the steady state activation curve for the Ca^{2+} channel. We measured tail currents at -50 mV since their decay is slow enough to allow for reliable measurements of their peak values. To assure that only ionic current through Ca^{2+} channels were included, the peak tail current measurements were done over traces that were subtracted from traces recorded in 2 mM Co^{2+}. The measurement of the G(V) relationship from peak tail currents is accurate since it is obtained at a constant potential with a constant single channel amplitude and it is not contaminated with outward ionic current.

Figure 1 shows G(V) curves for α_{1E} and α_{1E}+β_{2a} fitted by the sum of two Boltzmann distributions (upper panel) with the averaged values of the parameters obtained fitting individual experiments. (lower panel). Both G(V) curves can be empirically separated into two components described by the sum of two Boltzmann distributions. Averaged data points were fitted by: $G_m/G_{max}=G_1/G_{max}/\{1+\exp[z_1(V_{1half}-V_m)F/RT\}+G_2/G_{max}/\{1+\exp[z_2(V_{2half}-V_m)F/RT\}$ where G_m, G_1 and G_2 are the conductance values as a function of the prepulse potential V_m, and G_{max} is the limiting conductance with the largest prepulse potential (+132 mV).

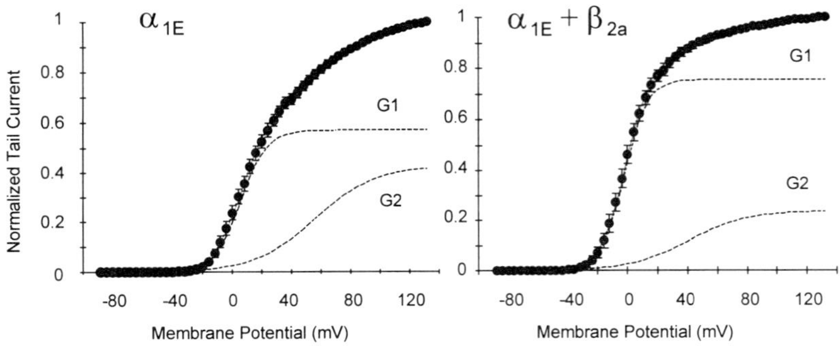

	G_1	V_{1half}	z_1	G_2	V_{2half}	z_2
α_{1E}	0.53 ± 1.2	-0.34 ± 2.1	2.9 ± 0.1	0.47 ± 1.2	52.4 ± 3.6	1.3 ±0.04
$\alpha_{1E}\beta_{2a}$	0.74 ± 1.72	-0.18 ± 0.1	3.6 ± 0.1	0.26 ± 1.7	41.5 ± 2.9	1.4 ± 0.1

Figure 1. The upper panel show data points of representative G(V) curves for α_{1E} (left) and $\alpha_{1E}+\beta_{2a}$ (right) from isocronic measurement of the peak tail current during repolarization to -50 mV. Data points are fitted by the sum of two Boltzmann distribution (continuous line) and the two components,G_1 and G_2 are displayed separately (dashed lines). In the lower panel the averages of parameters fitting the individual G(V) curves are shown. The coexpression of the β subunit seems to have little action on the half activation potential (V_{1half} ,V_{2half}) and the effective valences (z_1 ,z_2) of the components, while it clearly enhances the first, more negative and steeper component of the G-V. Error bars are the standard error of the mean (α_{1E}, n=15; $\alpha_{1E}\beta_{2a}$, n=15).

V_{1half} and $V_{2\ half}$ are the half activation potentials, z_1 and z_2 are the effective valences, and F, R and T (295°K) are the usual thermodynamic parameters. Similar values were obtained for the equivalent effective valences and half activation potentials for each component of the Boltzmann distribution fitting α_{1E} and $\alpha_{1E}+\beta_{2a}$.

The difference in the Boltzmann distributions that fitted the α_{1E} and $\alpha_{1E}+\beta_{2a}$ G(V) data, was found to reside in the relative proportion of their amplitude factors G_1/G_{max} and G_2/G_{max}. For α_{1E} G(V) curves, both components had a similar weight, while for $\alpha_{1E}+\beta_{2a}$ G(V) curves, the component with the more negative half activation potential and steeper voltage dependence became predominant, explaining the overall negative shift in the G(V) curve.

Note that the effective valence z obtained fitting the macroscopic conductance curve, does not correspond to the number of effective charges per channel which will be discussed later.

β_{2a} Subunit Increases the Limiting Open Probability

Besides the shift of the activation curve, another effect of the β_{2a} subunit on

α_{1E} currents is the increase in the absolute values of the limiting open probability.

This absolute value cannot be obtained by measuring peak tail currents, but can be estimated by non-stationary noise (variance) analysis of ionic currents. By fitting the experimental variance of consecutive current records, one can obtain the values of the single channel current and of the total number of channels activating in a patch. From these parameters and from the maximal current amplitude, the maximum value of the open probability (Po_{max}) can be extrapolated.

The results of this procedure are shown in Figure 2. The two graphs in A and B show plots and fits of the variance *vs.* mean current for α_{1E} and $\alpha_{1E}\beta_{2a}$. The Po_{max} reached by α_{1E} was low (0.55), while in the case of $\alpha_{1E}\beta_{2a}$ Po_{max} was higher (0.84).

Coupling Efficiency in α_{1E} and α_{1E}+β_{2a}

Figure 3 illustrates ionic and gating currents of α_{1E} and $\alpha_{1E}+\beta_{2a}$ Ca^{2+} channels. Coexpression of the β_{2a} subunit has no major effect on the activation time constant (τ). The time constant (τ) of the Ba^{2+} current activation for depolarizations to 16 mV was 3.1 ms and 2.7 ms for α_{1E} and $\alpha_{1E}+\beta_{2a}$, respectively. The trace to +60 mV is at the ionic current reversal potential and shows the ON gating current as a small upward current deflection.

Note the proportionally smaller charge movement in relation to the Ba^{2+} current for the $\alpha_{1E}+\beta_{2a}$ channel. Panels C and D show records of α_{1E} and $\alpha_{1E}+\beta_{2a}$ gating current elicited by voltage steps from -90 mV HP to various potentials

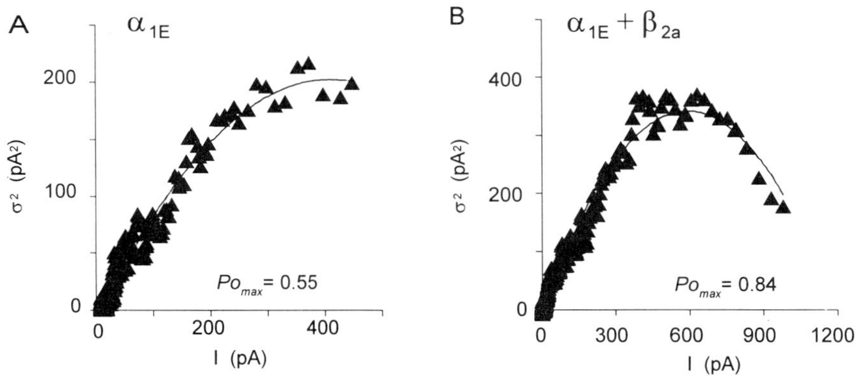

Figure 2. Variance-mean data and corresponding fitted curves to the theoretical function in α_{1E} and $\alpha_{1E}\beta_{2a}$. The fitted parameters and calculated Po_{max} were: i=1.00 pA, N=801, Po_{max}=0.55 in α_{1E} expressed alone; and i=1.14 pA, N=1033, Po_{max}=0.84 in $\alpha_{1E}\beta_{2a}$. The experimental variance is fitted to σ^2=Ii-I²/N Po_{max} was extrapolated as $Po_{max} = I_{max}/iN$. Experiments were performed at 30° C in order to reach a Po >0.5. This condition is necessary for fitting to the parabolic function whose maximum corresponds to Po=0.5.

and demonstrates that the time courses of the α_{1E} gating currents were not modified by β_{2a} coexpression.

In all batches of oocytes, coexpression of the β_{2a} subunit consistently induced a left shift of the G(V) curves, as shown in Figure 4. This action was independent of the expression level which was variable. For example, in batches of oocytes with low expression of α_{1E} (G_{max}=64±14 μS, n=5), the β_{2a} subunit *ca.* doubled ionic currents (G_{max}=154±31 μS, n=5), while in other batches of oocytes with high expression of α_{1E}, G_{max} had similar values for α_{1E} 123 ± 14 μS (n=20, from 10 batches) and for $\alpha_{1E}+\beta_{2a}$ 82 ± 19 μS (n=14, from 7 batches of oocytes). In the same group of oocytes with high α_{1E} expression, coexpression of β_{2a} reduced Q_{max}.

A better way to evaluate the action of the β_{2a} subunit on channel gating is to measure in the same oocyte the coupling efficiency between charge movement and pore opening by the G_{max}/Q_{max} ratio where Q_{max} is the total charge moved. Q_{max} was measured at the ionic reversal potential in 10 mM Ba^{2+} (Fig. 5A,B) and after blocking the Ba^{2+} currents with 2 mM Co^{2+}.

The G_{max}/Q_{max} ratio was significantly higher for $\alpha_{1E}+\beta_{2a}$ channels: $\alpha_{1E=}$9.22 x 10^5 ± 1.4 x 10^5 S/C (n=20) and $\alpha_{1E}+\beta_{2a}$= 21.95 x 10^5 ± 2.8 x 10^5 (n=14) (Fig. 3B). Assuming that the β_{2a} subunit does not change the number of effective charges

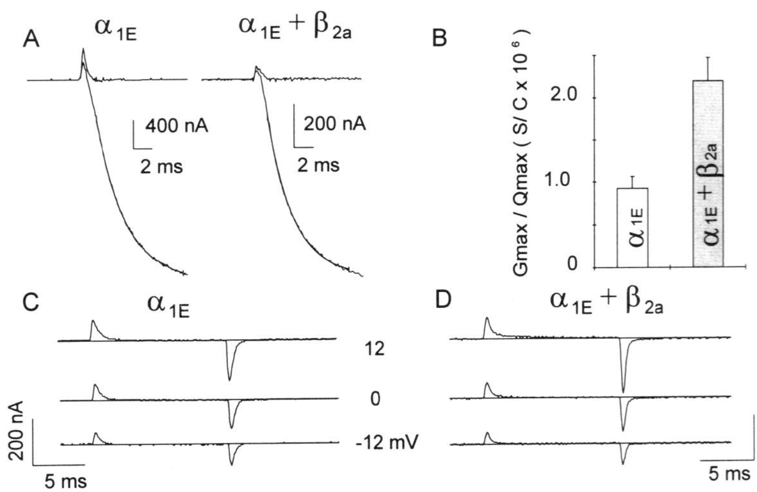

Figure 3. (A) Current traces in 10 mM external Ba^{2+} elicited by depolarization to 30 and 60 mV in oocytes expressing α_{1E} and $\alpha_{1E}+\beta_{2a}$. The activation time constants, obtained fitting to a single exponential function the very early phases of the ionic current during depolarization to 30 mV, is practically the same in the two clones (3.1 ms for α_{1E} and 2.7 ms for $\alpha_{1E}+\beta_{2a}$). Note the relative reduction in size of the charge movement which precedes the ionic current in the $\alpha_{1E}+\beta_{2a}$ compared to α_{1E} expressed alone. The ratio G_{max}/Q_{max} for α_{1E} and $\alpha_{1E}+\beta_{2a}$ is shown in (B). The coexpression of the β_{2a} improved the "charge movement-pore opening" coupling efficiency of α_{1E} channel . In C and D α_{1E} and α_{1E} + β_{2a} gating current recorded in 2 mM Co^{2+} are shown.

per channel, the higher coupling efficiency indicates that an equivalent number of channels measured by the Q_{max} elicits larger currents when co-expressed with the β_{2a} subunit. Since the single channel amplitude was not affected by the β_{2a} subunit (Noceti et al. 1996), the ionic current potentiation should arise from an increase in channel openings. The fact that the potentiated $\alpha_{1E}+\beta_{2a}$ currents have a time course that is very similar to that of the α_{1E} currents, together with the effect on the Po_{max} strongly suggests that the main action of the β_{2a} subunit is to reduce the proportion of null traces.

Coexpression of the β_{2a} Subunit Does Not Alter the Charge Movement of α_{1E}

To further evaluate the action of the β_{2a} subunit on pore opening of α_{1E}, we plotted Q and G *vs* the membrane potential (Fig. 4). The plots show that coexpression of the β_{2a} subunit induced an overall shift of the G(V) curves to more negative potentials, with the characteristics described in Figure 1, and that these changes in the G(V) curves were not associated with modifications in the Q(V) curve. The Q(V) relationships, obtained by plotting the time integral of the ON gating current *vs* the membrane potential, were empirically fitted by sums of two Boltzmann distributions: $Q_{on}/Q_{max}=Q_1/Q_{max}/\{1+\exp[z_{q1}(V_{q1half}-Vm)F/RT]\} + Q_2/Q_{max}/\{1+\exp[z_{q2}(V_{q2half}-Vm)F/RT]\}$ where Q_{max} is the limiting charge movement, Q_1 and Q_2 are the amplitudes of the charge components, z_{q1} and z_{q2} are the effective valences and V_{q1half} and V_{q2half} are the respective half activation potentials.

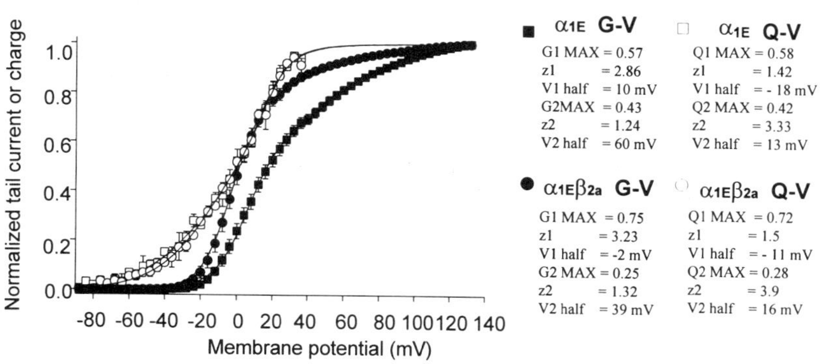

Figure 4. The figure shows averaged data points voltage dependence of the charge movement for α_{1E} (□), and $\alpha_{1E}+\beta_{2a}$ (O), and the Ba^{2+} conductance (■ α_{1E}, ● $\alpha_{1E}+\beta_{2a}$). Error bars are the standard error of the mean (n=3). The G(V) and Q(V) curves are fitted to the sum of two Boltzmann distributions. Parameters fitting the experimental data are shown on the side: z_1 and z_2 are the effective valence, $V1_{half}$ the $V2_{half}$ are half activation potentials, G_1 and G_2, Q_1 and Q_2 are the relative amplitude of the G(V) or Q(V) curves.

The charge movement of α_{1E} could be separated into two components with about the same amplitude: the first had a shallower voltage dependence, $z_{q1}=1.42$ with $V_{q1half}=-18mV$; the second component had a steeper voltage dependence $z_{q2}=3.3$ with $V_{q2half}=16$ mV. The Q(V) curve for α_{1C} showed only one component with a voltage dependence practically identical to that of the first component of α_{1E}. The coexpression of β subunits did not modify the voltage dependence of the charge movement.

The Q(V) curves are more negative in voltage axis than the G(V) curves indicating the existence of several closed states prior to channel opening.

The Number of Charges per Channel in Ca^{2+} Channels: Role of the β Subunit in its Determination.

The voltage-dependence of channel activation is strictly related to the size of the total charge per channel, as well as to its mobility and its coupling to pore opening. Therefore, one critical quantity to evaluate is the number of effective charges per channel, z. z is the number of elementary charges, z_0, times the fraction of the field, x, that the charge traverses across the membrane ($z=z_0x$). z can be derived from the conductance to G(V) relationship by its limiting slope value at a very low probability of opening (Almers, 1978; Almers and Armstrong, 1980; Bezanilla and Stefani, 1994; Hirschberg et al., 1995). Alternatively, z can be obtained from the ratio between the limiting charge (Q_{max}) and the number of channels, which may be evaluated by the variance method in the same patch (Schoppa et al., 1992), or by channel particle measurement in freeze fractured membranes (Zampighi et al., 1995).

We investigated the number of charge per channel and the potential role of the β subunit in its determination. The number of charge per channel (z), can be achieved through non-stationary noise analysis and through limiting slope analysis. As already discussed, the number N of channels in a patch can be evaluated through non-stationary noise analysis. By integrating gating currents in the same patch, one can estimate the total charge movement Q_{max}. The effective number of charges per channel is therefore obtained as

$$z = \frac{Q_{max}}{e_0 N}$$

The other method used to evaluate z is the limiting slope analysis, which is based on the assumption of dealing with a channel with one single open state preceded by a sequence of closed states. Under these conditions, the G(V) at very negative potential, i.e. for very low values of the open probabilities, can be approximated with a single exponential function, where z is one of the

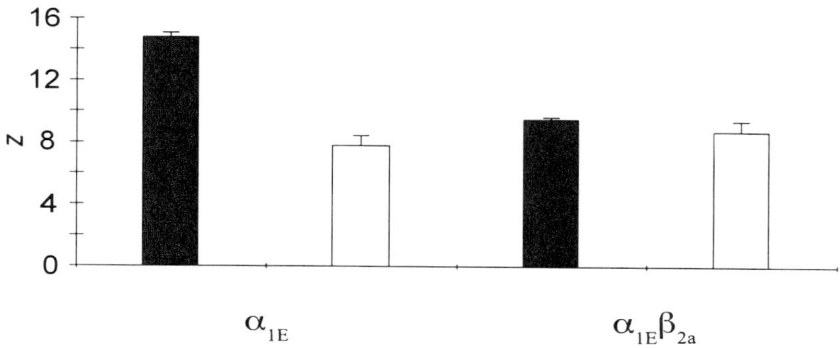

Figure 5. Summary plot of the results for α_{1E} Ca^{2+} and for the cardiac α_{1C} channels with or without different β subunits. The mean z values are plotted with their standard errors (SEM). The filled bars represent the results from the variance analysis, while the empty bars are the results from the limiting slope experiments. The average value of z obtained with the limiting slope experiments is: $z=8.58\pm0.18$.

parameters. It can be shown that the following is true:

$$z(V) = \frac{kT}{e_0}\frac{\mathrm{d}}{\mathrm{d}V}\ln G(V)$$

Figure 5 summarizes the number of charges obtained with the two methods for α_{1E} Ca^{2+} channels with and without β subunits. Both methods agreed for the $\alpha_{1E}\beta_{2a}$ combination, however, the variance method gave larger *z* values for α_{1E}.

This discrepancy may be explained by the correlation between a low value of the Po_{max} in α_{1E} and $\alpha_{1E}\beta_{1a}$ with more occurrence of traces without opening (*"nulls"*). Null traces have charge movement but do not contribute to the variance, causing an underestimation of *N*. The limiting slope method gave similar results, for all Ca^{2+} channel combinations tested, of *ca.* 9 e_0.

CONCLUSIONS

The transduction of the changes of membrane potential into structural modification of the channel protein is triggered by the voltage sensor which, behaving like an electrical dipole, reorients within the electric field and "drags" the rest of the protein through the different conformational states. How efficiently the movement of the voltage sensor will drive the rest of the molecule toward the open conformation will reflect on the biophysical properties of the channels such as the position of the G(V) curve relative to the Q(V) curve along the voltage axis.

We, and others, have previously shown that different β subunits (β_{1a}, β_{1b}, β_{2a}, β_{2b} and β_3) and chimeric constructs among them when co-expressed with the α_{1E} subunit or other Ca^{2+} channels, share the common action of facilitating pore opening by shifting the activation curve to more negative potentials. On the other hand, β subunits can have an opposite effect on the voltage dependent inactivation (Olcese et al.,1994).

The regulatory properties of β subunits have been studied in different expression systems and it is generally agreed that the expression of β subunit either potentiates α_1 ionic current (Singer *et al.,* 1991; Wei et al., 1991; Neely et al., 1993; Lacerda et al., 1991; Varadi et al., 1991) and/or increases the number of high affinity dihydropyridine (DHP) binding sites (Lacerda et al., 1991; Varadi et al., 1991; Perez-García, Kamp and Marban, 1995). The mechanism by which these changes occur may vary in different clones and expression systems. Increased high affinity binding and potentiation of ionic current could be due either to an increase of cell surface α_1 subunits or to stabilization of new conformations that show the potentiated ionic current and a shift in the equilibrium between low (undetectable) and high affinity DHP binding. Although the effect of β on the expression of α_1 was specifically addressed in two instances, the results reported were opposite. Upon β subunit coexpression, Nishimura and colleagues (1993) measured an increase in DHP binding to *Xenopus* membranes without changes in expression of α_1 assayed by Western blotting, while Josephson and Varadi (1996), studying the effect of β subunit on α_1 function in HEK 293 cells reported an increase in both ionic current and amount of α_1, assayed by measuring charge movement in α_1+β expressing *vs.* α_1 expressing cells. In other cases, it has been demonstrated that the enhancement of ionic current in *Xenopus* oocytes after coexpression of the accessory β subunit was not due to increased α_1 expression since the increase in the ionic current could be totally accounted for by an increase in the open probability of the expressed channels (Neely et al. 1993; Shistik et al. 1995). In comparison, the $\alpha_2\delta$ subunit alone has been shown to boost the expression of the α_1 subunit in *Xenopus* oocytes (Shistik et al. 1995).

We have shown that the coupling efficiency of the neuronal α_{1E} is further improved by the β subunit. The tighter coupling caused by the β subunit results from a shift in the G(V) curve to more negative potentials, which reduces the gap between the Q(V) and G(V) curves. However, this does not happen by heavily affecting the voltage sensitivity of the channel at negative potentials, since the feet of the G(V) curves for both α_{1E} and α_{1E}+β_{2a} are superimposed and the detection of the ionic current is close to -30 mV for both types of channels. The tighter coupling correlates from an increase in the relative amplitude of the first, steeper component of the two components of the G(V) curve, from ~53%

to ~75% of the total. Because of the equality of the number of charges per channel for α_{1E} and α_{1E}+β_{2a}, as measured from experiments estimating the limiting slope, the difference in the channel efficiency when the β is coexpressed must lie in the way the sensor is coupled to the channel opening. This is mostly reflected in the relative position of the G(V) curve along the voltage axis, because the properties of the voltage sensor, as evaluated by the Q(V), are similar: for the same displacement of charges, a higher open probability is achieved when the α_1 subunit is coupled to the β subunit.

In the present work, in batches of oocytes with very poor α_{1E} expression, β_{2a} promoted an increase in the current level which can be accounted by the increase in open probability (Noceti et al., 1996). On the other hand, in batches of oocytes with high α_{1E} expression, the β_{2a} subunit induced a reduction of the total charge movement, suggesting a lower expression level. One possible explanation for this observation may be that in oocytes in which the α_{1E} mRNA induces a saturating rate of protein synthesis, the coexpression of an additional protein, in this case β_{2a}, competes for the synthetic machinery and lowers the total number of α_1 subunits that are made.

The comparison of these results obtained for α_{1E} with the ones obtained for α_{1C} (Neely et al., 1994; Olcese et al., 1996) brings to interesting considerations: the coexpression of the β_{2a} subunit has a more remarkable effect on the α_{1C} than on the α_{1E} channel. In α_{1E} the shift to the left of the G(V) curve caused by β_{2a} coexpression is ~ -15 mV while in α_{1C} is ~ -50 mV. This difference could be related to the intrinsic differences in voltage separation between the Q(V) and G(V) curves (20 mV for α_{1E} and ca. 100 mV for α_{1C}). In fact, we can speculate that the coupling efficiency between charge movement and channel opening in α_{1E} is already high, making the regulation by the β_{2a} less evident, while the large voltage separation between G(V) and Q(V) of α_{1C} leaves more room for regulation. These findings allow to conclude that in the α_{1C} Ca^{2+} channel there are many closed states prior to channel opening and the charge movement is not tightly coupled to pore opening, while the opposite is true for α_{1E}. Related to this result is the marked change in the time course that the β_{2a} causes in the installation of the α_{1C} ionic current (Perez-Reyes et al., 1992), which it does not do when it is expressed with α_{1E} (Fig.3A).

We conclude that in the neuronal α_{1E} the charge movement is tightly coupled with the pore opening and that the β_{2a} subunit coexpression further improves this coupling.

REFERENCES

Almers, W., 1978, Gating currents and charge movements in excitatory membranes. *Rev of Physiol. Biochem and Pharmacol.* 82:96.

Almers, W., and. Armstrong, C. M., 1980, Survival of K^+ permeability and gating currents in squid axons perfused with K^+-free media. *J. Gen. Physiol.* 75:61.

Barish, M. E. , 1983, A transient calcium-dependent chloride current in the immature *Xenopus* oocyte. *J. Physiol.* 342:309.

Bean, B. P and Rios, E., 1994, Non linear charge movement in mammalian cardiac ventricular cell. Component from Na and Ca channels gating. . *J. Gen. Physiol* 94:65.

Bezanilla, F.and E. Stefani. 1994, Voltage dependent gating of ionic channels. *Ann Rev Biophys Biochem Chem* 23:819.

Catterall, W. A., 1991, Functional subunit structure of voltage-gated calcium channels. *Science* 253:1499.

Delbono, M., Gopalakrishnan, A., Neely, A. Olcese, R. Birnbaumer, L. and Stefani E., 1994, abstract. Skeletal muscle calcium channel domain I slows down $\alpha 1$ cardiac channel opening without changes in the voltage sensor. *Biophys. J.* 66:88.

De Waard, M. and Campbell, K. P., 1995, Subunit regulation of the neuronal alpha 1A Ca^{2+} channel expressed in Xenopus oocytes *J.Physiol.* 485:619

Hirschberg, B., A. Rovner, M. Lieberman and J. Patlak. 1995, Transfer of twelve charges is needed to open skeletal muscle Na^+ channels. *J Gen Physiol.* 106:1053.

Hofmann F., Biel M. and Flockerzi V., 1994, Molecular basis for Ca^{2+} channel diversity. *Annual Review of Neuroscience*, 17:399.

Hullin R., Singer-Lahat D., Freichel M., Biel M., Dascal N., Hofmann F. and Flockerzi V., 1992, Calcium channel beta subunit heterogeneity: functional expression of cloned cDNA from heart, aorta and brain. *Embo J.* 11:885.

Josephson, I. R. and Varadi, G., 1996, The β subunit increase Ca^{2+} current and gating charge movement of human cardiac L-type Ca^{2+} channels. *Biophys J.* 70:1285.

Kamp, T. J., Perez-Garcia, M. T., and Marban, E. M., 1996, Enhancement of ionic and charge movement by coexpression of calcium channel β_{1A} subunit with α_{1C} subunit in a human embryonic cell line. *J.Physiol.* 492:89.

Lacerda, A. E., Kim, H. S., Ruth, P., Perez-Reyes, E., Flockerzi, V., Hofmann, F., Birnbaumer, L., and Brown, A. M., 1991, Normalization of current kinetics by interaction between the α_1 and β subunits of the skeletal muscle dihydropyridine-sensitive Ca^{2+} channel. *Nature* 352:527.

Lory, P., Varadi, G., Slish, Df., Varadi, M. and Schwartz A., 1993, Characterization of beta subunit modulation of a rabbit cardiac L-type Ca^{2+} channel alpha 1 subunit as expressed in mouse L cells. *FEBS Letters* 315:167.

Neely, A., Olcese, R., Wei X., Birnbaumer, L. and Stefani E., 1994, Ca^{2+}-dependent inactivation of a cloned cardiac Ca^{2+} channel alpha 1 subunit, (α_{1C}) expressed in Xenopus oocytes. *Biophys. J.* 66:1895.

Neely, A. Wei, X. Olcese, R. Birnbaumer, L. and. Stefani. E, 1993, Potentiation by the β subunit of the ratio of the ionic current to the charge movement in the cardiac calcium channel. *Science*. 262:575.

Nishimura, S., Takeshima, H., Hofmann, F., Flockerzi, V. and Imoto K., 1993, Requirement of the calcium channel beta subunit for functional conformation. *FEBS Lett.* 324:283.

Noceti, F., Baldelli, P.,Wei, X., Qin, N., Toro, L., Birnbaumer, L. and Stefani, E., 1996, Effective gating charges per channel in voltage dependent K^+ and Ca^{2+} channels. *J.Gen.Physiol.* 108:143.

Olcese, R., Qin, N., Schneider, T., Neely, A., Wei, X., Stefani, E., and Birnbaumer, L., 1994, The amino termini of calcium β subunits set rates of inactivation independently of their effect on activation. *Neuron.* 13:1433.

Olcese, R., Neely, A., Qin, N., Wei, X., Birnbaumer, L. and Stefani, E., 1996 Coupling between charge movement and pore opening in vertebrate neuronal α_{1E} calcium channel. *J. Physiol. in press*

Perez-Garcia, M. T., Kamp, T. J. and Marban, E., 1995, Functional properties of cardiac L-type calcium channels transiently expressed in HEK293 cells. Roles of alpha 1 and beta subunits. *J.Gen.Physiol* 105:289.

Perez-Reyes, E., Kim, H. S., Lacerda, A. E., Horne, W., Wei, X., Rampe, D., Campbell, K. P., Brown, A. M. and Birnbaumer, L., 1989, Induction of calcium current by the expression of the α_1-subunit of the dihydropyridine receptor from skeletal muscle. *Nature* 340:233.

Perez-Reyes, E., Castellano, A., Kim, H. S., Bertrand, P., Baggstrom, E., Lacerda, A. E., Wei, X. and Birnbaumer, L., 1992, Cloning and Expression of Cardiac/Brain β Subunit of the L-type Calcium Channel. *J. Biol. Chem.* 267:1792.

Schneider, T., Wei, X., Olcese, R., Costantin, J. L., Neely, A., Palade, P., Perez-Reyes, E., Qin, N., Zhou, J., Crawford G. D., Smith, R. G., Appel, S. H. Stefani, E., and Birnbaumer, L., 1994, Molecular analysis and functional expression of the human type E neuronal Ca^{2+} channel α_1-subunit. *Receptors and Channels* 2:255.

Schoppa, N. E., Mccormack, K., Tanouye, M. A. and Sigworth, F. J., 1992, The size of gating charge in wild-type and mutant *Shaker* potassium channels. *Science* 255:1712.

Shistik, E. Ivanina, T. Puri, T. Hosey, M. and Dascal, N., 1995, Ca^{2+} channel enhancement by α_2/δ and β subunits in *Xenopus* oocytes: contribution of changes in channel gating and α_1 protein level. *J. Physiol.* 489:55.

Singer, D., Biel, M., Lotan, I., Flockerzi, V., Hofmann, F., and Dascal, N. (1991, The roles of the subunits in the function of the calcium channel. *Science* 253:1553.

Stefani, E., Toro, L., Perozo, E. and Bezanilla, F., 1994, Gating of Shaker K^+ channels: I. Ionic and gating currents. *Biophys. J.* 66:996.

Varadi G., Lory P., Schultz D., Varadi M. and Schwartz A., 1991, Acceleration of activation and inactivation by the beta subunit of the skeletal muscle calcium channel. *Nature* 352:159.

Wei, X. Y., Perez Reyes, E., Lacerda, A. E., Schuster, G. Brown A. M. and Birnbaumer, L., 1991, Heterologous regulation of the cardiac Ca^{2+} channel α_1 subunit by skeletal muscle β and γ subunits. Implication for the structure of cardiac L-type Ca^{2+} channels *J. Biol. Chem.* 266:21943.

Williams, M. E., Feldman, D. H., Mccue, A. F., Brenner, R., Velicelebi, G., Ellis, S. B. and Harpold, M. M., 1992, Structure and 2 functional expression of alpha 1, α_2, and β subunits of a novel human neuronal calcium channel subtype. *Neuron* 8:71.

Zampighi, G.A., M. Kreman, K.J. Boorer, D.D. Loo, F. Bezanilla, G. Chandy, J.E. Hall, and E.M. Wright., 1995. A method for determining the unitary functional capacity of cloned channels and transporters expressed in *Xenopus laevis* oocytes. *J Membr Biol* 148:65.

THERMODYNAMIC DOUBLE MUTANT CYCLES WITH THE SHAKER K^+ CHANNEL AND A PEPTIDE INHIBITOR DEFINE THE SPATIAL LOCATION OF PORE-FORMING RESIDUES

Patricia Hidalgo[1]
Department of Neurobiology
Harvard Medical School
Boston, MA 02115, USA

INTRODUCTION

K^+-selective channels comprise the most diverse family among ion channels. K^+ channels are gated in response to changes in membrane potential (voltage-activated K^+-channels), G-protein activation, or second messengers such as Ca^{2+} or cAMP. Despite the differences in their gating mechanisms, all K^+ channels share same conduction properties. The high selective permeability to K^+ ions is the most remarkable property displayed by K^+ channels considering that in order to achieve its selectivity K^+ channels must exclude Na^+ ions, which have a smaller crystal radius than K^+ ions (the permeability ratio Na^+ to K^+ is greatly than 1:100; Hille, 1973).

Full understanding of the molecular basis underlying selectivity in K^+ channels requires both, functional and structural studies. K^+ channels, as any other integral membrane protein, present difficulties for solving the structure at high atomic resolution.

[1]Present address: Biological Chemistry and Molecular Pharmacology, Harvard Medical School.

In this chapter an alternative approach for obtaining structural information on ion channels, based on thermodynamic double mutant cycle analysis, is described. Thermodynamic double mutant cycles have provided a formalism for studying energetic coupling between amino acids on the same polypeptide chain. A detailed description of mutant cycles applied to interactions between amino acids within a single protein has been given (Carter et al., 1984; Serrano et al., 1990; Horovitz and Fersht, 1990).

Thermodynamic double mutant cycles were applied to the voltage activated *Shaker* K^+ channel and the high affinity inhibitor agitoxin2 (AgTx2). AgTx2 belongs to a larger family of peptide toxins, derived from scorpion venom, which binds to the extracellular pore entryway of K^+ channels (Miller 1988, 1995; MacKinnon and Miller, 1988, 1989a; Candia et. al., 1992). The three-dimensional solution structure of AgTx2 has been solved at high resolution (Krezel et al., 1995). The basic idea is simple: the known distance between residues on the toxin can be translated into the distance separating two residues in the channel if such toxin residues interacts closely with known channel residues. Thus, the spatial location of the *Shaker* K^+ channel residues are defined with respect to AgTx2 structure by identifying pairwise interactions, through thermodynamic cycles, between amino acid residues on the channel and on the toxin (Hidalgo et al. , 1995; Ranganathan et al. , 1996). Because the same stretch of amino acids that determines the toxin binding site contributes to the formation of the ion conduction pathway (pore), the structural information obtained pertains to part of the functionally most important region of K^+ channels.

Further spatial constraints on the structure of the toxin binding site of the *Shaker* K^+ channel was obtained by extending the thermodynamic double mutant cycle analysis to examine the dependence of a pairwise toxin-channel interaction on a second channel residue.

Altogether, the results place the amino-terminal end of the pore forming region of the *Shaker* K^+ channel 12 to 15 Å from the central axis of the channel and suggest a helical conformation for a stretch of amino acids constituting the AgTx2 binding site.

METHODS

Mutagenesis and Expression of the *Shaker* K^+ Channel and AgTx2

All experiments were carried out using an inactivation-removed *Shaker* K^+ channel clone expressed in Xenopus oocytes (Hoŝhi et al., 1990).

Oocytes were prepared and injected with in vitro transcribed cRNA as previously described (Lu and MacKinnon, 1994).

Recombinant toxin was prepared as described (Park et al., 1991; Garcia et al., 1994). An extinction coefficient of 8.6 mM^{-1} cm^{-1} at 235 nm was used to calculate toxin concentration. Mutants were produced in the toxin gene using PCR and were confirmed by dideoxy DNA sequencing (Sanger et al., 1977) across a Sal I-Hind III cassette containing the coding sequence.

Electrophysiological Recording

K^+ currents were recorded with a two-electrode voltage clamp amplifier (Warner Instruments). Recording solution contained: 95 mM NaCl, 2 mM KCl, 0.3 mM $CaCl_2$, 1 mM $MgCl_2$, 5 mM HEPES, pH 7.6. Currents were elicited with a 50 ms step to 0 mV every 5 to 25 seconds, depending on the kinetic of the toxin blockade, from a holding potential of -80 mV. Current magnitudes ranged from 0.5 to 1.5 μA.

Perfusion with toxin-free and toxin-containing solution was controlled with separate lines using computer-driven solenoids valves. Bovine serum albumin at a concentration of 50 μg/ml was used to prevent non-specific adherence of the toxin to the perfusion lines.

The time course of toxin blockade was monitored to ensure that equilibrium inhibition was achieved and to confirm that the control current level was the same before addition and after removal of toxin (figure 1). From 3 to 8 experiments were made on separate oocytes.

The equilibrium inhibition constant (Ki) was calculated according to the following equation: $Ki = [AgTx2](I/I_0)/(1-(I/I_0))$, where I_0 is the control current level and I is the current in the presence of AgTx2 concentration [AgTx2].

PRIMARY STRUCTURE OF THE PORE FORMING REGION OF THE *SHAKER* K^+ CHANNEL

The predicted membrane topology based on the hydropathy profile of the deduced amino acid sequence of the *Shaker* K^+ channel is shown in figure 2 (Catterall, 1988; Tempel et al., 1987). The channel comprises six putative transmembrane segments, named S1 to S6. The amino and carboxy termini are both located on the cytoplasmic side of the membrane. The channel presumably is formed by the assembly of four identical subunits (MacKinnon, 1991), each

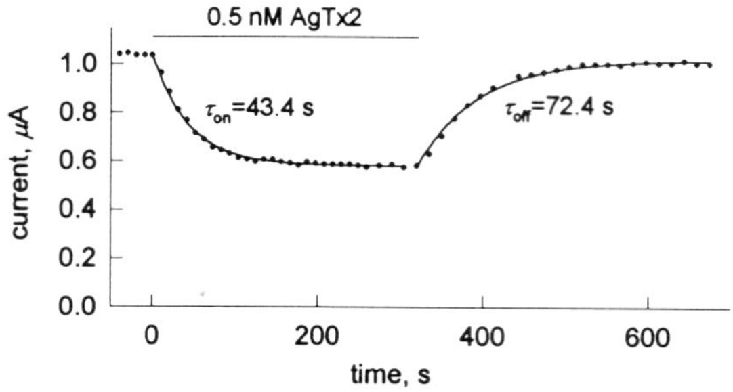

Figure 1. Inhibition of *Shaker* K^+ channels by AgTx2. *Shaker* K^+ channels were expressed in *Xenopus* oocytes. Current elicited by voltage steps of 50 ms duration from a holding potential of -80 mV to 0 mV are recorded by two-electrode voltage clamp. After 5 traces were recorded in the absence of toxin (control solution), solution containing 0.5 nM AgTx2 was perfused. The switch of perfusion lines was done by computer controlled solenoides valves. Isochronal current values upon maximal channel opening are plotted against time. Once equilibrium blockade was reached toxin was removed by perfusing with control solution. The inhibition is fully reversed upon removal of the toxin. The equilibrium inhibition constant (Ki) was calculated according to $Ki = [AgTx2](I/I_0)/(1-(I/I_0))$ where I_0 is the control current level and I is the current in the presence of AgTx2 concentration [AgTx2]. The solid curves correspond to single exponential fits with t_{on} = 43.4 sec and t_{off} = 72.4 sec.

containing the six transmembrane segments, and arranged about a central pore (a four-fold symmetry axis). The linear amino acid sequence between residues 431 and 449 on the S5-S6 linker, usually referred as the P-region (pore-forming region), of each subunit contribute to the formation of the ion conduction pathway (MacKinnon and Miller, 1989b; MacKinnon et al., 1990; MacKinnon and Yellen, 1990; Yellen et al., 1991; Yool and Schwarz, 1991; Hartmann et al., 1991; Heginbotham and MacKinnon, 1992; Heginbotham et al., 1992; Kavanaugh et al., 1991; Kavanaugh et al., 1992; Kirsch et al., 1992; De Biasi et al., 1993). A smaller subset of amino acids on the P-region, named the K^+ channel signature sequence, has been postulated to form, although in an unknown way, the selectivity filter (Heginbotham et al., 1992, 1994). Other regions of the voltage-activated K^+ channels, such as the S4-S5 loop and the S6 segment has also been shown to affect K^+ permeation (Jan and Jan, 1994; Choi et al., 1993; Slesinger et al., 1993; Lopez et al., 1994), although most of the evidence points to the P-region as the main determinant of K^+ selectivity.

The binding site for pore blocking peptides purified from scorpion venom, such as AgTx2 also resides in part of the P-region of K^+ channels (MacKinnon and Miller 1989b; MacKinnon et al., 1990; Candia et al., 1992; Giangiacomo et al., 1992; Gross et al., 1994, Ranganathan et al. , 1996). Mutational analysis of the S5-S6 linker of the *Shaker* K^+ channel identified several residues that affect AgTx2 binding in one of two different ways, through electrostatic mechanisms or through possible steric effects (MacKinnon and Miller, 1989b; MacKinnon et al., 1990, see also Garcia et al., 1994).

THREE-DIMENSIONAL STRUCTURE OF AGTX2

AgTx2 was originally purified from scorpion venom and described as a potent inhibitor of the *Shaker* K^+ channel and other related voltage activated K^+ channels (Garcia et al., 1994). The toxin is a highly basic, 38 amino acids long and an extremely structured peptide (figure 3a). Most of AgTx2 linear sequence is involved in the formation of secondary structure, a triple stranded antiparallel

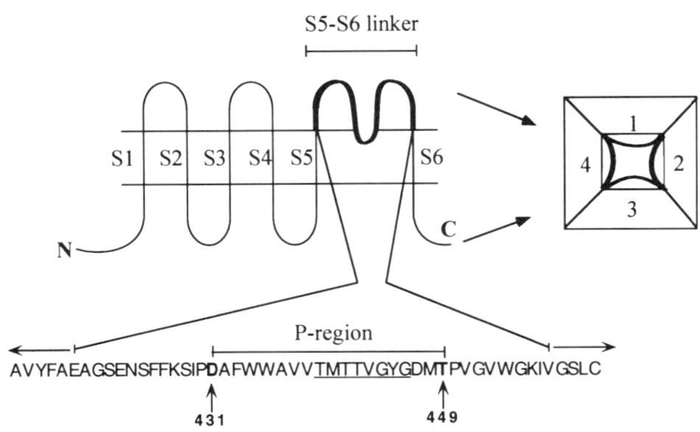

Figure 2. Current model of the membrane topology and partial amino acid sequence of the *Shaker* K^+ channel. The hydropathy profile predicts six transmembrane spanning regions (S1 to S6). Structure-function studies place the segment connecting the S5 and S6 transmembrane domains (S5-S6 linker, shown in bold) as partially dipping into the plane of the membrane. Channel residues 431 and 449 delimit the pore-forming region (P-region). The K^+ channel signature sequence is underlined. The channel (left panel) presumably has four-fold symmetry. Each subunit (numbered 1 to 4) corresponds to one polypeptide chain with its six transmembrane spanning regions as shown.

A

1 5 10 15 20 25 30 35

GVPINVS**C**TGSPQ**C**IKP**C**KDAGMRFGK**C**MNRK**C**H**C**TPK

B

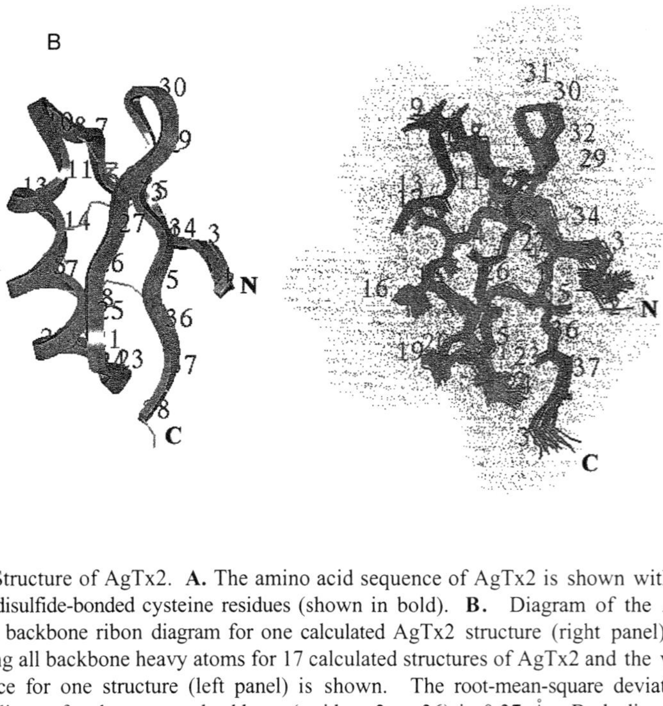

Figure 3. Structure of AgTx2. **A.** The amino acid sequence of AgTx2 is shown with lines connecting disulfide-bonded cysteine residues (shown in bold). **B.** Diagram of the AgTx2 structure. A backbone ribon diagram for one calculated AgTx2 structure (right panel) and a view showing all backbone heavy atoms for 17 calculated structures of AgTx2 and the van der Waals surface for one structure (left panel) is shown. The root-mean-square deviation of atomic coordinates for the average backbone (residues 3 to 36) is 0.37 Å. Both diagram are shown with the same AgTx2 orientation. The three disulfide bonds are also shown. The numbering correspond to the amino acid sequence shown in panel A.

β-sheet and single helix (figure 3b). The toxin appears to be a "tightly packaged" molecule as a consequence of three disulfide bonds, two of which bring together the β-sheet and the helix. AgTx2 has an asymmetric shape that resembles an ellipse, with a long axis of about 30 Å and two short axes of about 20 Å. Most of the amino acid residues are exposed to the solvent, except for the six cysteines, Gly26 and the side chain of Met23, which form the core of the molecule (Krezel et al., 1995).

THERMODYNAMIC DOUBLE MUTANT CYCLES ANALYSIS WITH THE SHAKER K⁺ CHANNEL AND AGTX2

To organize channel residues in space with respect to the AgTx2 structure, specific pairwise interactions between residues on the toxin and on the channel must be identified. The assignment of pairwise interactions is done through thermodynamic double mutant cycle analysis by mutating both residues under study, one on the toxin and one on the channel. Each corner of the thermodynamic double mutant cycle (figure 4) corresponds to the inhibition equilibrium constant for the indicated toxin-channel pair, (wild-type toxin/wild-type channel; mutant toxin/wild-type channel; wild-type toxin/mutant channel and mutant toxin/mutant channel). For each cycle, the horizontal arrows

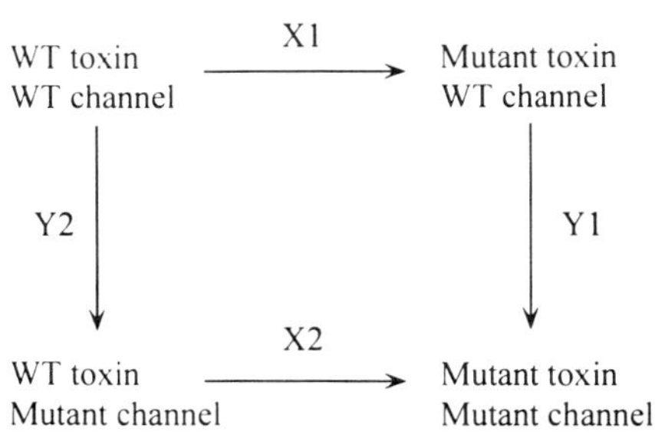

Figure 4. Thermodynamic double mutant cycle with the *Shaker* K^+ channel and AgTx2. Equilibrium inhibition constants for the affinity of wild type (WT) toxin and mutant toxin for a wild type channel and mutant channel define a thermodynamic double mutant cycle where X1, X2, Y1 and Y2 represent fold change in affinity along the arrow (inhibition constant at the tail of the arrow divided by that at the head). [Adapted from "Revealing the Architecture of a K^+ channel pore through mutant cycles with a peptide inhibitor". Hidalgo and Mackinnon. Science, 14 April 1995, Volume 268, pp. 307-310].

connecting a given toxin-channel pair indicate mutation of the toxin and the vertical arrows indicate mutation of the channel. The effect of the toxin mutation on the affinity for the wild-type channel is reported as X1 (top horizontal bar) and the effect of the same toxin mutation on the affinity for a channel with a point mutation is reported as X2 (bottom horizontal arrow). The number assigned to each arrow corresponds to the factor by which affinity changes along the direction of the arrow (i.e. X1 = Ki(wt:wt)/Ki(mut:wt), where Ki(toxin:channel) refers to the equilibrium inhibition constant for the indicated pair). Because the overall affinity change in going from the wild-type toxin/wild-type channel interaction (upper left) to the mutant toxin/mutant channel interaction (lower right) must be the same regardless of the pathway taken, the following thermodynamic relationship is fulfilled

$$X1 \cdot Y1 = X2 \cdot Y2 \qquad (1)$$

Dividing both sides of Eq. 1 by X2·Y1 leads to the coupling coefficient, Ω, defined as,

$$\Omega = \frac{X1}{X2} = \frac{Y2}{Y1} = \frac{Ki(wt{:}wt)\cdot Ki(mut{:}mut)}{Ki(wt{:}mut)\cdot Ki(mut{:}wt)} \qquad (2)$$

Because Ω is a ratio of a ratio of equilibrium constants, Ω is related to the difference in the difference in free energy change brought about by the mutation on the toxin and the channel. Thus, in energetic terms, Ω reports the degree of interaction, or *coupling energy brought about by the double mutation.*

The value of Ω reflects whether or not the two residues under study interact between them. Consider two cases: 1) the mutated residues, one on the toxin and one on the channel do not interact and 2) the mutated residues do interact. The outcome for these two cases is radically different. In the first case (mutated residues do not interact) the mutation of the toxin is not "sensed" by the mutation of the channel; therefore, the toxin mutation alters the affinity by the same factor for both the wild type and the mutant channel (i.e. X1 = X2). Thus, if the two mutated residues are independent of one another Ω will be unity. In the second case (mutated residues interact) the toxin mutation has a given effect on the affinity for the wild-type channel. Since the mutated residue on the toxin interacts with that on the channel, the effect on the affinity for the mutant channel depends on the amino acid residue at that particular position on the channel (i.e. X1 $\neq$ X2). Therefore, if the residues on the channel and on the toxin interact with each other and the mutations alter their interaction Ω will deviate from unity. Since the goal of applying thermodynamic double mutant cycle to the Shaker K^+ channel and AgTx2 is to identify pairwise interactions between residues on both proteins, the present study focuses on Ω values that differ from unity. It is worth it to mention that it is possible, although unlikely, for double mutations to alter an interaction between both residues in such a way that there is energetic compensation giving rise to an Ω value of unity.

To allow easy comparison of magnitudes, throughout this study Ω is reported as a number greater than unity (if Ω is less than unity according to Eq. 2 then its reciprocal is given).

THE COUPLING BETWEEN TWO TOXIN RESIDUES AND A SINGLE CHANNEL RESIDUE, ASP431, DEFINES THE DISTANCE BETWEEN TWO ASP431 IN DIAGONALLY-OPPOSED CHANNEL SUBUNITS

Electrostatic interactions are known to mediate binding of scorpion toxins to K^+ channels (MacKinnon and Miller, 1989b; MacKinnon et al., 1990, also see Garcia et al., 1994), accordingly, this study focuses on acidic and basic amino acid residues (Figure 5a,b).

An Ω plot for charge-neutralizing mutations of the eight basic residues on the toxin with charge-altering mutations at positions 422, 427, and 431 on the channel is shown in figure 5c. The large Ω value for the pair R24Q (toxin)/D431N (channel), which stands out above all others, results from the R24Q mutation reducing toxin affinity for the wild type channel about 1000-fold (X1 = 977.03), but having little effect on toxin binding to a D431N mutant channel (X2 = 1.33) (table 1). Mutation of either residue alone disrupts the interaction so that subsequent mutation of the second residue has little further effect. In contrast, for R24Q/K427E pair the channel mutation and the toxin mutation both have very large effects on affinity, but Ω for this pair is near unity. From the Ω definition, the large Ω value for the pair R24Q/D431N, corresponding to a coupling energy of about 4 kcal $mole^{-1}$, is expected if Arg24 and Asp431 interact strongly and specifically with each other. The strong coupling between the R24Q/D431N pair must come about through their local interaction and not through a global structural change of either the toxin or the channel. If the R24Q mutation altered the toxin structure, or caused the toxin to change its position in the binding site, then R24Q would in general be coupled

Table 1. Inhibition constants and Ω values for channel/toxin mutants pairs.

	WT (nM)	**E422K** (nM)	Ω	**K427E** (nM)	Ω	**D431N** (nM)	Ω
WT	0.74 ± 0.04	9.1 ± 0.7		0.007±0.001		2233 ± 60	
K16M	1.42 ± 0.03	11.1 ± 0.8	1.6 ± 0.2	0.015±0.001	1.1 ± 0.2	4092 ± 302	1 ± 0.1
K19M	2.7 ± 0.2	9.9 ± 0.5	3.4 ± 0.4	0.014±0.001	1.8 ± 0.3	3791 ± 107	2.1 ± 0.2
R24Q	723 ± 43	3486 ± 264	2.6 ± 0.3	9.0 ± 0.7	1.3 ± 0.2	2976 ± 396	733 ± 116
K27M	575 ± 42	8400 ± 1000	1.2 ± 0.2	5.4 ± 0.2	1.0 ± 0.2	NB	....
R31Q	4.3 ± 0.3	32.7 ± 1.7	1.6 ± 0.2	0.21 ± 0.01	5.2 ± 0.9	2619 ± 213	5.0 ± 0.6
K32M	4.4 ± 0.3	20.0 ± 0.7	2.7 ± 0.2	0.025 ± 0.003	1.7 ± 0.3	4522 ± 223	3.0 ± 0.3
H34F	0.8 ± 0.1	24.9 ± 0.3	2.5 ± 0.5	0.004±0.001	1.9 ± 0.6	3724 ± 174	1.5 ± 0.3
K38M	0.91 ± 0.03	9.1 ± 1.0	1.2 ± 0.2	0.019±0.001	2.2 ± 0.3	2084 ± 160	1.3 ± 0.1

The toxin mutations (left) were designed to neutralize basic amino acids. The channel mutations (top) altered the charge at three positions (422, 427 and 431) in the S5-S6 linker. Each Ki (nM) for a given toxin/channel pair shown was measured as described in methods and is the mean ± SEM of 3 to 8 separate measurements. NB refers to no inhibition at 2 μM. (If the K27M and D431N mutations are independent then the expected Ki for this pair is 1.7 mM). Ω was calculated according to Eq. 2 but is reported as a number greater than unity (see text). [Adapted from "Revealing the Architecture of a K^+ channel pore through mutant cycles with a peptide inhibitor". Hidalgo and Mackinnon. Science, 14 April 1995, Volume 268, pp. 307-310].

to many other channel mutations. Likewise, if the D431N mutation produced a global alteration of the channel structure then strong coupling to multiple toxin mutations would be expected. Thus, examination of the Ω plot constitutes by itself an internal control for the structural integrity of both mutated molecules and reinforces the conclusion that Asp431 interacts locally with Arg24.

Another, smaller but significant Ω is observed with the same Asp431 channel residue and Arg31 on the toxin (see figure 5c). Two toxin residues (Arg24 and Arg31) separated by approximately 25 Å from each other (figure 6) which interact with the same channel residue (Asp431) does not contradict the above statement of a local interaction between residues 24 and 431. The *Shaker* K^+ channel is formed by the assembly of four identical subunits (MacKinnon,1991); therefore the channel exposes four Asp431 to the toxin.

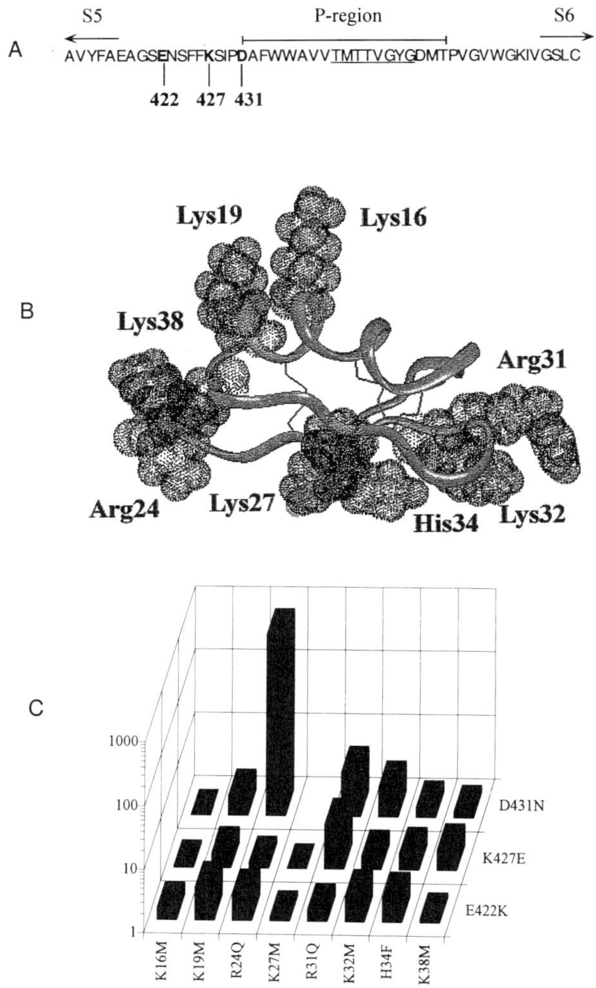

Figure 5. Thermodynamic double mutant cycles identify pairwise interactions between the *Shaker* K^+ channel and AgTx2. **A.** Linear amino acid sequence of the S5-S6 linker of the *Shaker* K^+ channel. The three toxin-interacting residues under study, 422, 427 and 431 are shown in bold. The K^+ channel signature sequence is underlined. **B.** AgTx2 structure is shown with a gray backbone ribbon and the eight basic amino acids under study are shown with their respectives van der Waals surface. **C.** The Ω plot graphs Ω values calculated according to Eq. 2 (see text) for the paired mutations shown in single letter code. [Adapted from "Revealing the Architecture of a K^+ channel pore through mutant cycles with a peptide inhibitor". Hidalgo and Mackinnon. Science, 14 April 1995, Volume 268, pp. 307-310]. Our inability to determine Ω for the K27M/D431N pair is compatible with their independence (see table 1). Single letter code for amino acid abbreviation.

Moreover, the interaction between Arg31 and Asp431 can be predicted considering both, the four-fold symmetry of the *Shaker* K^+ channel and the toxin's dimensions (Krezel et al., 1995), as described below. Previous biophysical studies demonstrated that Lys27 on ChTx interacts with K^+ ions inside the Ca^{2+}activated K^+ channel pore (Park and Miller, 1992; MacKinnon and Miller, 1988). Studies on the interaction between ChTx and the *Shaker* K^+ channel with the F425G mutation, which confers sensitivity to ChTx, suggest that Lys27 is also located over the pore (Goldstein and Miller, 1993). Lys27 together with the six cysteine residues, are the only fully conserved residues among the K^+ channel inhibitors from scorpion venom (Garcia et al., 1994). If Lys27 on AgTx2 is located centrally over the pore, then the strong interaction between Arg24 and Asp431 predicts that Arg31, located at the opposite side of the toxin from Arg24, should also interact with the Asp431 residue located in the diagonally-opposed channel subunit.

The assumption that Lys27 on AgTx2 resides over the pore of the *Shaker* K^+ channel has been validated by recent results. It was shown, by means of double mutant cycle analysis, that Lys27 on AgTx2 interacts with a residue (Tyr445) located on the proposed selctivity filter of the *Shaker* K^+ channel (see figure 2) and the interaction is dependent on extracellular K^+ concentration (Ranganathan et al., 1996).

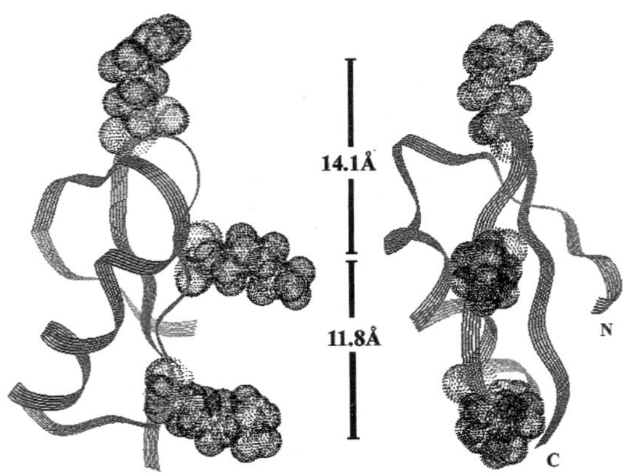

Figure 6. Ribbon diagram of AgTx2 with the van der Waals surface of the side chains of Arg24 (bottom), Lys27 (middle) and Arg31 (top). The distances between the γ-carbons of Arg24 and Arg31 to the the γ-carbon of Lys27 are indicated. Right panel is a 90° rotation of the molecule shown in the left panel.

Arg31 and Arg24 are not exactly symmetric with respect to Lys27 (figure 6) and, therefore, their interactions with Asp431 residues on diagonal subunits are not expected to be identical. Nevertheless, the strong coupling between R24/D431N pair, which tightly constrains the location of Asp431 on one subunit with respect to Arg24, would place Asp431 on the diagonal subunit within a few angstroms of Arg31; thus a through-space electrostatic interaction is likely to occur between residues Arg31 on the toxin and Asp431 on the channel. The Ω value for the R31Q/D431N pair, 5 ± 0.6 (table 1) corresponds to a coupling energy of approximately 1 kcal $mole^{-1}$. The fact that many Ω values are close to unity in the Ω plot reflects a rapid decay of the electric potentials across space and therefore supports the proposal that Asp431 and Arg31 must be close in space.

A Network of Thermodynamic Double Mutant Cycles Defines a Through Space Electrostatic Interaction Between a Toxin (Arg31) and a Channel (Asp431) Residue

To further examine an electrostatic interaction between residues 31 on the toxin and 431 on the channel, a network of cycles was constructed by making multiple amino acid substitutions at both positions (figure 7a). The Arg31 was substituted with Lys (R31K), to preserve the positive charge at that position, or by a Gln (R31Q), to neutralize the charge. Asp431 was substituted by Glu (D431E, to preserve charge) or an Asn (D431N, to neutralize the charge). The combination of three different residues (Arg, Lys, Gln) at position 31 on the toxin and three different residues (Asp,Glu,Asn) at 431 on the channel give rise to nine possible thermodynamic cycles for which the Ω value was calculated (figure 7a, bottom panel). A through-space electrostatic mechanism predicts that mutated pairs will show coupling (i.e. Ω deviate from unity) only if charge is altered on both residues. If charge is conserved at the mutated site on either or both proteins then the electrostatic interaction is not altered and thus no coupling should be observed (Ω = 1). The pattern of Ω values, larger only for cycles where charge perturbation occurs on both the toxin and the channel (cycles shown in bold, figure 7a bottom panel) fulfills the expected outcome for a through-space electrostatic mechanism and establishes the electrostatic nature of the interaction between residues 31 (toxin) and 431 (channel). Accordingly, residues 31 and 431 must be within a few angstroms of each other as was anticipated based on channel's geometry and toxin's dimensions considerations.

Figure 7b shows the cycle network analysis to characterize the strong interaction between Arg 24 on the toxin and Asp431 on the channel. The same substitutions as for the 31/431 pair were made: Arg24 on the toxin was substituted with Lys and Gln and Asp431 on the channel was substituted with Glu and Asn. In contrast to the pattern observed for the 31/431 pair, the 24/431 pair shows large Ω values for all cycles independent of the charge perturbation at both positions (figure 7b, bottom panel). The results obtained for the 31/431 pair suggests that the chemical nature rather than the charge of the residue at the position under study is important for the interaction. Residues 24 (toxin) and 431 (channel) must interact through shorter range molecular forces rather than through-space electrostatics. Arg24 and Asp431 apparently come into intimate contact either sterically (with an electrostatic component) or, more likely, through the formation of an ionized hydrogen bond (salt bridge). Frequently, the side chain of arginine residues serves as hydrogen-bond donors while the side chain of aspartate residues serves as hydrogen-bond acceptors (Branden and Tooze, 1991).

In summary, the results presented in this paper demonstrated that Arg24 and Arg31 on the toxin are in physical proximity to Asp431 on the channel. Therefore, the known distance separating Arg24 and Arg31 on the toxin can be translated into the distance separating two Asp431 residues on different channel subunits. The toxin's dimensions constrain the two Asp431 residues to be on diagonally-opposed channel subunits.

THREE-WAY INTERACTION BETWEEN A SINGLE TOXIN RESIDUE, ARG31, AND TWO CHANNEL RESIDUES

R31Q appears to be coupled to two different channel mutations, at position 431, which has been described in detail in above section, and at position 427 (Ω value of 5.2 ± 0.9 for the R31Q/K427E pair, see figure 5c, table 1). Two channel residues interacting with a single toxin residue suggest a spatial proximity between those channel residues. If channel residues 427 and 431 are indeed close to each other in space, an interaction between them is likely to occur. The binding of the toxin can be used to assess the dependence of channel residues 427 and 431 by inspecting the additivity of their individual contributions to the binding of AgTx2 (Horovitz, 1987).

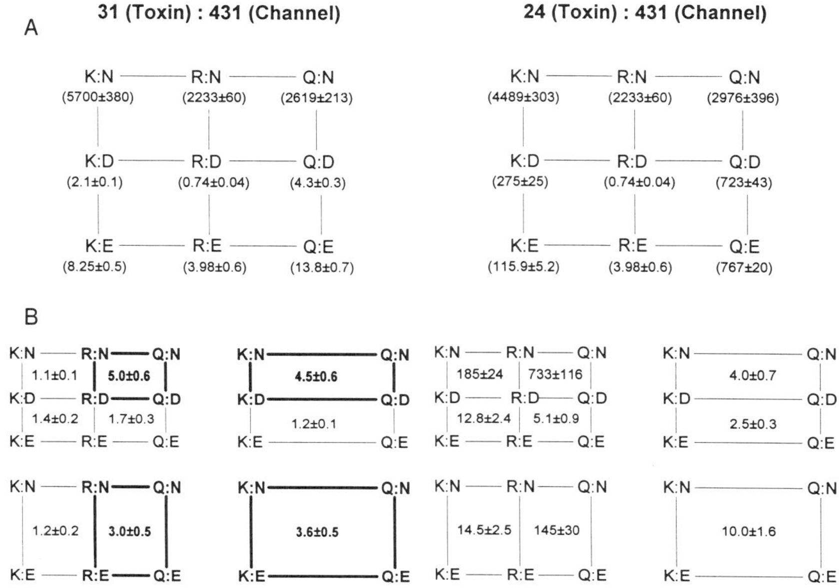

Figure 7. A network of thermodynamic double mutant cycles examine the nature of the interaction between mutated residues. **A.** The network contains nine toxin-channel pairs corresponding to all combinations of three residues at position 31 (toxin) with three residues at position 431 (channel). At each node, the amino acid (single letter code) on the left refers to the toxin residue and that on the right refers to the channel residue. Inhibition constants (nM, mean ± SEM of 3 to 8 measurements) for each toxin-channel pair are shown in parenthesis (top panel). The value of Ω for every possible cycle defined by the network is shown (bottom panel). Ω for each cycle (reported as a number greater than unity) was calculated according to Eq. 2 using Ki values corresponding to the interaction pairs at the four corners of the cycle. Cycles with larger Ω values (3 to 5) are shown in bold. Note that Ω for the four smallest cycles fully determine Ω for the remaining five larger cycles. **B.** Network cycle analysis for the interaction between mutated pairs at position 24 (toxin) and 431 (channel). [Adapted from "Revealing the Architecture of a K^+ channel pore through mutant cycles with a peptide inhibitor". Hidalgo and Mackinnon. Science, 14 April 1995, Volume 268, pp. 307-310].

The effect of the channel mutation K427E on AgTx2 affinity was tested in the presence of an Asp (WT channel) or an Asn (D431N) at position 431. When the channel residues at positions 427 and 431 are both mutated, the magnitude of the equilibrium inhibition constant for toxin inhibition is 30 times lower than expected if residues 427 and 431 were independent from each other. Figure 8 shows the thermodynamic double mutant cycle for the K427E/D431N pair. The

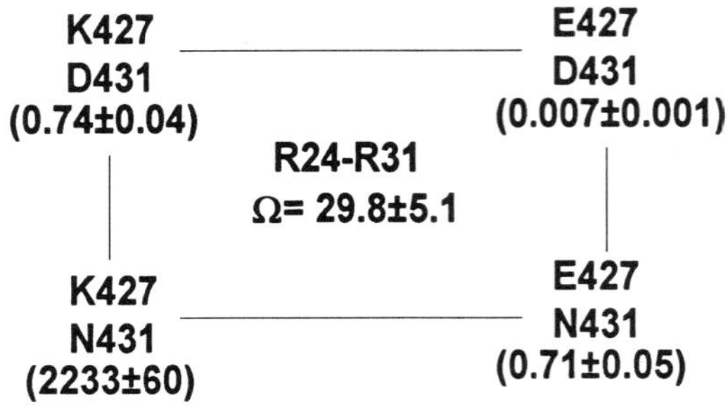

Figure 8. Thermodynamic double mutant cycle for the *Shaker* K^+ channel-channel mutated pair K427E/D431N. The Ω value was calculated according to Eq. 2. The Ki of AgTx2 (WT toxin, R24-R31) for each channel/channel pair is shown in parentheses (SEM for 4 to 8 separate measurements).

Ω value obtained, approximately 30, reflects an interaction between channel residues 427 and 431 in the context of our toxin blocking assay.

The coupling observed for the K427E/D431N pair predicts that the toxin-channel interaction mediated by the toxin residues 24 and 31 (which interact with residue 431 on diagonally-opposed channel subunits) will reciprocally depend on channel residues at position 427 and 431. Figure 9a (see also table 2) shows an Ω plot for seven charge-neutralizing mutations of AgTx2 paired with K427E channel in the presence of an Asp or Asn at channel position 431. The degree of interaction between the R31Q/K427E pair and the R24Q/K427E pair is altered depending on the residue at position 431 (Asp or Asn). The coupling for others mutant toxin-K427E channel pairs are independent of the residue at position 431. As expected, the degree of interaction between the R24Q/D431N pair and the R31Q/D431N pair is also influenced by the residue at position 427 (Lys or Glu) (figure 9b, table 2).

The above results indicate that channel residues 427 and 431 interact with each other, as detected by the toxin inhibition assay. The necessity of placing these two residues close to each other in space constrains the folding pattern formed by the stretch of amino acids between channel residues 427 and 431. At least, an extended conformation, such as β-sheet is discarded since it would place channel residues 427 and 431 far apart in space to account for the observed interaction (Hidalgo, 1995 and see discussion).

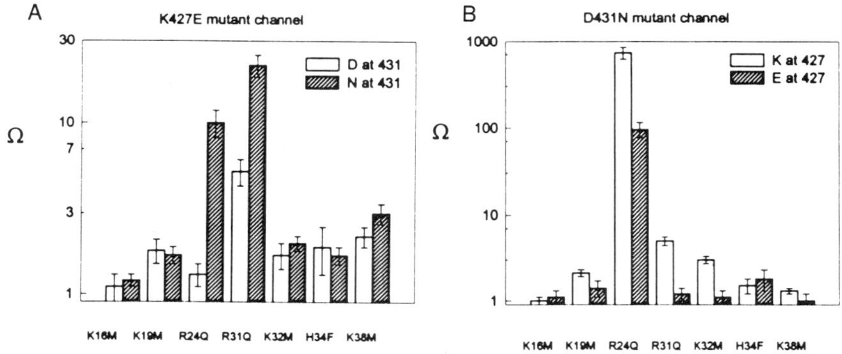

Figure 9. Toxin-channel interaction mediated by toxin residues 24 and 31 depends on channel residues at position 427 and 431. **A.** Plot of Ω values for seven AgTx2 charge mutants paired with the K427E channel, in the presence of an Asp (white bars), or an Asn (cross-hatched bars) at 431. **B.** Plot of Ω values for seven AgTx2 charge mutants paired with the D431N channel, in the presence of a Lys (white bars) or a Glu (cross-hatched bars) at 427.

Table 2. Ω values for seven charge-neutralizing toxin mutations with K427E, D431N and K427E/D431N channels.

	K16M	K19M	R24Q	R31Q	K32M	H34F	K38M
K427E (D431)	1.1±0.2	1.8±0.3	1.3±0.2	5.2±0.9	1.7±0.3	1.9±0.6	2.2±0.3
K427E (N431)	1.2±0.1	1.7±0.2	10.0±1.8	21.5±3.1	2.0±0.2	1.7±0.2	3.0±0.4
D431N (K427)	1.0±0.1	2.1±0.2	733±116	5.0±0.6	3.0±0.3	1.5±0.3	1.3±0.1
D431N (E427)	1.1±0.2	1.4±0.3	96.1±18.9	1.2±0.2	1.1±0.2	1.8±0.5	1.0±0.2

The Ω was obtained as described in methods and table1. Ki values were taken from table 1. The others Ki values for calculation of Ω for the K427E/D431N channel with the corresponding toxin mutant are in nM: K16M, 1.61±0.09; K19M, 2.0±0.1; R24Q, 9.5±0.8; R31Q, 17.9±1.6; K32M, 2.9±0.2; H34F, 0.71±0.07 and 2.0±0.1.

STRUCTURAL CONSTRAINS TO THE SHAKER K^+ CHANNEL

The relationship between the Asp431 residues on four channel subunits with respect to the toxin is shown in figure 10. Asp431, corresponding to the amino-terminal limit of the P-region, is located 12 to 15 Å away from the central axis. The range is derived from the range in the distance separating the guanidinium carbons of Arg24 and Arg31 in different rotomers of a set of 17 calculated AgTx2 NMR structures (Krezel et al., 1995). The distance of closest approach is 24 Å and of greatest separation is 31 Å. The distance (12 to 15 Å) is nearly ten times greater than the radius of a K^+ channel selectivity filter, estimated by functional measurements to be about 1.5 Å (Hille, 1975). The distant radial location of Asp431 raises the possibility that a significant segment of the P-region could be oriented parallel rather than perpendicular to the plane of the membrane.

The coupling between channel residues 427 and 431, in the context of the toxin inhibition assay, constrains the folding pattern formed by the stretch of amino acids between them. The results exclude a β-sheet conformation for this

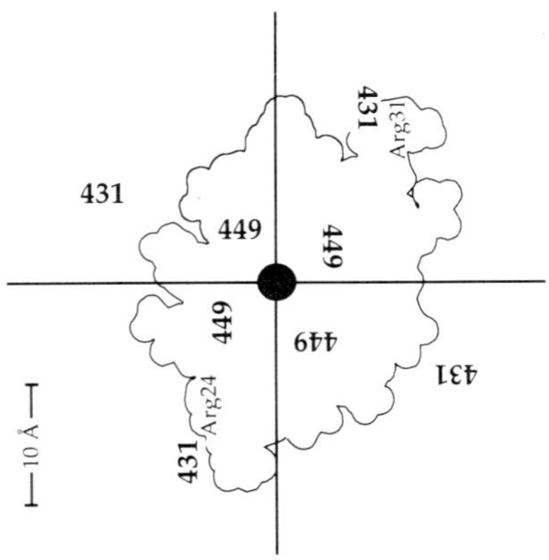

Figure 10. Channel residue Asp431 is located between 12 and 15 Å from the central axis of the K^+ channel pore. The scalloped outline shows the shadow of AgTx2, oriented with Lys27 over the pore (black circle) at the interface between four subunits (defined by axes). The radial positions of channel residues 431 and 449 are shown in bold. Residue 431 was placed near toxin residue Arg24 and then rotated by 90° about the central axis. Residue 449 was placed 5 Å from the center. The relative positions of 431 and 449 with respect to the subunit borders is arbitrary. [Reprinted with permission from "Revealing the Architecture of a K^+ channel pore through mutant cycles with a peptide inhibitor". Hidalgo and Mackinnon. Science, 14 April 1995, Volume 268, pp. 307-310. Copyright 1997 American Association for the Advancement of Science].

segment because the axial distance between adjacent amino acids in a β-sheet structure is approximately 3.5 Å (Creighton, 1993; Branden and Tooze, 1991), which gives a separation of 14 Å between residues 427 and 431. This distance is too large to account for the interaction observed between channel residues 427 and 431. Although our results can not exclude other foldings patterns, for example a β-turn, they are consistent with a helix conformation, which would place amino acid residues that are three or four residues apart in the linear sequence spatially close to one another. A helical conformation places residues 427 and 431 approximately 6 Å apart (1.5 Å axial distance between adjacent residues, Creighton, 1993; Branden and Tooze, 1991), and importantly, facing the same side of the helix. The amino acids between residues 427 and 431, Ser428-Ile429-Pro430, are well conserved among the voltage-activated K^+ channels. A further speculation would be that in a helical conformation Ile429 will lie on the opposite face of residues 427 and 431 facing the membrane plane or the the core of the channel protein. This location is consistent with its hydrophobic nature.

Mutagenesis and site-directed chemical modification studies on the interaction between AgTx2 and the *Shaker* K^+ channel imply a helical conformation for a stretch of amino acids between residues 427 and 438 of the S5-S6 linker (Gross and Mackinnon, 1996).

CONCLUSION

Thermodynamic double mutant cycle analysis with the *Shaker* K^+ channel and AgTx2 placed structural constraints to the pore forming region of the *Shaker* K^+ channel. The amino-terminal residue of the P-region, Asp 431, is placed 12 to 15 Å from the central axis of the K^+ channel (Hidalgo and MacKinnon, 1995). Thus, the amino-terminal end of the P-region cannot form the narrow part of the ion conduction pathway.

Current transmembrane topology models of K^+ channels depict the P-region as a hairpin that dips symmetrically into the plane of the membrane. The position assigned to Asp431 by the present work contrasts with those models and suggests that part of the P-region at the amino terminal end runs fairly parallel to the membrane plane.

Channel residues 427 and 431 in the in the S5-S6 linker of the *Shaker* K^+ channel, were shown to be close to each other in space. Their physical proximity and the number of amino acids separating them suggest a helical conformation for this segment. Thus, the segment of the P-region that runs parallel to the plane of the membrane may adopt a helical conformation, at least over some stretch.

The thermodynamic double mutant cycle approach constitutes a systematic method for mapping protein surfaces with a ligand of known structure. This approach is of general applicability to other ion channels where high affinity ligands are available.

NOTES

Abbreviations. AgTx2, agitoxin2; cAMP, cyclic adenosine monophosphate; NMR, nuclear magnetic resonance; PCR, polymerase chain reaction. A, Ala, alanine; C, cys, cysteine; D, asp, aspartate; E, glu, glutamate; F, phe, phenylalanine; G, gly, glycine; H, his, histidine; I, ile, isoleucine; K, lys, lysine; L, leu, leucine; M, met, methionine; N, asn, asparagine; P, pro, proline; Q, gln, glutamine; R, arg, arginine; S, ser, serine; T, thr, threonine; V, val, valine; W, trp, tryptophan; Y, tyr, tyrosine. Substituted proteins are named with a shorthand designation consisting of *(wild type amino acid) (position number) (substituted amino acid)*.

Acknowledgments. I would like to acknowledge Dr. Roderick Mackinnon in whose laboratory the present work was developed. This work would not have been possible without him.

REFERENCES

Branden, C. and Tooze, J. 1991. Introduction to Protein Structure. Garland Publishing Inc., New York:.

Candia, S., García, M.L., and Latorre, R. 1992. Mode of action of Iberiotoxin, a potent blocker of the large conductancer Ca^{2+}-activated K^{+}channel. *Biophys. J. 63*, 583.

Carter, P.J., Winter, G., Wilkison, A.J., and Fersht, A.R. 1984. The use of double mutants to detect structural changes in the active site of the tyrosyl-tRNA synthetase (*Bacillus stearothermophilus*). *Cell 38*, 835.

Catterall, W.A. 1988. Structure and function of voltage-sensitive ion channels. *Science 242*, 50.

Choi, K.L., Mossman, C., Aube, J., and Yellen, G. 1993. The internal quaternary ammonium receptor site of *Shaker* potassium channels. *Neuron 10*, 533.

Creighton, T. E. 1993. Proteins: Structure and Molecular Properties. W.H. Freeman Company, second edition, New York.

De Biasi, M., Kirsch, G.E., Drewe, J.A., Hartmann, H.A., and Brown, A.M. 1993. Cesium selectivity conferred by histidine substitution in the pore of the potassium channel Kv2.1. *Biophys. J. 64*, A341.

García, M.L., García-Calvo, M., Hidalgo, P., Lee, A., and MacKinnon, R. 1994. Purification and characterization of three inhibitors of voltage-dependent K^+ channels from *Leiurus quinquestriatus* var. *hebraeus* venom. *Biochemistry 33*, 6834.

Giangiacomo, K.M., García, M.L., and McManus, O.B. 1992. Mechanism of Iberiotoxin block of the large conductance calcium-activated potassium channel from bovine aortic smooth muscle. *Biochemistry 31*, 6719.

Goldstein, S. and Miller, C. 1993. Mechanism of Charybdotoxin block of a voltage-gated K^+ channel. *Biophys. J. 65*, 1613.

Gross, A., Abramson, T., and MacKinnon, R. 1994. Transfer of the scorpion toxin receptor to an insensitive potassium channel. *Neuron 13*, 961.

Gross, A and MacKinnon, R. 1996. Agitoxin footprinting of the Shaker potassium channel. *Neuron, 16,* 399.

Hartmann, H.A., Kirsch, G.E., Drewe, J.A., Taglialatela, M., Joho, R.H., and Brown, A.M. 1991. Exchange of conduction pathways between two related K^+ channels. *Science 251*, 942.

Heginbotham, L. and MacKinnon, R. 1992. The aromatic binding site for tetraethylammonium ion on potassium channels. *Neuron 8*, 483.

Heginbotham, L., Abramson, T., and MacKinnon, R. 1992. A functional connection between the pores of distantly related ion channel as revealed by mutant K^+ channels. *Science 258*, 1152.

Heginbotham, L., Lu, Z., Abramson, T., and MacKinnon, R. 1994. Mutations in the K^+ channel signature sequence. *Biophys. J. 66*, 1061.

Hidalgo, P. 1995. Obtaining structural data of the pore-forming region of the *Shaker* K^+ channel using the scorpion toxin Agitoxin2. Facultad de Ciencias Universidad de Chile, Santiago, Chile.

Hidalgo, P. and MacKinnon, R. 1995. Revealing the architecture of a K^+ channel pore through mutant cycles with a peptide inhibitor. *Science 268,* 307.

Hille, B. 1973. Potassium channels in myelinated nerve. Selective permeability to small cations. *J. Gen. Physiol. 61*, 669.

Hille, B. 1975. Ionic Selectivity of Na and K Channels of Nerve Membranes: Marcel Dekker, New York.

Horovitz, A. 1987. Non-additivity in protein-protein interactions. *J. Mol. Biol. 196*, 733.

Horovitz, A. and Fersht, A.R. 1990. Strategy for analysing the co-operativity of intramolecuar interactions in peptides and proteins. *J. Mol. Biol. 214*, 613.

Hoshi, T., Zagotta, W.N., and Aldrich, R.W. 1990. Biophysical and molecular mechanisms of *Shaker* potassium channel inactivation. *Science250*, 533.

Jan, L.Y. and Jan, Y.N. 1994. Potassium channels and their evolving gates. *Nature 371*, 119.

Kavanaugh, M.P., Hurst, R.S., Yakel, J., Varnum, M.D., Adelman, J.P., and North, R.A. 1992. Multiple subunits of a voltage-dependent potassium channel contribute to the binding site for tetraethylammonium. *Neuron 8*, 493.

Kavanaugh, M.P., Varnum, M.D., Osborne, P.B., Christie, M.J., Busch, A.E., Adelman, J.P., and North, R.A. 1991. Interaction between tetraethylammonium and amino acid residues in the pore of cloned voltage-dependent potassium channels. *J. Biol. Chem. 266*, 7583.

Kirsch, G.E., Drewe, J.A., Hartmann, H.A., Taglialatela, M., de Biasi, J., Brown, A.M., and Joho, R.H. 1992. Differences between the deep pores of K^+ channels determined by an interacting pair of nonpolar amino acids. *Neuron 8*, 499.

Krezel, A., Khasibhatla, C., Hidalgo, P., MacKinnon, R., and Wagner, G. 1995. Solution structure of the potassium channel inhibitor Agitoxin 2: Calipers for probing channel geometry. *Protein. Science 4,* 1478.

López, G.A., Jan, Y.N., and Jan, L.Y. 1994. Evidence that the S6 segment of the *Shaker* voltage-gated K^+ channel comprises part of the pore. *Nature 367*, 179.

Lu, Z. and MacKinnon, R. 1994. A conductance maximum observed in an inward-rectifier potassium channel. *J. Gen. Physiol. 104*, 477.

MacKinnon, R. 1991. Determination of the subunit stoichiometry of a voltage-activated potassium channel. *Nature (London) 350*, 232 .

MacKinnon, R. and Miller, C. 1988. Mechanism of charybdotoxin block of Ca^{2+}-activated K^+ channels. *J. Gen. Physiol. 91*, 335 .

MacKinnon, R. and Miller, C. 1989a. Functional modification of a Ca2+-activated K^+ channel by trimethyloxonium. *Biochemistry 28,* 8087.

MacKinnon, R. and Miller, C. 1989b. Mutant potassium channels with altered binding of Charybdotoxin, a pore-blocking peptide inhibitor. *Science 245*, 1382.

MacKinnon, R. and Yellen, G. 1990. Mutations affecting TEA blockade and ion permeation in voltage-activated K^+ channels. *Science 250*, 276.

MacKinnon, R., Heginbotham, L., and Abramson, T. 1990. Mapping the receptor site for Charybdotoxin, a pore-blocking potassium channel inhibitor. *Neuron 5*, 767.

Miller. C. 1995. The charybdotoxin family of K^+ channel-blocking peptides. *Neuron 15*, 5.

Park, C.S. and Miller, C. 1992. Interaction of Charybdotoxin with permeant ions inside the pore of a K^+ channel. *Neuron 9*, 307.

Park, C.S., Hausdorff, S.F., and Miller, C. 1991. Design, synthesis, and functional expression of a gene for Charybdotoxin, a peptide blocker of K^+ channels. *Proc. Natl. Acad. Sci. U. S. A. 88*, 2046.

Ranganathan, R., Lewis, J.H., and MacKinnon, R. 1996. Spatial localization of the K^+ channel selectivity filter by mutant cycle-based structure analysis. *Neuron 16*, 131.

Sanger, F., Nicklen, S., and Coulson, A.R. 1977. DNA sequencing with chain-terminating inhibitors. *Proc. Natl. Acad. Sci. U. S. A. 74*, 5463.

Serrano, L., Horovitz, A., Avron, B., Bycroft, M., and Fersht, A.R. 1990. Estimating the contribution of engineered surface electrostatic interactions to protein stability by using double-mutant cycles. *Biochemistry 29*, 9343.

Slesinger, P.A., Jan, Y.N., and Jan, L.Y. 1993. The S4-S5 loop contributes to the ion-selective pore of potassium channels. *Neuron 11*, 739.

Stampe, P., Kolmakova-Partensky, L., and Miller, C. 1994. Intimations of K^+ channel structure from a complete functional map of the molecular surface of Charybdotoxin. *Biochemistry 33*, 443.

Tempel, B.L., Papazian, D.M., Schwarz, T.L., Jan, Y.N., and Jan, L.Y. 1987. Sequence of a probable potassium channel component encoded at *Shaker* locus of *Drosophila*. *Science 237*, 770.

Yellen, G., Jurman, M., Abramson, T., and MacKinnon, R. 1991. Mutations affecting internal TEA blockade identify the probable pore-forming region of a K^+ channel. *Science 251*, 939.

Yool, A.J. and Schwarz, T.L. 1991. Alteration of ionic selectivity of a K^+ channel by mutation of the H5 region. *Nature (London) 349*, 700.

BARIUM AS A PROBE OF THE MOLECULAR ARCHITECTURE OF THE PORE OF K^+ CHANNELS

Ramon Latorre, Raymond Hurst, Felipe Diaz, Ligia Toro, and Enrico Stefani

Centro de Estudios Cientificos de Santiago, Departamento de Biologia, Facultad de Ciencias, Universidad de Chile, Santiago, Chile and Department of Anesthesiology, UCLA, Los Angeles, CA.

INTRODUCTION

A molecular picture of the ion channel pore is emerging from the characterization of the structural determinants involved in K^+ conduction (MacKinnon, 1995). The hallmark of potassium channels is their high selectivity for K^+ over Na^+.[1] However the molecular nature of the site(s) where the process of ion discrimination takes place is unknown. Actually, several potassium binding sites are expected to exist since phenomena usually associated with channels that allow several ions to coexist, take place in the conduction machinery of K^+ channels (Hodgkin and Keynes, 1955; Armstrong, 1975; Adelman and French, 1978; Hille, 1978; Yellen, 1984; Eisenman et al., 1986; Cecchi et al., 1987; Neyton and Miller, 1988b; Stampe and Begenesich, 1996). Barium strongly blocks most potassium channels and the combination of this divalent cation with its receptors forms very stable complexes. For example, Ba^{2+} leaves the closed delayed rectifier channel from the squid axon or the closed Shaker K^+ channel with a time constant of several minutes (Armstrong and Taylor, 1980; Hurst et al.

[1] The situation, however, is more complex since Korn and Ikeda (1995) showed that a class of K^+ channels (Kv2.1) are able to transport Na^+ in the absence of K^+. In this regard this type of K^+ channels showed a competition mechanism -low $[K^+]$ inhibited Na^+ conductance- similar to that of Ca^{2+} channels.

From Ion Channels to Cell-to-Cell Conversations
Edited by Latorre and Sáez, Plenum Press, New York, 1997

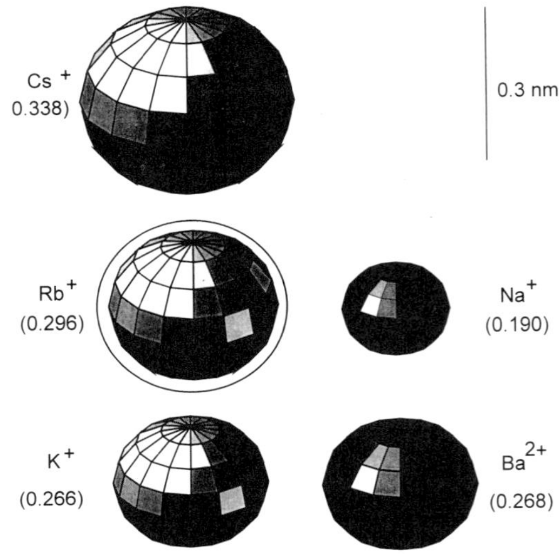

Figure 1. Crystal size of various alkali cations and barium ion. The circumference surrounding the rubidium ion has a diameter of 0.33 nm. This diameter corresponds to a selectivity filter large enough to let rubidium to go through, but not cessium.

The stability of the complexes formed by Ba^{2+} and K^+ channels makes this divalent cation an excellent probe of their ion conduction system.

The largest permeant ion in K^+ channels is Rb^+ (0.296 nm), Cs^+ conductance is about a 100-fold lower than that of K^+ (Heginbothan and MacKinnon, 1993) and methyl groups (diameter = 0.4 nm) do not pass. On this basis, Hille (1973) and Bezanilla and Armstrong (1972) proposed the existence of an specialized region in the K^+ channel pore, the selectivity filter, with a sectional area corresponding to a circumference with a diameter of 0.33 nm (Fig. 1). Barium with a crystal radius of 0.134 nm differs from that of potassium only by 0.001 nm. Therefore, if the picture of the selectivity filter is correct, Ba^{2+} is ideally suited to interact with it. Moreover, due to the divalent nature of the cation, Ba^{2+} can establish strong coulombic interactions with negative charges located in or in the neighborhood of the selectivity filter.

The similarity in size between Ba^{2+} and K^+, and the divalent nature of Ba^{2+}, makes very likely the possibility that Ba^{2+} strongly binds to sites used by K^+ when traveling across a K^+ channel. Therefore, this divalent cation constitute an ideal probe to inquire about the still unknown molecular nature of such sites in K^+ channels. In this chapter, we will review what is known about the mode of action of Ba^{2+} in different K^+ channels from a mechanistic point of view. We will attempt to describe in each case what we have learned using this divalent cation

1995). The stability of the complexes formed by Ba^{2+} and K^+ channels makes this divalent cation an excellent probe of their ion conduction system.

PHENOMENOLOGY

Although it was known for a long time that Ba^{2+} was a potent blocker of K^+ channels (Werman and Grundfest, 1961), Armstrong and his colleagues (Armstrong and Taylor, 1980; Armstrong et al. 1982) were the first to realize fully the potential of Ba^{2+} as a probe of the gross architecture of K^+ channels.

Internal Ba^{2+} block.

Eaton and Brodwick (1975, 1980) and Armstrong and Taylor (1980) showed that Ba^{2+} applied from the cytoplasmic side is able to block the squid axon delayed rectifier K^+ channel only when the channel is open. Internally applied Ba^{2+} causes a decay with an exponential time course that is similar to an inactivation process. The blockade kinetic is well described by a bimolecular reaction:

$$Ba^{2+} + \text{open channel} \underset{\beta}{\overset{\alpha}{\rightleftharpoons}} \text{blocked channel} \qquad \text{(R1)}$$

and hence, the fraction of blocked channels at equilibrium, f_{block} is well described by an expression of the type:

$$f_{block} = \alpha[Ba]/(\alpha[Ba] + \beta) \qquad (1)$$

implying that only one Ba^{2+} is enough to block a K^+ channel. The reaction shown in scheme 1 is voltage-dependent, large depolarizing voltages speed up blockade and increase the probability of finding the channel in the blocked state. Blockade is relieved by increasing the *external* K^+ concentration (Fig. 2). These last two observations are compatible with a Ba^{2+} binding site located deep within the pore. In particular, the relief of block by external K^+ suggests that external K^+ when entering the channel can push Ba^{2+} toward the cytoplasm. The apparent dissociation constant for the blocking reaction (scheme R1), $K^+_{0.5}$, is well described by:

$$K_{0.5}(V) = K_{0.5}(0)\exp(-AV) \qquad (2)$$

For single-ion channels $A = z\delta F/RT$ where z is the valence of the blocking ion and δ the distance the ion must travel along the electric field before reaching its site.

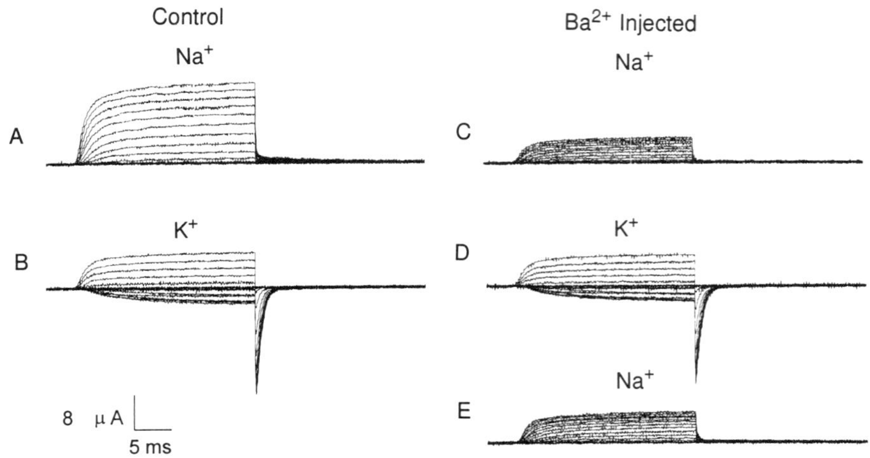

Figure 2. External K^+ relieves internal Ba^{2+} block of *Shaker* H4IR K^+ channels. Potassium currents were measured in the presence of (A) external Na^+ or in the presence of external K^+ (**B**). Injecting the oocyte with Ba^{2+} induce a marked decrease in the ionic currents in the presence of external Na^+ (**C**). However, replacement of external Na^+ by K^+ (**D**) restores them almost to the control level (**A**). Currents were measured using the cut-open oocyte Vaseline gap voltage clamp. Test voltages were from a holding voltage of -90 mV in 10 mV steps up to 40 mV.

Given the multi-ion pore nature of most K^+ channels, the actual fractional electrical distance that Ba^{2+} travels before binding to its site is uncertain. In several different types of K^+ channels, the apparent dissociation constant $K_{0.5}$ for Ba^{2+} block varies an *e*-fold per 12-25 mV (Table 1). In multi-ion channels the interpretation of this is not a unique. For example, an *e*-fold/25 mV could mean that Ba^{2+} travels half of the way through the electric field before blocking or Ba^{2+} can travel a quarter of the way and two K^+ each moves a quarter of the way.

Armstrong et al. (1980) showed that Ba^{2+} can exit closed channels albeit very slowly indicating that the complex formed between a Ba^{2+} and a closed channel is very stable. This observation together with the fact that a closed channel cannot be blocked from the internal side suggests that the gate is located toward the cytoplasmic side of the channel (see Fig. 3).

External Ba^{2+} Block.

External Ba^{2+} can block open *and* closed channels (Armstrong et al. 1982; Hurst et al., 1995; Wollmuth, 1994). This observation gives further support to the hypothesis that the gate in this class of channel is at the inner end of the channel (Fig. 3).

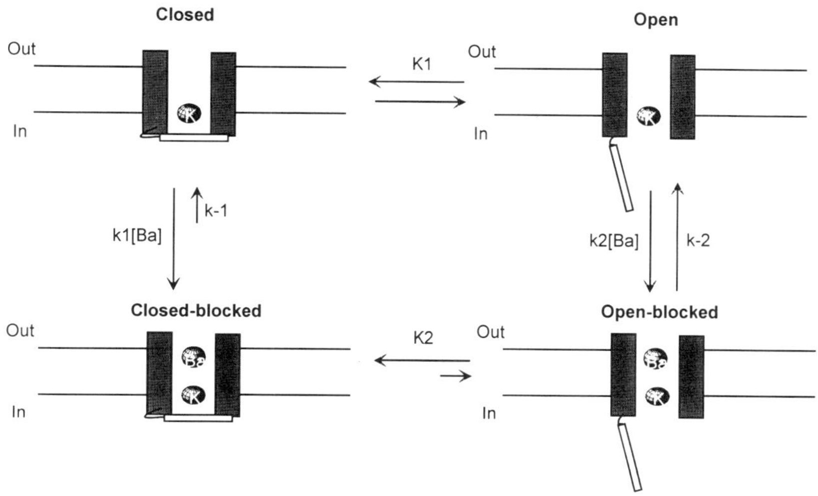

Figure 3. External Ba^{2+} can enter closed and open voltage-dependent K^+ channels. The cartoon shows that Ba^{2+} stabilizes the closed configuration since the equilibrium described by the constant K2 is much displaced toward the left. Based on the fact that external but not internal Ba^{2+} can block a closed channel, the cartoon emphasizes the cytoplasmic location of the gate. A potassium ion is also located in the conduction machinery of the channel to stress the multi-ion nature of K^+ channels.

Table 1
Equilibrium Characteristics of Internal Ba^{2+} Block

Channel	$K_{0.5}(0)$ (M)	A (mV^{-1})	Reference
Squid K^+	-	12	1
Squid K^+	8.4×10^{-7}	12.5	2
Skeletal muscle K_{Ca}	1.8×10^{-5}	19	3
Sh B IR	6.3×10^{-4}	25	4
Kv2.1	1.3×10^{-5}	33	5
h*slo*	3.6×10^{-4}	13.6	6

1. Armstrong and Taylor, 1980; 2. Eaton and Brodwick. 1980; 3. Vergara and Latorre, 1983; 4. Slensiger et al. 1993; 5. Taglialatela et al. 1993; 6. Diaz et al. 1996

A detailed study of the external Ba^{2+} block of the *Shaker*H4 IR K^{+} channel has been done recently by Hurst et al. (1995). When Ba^{2+} is added to the external side and the channel is mantained at hyperpolarized voltages, a short positive test voltage pulse indicates that channels progressively enter into a closed-blocked state.

In this channel, the blockade has a fast and a slow component. The slow component has an exponential time course (τ = 120 s at a [Ba^{2+}] = 10 mM) that is speeded up by increasing [Ba^{2+}]. Recovery from block is also well described by a single exponential with a [Ba^{2+}]-independent time constant. Both components are voltage-dependent; the fast component is well described with a $K_{0.5}$ = 27 mM and a δ = 0.14 whereas for the slow component the same parameters are 1.3 mM and 0.25, respectively. This bi-phasic time course suggests the existence of two external blocking sites on *Shaker* H4-IR and the magnitude of the electrical distances suggest that they are in series. Moreover, the reaction scheme R1 predicts that the reciprocal value of the blocking time constant (τ_{on}) must be a linear function of [Ba^{2+}]. Contrary to this expectation, τ_{on} was found to be a saturating function of [Ba^{2+}]. Reaction scheme R1 must be then modified to satisfy the data. The most economical model is one with two sequential sites:

$$Ba^{2+} + Site1\text{-}Site2 \underset{k_{-1}}{\overset{k_{1}}{\rightleftharpoons}} Ba^{2+}\text{:}Site1\text{-}Site2 \underset{k_{-2}}{\overset{k_{2}}{\rightleftharpoons}} Site1\text{-}Site2\text{: }Ba^{2+} \quad (R2)$$

where Site1 and Site2 are the two Ba^{2+} binding sites, Ba^{2+} binding to Site2 produces the slow block. The reaction scheme R2 predicts:

$$1/\tau = k_2\{[Ba]/(Ba] + K_1)\} + k_{-2} \qquad (3)$$

where K_1 is the dissociation constant of site 1 and Ba]/([Ba] + K_1) is the fractional time the channel has a Ba^{2+} in Site1. In agreement with the data, eq. 3 predicts that $1/\tau$ is a saturating function of [Ba^{2+}].

Armstrong et al. (1982) argued that Ba^{2+} not only blocks the channel but it also modifies channel gating. They based their contention on the fact that recovery from block has a fast and a very slow component. The fast component was clearly manifested as a hook in the current tails that are concomitant with the off of the voltage pulse. The hook is a consequence of Ba^{2+} ions clearing open clannels. As discussed above, once channels close, Ba^{2+} forms a very stable complex with its site and leaves the channel very slowly. This closed-blocked state is more stable than the closed state as implied by the decrease in amplitude of the tail currents with time after a large depolarizing pulse in the presence of internal Ba^{2+}. A decrease in the tail current implies that the same depolarizing pulse is able to open fewer channels as they become progressively blocked and hence *barium ions must stabilize the closed configuration of the squid potassium channel.*

Taking advantage of the large expression of *Shaker*H4 in *Xenopus* oocytes, the gating currents induced by this K^+ channel have been extensively studied (for a review, see Bezanilla and Stefani, 1994). In *Shaker* H4 IR gating charge movement has two components. In one component, the charge (Q1) moves in the hyperpolarized region; whereas in the second, the charge (Q2) moves in the depolarized voltage range and has a larger apparent valence. These two components can also be differentiated kinetically. For small depolarizations, the decay of the gating current is fast (i.e., charge during the OFF returns quickly). This charge movement corresponds to Q1 and occurs within closed conformations of the channel. At intermediate depolarizations, gating current decay has a double exponential time course and at more positive voltages the slow component predominates. The slow component has charge moving between closed states preceding the open state and corresponds to the displacement of Q2 (Bezanilla et al. 1994).

Measurements of gating currents in a wider range of voltages are more easily done using a nonconducting mutant of the *Shaker* K^+ channel. In this particular mutant, the tryptophan (W) 434 residue located in the pore region (Table 2) is changed to a phenylalanine (F). Taking advantage of the W434F mutant, Hurst et al. (1996a) explored the effect of external Ba^{2+} on gating currents. Fig. 4B shows that external Ba^{2+} increases the rate of charge return compared to the control taken in 2 mM Ca^{2+} (Fig. 4 A).

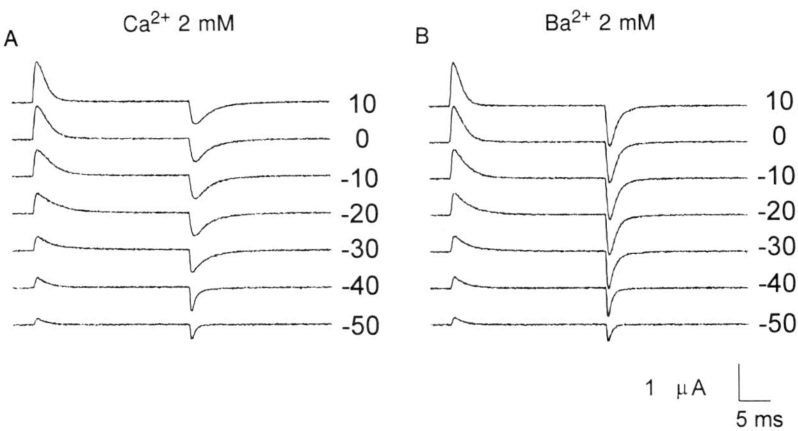

Figure 4. Barium modifies the off component of gating currents of the W434F *Shaker* H4 IR mutant. **A**. Control records of gating currents were obtained in 2 mM Ca^{2+}. Notice that the time course of the OFF response becomes slower as voltage is increased. Holding voltage was -90 mV. (**B**). 2 mM Ba^{2+} speeds the return of the charge. The Ba^{2+} effect is only noticeable at voltages large than - 55 mV. Stefani (unpublished results).

A detailed study of the Ba^{2+} effects on the *Shaker* H4 IR gating currents shows that the off gating charge kinetics in the presence of the divalent cation is not modified if depolarizations are negative to -55 mV (Hurst et al. 1995, 1996). However, at larger voltages, Ba^{2+} increases the rate of charge return compared to the control taken in 2 mM Ca^{2+}. Barium alters specifically the movement of the Q2 component or, in other words, the movement of the charge that occurs in closed states just preceding the open state. This result can be explained as an increase in the backward rate of some rate limiting transition between closed states in the neighborhood of the open state (Hurst et al. 1996a).

The effect of Ba^{2+} is contrasted with that of internal tetraethylammonium (TEA) in Fig. 5B. TEA slows down charge return, a phenomenon known as "charge immobilization" (Bezanilla et al. 1991). TEA stabilizes the open-blocked

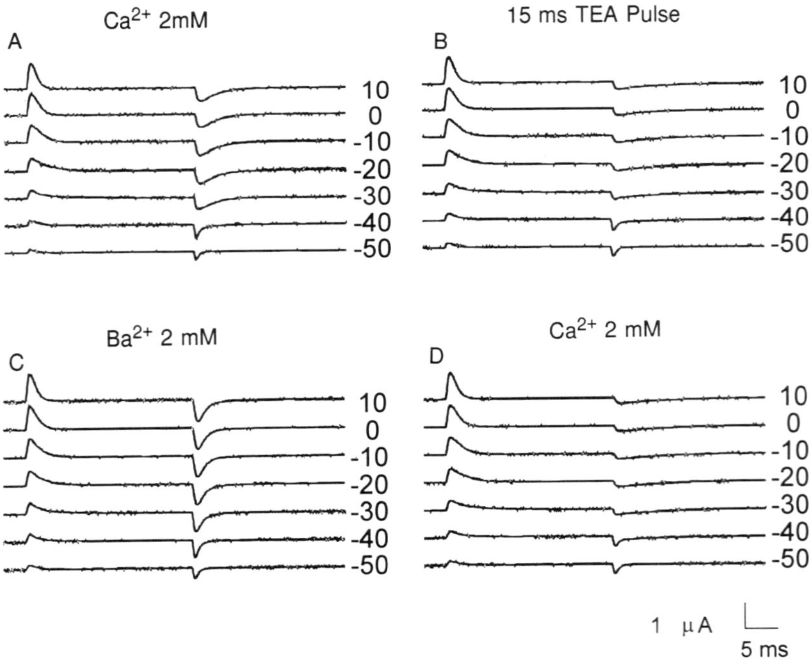

Figure 5. Charge immobilization induced by internal TEA is relieved by external Ba^{2+} in the W434F *Shaker* mutant. **A**. Control. Gating induced by *Shaker* H4IR K^+ channels in *Xenopus* oocytes. **B**. Injection of TEA causes a dramatic decrease in the rate at which charge returns during the OFF . This phenomenon is known as "charge immobilization" (Bezanilla et. al 1991). C. Charge immobilization is relieved by the presence of 2 mM external Ba^{2+} but no by 2 mM Ca^{2+} (D). Currents measured using the cut-open vaseline-gap oocyte voltage clamp. Stefani unpublished results.

Figure 6. A cartoon giving a simple view of the effects of Ba^{2+} on the kinetics of the off response. Depolarizing voltages induces a displacement of gating charge represented by the positive charge in the in-out direction promoting the opening of the channel. In this picture the configuration energy is given by the spring compliance that at hyperpolarized voltages tends to keep the charge toward the cytoplasmic side inducing channel closing. In the blocked configuration, Ba^{2+} by electrostatics reasons (the most economical assumption), stabilized the closed configuration by making the charge return faster (For more details see text).

configuration of the channel. The effect of internal TEA can be overcome by external Ba^{2+} as shown in Fig. 5C. The simplest interpretation of this result is that by stabilizing the closed configuration of the channel, Ba^{2+} does not allow TEA to bind to its receptor. Notice that external Ca^{2+} is not able to relieve the effect of internal TEA (Fig. 5D).

The cartoon showed in Fig. 6 is based on a simulation of the effect of Ba^{2+} on gating currents using the linear activation model proposed by Bezanilla et al. (1994) (Hurst et al. 1996a). In this model, comprising seven closed (C_0, C_1,...C_6) and one open state connected with C6, the transition between C5 and C6 carries a large charge (see also Sigg et al. 1994) and the backward transition C6-C5 is rate limiting producing the slow falling phase of the gating current. It is enough for Ba^{2+} to modify this transition to reproduce most of the effects of the divalent cation on the OFF gating current time course.

TRAPPING Ba^{2+} INSIDE POTASSIUM CHANNELS

Miller (1987) and Miller et al (1987) showed that Ca^{2+}-activated K^+ channels (K_{Ca}) of large conductance can trap Ba^{2+} once they closed (see also Grissmer and Cahalan, 1989). In this regard, K_{Ca} and *Shaker* K^+ channels, despite the fact that they belong to the same S4 superfamily (Jan and Jan, 1990), behave qualitatively different when confronted to Ba^{2+}. K_{Ca} channels cannot be blocked by internal or external Ba^{2+} once they have closed and when closed the divalent cation cannot exit the channel. It is also noticeable, that in the case of K_{Ca} channels, Ba^{2+} *destabilizes* the closed channel configuration.

Barium blocks voltage-dependent K^+ channels with higher potency when applied internally (Eaton and Brodwick, 1982; Armstrong et al., 1982). The same is true for K_{Ca} channels. Vergara and Latorre (1983) found that external Ba^{2+} block is weaker because the rate constant for association is sluggish (about 3,000-fold slower than from the inside). In symmetrical K^+ the dissociation rate constant is identical for Ba^{2+} added to either side. This is because Ba^{2+}, due to the large external energy barrier, always leaves the channel toward the internal side. These observations led us to the conclusion that the same Ba^{2+} binding site(s) can be reached from both sides of the membrane. Ba^{2+} can permeate K_{Ca} channels, but that exit to the external side is strongly hindered.

Neyton and Miller (1988a) showed that removal of external potassium induces a notable decrease in the mean blocked time[2]. An external K^+ site that when occupied does not allow Ba^{2+} to leave the channel toward the external side was proposed by Neyton and Miller (1988a) to explain this result. This site was dubbed the external lock-in-site since occupancy of this site by K^+ increases the dwell time of Ba^{2+} inside the pore. Barium in high external K^+ is forced to dissociate only to the internal solution (Vergara and Latorre, 1983; Miller et al. 1987). The fact that in the absence of external K+, depolarizing voltages increase *both* the on and the off rates for internal Ba^{2+} block implies that the blocker leaves the channel preferentially toward the external side. In view of these results, it is tempting to suggest that in K_{Ca} channels, as in voltage-dependent channels, the gate is located in the cytoplasmic neighborhood of the channel. In K_{Ca} channels, Ba^{2+} cannot leave a closed channel because on the external side of it resides a K^+ very tightly bound to the channel.

MOLECULAR DETERMINANTS OF Ba^{2+} BLOCK

All the results presented fit into a picture of a Ba^{2+} binding site located within the conduction pore of K^+ channels. Mutagenesis approaches have revealed that the amino acids that may line the ion channel pore are well conserved in different K^+ channel families. The cloning of the *Shaker* K^+ channel (Baumann et al. 1988; Kamb et al. 1987; Tempel et al. 1987) was the first step toward the identification of several subfamilies of K^+ channel-forming proteins characterized by having six transmembrane domains. The fourth (S4) domain contains up to eight positively charged amino acids at every third position with nonpolar residues in between. Because of its peculiar structure, the S4 segment was postulated to be

---

[2] This observation is incompatible with a single-ion channel model since in a channel that can accept only one ion at a time K^+ should affect only the rate of Ba^{2+} entry.

part of the voltage sensor (e.g., Bezanilla and Stefani, 1994)[3]. In these types of channels, the conduction machinery appears to be the hairpin shaped region comprised between segments S5 and S6 dubbed H5 or P region, the S4-S5 loop and the S6 segment (for a review, see Kukuljan et al. 1995).

As discussed earlier, external Ba^{2+} blocks *Shaker* H4 IR K^+ channels and the results suggest the presence of two binding sites (Hurst et al. 1995). The slow component of this block is modified by mutations at two positions in the P region: D447 and T449 (Hurst et al. 1966b). Substitution of T449 by A, V, or Y, greatly increases k_2, the blocking rate for the slow component (scheme R2) but does not affect the off rate k_{-2}. The simplest interpretation of this result is that these mutations raise the energy barrier that gives access to the second site located deeper into the pore. In other words, position T449 modulates the passage of Ba^{2+} to that site. Mutation at position D447 in *Shaker* or D378 in Kv2.1 are not tolerated, but Kirsch et al. (1995) successfully mutated D378 to T in a tandem dimer. Since K^+ channel are tetramers (MacKinnon, 1991), the mutated channels contain only two aspartic residues instead of four. In this construct the permeability ratio between Na^+ and K^+ increases five fold indicating that this negative residue is

Table 2

Sequence Alignment of the P region of Different Cloned K^+ Channels

				434							441						447		449
Sh	D	A	F	W	W	A	V	V	T	M	T	T	V	G	Y	G	D	M	T
													374				378		
Kv2.1	A	S	F	W	W	A	T	I	T	M	T	T	V	G	Y	G	D	I	Y
											121								
IRK1	A	A	F	L	F	S	I	E	T	Q	T	T	I	G	Y	G	F	R	C
ROMK2	S	A	F	L	F	S	L	E	T	Q	V	T	I	G	Y	G	F	R	F

Single-letter amino acid code was: A, Ala; R, Arg; D, Asp; N, Asn; C, Cys; Q, Gln; H, His; I, Ile; L, Leu; K^+, Lys; M, Met; F, Phe; P, Pro; S, Ser; T, Thr; Y, Tyr; W, Trp; V, val; E, Glu; G, Gly.

an important part of the selectivity filter. In view that Ba^2 can be interacting with these negatively charged residues, Hurst et al. (1996b) mutated D447 in the *Shaker* protein to N. The mutant D447N does not conduct but induces gating currents like the nonconducting W434F clone discussed earlier in this chapter. The effect of external Ba^{2+} in these two clones was compared. The charge neutralization abolishes the ability of Ba^{2+} to speed up Q_{off} (Fig. 7).

More recently, Mannuzzu et al. (1996) presented evidence that during activation of *Shaker* K^+ channels a stretch of at least seven amino acids of S4 is moved from a buried position to the extracellular side of the membrane. It is possible tha external Ba^{2+} once in the pore, accelerates the return of the exposed S4 segment to its buried position.

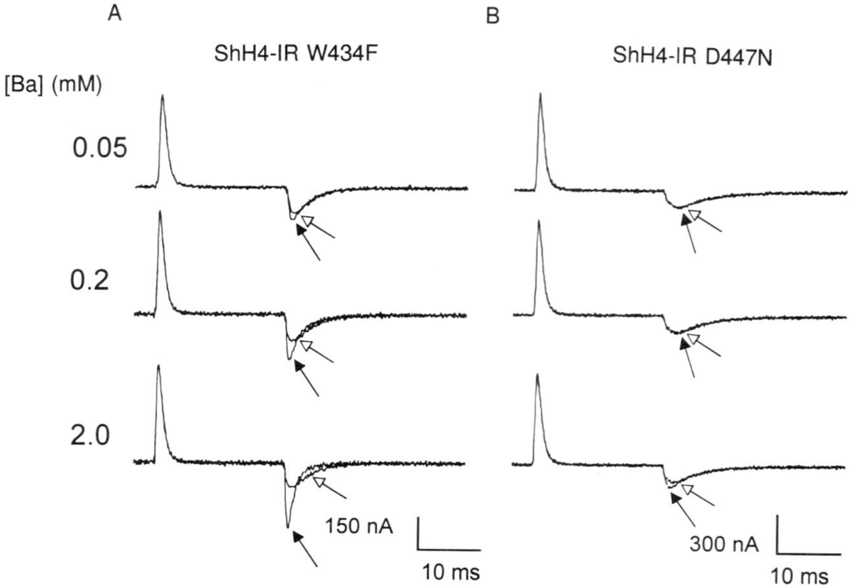

Figure 7. Aspartate 447 modulates the effect of Ba^{2+} on gating charge return. **A**. Barium speeds charge return in a concentration-dependent fashion in *Shaker* H4 IR W434F mutant. Records taken in the absence or in the presence of Ba^{2+} are indicated by the open and solid arrows, respectively. Notice that Ba^{2+} does not affect the ON component of the gating current. The effect of Ba^{2+} on the off gating current is almost absent in the mutant D447N *Shaker* mutant (**B**) and in the double mutant W434F, D447N (not shown). Gating current records were obtained by applying a depolarizing test pulse to 30 mV from a holding voltage of -90 mV. The effect of Ba^{2+} was measured by applying a 30 mV test pulse every 5 s. Modified from Hurst et al. (1996).

These results provide evidence, albeit indirect, that the ring of carboxyl groups of the aspartate residues may provide the Coulombic interactions that stabilize the Ba^{2+}-channel complex. Barium was used to probe the pore structure of the cloned delayed rectifier DRK1 or Kv2.1 (Taglaliatela et al. 1993). Like other K^+ channels, Kv2.1 is blocked by internally and externally added Ba^{2+}. External Ba^{2+} block is characterized by its low affinity ($K_{0.5}$ = 30 mM) and it is voltage independent. On the other hand, Ba^{2+} encounters a high affinity site ($K_{0.5}$ = 13 μM; Table 2) when entering the channel from the cytoplasmic side and binding occurs at a fractional electrical distance of 0.39. However, internal Ba^{2+} block does not display the kinetic properties of a simple bimolecular reaction as found for the squid delayed rectifier (Armstrong and Taylor, 1980; Eaton and Brodwick, 1980) or for the K_{Ca} channel (Vergara and Latorre, 1983). In the case of the Kv2.1, the inhibition curve of the internal block was best fitted using a Hill coefficient of 3 and the reciprocal of the forward (on) rate constant was not a linear function of the Ba^{2+} concentration (the relationship between $1/\tau_{on}$ and $[Ba^{2+}]$ is superlinear). To fit the data Taglialatela et al (1993) proposed a model containing two binding sites outside the internal channel vestibule (binding to those site did not lead to a blocked state) from where Ba^{2+} can jump to a site located at 39% of way across the membrane field and block the channel.

Previous studies done by Brown and his colleagues (Kirsch et al. 1992; Taglialatela et al. 1993) showed that the P region amino acid residue valine 374 (Table 2) regulates K^+ and Rb^+ conductances. Normally, the Kv2.1 channels has a larger single-channel conductance for Rb^+ than for K^+, but when the 374 residue is mutated to a threonine (V374T) or to a serine (V374S) the channel switches its ion selectivity and becomes K^+ selective. In the case of the V374S mutant, K^+ single channel conductance increases 2.4-fold and Rb^+ conductance decreases 2-fold. These two mutations produced a significant decrease in the rate of dissociation of internal Ba^{2+} from the pore.

Inward rectifier K^+ channels have only two transmembrane domains (M1 and M2), but the segments that connect M1 to M2 has a high degree of identity with the P region of voltage dependent K^+ channels (Table 2). The case of the IRK1 and ROMK2 are worthwhile to discuss in the present context because they are both blocked by external Ba^{2+}. However, the dissociation constant for Ba^{2+} block is 4.7 mM for ROMK2 (Zhou et al. 1996) and is in the μM range for IRK1 (Kubo et al. 1993). The P regions of ROMK2 and IRK1 differ in only two amino acid residues. The leucine 117 in ROMK2 becomes an isoleucine in IRK1 and the valine 121 in ROMK2 is mutated to a threonine in IRK1. Zhou et al. (1996) found that it was enough to change the valine 121 to a threonine in ROMK2 to recover the Ba^{2+} sensitivity of IRK1. The dissociation constant decreased from 4.7 mM in the wild type channel to 70 μM in the V121T mutant.

It is surprising that in both Kv2.1 and ROMK2 a point mutation of a V to T increases Ba^{2+} sensitivity. Notice that the alignment sequences show that V121 in ROMK2 and V374 in Kv2.1 are only two amino acid residues apart in the P region (Table 2) and toward the cytoplasmic side of the loop[4]. It is also amazing that in ROMK2 the mutation increases sensitivity to *external* Ba^{2+} block whereas the V374T or V374S mutations affect internal Ba^{2+} block without modifying external Ba^{2+} block. We do not know at present the molecular details of the interaction between residues 374 in Kv2.1 and 121 in ROMK2 and Ba^{2+}. Nevertheless, it is possible that substitution of valine at these positions with residues having polar hydroxyl groups provides the adequate shelter for stabilizing Ba^{2+} in the pore (Taglialatela et al. 1993; Zhou et al. 1996). Any molecular interpretation of the

---

[4] Residues T441 in *Shaker* K^+ and T372 in Kv2.1 form part of the internal tetraethylamonium receptor (Yellen et al. 1991; Taglialatela and Brown, 1994).

results has the caveat of the multi-ion nature of K^+ channels. It may well be that changes in ion affinity induced by a given mutation are only reflecting changes in ion occupancy of the channel pore.

Internal Ba^{2+} block in *Shaker* B IR was characterized by Slesinger et al. (1993). In this channel, internal Ba^{2+} induces a voltage- and time dependent block similar to that found in the squid delayed rectifier (Armstrong and taylor, 1980) and in K_{Ca} channels (Diaz et al. 1996). The steady state block is well described by a $K_{0.5}(0) = 0.6$ mM and a fractional electrical distance (δ)of 0.5 (Table 1). Slesinger et al. (1993) found that two point mutations in the loop joining together the transmembrane segments S4 and S5 altered the characteristics of the internal Ba^{2+} block. The mutations L385A and S392C increased affinity *and* voltage dependence of block. In both cases $K_{0.5}(0)$ decreased from 0.6 to about 0.3 mM and δ from 0.5 to about 0.15. These mutations in the S4-S5 loop also affected the binding of TEA suggesting that this portion of the channel forming protein line some internal region of the K^+ channel conduction machinery. As discussed earlier in the chapter, changes in the voltage dependence of block induced by a given mutation are difficult to interpret because it can be reflecting a change in the position of the site in the electric field and/or a change in the ion occupancy of the ion channel pore.

The use of chimaeras has also helped us to locate possible Ba^{2+} sites in K^+ channels. Lopez et al. (1994) noticed that the internal Ba^{2+} block of the NGK2 channels was faster and more potent than the Ba^{2+} block of the *Shaker* B IR. Moreover, the voltage dependence of the Ba^{2+} block of the NGK2 channel is more pronounced than that of the *Shaker* B IR K^+ channel. Substitution of the S6 segment of *Shaker* B IR by the S6 segment of NGK2 produces a chimaera with the Ba^{2+} voltage-dependent block with most of the characteristics of Ba^{2+} block of the NGK2 channels. This result indicates that the S6 contains a Ba^{2+} binding site or exposes a hidden one (Lopez et al. 1994).

CONCLUSIONS

The main conclusions of this chapter are: a) Ba^{2+} has several binding sites in the pore of K^+ channels; b) external, but not internal Ba^{2+} can block closed voltage-dependent K^+ channels. This finding suggest that a cytoplasmic portion of the channel-forming protein undergoes the conformational change that closes the channel. However, closed K_{Ca} channels cannot be blocked by internal *or* external Ba^{2+}. This result implies that in these channels the closed state posses a large barriers to the ion permeation from both sides.; and c) Ba^{2+} stabilizes the closed configuration of the channel. Gating current experiments clearly show that Ba^{2+} speeds charge return. Kinetic modeling of this effect tells us that this can be the consequence of an increase in the backward rate constant in a closed-closed transition preceding opening. The molecular nature of the Ba^{2+} sites remains a

mystery. However, some clues have been obtained using site-directed mutagenesis and electrophysiological measurements. Given its negative nature, a well conserved aspartate residue in the P region of K^+ channels can be or be part of a Ba^{2+} binding site. One interesting, but not conclusive evidence in this regard is that mutation of this aspartate residue to asparragine in *Shaker* K^+ channels eliminates the effect of Ba^{2+} on the charge return. Also in the P region, a conserved valine appears to play a role in modulating Ba^{2+} binding. Substitution of this valine for threonine in Kv2.1 channels increases Ba^{2+} binding potency about 10-fold. Barium binding sites can also exist in the S4-S5 linker and in the S6 transmembrane domain. Due the similarity in crystal radius of K^+ and Ba^{2+}
and the divalent nature of the latter, Ba^{2+} is an ideal probe to search for ion binding sites in K^+ channels.

ACKNOWLEDGEMENTS

This work was supported by grants HL54970 (to LT), GM50550 (to ES), FNI 2950028 (FD), and FNI 940227, Catedra Presidencial and a Group of Chilean private companies (CMPC, CGE, COPEC, Minera Escondida, NOVAGAS, Bussiness Design Ass., and XEROX Chile) (to RL).

REFERENCES

Adelman, W.J. and French, R.J. 1978. Blocking of the squid axon potassium-channel by external caesium ions. *J. Physiol.* 276:13-25.

Armstrong, C.M. 1975. Evidence for ionic pores in excitable membranes. *Biophys. J.* 15:932-933,.

Armstrong, C.M., Swenson, R.P., and Taylor, S.R. Block of squid axon K channels by internally and externally applied barium ions. *J. Gen.Physiol.* 80:663-682, 1982.

Armstrong, C.M. and Taylor, S.R. Interaction of barium ions with potassium channels in squid giant axons. *Biophys. J.* 30:473-488, 1980.

Bezanilla, F. and Armstrong, C. 1972. Negative conductance causes by entry of sodium and cesium ions into the potassium channels of squid axon. *J. Gen. Physiol.* 60:588-608.

Bezanilla, F., Perozo, E., Papazian, D.M., and Stefani, E. 1991. 1972. Molecular basis of gating charge immobilization in Shaker potassium channels. *Science* 254:679-683.

Bezanilla, F., Perozo, E., and Stefani, E. 1994. The gating of Shaker K+ channels. II. The components of gating currents. *Biophys. J.* 66:1011-1021.

Bezanilla, F. and Stefani, E. Voltage dependent gating of ionic channels. *Ann. Rev. Biophys. Biochem. Chem.* **23**:819-846, 1994.

Cecchi, X., Wolff, D., Alvarez, O., and Latorre, R. 1987. Mechanisms of Cs+ blockade in a Ca2+-activated K+ channel from smooth muscle. *Biophys. J.* 52:707-716.

Diaz, F., Wallner, M., Stefani, E., Toro, L., and Latorre, R. 1996. Interaction of internal Ba^{2+} with a cloned Ca^{2+}-activated K^{+} (*hslo*) channel from smooth muscle. *J. Gen. Physiol.* 107:399-407.

Eaton, D.C. and Brodwick, M.S. 1980. Effects of barium on the potassium conductance of squid axon. *J. Gen. Physiol.* 75:727-750.

Eisenman, G., Latorre, R., and Miller, C. 1986. Multi-ion conduction and selectivity in the high-conductance Ca^{++}-activated K^{+} channel from skeletal muscle. *Biophys. J.* 50:1025-1034.

Grissmer, S. and Cahalan, M.D. 1989. Divalent cation trapping inside potassium channels of human T lymphocytes. *J. Gen. Physiol.* 93:609-630.

Heginbotham, L. and MacKinnon, R... 1993. Conduction properties of the cloned *Shaker* K^{+} channel. *Biophys. J.* 65:2089-2096.

Hille, B. 1973. Potassium channels in myelinated nerve. Selective permeability to small cations. *J. Gen. Physiol.* 61:669-686,.

Hille, B. and Schwarz, W. 1978. Potassium channels as multi-ion single-file pores. *J. Gen. Physiol.* *72*:409-472.

Hurst, R.S., Latorre, R., Toro, L., and Stefani, E. 1995. External barium block of *Shaker* potassium channels:evidence for two binding sites. *J. Gen. Physiol.* 106:1069-1087.

Hurst, R.S., Roux, M., Toro, L., and Stefani, E. 1996a . External Barium influences the voltage sensing mechanism of *Shaker* K^{+} channels. *Biophys. J.*. In press.

Hurst,R.S, Toro, L., and Stefani,E. 1986b . Amino acid residues involved in external barium block of *Shaker* potassium channels. *FEBS Lett.* In press.

Jan, L.Y. and Jan, Y.N. 1990. A superfamily of ion channels. *Nature (London)* 345:672.

Kamb, A., Tweng-Drank, J., and Tanouye, M.A. 1988. Multiple products of the *Drosophila Shaker* gene may contribute to poasssium channel diversity. *Neuron* 1:421-430.

Kirsch, G.E., Pascual, J.M., and Shieh, C. 1995. Functional role of a conserved aspartate in the external mouth of voltage-gated potassium channels. *Biophys. J.* 68:1804-1813,.

Korn, S.J. and Ikeda, S.R. 1995. Permeation selectivity by competition in a delayed rectifier potassium channel. *Science* 269:410-412.

Kubo, Y., Baldwin, T.J., Jan, Y.N., and Jan, L.Y. 1993. Primary structure and functional expression of a mouse inward rectifier potassium channel. *Nature* 362:127-133.

Kukuljan, M., Labarca, P., and Latorre, R. 1995. Molecular determinants of ion conduction and inactivation in K+ channels. *Am. J. Physiol.* 37:C535-C556.

Lopez, G.A., Jan, Y.N., and Jan, L.Y. 1994. Evidence that the S6 segment of the *Shaker* voltage-gated K^+ channel comprises part of the pore. *Nature* 367:179-182.

MacKinnon, R. 1991. Determination of the subunit stiochiometry of a voltage-activated potassium channel. *Nature* 350:232-235.

MacKinnon, 1995. R. Pore loops: am emerging theme in ion channels structure. *Neuron* 14:889-892.

Mannuzzu, L.M., Moronne, M.M., and Isacoff, E.Y. 1996. Direct physical measure of conformational rearrangement underlying potassium channel gating. *Science* 271:213-216.

Miller, C. 1987Trapping single ions inside single ion channels. *Biophys. J.* 52:123-126.

Miller, C., Latorre, R., and Reisin, I. 1987. Coupling of voltage-dependent gating and Ba^{++} block in the high conductance Ca^{++}-activated-K^+ channel. *J. Gen. Physiol.* 90:427-449,.

Neyton, J. and Miller, C. 1988a . Potassium blocks barium permeation through a calcium-activated potassium channel. *J. Gen. Physiol.* 92:549-568.

Neyton, J. and Miller, C. 1988b . Discrete Ba^{2+} as a probe of ion occupancy and pore structure in the high-conductance Ca^{2+} activated K^+ channel. *J. Gen. Physiol.* 92:569-586.

Sigg, D., Stefani, E., and Bezanilla, F. 1994. Gating current noise produced by elementary transitions in *Shaker* potassium channels. *Science* 264:578-582.

Slesinger, P.A., Jan, Y.N., and Jan, L.Y. 1993. The S4-S5 loop contributes to the ion-selective pore of potassium channels. *Neuron* 11:739-749.

Stampe, P., and Begenisich, T. 1996. Unidirectional K^+ fluxes through recombinant *Shaker* potassium channels expressed in single *Xenopus* oocytes. *J. Gen. Physiol.* 107:449-457.

Stuhmer, W., Stocker, M., Sakmann, B., Seeburg, P., Baumann, A., Grupe, A., and Pongs, O. 1988. Potassium channels expressed from rat brain cDNA have delayed rectifier properties. *FEBS. Lett.* 242:199-206,.

Taglialatela, M., Drewe, J.A., and Brown, A.M. 1993. Barium blockade of a clonal K^+ channel and its regulation by a critical pore residue. *Mol. Pharmacol.* 44:180-190.

Tempel, B.L., Papazian, D.M., Schwarz, T.L., Jan, Y.L., and Jan, L.Y. 1987. Sequence of a probable potassium channel component encoded at *Shaker* locus of *Drosophila*. *Science* 237:770-775.

Vergara, C. and Latorre, R. 1983. Kinetics of Ca^{2+}-activated K^+ channels from rabbit muscle incorporated into planar bilayers. Evidence for a Ca^{2+} and Ba^{2+} blockade. *J. Gen. Physiol.* 82:543-568.

Werman, R. and Grundfest, H. 1961. Graded and all-or-none electrogenesis in arthropod muscle. *J. Gen. Physiol.* 44:997-1027.

Wollmuth, L. P. 1994. Mechanism of Ba^{2+} block of M-like K^{+} channels of rod photoreceptors of tiger salamanders. *J. Gen. Physiol.* 103:45-66.

Yellen, G. 1984. Relief of Na+ block of Ca^{2+}-activated K^{+} channels by external cations. *J. Gen. Physiol.* 84:187-199.

Zhou, H., Chepilko, S. Schutt, W., Choe, H., Palmer, L. G., and Sackin, H. 1996. Mutations in the pore region of ROMK enhance Ba^{2+} block. *Am. J. Physiol.* In press.

POSSIBLE PARTICIPATION OF A cAMP REGULATED K^+ CHANNEL FROM THE SEA URCHIN SPERM IN THE SPERACT RESPONSE

Pedro Labarca#, Celia Santi, Otilia Zapata, Carmen Beltrán, Arturo Liévano, Yolanda Sandoval and Alberto Darszon

#Centro de Estudios Científicos de Santiago and Facultad de Ciencias, Universidad de Chile; Santiago, Chile. Dept. Genética y Fisiología Molecular, Instituto de Biotecnología, Universidad Nacional Autónoma de México; Apartado. 510-3, Cuernavaca, Morelos 62271, México.

INTRODUCTION

Ion channels are key elements in the activation of respiration and motility, in chemotaxis, and in triggering the acrosome reaction (AR) in sperm (Schackmann, 1989; Ward and Kopf, 1993; Darszon *et al.*, 1994). The small size of these cells (head diameter ~ 2 μm) has precluded the careful characterization of their electrophysiological properties. Because of this we have used planar bilayers with incorporated sperm plasma membrane components, to detect single channels from sea urchin sperm. Although with a low success rate we have also succeeded in patch clamping sea urchin sperm (Guerrero *et al.*, 1986; Babcock *et al.*, 1992). Using these techniques, together with studies of membrane potential, $[Ca^{2+}]$i and pHi in whole sperm, we have established the presence of K^+, Ca^{2+} and Cl^- channels in this cell and are exploring their regulation and participation in chemotaxis and in the AR (Darszon *et al.*, 1994).

Sperm respond to the outer layer of eggs with profound changes in membrane permeability, metabolic activity, swimming behavior and second messenger levels ($[Ca^{2+}]i$, cyclic nucleotides, IP_3 levels) (reviewed in Schackmann, 1989; Garbers, 1989; Ward and Kopf, 1993). It is known that ion channels are subject to multiple forms of regulation (Brown, 1993), and cyclic nucleotides can directly or indirectly modulate them (Latorre *et al.*, 1991; Zufall *et al.*, 1994). Thus, we are now exploring the regulation of the sea urchin sperm ion channels by cyclic nucleotides. Here, we present a study of a cAMP regulated K^+ channel from isolated *S. purpuratus* flagellar membranes incorporated into Müeller-Rudin planar bilayers. Because this channel is only 4.9 fold more selective for K^+ than for Na^+, its opening would allow a significant Na^+ current into sea urchin sperm in sea water, causing a depolarization.

Speract, a decapeptide from *S. purpuratus* and *H. pulcherrimus* (Hansbrough and Garbers 1981, Garbers *et al.*, 1982; Suzuki *et al.*, 1981; Garbers 1989) egg jelly, activates respiration and lipid oxidation in the pM range. It transiently stimulates guanylyl cyclase, and binds to a 77 kDa protein resembling the low density lipoprotein receptor (Dangot *et al.*, 1989). Nanomolar concentrations of this peptide elevate the levels of cGMP, cAMP, and induce permeability changes in homologous sperm (Schackmann, 1989, Garbers, 1989, Ward and Kopf, 1993). It has been argued that these egg peptides in addition to regulating respiration and sperm motility (Ward *et al.*, 1985; Cook *et al.*, 1994) cooperate with other factors in the egg-jelly to promote fertilization (Suzuki *et al.*, 1984; Yamaguchi *et al.*, 1988; Shimizu *et al.*, 1990).

Since in swollen sea urchin sperm nM speract induces a Na^+ dependent repolarization (Reynaud *et al.*, 1993), the speract responses of normal sperm (Babcock *et al.*, 1992; Cook and Babcock 1993a, b) were further characterized to explore if the cAMP-regulated K^+ channel could be involved in them. The speract-induced pHi and membrane potential changes were simultaneously or separately measured in sperm suspensions using pHi and membrane potential sensitive fluorescent dyes. We found that in normal sperm the depolarization induced by nM speract is only partly dependent on external Ca^{2+}, suggesting that it must be Na^+ -dependent. We show that indeed this is the case, and that this dependence is direct and not via pHi. Taking this into account, together with other results, we conclude that the cAMP-regulated K^+ channel detected in planar bilayers could participate in the speract-induced depolarization of sea urchin sperm.

METHODS AND MATERIALS

Materials: *S. purpuratus* sea urchins were obtained from Pacific Bio-Marine Laboratories (Venice, CA) or from Marinus (Long Beach, CA). Valinomycin, and tetraethylammonium (TEA^+) were from Sigma (St. Louis, Mo). Phospholipids for planar bilayer experiments were from Avanti Polar Lipids (Birmingham, AL). Speract was from Peninsula Laboratories (Belmont, CA). Dipropylthiodicarbocyanine (DiS-C_3-(5)) and 2', 7'- Bis (2-carboxyethyl)-5(6)- carboxyfluorescein acetoxymethyl ester (BCECF) were from Molecular Probes (Eugene, OR).

Gamete collection: Sperm were obtained from *S. purpuratus* sea urchins by intracoelomic injection of 0.5 M KCl and collected with a Pasteur pipette directly from the gonopores (Darszon *et al.*, 1994). The cell suspension was kept on ice until used to obtain flagellar plasma membranes or for membrane potential and pHi measurements (see bellow). ASW was in mM: 486 NaCl, 10 KCl, 29 $MgSO_4$, 27 $MgCl_2$, 10 $CaCl_2$, 2.4 $NaHCO_3$, 0.1 EDTA, 10 HEPES, pH 8.0. The experiments at various Na^+ concentrations were done using in $0SO_40CO_3$ ASW where Na^+ was replaced (mole per mole) by choline$^+$, and $MgSO_4$ by $MgCl_2$, and $NaHCO_3$ was removed. Because Ba^{2+} precipitates in the presence of CO_3^{2-} and SO_4^{2-} , $0SO_40CO_3$ ASW was used to test the blocking effect of this divalent.

Isolation of plasma membranes and formation of vesicles: Plasma membranes were isolated as described by (Darszon *et al.*, 1984). The membranes were preloaded with 0.6 M sucrose by sonication (2 min) with a water bath sonicator (Bransonic Heat Systems Ultrasonics, Plainview, NY, model B12) or allowing them to stand overnight in the sucrose solution, to provide an osmotic gradient which favors fusion with planar bilayers (Darszon *et al.*, 1994). Thereafter, the vesicles were stored in aliquots (100 µl) at
-70 °C.

Planar bilayer experiments: Planar lipid bilayers were formed according to Müller and Rudin (1969) from a synthetic lipid, diphytanoylphosphatidylcholine (DPPC) or a mixture of DPPC: phosphatidylethanolamine:phos-phatidylserine (70:15:15, 20 mg/ml, total concentration in n-decane). Bilayers were formed on a hole (200-300 µm diameter) made on the plastic septum of a chamber having two 300 µl compartments. Each compartment was connected through an agar-Ag/AgCl electrode to a home-made I-V transducer (10 GΩ resistor in the feedback loop, *trans* at virtual ground) or a Dagan 3900A patch-clamp amplifier (Dagan Corp., Minneapolis, MN). After formation DC pulses of ±100 mV were applied during 30 sec to ensure that the bilayer did not show any single channel-like activity. After this, 5-10 μl of membrane vesicles (5 μg of protein) were added to *cis* side, with a holding potential of 25-50 mV. The fusion of vesicles containing ionic channels was usually detected within 1 - 10 min by the appearance of single channel transitions.

The solutions used were prepared with deionized and glass distilled water filtered through 0.22 μm Millipore sieves. Their composition was: 300-600 mM KCl in *cis* and 100 mM in *trans*, both containing 5 mM HEPES pH 8.0 and 100 μM-5 mM $CaCl_2$. All reagents were of the best quality commercially available.

Data storage and analysis: Single channel currents were recorded unfiltered and stored using a modified VCR (Bezanilla, 1985). For single channel current analysis, the records were low-pass filtered at 500 Hz (four pole Bessel filter, Warner Inst. Corp., Handem, CT), acquired at 5 kHz, and analyzed using pClamp 6.0 software (Axon Instruments Inc., Foster City, CA). Dwell time distributions were fitted by least square minimization methods using pStat. Curve fitting was performed by least square minimization methods using Axum 2.0 (Trimetrix Inc., Seattle, WA).

Biionic conditions were used to determine the selectivity as follows: after detecting the channels in a KCl gradient, the *cis* compartment was perfused with 100 mM KCl to have a control record under symmetrical conditions. Thereafter, the solution in *cis* was changed perfusing at least 10 chamber volumes with another monovalent cation (100 mM of RbCl, NaCl or LiCl). Reversal potentials (E_R) were estimated from the I-V curves of the single channel transitions (Darszon *et al.*, 1994).

Membrane potential and pHi measurements in sea urchin spermatozoa: Membrane potential was monitored with the fluorescent, positively charged membrane potential sensitive dye DiS-C_3 -(5) (0.5 μM final, 1 mM stock in DMSO) at the 620/670 excitation/emission wavelength pair, as described in (González-Martínez *et al.*, 1992, Reynaud *et al.*, 1993). The dye response was calibrated and the resting membrane potential determined after treatment with the mitochondrial uncoupler bis(hexafluoroacetonyl)acetone (5 μM) as in (Reynaud *et al.*, 1993) using valinomycin and K^+ additions with each of the solutions employed. Measurements of pHi were made using the intracellular fluorescent probe BCECF monitored at 500/550 nm excitation/emission wavelength pair, respectively and loaded and calibrated as in (Guerrero and Darszon, 1989, Reynaud *et al.*, 1993). The loaded cells were then diluted 10 times at 4 °C, in 1 mM $CaCl_2$ ASW pH 7.0, pelleted (6 min, 1000 x g) and resuspended in the same media. Thereafter, 10 - 20 μl of this dilution was added to a round cuvette containing 700 μl of ASW (10^6 cells/ml final) for fluorescence measurements at 14 °C. Membrane potential and pHi were simultaneously measured using a magnetically stirred, temperature regulated Hansatech Mk II fluorescence cell (Norfolk, England) coupled to a Schott lamp (KL 1500) through a bifurcated optic fiber. Each arm of the fiber had the corresponding interference filter for excitation attached. The emission was determined with two Hansatech fluorescent detectors (FDC), each with its corresponding interference filter. In these experiments each dye was calibrated in the presence of the other, as described (Reynaud *et al.*, 1993). Initial experiments were confirmed measuring membrane potential and pHi separately, and also with a Perkin Elmer LS 3 spectrofluorometer. All determinations were made at 14 °C under constant stirring. Membrane potential and pH experiments were performed at least 4 times with different batches of sperm.

RESULTS

A weakly voltage-dependent K^+-selective channel from sea urchin plasma membranes.

A cation-selective channel exhibiting a low probability of opening (Po= 0.80 ± 0.20 %, n=8, at zero-applied voltage) was detected upon fusion of plasma membrane fragments purified from the sea urchin *S. purpuratus* to planar lipid bilayers, as described in Methods (Fig.1A). Incorporation of this channel to planar bilayers can be achieved reproducibly in all membrane preparations tested (>15).

The fraction of time the channel spent in the open state depended, although weakly, on the applied voltage (Fig. 1B). At more negative-applied potentials, the mean open times, estimated by averaging open time durations, became longer (Fig.1C). Fig. 1C also shows the line joining the experimental points corresponding to the best fit of the function:

$$\tau(V) = \tau_0 \exp -V/A \quad (1)$$

to the data. In this equation, τ_0 denotes the mean open times estimated at zero-applied voltage, and A is a constant, in mV, calculated here at 174 mV. In the ± 80 mV range explored, the I-V relation for the channel was linear, with a slope conductance of 144 pS in 0.6 M KCl *cis*/0.1 M KCl *trans*. The reversal

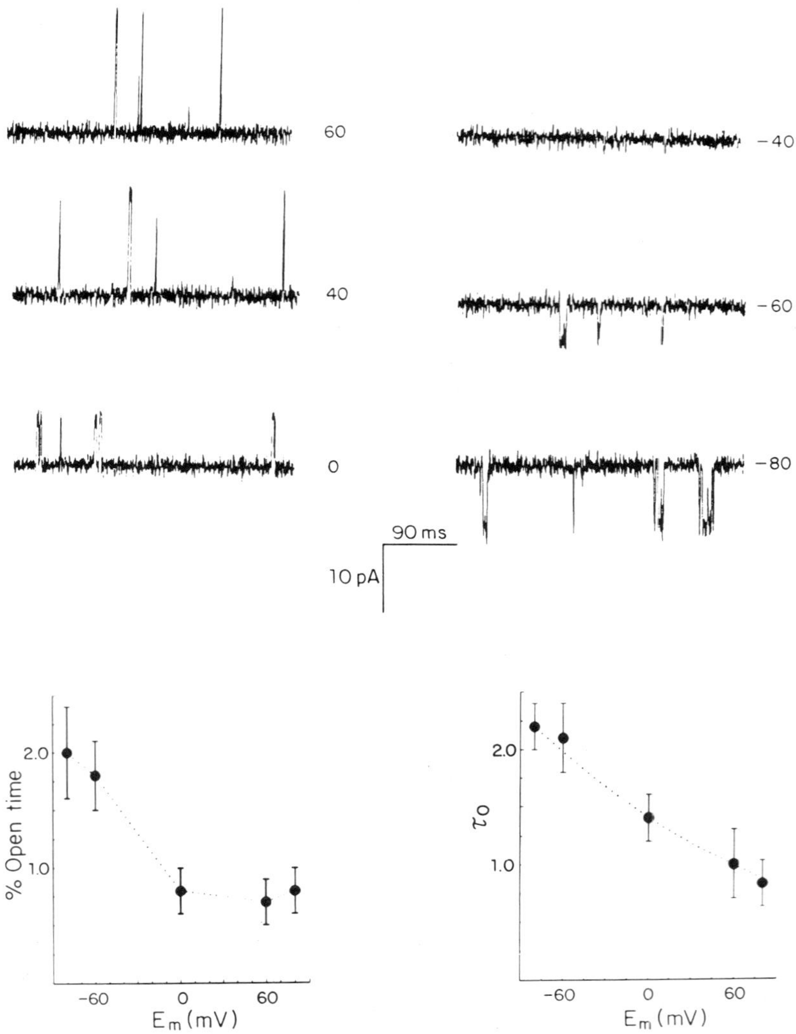

Figure 1. Characteristics of a K^+-selective channel from purified plasma membrane of *S. purpuratus* in planar bilayers. A. Channel incorporation to planar lipid bilayers was achieved as explained in Methods having 600 mM KCl in the *cis* chamber and 100 mM KCl in the *trans*. Single channel currents from the same planar bilayer, low-pass filtered at 500 Hz, are shown at different applied-voltages. Near -40 mV the current became zero, indicating that the channel is cation-selective. **B**. Voltage-dependence of the open probability. Computer analysis (see Methods) was used to obtain the fraction of open time (% Open tiem). Each point is the average from eight different experiments performed as in A and the bars are SEM. **C**. Voltage-dependence of the open time durations (τ_o). All open times in an experimental record were averaged to estimate the durations. Each point represents the average of eight different experiments over the voltage range shown, under the conditions of A and the bars are SEM. The line is best fit of equation 1 to the experimental points.

Table 1. Reversal potentials and relative permeability ratios for monovalent cations of sperm K^+-selective channel. Biionic experiments were performed as explained in Methods. Reversal potentials (E_R) were estimated as the zero-current voltage, determined from single channel current *vs* voltage relations. Each E_R value in the Table corresponds to the average ± SEM from three different experiments, from which the P_{K+}/P_{X+} were calculated by using the Goldman-Hodgkin-Katz equation, and ion concentrations were corrected for activity coefficients.

K^+ *vs:*	K^+	Rb^+	Na^+
E_R (mV):	0	25 ± 5	40 ± 8
PK^+/PNa^+:	1	2.7	4.9

potential for the channel under these experimental conditions was -40 mV, close to the Nernst potential predicted for a perfectly cation selective pore (Fig. 1A and 5B). The channel's slope conductance in symmetrical 0.1M KCl, was 103 pS (Fig. 2). Fig. 2 and Table I show that the channel followed the selectivity sequence $P_K > P_{Rb} > P_{Na} > P_{Cs}$, estimated from the reversal potentials obtained from current-voltage relations under a bionic regime.

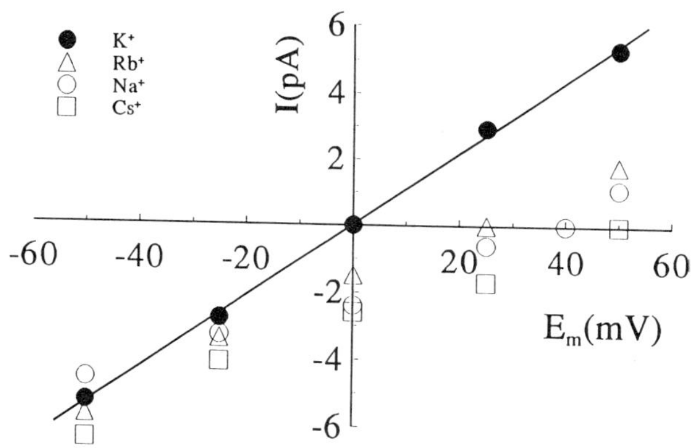

Figure 2. Selectivity of the sea urchin sperm K^+ channel. Channel incorporation to planar bilayers was performed under symmetrical 0.1 M KCl, thereafter *cis* was extensively perfused with 0.1 M of the monovalent cation of interest. Single channel currents were recorded at different applied-voltages to estimate the zero-current voltage under bionic conditions. The experimental points are the averages from three different studies with SEM being enclosed by symbols

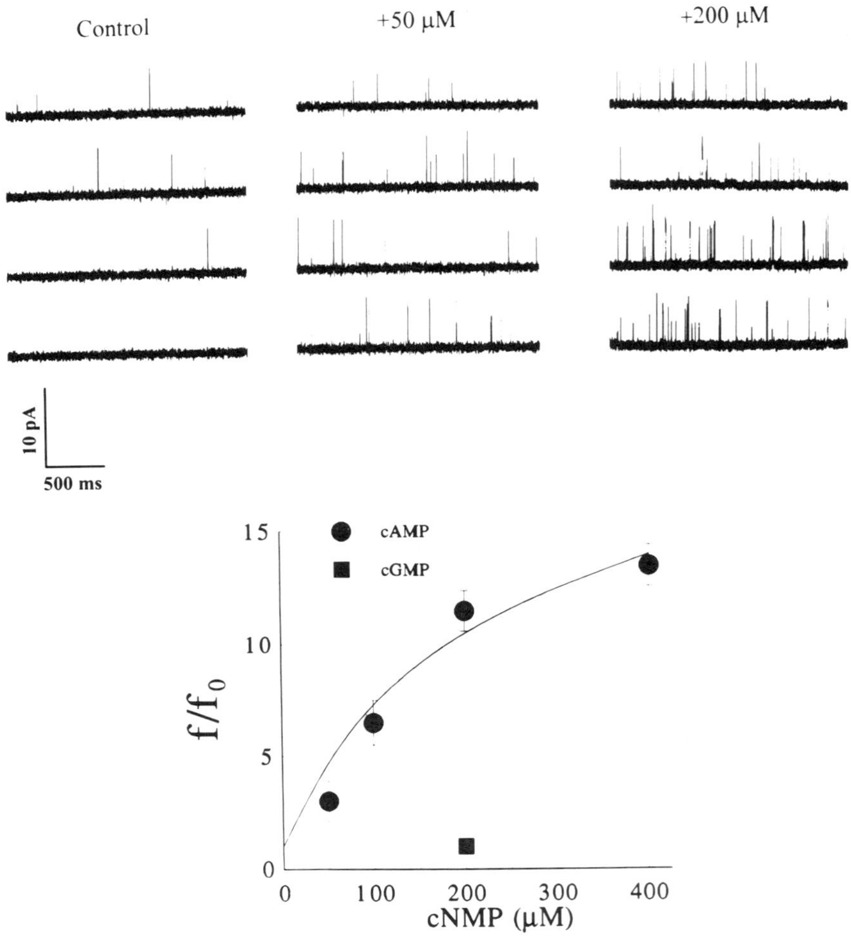

Figure 3. Upward modulation by cAMP of the sea urchin sperm K^+ channel. **A**. The open probability of the channel is increased by cAMP. Single channel curents were recorded at zero-applied voltage using the experimental conditions described in Fig. 1A. Records from a bilayer, low pass filtered at 500 Hz, exposed to the indicated [cAMP]. **B**. Channel open probability vs [cAMP]. The ratio f/f_0, where f is the fraction of open time at any [cAMP] and f_0 the fraction of open time without cAMP, was used as a normalized parameter to assay the effects of cAMP on open probability (filled circles). Each point represents the average open probability from at least four different studies where a minimum of three different [cAMP] were tested in the same experiment, and the bars are ± SEM. The filled square is the average f/f_0 of three experiments done with 200 μM cGMP. The best fit to equation (2) was used to generate the line.

Stimulation by cAMP. The planar lipid bilayer assay was used to determine whether agents relevant to sperm physiology, such as cGMP and cAMP (Garbers 1989), were able to modulate the probability of opening of the K^+-selective channel derived from *S. purpuratus* sperm. The channel's open probability was not affected by up to 200 μM cGMP *cis* or *trans* (Fig. 3B). In contrast, channel exposure to cAMP from the *cis* side, at concentrations above 50 μM, caused a dose-dependent increase in channel open probability in 80 % of assays (n=23). Fig. 3A illustrates the increase in channel activity induced by cAMP. Upward modulation by cAMP was only seen in *cis* and could be reversed upon perfusing this bilayer chamber with agonist-free buffer. The results from 5 bilayers where at least 3 cAMP concentrations were assayed are shown in Fig. 3B. At zero-applied voltage, addition of 200 μM cAMP increased channel open probability in average by 11-fold, from 0.5 ± 0.2 % to 6.1 ± 2 %. The dotted line in Fig. 3B is the best fit of the function:

$$\mathbf{f/f_0} = 1 + (\mathbf{f_{max}}\ ([\mathrm{cAMP}]/[\mathrm{cAMP}] + K_D)\ (2)$$

to the experimental data. Here, **f** is the fraction of open time at any cAMP concentration, $\mathbf{f_0}$ the fraction of open time in the absence of cyclic nucleotide, $\mathbf{f_{max}}$ is the value of $\mathbf{f/f_0}$ at very large [cAMP], and K_D is the apparent dissociation constant. In Fig. 3B $\mathbf{f_{max}} = 20$ and K_D was estimated at 214 μM. These results indicate that the effect of cAMP could be explained, in principle, considering a single binding site in the *cis* face of the K^+-selective channel whose occupancy by the cyclic nucleotide increases the probability of reaching the channel open conformation. The *cis* side was assigned as the cytoplasmic face of the channel since upward modulation occurred from this side.

Further analysis of the results indicated that, in the range of cAMP concentrations tested, mean open times were not altered by the presence of the cyclic nucleotide (Fig. 4B, left). On the other hand, mean closed times became shorter as cAMP was raised (Fig. 4B). In the absence and in the presence of cAMP, open times could be accounted for in terms of a single exponential component (Fig. 4C, left histogram). In turn, closed times could be fitted to two exponential components (Fig. 4C, right histogram). The results would be consistent with a linear kinetic scheme to account for the upward modulation by cAMP (Moczydlowski, 1985):

$$\mathrm{closed}_1 \underset{K_{-1}}{\overset{K_1}{\longleftrightarrow}} \mathrm{Closed}_2\text{-cAMP} \underset{\alpha}{\overset{\beta}{\longleftrightarrow}} \mathrm{open}$$

In agreement with the experimental data the scheme predicts that the distribution of open times should be single-exponential and independent of

[cAMP], while there should be two components of closed times. Binding of cAMP to the $closed_1$ conformation shifts the equilibrium towards the $closed_2$ state. Channel opening from $closed_2$ would then proceed through a

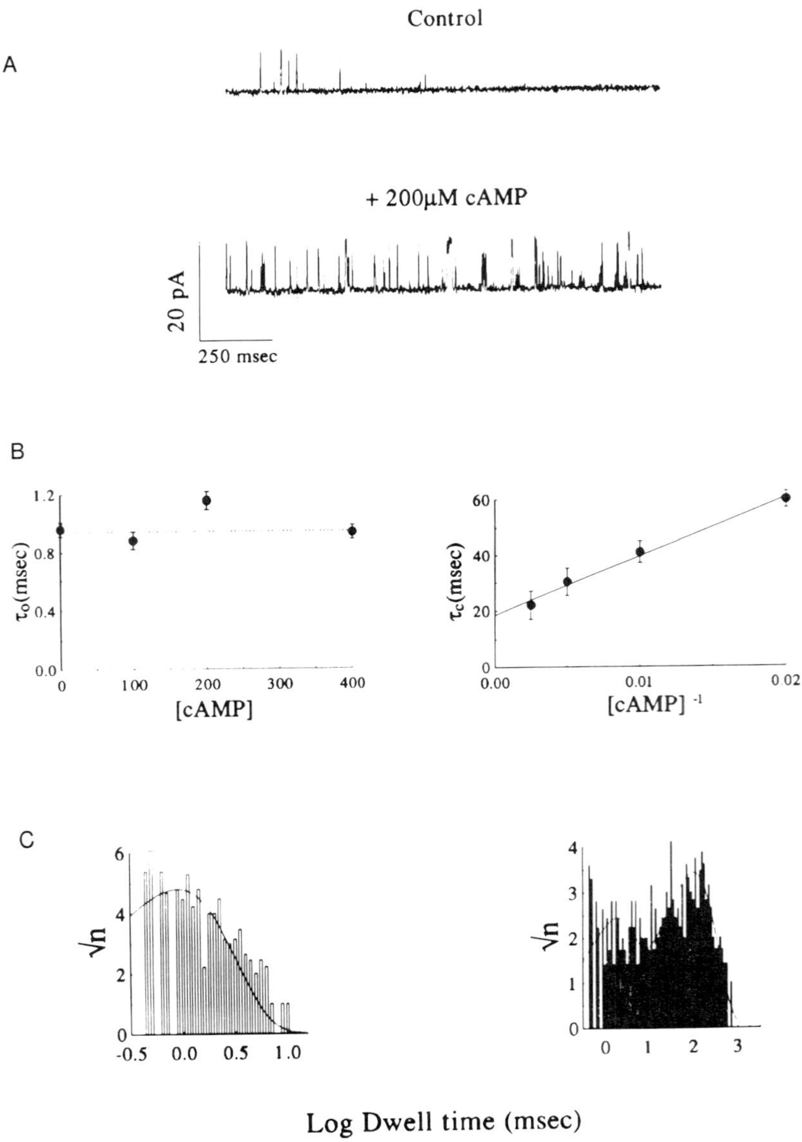

conformational transition. Only the transition leading away from the $closed_1$ into the $closed_2$ conformation involves a cAMP-binding step. The scheme predicts that mean closed times, obtained by averaging closed durations, should vary linearly as a function of $[cAMP]^{-1}$ (Moczydlowski, 1985). The right panel of Fig. 4B illustrates that within the range of cyclic nucleotide concentration tested this prediction was fulfilled by the experimental data. The function used to fit the experimental data in 4B is (Moczydlowski, 1985):

$$\bar{\tau}_c = A\,(1 + K_D/[cAMP]) \quad (4)$$

where $\bar{\tau}_c$ is the mean closed time calculated by averaging the closed durations, A is a constant and K_D is the apparent dissociation constant. Consistent with the K_D (214 μM) estimated from the cAMP dependence of open channel probability (Fig. 3B), a K_D of 120 μM resulted from the slope of the straight line obtained from the results.

TEA^+ and Ba^{2+} block the channel in a voltage-dependent manner. Two K^+ channel blockers, TEA^+ and Ba^{2+} (Latorre and Miller, 1983; Rudy, 1988), were tested on the K^+-selective channel from *S. purpuratus*. Single channel currents were not altered by up to 10 mM TEA^+ in the *cis* chamber (not shown). In contrast, channel exposure to 10 mM TEA^+ in *trans* reduced channel open current amplitude at negative-applied voltages (Fig. 5A). The channel was blocked by 10 mM Ba^{2+} in *trans,* even at zero-applied voltage; blockade became more dramatic at negative applied voltages (Fig. 5B).

The dose and voltage dependence of the Ba^{2+} blockade was further investigated. At all voltages explored, the decrease in channel open currents as

Figure 4. Modulation of open and closed times by [cAMP]. A. Single channel currents at zero-applied voltage from a bilayer under control conditions (upper record) and after exposure to 200 μM cAMP *cis* (lower record); filtering was as in Fig. 4. **B.** Closed times (left) but not open times (right) were modulated by cAMP. Each point represents the average of at least five experiments in which a minimum of three different [cAMP] were tested, the bars are SEM. Mean open times (τ_o) were obtained as in Fig. 1B. The line is an average open time derived from open durations in all experiments. Mean closed times were obtained by averaging all closed durations vs $[cAMP]^{-1}$. Equation 4 was used to fit the line to the results. **C.** Open (left) and closed (right) dwell time distributions built from 1000 single channel transitions in the presence of 200 μM cAMP were analyzed as described in Materials and Methods. The continuous line in the left histogram is the best fit to a single exponential component of open dwell times with a τ_o = 1.6 msec. The continuous line on the right histogram is the best fit to the closed dwell times with two exponential components with τ_1= 20 mec and τ_2= 1 msec. The relative amplitude of the slow component was 0.51.

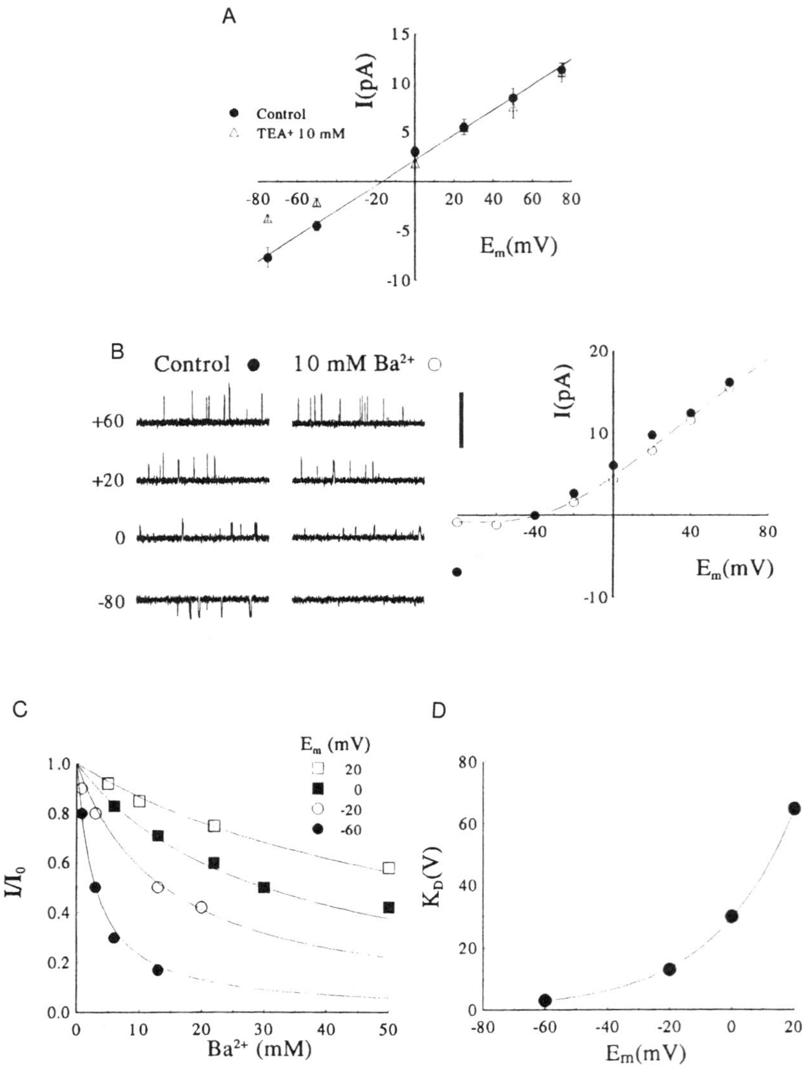
A
I(pA)
15
10
5
-5
-10
-80
-60
-20
20
40
60
80
Em(mV)
Control
TEA+ 10 mM
B
Control
10 mM Ba2+
+60
+20
0
-80
I(pA)
20
10
-10
-40
40
80
Em(mV)
C
I/I0
1.0
0.8
0.6
0.4
0.2
0.0
0
10
20
30
40
50
Ba2+ (mM)
Em (mV)
20
0
-20
-60
D
KD(V)
80
60
40
20
0
-80
-60
-40
-20
0
20
Em(mV)

a function of blocker concentration could be fitted with a single Langmuir isotherm (Fig. 5C; Moczydlowski, 1985).

$$i/i_0 = (1+ [Ba^{2+} trans]/K_D)^{-1} \quad (5)$$

where i is the single channel current amplitude measured at any given $[Ba^{2+}]$, i_0 is the current measured in the absence of the divalent and K_D is the apparent dissociation constant at any applied voltage. The dissociation constant estimated by fitting (5) to the experimental data at zero-applied voltage (K_D0) yielded a value of 30 mM. The effect of *trans* Ba^{2+} on channel open currents became less evident at +20 mV, showing an apparent dissociation constant of 65 mM (Fig. 5C). In contrast, at -20 and -60 mV the block by Ba^{2+} became more effective, yielding estimates for the apparent K_D's of 14 and 3 mM, respectively (Fig. 5C).

The previous results indicate that Ba^{2+} could partially permeate the open channel from the *trans* side, blocking ion passage through it. Thus, the divalent would be expected to sense a fraction of the voltage drop across the bilayer, accounting for the observed voltage-dependence of the apparent dissociation constant (Hille, 1992). It should be possible to describe the voltage-dependence of K_D using the following equation (Hille, 1992):

$$K_D(V) = K_D0 \exp z\delta FV/RT \quad (6)$$

Figure 5. Open channel blockade by TEA^+ and Ba^{2+}. A. Single channel currents were monitored in 300 mM KCl *cis*, 100 mM KCl, *trans* in the absence (filled circles) and in the presence of 10 mM TEA *trans* (triangles). The points are averages of single channel current amplitudes from three different studies at the indicated voltages. The ± SEM are enclosed within the symbols. The straight lines across the experimental points in 5A and B were drawn by eye. **B.** Single channel currents recorded in 600 mM KCl *cis*/100 mM KCl *trans* in the absence (filled circles) and presence of 10 mM Ba^{2+} *trans*. The left panel shows single channel currents from a bilayer at the indicated voltages under control conditions (left) and in 10 mM *trans* Ba^{2+} (right). The right panel illustrates the current vs voltage relation of single channel currents in control (filled circles) and in 10 mM Ba^{2+} (open circles). The points (± SEM enclosed within them) are averages from three different studies. The curve joining the experimental points in the presence of Ba^{2+} (10 mM) is the best fit to Eq. (7). Each trace is 150 msec long, and the current scale bar is 20 pA. The inset illustrates single channel currents obtained from a same planar bilayer at the indicated voltages under control and in 10 mM *trans* Ba^{2+}. **C.** Dose-dependence of Ba^{2+} blockade at different applied voltages. Single channel currents wererecorded under 600 mM KCl *cis*/100 mM KCl *trans* at the indicated voltages and over the range of $[Ba^{2+}]$ shown. The points are the average currents from three different studies and ± SEM are enclosed within symbols. The lines joining experimental points are best fits to Eq. (5). The apparent K_D's were 65 mM (V= 20 mV), 30 mM (0 mV), 14 mM (-20 mV) and 3 mM (-60 mV). **D.** K_D's vs applied voltage (E_m). Filled circles are K_D values derived from data shown in **C** at the indicated voltages. The line joining the points is the best fit to Eq. (6) with $\delta = 0.49$ and K_D0= 30 mM.

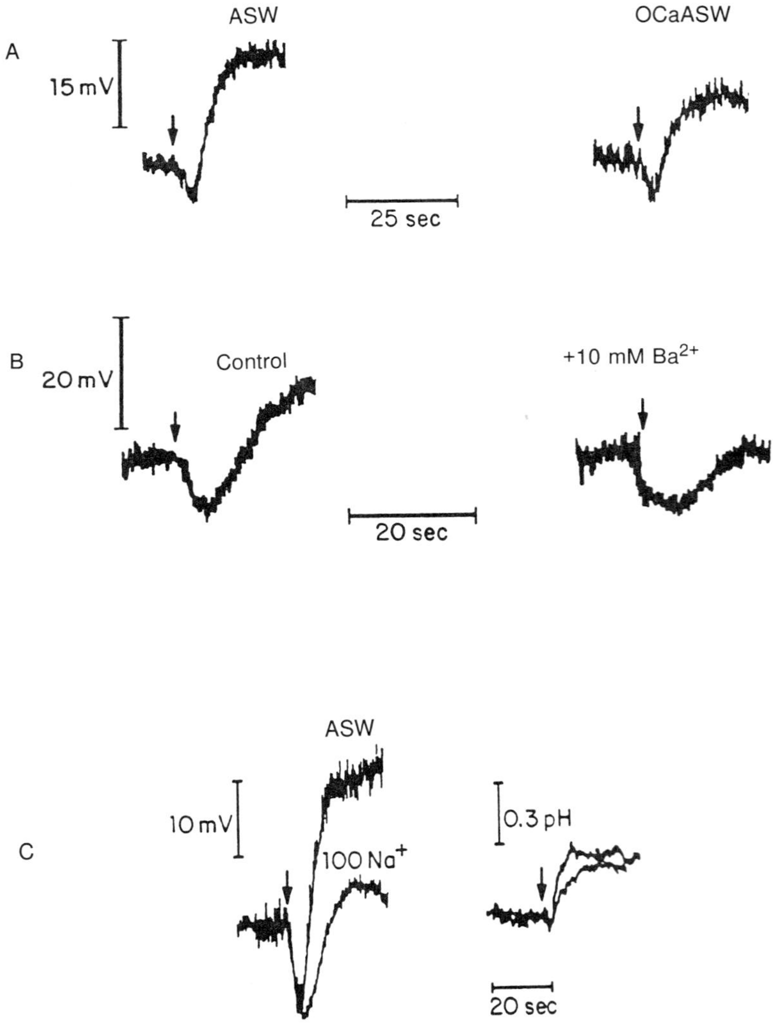
ASW
OCaASW
A
15 mV
25 sec
B
20 mV
Control
+10 mM Ba2+
20 sec
ASW
10 mV
0.3 pH
C
100 Na+
20 sec

where K_D0 is the apparent dissociation constant at zero-applied voltage, z is the valence of the blocker, δ is the fraction of the electric field sensed by the blocker, V is the applied-voltage, and F, R and T have their usual meaning. A K_D0 of 29.6 mM and δ= 0.49 were obtained from fitting the results (Fig. 5D) to Eq. 6. The estimated value of δ indicates that the divalent exerted its effect by occupying a site deep into the open channel pore. Figure 5B shows that combining (5) and (6) yields an expression:

$$i = i_0 \left(1+ [Ba^{2+}\ trans]\ \exp -(z\delta FV/RT/K_D0)\right)^{-1} \quad (7; \text{Hille, 1992})$$

that accounts for the I-V relation, in the presence of 10 mM *trans* Ba^{2+}.

Possible Participation of the cAMP-Dependent Channel in the Speract-Induced Membrane Potential Changes in Sea Urchin Sperm.

Considering the PK^+/PNa^+ ratio of the cAMP-dependent channel it is predicted, using the Goldman-Hodgkin-Katz equation, that its opening in normal sea water (ASW) would actually depolarize sea urchin sperm. Speract, a decapeptide from egg jelly, induces increases in cGMP, and cAMP (Garbers, 1989; Suzuki and Yoshino, 1992), and triggers membrane potential changes (Lee and Garbers, 1986; Babcock *et al.*, 1992 ; Reynaud *et al.*, 1993; Cook and Babcock, 1993a, b). The possibility that a K^+-selective channel, alike that identified in planar bilayers, might participate in the sperm responses triggered by this peptide was tested.

Figure 6. Speract-induced changes in sperm membrane potential and pHi; inhibition by Ba^{2+} and dependence on extracelular Na^+. Membrane potential measurements were made with the membrane potential sensitive fluorescent dye DiS-C_3-(5). Loading of BCECF for pH_i and the simultaneous recording of membrane potential and pH_i were as indicated in Methods. Pre-equilibration (3 - 4 min) with DiS-C_3-(5) (0.5 μM), the effect of the mitochondrial uncoupler (~ 40 sec) and the final calibration with valinomycin and K^+ are not shown. An upward deflection indicates a depolarization in the membrane potential traces and an alcalinization in those of pH_i. The arrow on top of all traces corresponds to the addition of 100 nM speract. **A.** Speract-induced membrane potential changes. Left trace: in normal artificial sea water (ASW) ; right trace: in 0Ca ASW (0CaASW). **B.** Antagonizing effect of Ba^{2+} on the speract-induced depolarization. Left trace: control in 0CaASW; right trace: in 10 mM external Ba^{2+}. **C.** External $[Na^+]$ dependence of the speract-induced changes in membrane potential and pH_i. The experiments were done in the presence of 300 μM $NiCl_2$. ASW with various concentrations of Na^+ was prepared replacing NaCl (mole per mole) with choline chloride, $MgSO_4$ with $MgCl_2$ and removing $NaHCO_3$. Experiments with ASW with various Na^+ concentrations gave similar results. The records on the left display the membrane potential changes induced by speract in ASW (488 mM NaCl) and in $100Na^+$ ASW (388 mM choline chloride, 100 mM NaCl ASW). The records on the right show that the speract-induced pH_i changes in spermatozoa suspended in ASW or 100NaASW are basically the same.

The membrane potential changes induced by nM speract in normal sperm suspended in ASW have not been fully characterized (see Babcock *et al.*, 1992; Cook and Babcock, 1993a, b). Fig. 6A, left trace, illustrates how nM speract induces a hyperpolarization (-7.4 ± 4 mV, n= 20) which reaches its maximum in 2.6 ± 1.2 sec (n= 20). This was followed by a fast repolarization to the resting membrane potential (-48 ± 6.5 mV; n= 9) and a depolarization (15.1 ± 4.2 mV; n= 20) which plateaus at 13.6 ± 4.7 sec (n= 20). The repolarization plus the net depolarization will be simply referred to as depolarization (22.5 ± 4.2 mV; n= 20). Speract depolarized normal sperm in the absence of external Ca^{2+} (15.5 ± 6.5 mV, n= 9, Fig. 6A, right trace) and even in the presence of Ca^{2+}-channel blockers (Ni^{2+} and Co^{2+}, 100-300 μM, see Fig. 6C). Figure 6B documents that 10 mM Ba^{2+} inhibited the speract-induced depolarization. A similar effect was observed in the presence of TEA (>30 mM, not shown). The speract-induced depolarization of normal sperm suspended in $0SO_40CO_3$ ASW was dependent on external [Na^+] (Fig. 6C). This occurred in the presence of 300 μM Ni^{2+} which blocks the Ca^{2+}-dependent depolarization in swollen sperm (Reynaud *et al.*, 1993; Cook and Babcock, 1993b). As shown in figure 6C, under these experimental conditions the peptide triggers a larger depolarization in ASW (488 mM NaCl, 39 ± 4.7 mV; n= 6) than in 100NaASW (19.5 ± 2.4 mV; n= 5, Na^+ replaced by choline). On the other hand, there is less difference in the hyperpolarization (ASW, -14.9 ± 1.1 mV; n= 5; 100NaASW, -10.6 ± 1.1 mV; n= 5) and basically no change in pH_i (n= 3; Fig 6C) at both Na^+ concentrations.

DISCUSSION AND CONCLUSIONS

This report offers evidence, for the first time, that sea urchin sperm posses K^+-selective channels modulated by cAMP. The sea urchin sperm-flagellar plasma membranes from which the K^+ channel described here is derived have been shown to be quite pure (Cross, 1983; Darszon *et al.*, 1984; García-Soto *et al.*, 1988). This channel studied in planar lipid bilayers, adds to the growing list of ion channels displaying direct modulation by cyclic nucleotides. In addition to the K^+-selective channel described here, they include, cGMP-activated, cation-selective channels from photoreceptors cells (Bonigk *et al.*, 1993), cyclic nucleotide-activated, cation-selective channels from olfactory cells (Dhallan *et al.*, 1990), a K^+-selective channel from *Drosophila* larval muscle (Delgado *et al.*, 1991), the K^+-selective channel encoded by the *eag* gene (Brüggema *et al.*, 1993), a K^+-selective channel from olfactory cilia (Jorquera *et al.*, 1995) and a cGMP-activated K^+-selective channel in bay scallop *Pecten irradian* ciliary photoreceptors (Gomez and Nassi, 1995). Recently a cation-selective channel (P_{K^+}/P_{Na^+} ~1.3) modulated by cGMP found in human sperm has been recorded and cloned (Weyand *et al.*, 1994). It has 65 % sequence homology with the photoreceptor channel, and with one cloned from aorta (Biel *et al.*, 1993). The K_D of the sea urchin sperm K^+ channel for cAMP (214 μM) was higher than that of cation selective channels in olfactory cells (3–70 μM, Dhalla *et al.*, 1990) and the K^+ channel from *Drosophila* larval muscle (54 μM, Delgado *et al.*, 1991) but, most probably, lower than the *eag* channel. In this last case, upward modulation was reported to require mM cAMP (Brügemann *et al.*, 1993). The affinity of the sea urchin K^+ channel for cAMP is also lower than that reported for cation channels from olfactory neurons (Kaupp, 1991). At the present time we do not have an explanation for this lower apparent affinity. Perhaps channel insertion into an extraneous environment modifies the channel's affinity for the cyclic nucleotide. In fact, cyclic nucleotide gated-channels from photoreceptors and olfactory neurons, expressed from cDNA in heterologous systems, display lower affinities for their agonists than the same channels *in situ.* A recently cloned subunit of cyclic nucleotide-gated channels in olfactory neurons was shown to modulate their affinity for cyclic nucleotides (Liman and Buck, 1994).

We think that cAMP acts directly on the channel since up-regulation is seen in the absence of added ATP. It is difficult to believe that the purified sea urchin sperm plasma membranes contain enough residual ATP to support a kinase dependent phosphorylation. On the other hand, since the cAMP-induced changes in single channel behavior are reversible in the absence of ATP and dose-dependent, a cAMP-dependent phosphatase can be ruled out. The experimentally observed upward modulation by cAMP of the K^+ channel from sea urchin sperm could be accounted for by a simple linear kinetic model. The channel displayed one open and two closed states. Channel open times were independent of [cAMP] while closed times varied linearly with the reciprocal [cAMP].

The sperm K^+ channel exhibited a $P_{K^+}/P_{Na^+} = 4.9$, lower than the cAMP-modulated K^+ channel from *Drosophila* larval muscle and the *eag* K^+ channel. More so, the sea urchin sperm K^+ channel displayed a selectivity sequence $K^+ > Rb^+ > Na^+ > Cs^+$, different to that of the *eag* channel ($K^+ > Rb^+ > Cs^+ > Na^+$ (Brüggemann *et al.*, 1993). It will be interesting to determine the selectivity sequence for other cyclic nucleotide-modulated K^+ channels to examine if cyclic nucleotide modulation of K^+ channels might not be restricted to a particular K^+ channel class, defined in terms of selectivity sequences. Apparently Ca^{2+} permeates through the *eag* K^+ channel expressed in *Xenopus laevis* oocytes (Brüggemann et al., 1993). We have not seen divalent cation permeation through the sea urchin sperm, cAMP modulated K^+ channel. Moreover, we have shown that *trans* Ba^{2+} blocked this channel in a voltage-dependent fashion. Ca^{2+} in *trans* (100 mM) also caused a significant reduction in channel open currents at zero and negative applied-voltages. No sign of a reversal of open currents was seen up to -100 mV applied-voltage thus, Ca^{2+} does not seem to permeate the K^+ channel from sea urchin sperm.

The opening probability of the sea urchin sperm K^+ channel in the absence of cyclic nucleotide, was below 1% at zero-applied voltage. This probability increased ~10 fold at 200 μM cAMP. Close to the resting sperm membrane potential (E_R = -60 mV), the opening probability without cAMP was 2 %. Thus under these conditions, an upper limit for channel opening probability in the presence of 200 μM cAMP can be estimated at 10-20 %. Activation of 1 or 2 cAMP regulated K^+ channels in the sperm plasma membrane, with a conductance of 100 pS and opening probability of 10-20 %, would add significantly to the sperm conductance, estimated to be ~10 pS (Guerrero *et al.*, 1987).

Sea water contains 480 mM Na^+ and 10 mM K^+, while sperm intracellular $[K^+]$ is ~180 mM (Babcock *et al.*, 1992). Given that the channel described here has a P_{K^+}/P_{Na^+} = 4.9, its reversal potential in ASW can be estimated at -13 mV. This value is at least 30 mV more positive than the sea urchin sperm resting membrane potential in sea water (-48 ± 6.5 mV; n=9, this work and Reynaud *et al.*, 1993). Thus, because of the salt composition of sea water and the permeability properties of this K^+ channel, its opening would actually depolarize sea urchin sperm.

Sperm activation by speract, a decapeptide from egg-jelly or during the AR, involves increases in the cAMP levels of up to 300 μM (Ward and Kopf, 1993; Garbers, 1989; and C. Beltrán and A. Darszon, unpublished). The observations discussed above, derived from bilayer studies, together with the fact that speract can cause increases in cAMP, Na^+ uptake and a depolarization (Schackmann, and Chock, 1986, Garbers, 1989, Labarca *et al.*, 1995), led us to explore the possible participation of cAMP-regulated K^+ channels in the depolarization triggered by nM speract in sperm. We found that sperm in sea water (480 mM Na^+) depolarize upon exposure to speract, even in the absence of external Ca^{2+} (15 mV), suggesting that Na^+ contributes to this membrane potential change. Indeed, the speract induced depolarization decreases when external Na^+ is replaced by choline$^+$, in a $[Na^+]$ range where pHi is not affected. Therefore, Na^+ directly and not through its effect on Na^+/H^+ exchange affects the speract induced depolarization.

In swollen sperm, where cAMP has been implied in the speract-induced depolarization mediated by Ca^{2+} channels (Cook and Babcock, 1993b), this depolarization depends on external Ca^{2+}, and is completely blocked by 10 μM Ni^{2+} (Reynaud *et al.*, 1993; Cook and Babcock, 1993b). In normal sperm >100 μM Ni^{2+} did not significantly alter the speract response in 0CaASW, but was antagonized by K^+-channel blockers (TEA^+ and Ba^+). Altogether, these observations suggest that in normal sperm two different ion channels, a Ca^{2+}-selective one and a channel permeable to Na^+ and sensitive to TEA^+ and Ba^{2+}, but refractory to Ca^{2+} channel blockers, might operate during the depolarization induced by speract. This latter channel could be of the type identified here in planar bilayers, a cAMP-regulated K^+ channel with a PK^+/PNa^+ of 4.9 which is blocked by TEA^+ and Ba^{2+}.

ACKNOWLEDGMENTS: This work was partially supported by CONACyT, DGAPA -UNAM. and Fundacion Andes. A.D. is an International Scholar of the Howard Hughes Medical Institute. P. L. held a Catedra Patrimonial from CONACyT, a Catedra Presidencial en ciencias and a John Guggenheim Fellowship.

REFERENCES

Babcock, D. F., Bosma, M. M., Battaglia, D. E. and Darszon, A., 1992. Early perrsistent activation of sperm K^+ channels by the egg peptide speract. *Proc. Natl. Acad. Sci. USA* **89:**6001.

Bezanilla, F., 1985, A high capacity data recording device based on a digital audio processor and a video cassete recorder. *Biophys. J.* **47:**437.

Biel, M., Altenhofen, W., Hullin, R., Ludwig, J., Freichel, M., Flockerzi, N., Dascal, N., Kaupp, U. B. and Hofmann, F., 1993, Primary structure and functional expression of a cyclic nucleotide-gated channel from rabbit aorta. *FEBS lett.* **329:**134.

Bonigk, W., Altenhofen, W., Muller, F., Dose, A., Illing, M., Molday, R. S. and Kaupp, U. B., 1993, Rod and cone photoreceptor cells express distinct genes for cGMP-gated channels. *Neuron* **10:**865.

Brown, A. M., 1993, Membrane-delimited cell signaling complexes: direct ion channel regulation by G proteins. *J. Membr. Biol.* **131:**93.

Brüggemann, A., Pardo, L. A., Stühmer, W. and Pongs, O., 1993, Ether-á-go-go encodes a voltage-gated channel permeable to K^+ and Ca^{2+} and modulated by cAMP. *Nature* **365:**445.

Cook, S. and Babcock, D. P., 1993a, Selective modulation by cGMP of the K^+ channel activated by speract. *J. Biol. Chem.* **268:**22402.

Cook, S. and Babcock, D. P., 1993b, Activation of Ca^{2+} permeability by cAMP is coordinated through the pH_i increase induced by speract. *J. Biol. Chem.* **268:**22408.

Cook, S. P., Brokaw, C. J., Muller, C. H., and Babcock, D. F., 1994, Sperm chemotaxis: egg peptides control cytosolic calcium to regulate flagellar responses. *Dev. Biol.* **165:**10-19.

Cross, N. L., 1983, Isolation and electrophoretic characterization of the plasma membrane of sea urchin sperm. *J. Cell Sci.* **59:**13.

Dangot, L. J., Jordan, J. E., Bellet, R. A. and Garbers, D. L., 1989, Cloning of t the mRNA of the protein that crosslinks to the egg peptide speract. *Proc. Natl. Acad. Sci. USA* **86:**2128.

Darszon, A., Gould-Somero, M., de De La Torre, L. and Vargas, I., 1984, Response of isolated sperm plasma membranes from sea urchin to egg jelly. *Eur. J. Biochem.* **144:**515.

Darszon, A., Labarca, P., Beltrán C., García-Soto, J., and Liévano, A., 1994, Sea urchin sperm: An ion channel reconstitution study case. *Methods: A Companion to Meth. Enzym.* **6:**37.

Delgado, R., Hidalgo, P., Diaz, F., Latorre, R. and Labarca, P., 1991, A cyclic AMP-activated K^+ channel in *Drosophila* larval muscle is persistently activated in dunce. *Proc. Natl. Acad. Sci. USA* **88:**557.

Dhallan, R. S., Yau, K. W., Schrader, K. A. and Reed, R. R., 1990, Primary strucure and functional expression of a cyclic nucleotide-activated channel from olfactory neurons. *Nature* **347:**184.

Garbers, D. L., 1989, Molecular basis of fertilization. *Annu. Rev. Biochem.* **58:**719.

Garbers, D. L., Watkins, H. D., Hansbrough, J. R., Smith, A., and Misono, K. S., 1982, The amino acid sequence and chemical synthesis of speract and of speract analogs. *J. Biol. Chem.* **257:**2734.

García-Soto, J., Mourelle, M. I., Vargas, I., de De la Torre, L., Ramírez, E., López-Colomé, A. M. and Darszon, A., 1988, Sea urchin sperm head plasma membranes: characteristics and egg-jelly induced Ca^{2+} and Na^+uptake. *Biochim. Biophys. Acta* **944:**1.

Gomez, M. P. and Nassi. E., 1995, Involvement of cyclic GMP in transduction in hyperpolarizing invertebrate photoreceptors. *Biophys. J.* **68**:A19.

González-Martínez, M. T., Guerrero, A., Morales, E., de De La Torre, L. and Darszon, A., 1992, A depolarization can trigger Ca^{2+} uptake and the acrosome reaction when preceded by a hyperpolarization in *L. Pictus* sea urchin sperm. *Dev. Biol.* **150:**193.

Guerrero, A., Sánchez, J. A. and Darszon, A., 1987, Single-channel activity in sea urchin sperm revealed by the patch-clamp technique. *FEBS Lett.* **220:**295.

Guerrero, A., and Darszon, A., 1989, Evidence for the activation of two different Ca^{2+} channels during the egg jelly induced acrosome reaction of sea urchin sperm. *J. Biol. Chem.* **264**:19593.

Hansbrough, J. R., and Garbers, D. L., 1981, Speract. Purification and characterization of a peptide associated with eggs that activate spermatozoa. *J. Biol. Chem.* **256:**1447.

Hille, B., 1992, "Ion channels of excitable membranes," 2nd. edition. Sinauer Associates Inc., Sunderland MA.

Jorquera, O., Latorre, R., Labarca, P., 1995, Ion channel classes in purified olfactory cilia: planar lipid bilayer studies. *Am. J. Physiol.* (in press).

Kaupp, B. U., 1991, The cyclic nucleotide-gated channels of vertebrate receptors and olfactory epithelium. *Trends in Neurosciences* **14**:150.

Labarca, P., Santi, C., Morales, E., Zapata, O., Beltrán, C., Liévano, A. and Darszon, A., 1995, Possible participation of a cAMP regulated K^+ channedetected in planar bilayers in the speract-induced repolarization of sea urchin spermatozoa. *Biophys. J.* **68:**A224.

Latorre, R., Bacigalupo, J., Delgado, R., and Labarca, P., 1991, Four cases of direct ion channel gating by cyclic nucleotides. *J. Bioenerg. Biomemb.* **23:**577.

Latorre, R. and Miller, C., 1983, Conduction and selectivity in potassium channels. *J. Membr. Biol.* **71:**11.

Lee, H. C. and Garbers, D. L., 1986, Modulation of the voltage sensitive Na^+/H^+ exchange in sea urchin spermatozoa through membrane potential changes induced by the egg peptide speract. *J. Biol. Chem.* **261:**16026.

Liman, E-R. and Buck, L. B., 1994, A second subunit of the olfactory cyclic nucleotide-gated channel confers high sensitivity to cAMP. *Neuron* **13**:611.

Moczydlowski, E., 1985, Single-channel enzymology, *in,* "Ion Channel Reconstitution," C. Miller, ed., pp.75-114. Plenum press, New York, .

Müeller, P. and Rudin, D. O., 1969, Bimolecular lipid membranes: techniques of formation, study of electrical properties, and induction of gating phenomena. *In* "Laboratory Techniques of Membrane Biophysics," M. Passow and R. Stampfli, eds., pp. 141-156. Springer-Verlag, Berlin.

Reynaud, E., de De La Torre, L., Zapata, O., Liévano, A. and Darszon, A., 1993, Ionic bases of the membrane potential changes induced by speract in swollen sea urchin sperm. *FEBS lett.* **329:**210.

Rudy, B., 1988, Diversity and ubiquity of K channels. *Neuroscience* **25:**729.

Schackmann, R.W., 1989, Ionic regulation of the sea urchin sperm acrosome reaction and stimulation by egg-derived peptides, *in* "The cell biology of fertilization," Schatten, H. and Schatten, G., eds.., pp. 3-28. Academic Press, San Diego, CA.

Schackmann, R. W., and Chock, P. B.,1986, Alteration of intracellular $[Ca^{2+}]$ in sea urchin sperm by the egg peptide speract. *J. Biol. Chem.* **261:**8719.

Shimizu, T., Kinoh, H., Yamaguchi, M. and Suzuki, N., 1990, Purification and characterization of the egg jelly macromolecules, sialoglycoprotein and fucose sulfate glycoconjugate, of the sea urchin *Hemicentrotus pulcherrimus. Dev. Growth Differ.* **32:**473.

Suzuki, N., Shimomura, H., Radany, E. W., Ramarao, C. S., Ward, G. E., Bentley, J. K., and Garbers, D. L., 1984, A peptide associated with eggs causes a mobility shift in a major plasma membrane protein of spermatozoa. *J. Biol. Chem.* **10:**14874.

Suzuki, N. and Garbers, D. L., 1984, Stimulation of sperm respiration rates by speract and resact at alkaline extracellular pH. *Biol. reprod.* **30:**1167.

Suzuki, N., Nomura, K., Ohtake, H., and Isaka, S., 1981, Purification and primary structure of sperm-activating peptides from the jelly coat of sea urchin eggs. *Biochem. Biophys. Res. Comun.* **99:**1238.

Suzuki, N., Shimomura, H., Radany, E. W., Ramarao, C. S., Ward, G. E., Bentley, J. K., and Garbers, D., 1984, A peptide associated with eggs causes a mobility shift in a major plasma membrane in a major plasma membrane protein of spermatozoa. *J. Biol. Chem.* **259**:14874.

Suzuki, N. and Yoshino, K. I., 1992, The relationship between amino acid sequences of sperm-activating peptides and the taxonomy of equinoids. *Comp. Biochem. Physiol.* **102B:**679.

Ward, G. E., Brokaw, C. J., Garbers, D. L., and Vacquier, V. D., 1985, Chemotaxis of *Arbacia punctulata* spermatozoa to resact, a peptide from the egg jelly layer. *J. Cell. Biol.* **101**:2324.

Ward, C. R. and Kopf, G., 1993,. Molecular events mediating sperm activation. *Devel. Biol.* **158:**9.

Weyand, I., Godde, M., Frings, S., Weiner, J., Müller, F., Altenhofen, W., Hatt, H. and Kaupp, B., 1994, Cloning and functional expression of a cyclic-nucleotide-gated channel from mammalian sperm. *Nature* **368:**859.

Yamaguchi, M., Niwa, T., Kurita, M. and Suzuki, N. 1988, The participation of speract in the acrosome reaction of *Hemocentrotus pulcherrimus* spermatozoa by the egg jelly molecule, fucose-rich glycoconjugate and sperm-activating peptide I. *Dev. Growth Differ.* **30:**159.

Zufall, F., Firenstein, S. and Shepherd, G. M., 1994, Cyclic nucleotide-gated ion channels and sensory transduction in olfactory receptor neurons. *Annu. Rev. Biophys. Biomol. Struct.* **23:**577.

GATING AND SELECTIVITY PROPERTIES OF A cAMP-ACTIVATED K^+ SELECTIVE CHANNEL FROM *Drosophila* LARVAL MUSCLE

Pedro Labarca, Ricardo Delgado, Orlando Jorquera, and Carlos Maureira

Centro de Estudios Cientificos de Santiago,

Departamento de Biologia, Facultad de Ciencias, Universidad de Chile.

Correspondence to: Dr. Pedro Labarca, Centro de Estudios Científicos de Santiago, Casilla 16441, Santiago 9, Chile. Tel. # 562-233-8342. Fax # 562-233-8336. Email: plabarca@cecs.cl

I. INTRODUCTION

For a while, direct modulation of ion channels by cyclic nucleotides was thought to be confined to those cation-selective ones found in photoreceptors and in vertebrate olfactory neurons (Kaupp, 1991; Zufall, 1994). This idea was fortified by structural and functional similarities exhibited by cyclic nucleotide-gated channels in olfactory neurons with those of photoreceptor cells. The presence of a ciliary axonema in photoreceptors cells pointed out to a unique evolutionary relationship between visual and olfactor y sensory cells. However, cyclic nucleotide activated cation-selective channels have been reported now in a variety of animal cells, including retinal ganglion cells, aorta, testis, kidney, heart and sperm (Yau, 1994). Furthermore, there is increasing evidence for the presence in animals of K^+-selective channels that are directly gated by cyclic nucleotides (Delgado et al., 1991; Michael and Asche, 1992; Bruggemann et al., 1993; Gotow, et al., 1994; Gomez and Nassi, 1995; Jorquera et al., 1995; Labarca et al., 1995).

Cyclic nucleotide-activated, cation-selective channels in photoreceptors, olfactory neurons, and other cell types, have been cloned and expressed in heterologous systems, allowing for the establishment of structure-function correlates (see, for example: Biel et al., 1994; Goulding et al., 1994; Weyand et al., 1994; Gordon and Zagotta, 1995). In contrast, to the best of our knowledge, the only K^+-selective channel featuring direct modulation by cyclic nucleotides that has been cloned is the *eag* K^+ channel, from *Drosophila* (Warmke et al., 1991). *eag* belongs to the S4 superfamily of channels and posseses domains with a high degree of homology to cyclic nucleotide-binding proteins (Guy et al., 1991).

Currently, most of our knowledge on K^+-selective channels directly activated by cyclic nucleotides derives from electrophysiological studies (Delgado et al., 1991; Michael and Asche, 1992; Gotow, et al., 1994; Gomez and Nassi, 1994, 1995; Jorquera et al., 1995; Labarca et al., 1996). The first evidence for direct modulation of K^+ channels by cyclic nucleotides was obtained in *Drosophila* larval muscle (Delgado et al., 1991). Here, gating and selectivity properties of this cAMP-activated K^+-selective channel are reviewed, and compared to those of cyclic nucleotide-gated K^+ channels found in other animal cells.

II. GATING PROPERTIES

cAMP Activation. In excised patches, the cAMP-activated K^+ selective channel from *Drosophila* larval muscle exhibits, at rest, a small, albeit sizeable, probability of opening (P_o) with a weak, if any, dependence on the applied voltage (Fig. 1A). Exposure of its cytoplasmic face to micromolar amounts of cAMP, via chamber perfusion, causes a clear increase in P_o(Fig. 1B). Activation is dose-dependent and the sigmoid P_o vs [cAMP] curve can be well accounted for in terms of a Hill equation with an apparent dissociation constant of 60 μM, and an exponent of ~3 (Fig. 1B).

Sigmoidal dose-dependences, with a similar Hill coefficient, have been reported in cyclic nucleotide-gated, cation-selective channels (Kaupp, 1991; Biel et al., 1994; Weyand et al., 1994; Zufall et al., 1994;) and in a K^+-selective channel, derived from olfactory cilia (Jorquera et al., 1995). In contrast, activation of a K^+-selective channel from sea urchin sperm membranes, monitored in planar lipid bilayers, was reported to be Michaelian (Labarca et al., 1995).

The apparent affinity for its agonist of the K^+-selective channel in *Drosophila* larval muscle is lower than that reported for cation-selective channels in photoreceptors and olfactory cells (Kaupp, 1991; Zufall, 1994), similar to that of a K^+ channel derived from olfactory cilia (Jorquera et al., 1995), but higher than that reported for a cAMP-activated K^+-selective channel from sea urchin sperm (Labarca et al., 1996). We lack information on the dose-dependent activation properties of other K^+-selective channels activated directly by cyclic nucleotides.

Activation by cAMP is reversed upon perfusion of the experimental chamber with a cAMP-free solution (Fig. 1B). The reversibility of activation, added to the smooth dose-dependent increase in P_o as a function of [cAMP], indicates that activation is direct, rather than through a covalent modification of the channel, via phosphorylation. Indeed, the increase in P_o induced by cAMP in excised patches is observed in the absence of ATP.

Exposure of larval muscle to 200 μM of the permeable, slowly hydrolizing agonist analog 8-BrcAMP, allows to obtain records of this K^+ channel in cell-attached patches, under conditions of elevated intracellular cAMP. Under such

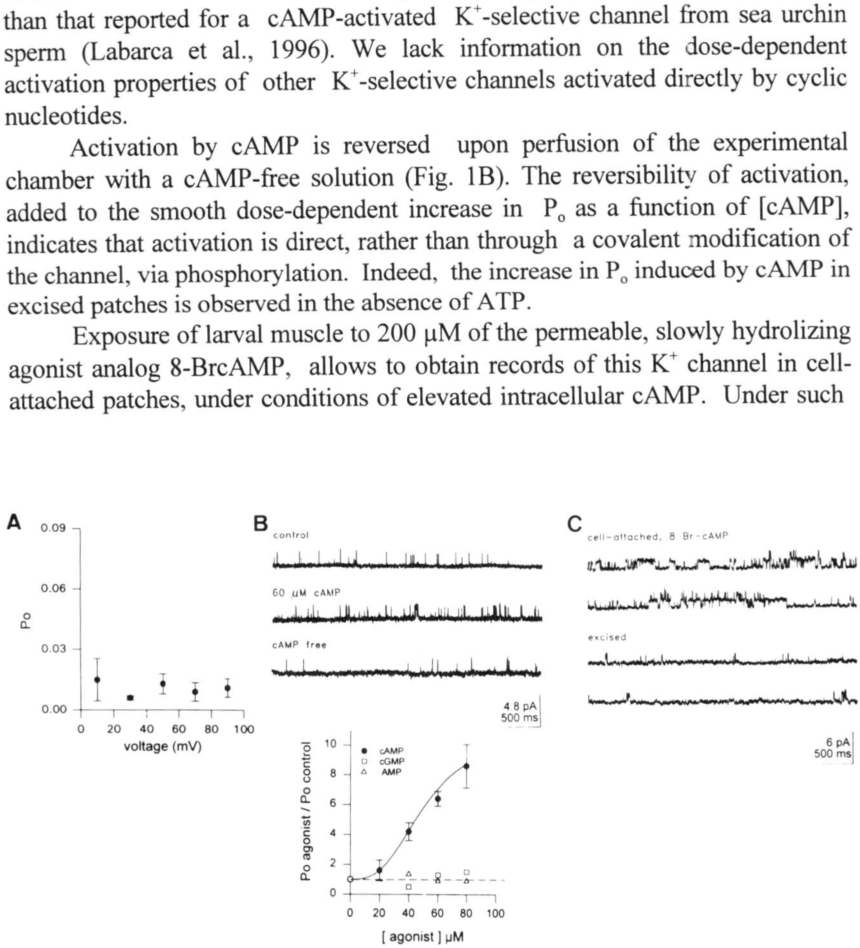

Figure 1. Activation by cAMP of a K+-selective channel from *Drosophila* larval muscle. **A.** Voltage-independence of channel open probability (P_o). **B.** cAMP activation in excised, inside-out patches. Upper record: control; middle record: after patch exposure to 60 μM cAMP; lower record: after perfusion with cAMP-free solution. **C.** Upper record: single channel current activity recorded in cell-attached patch, after muscle exposure to 200 μM 8-BrcAMP; lower record: channel activity after excising the patch into a cAMP-free solution. Insert: P_o vs [cAMP]. The continuous line is best fit to Hille's Eq. With n=3, and K_d (apparent dissociation constant)= 60 μM,

conditions, the channel exhibits an elevated P_o. Excising the patch into a cAMP-free solution causes an inmediate diminishment in channel activity (Fig. 1C), providing further support to the idea that cAMP activates this channel by direct binding. Channel activation could not be obtained in the presence of other potential agonists like cGMP, AMP or ATP, up to 200 μM (Delgado et al., 1991; Fig. 1B).

Distribution of open dwell times. The analysis of the distribution of open dwell times in the absence and in the presence of cAMP, under steady-state conditions, allowed to obtain information on the mechanisms by which the cyclic nucleotide causes increases in open probability. Under control conditions, in the absence of agonist, open dwell times distributed along two exponential components, a fast one and a slow component with time constants of 1.1 ± 0.10 ms (n=4) and 4.9 ± 1.3 ms (n=4), (Fig. 2A). The relative amplitude of the fast component amounted to 0.71 ± 0.10 (n=4), Delgado et al., 1995.

After patch exposure to cAMP, the distribution of open dwell times was fitted also by two exponentials components (Fig. 2A). Interestingly, neither the

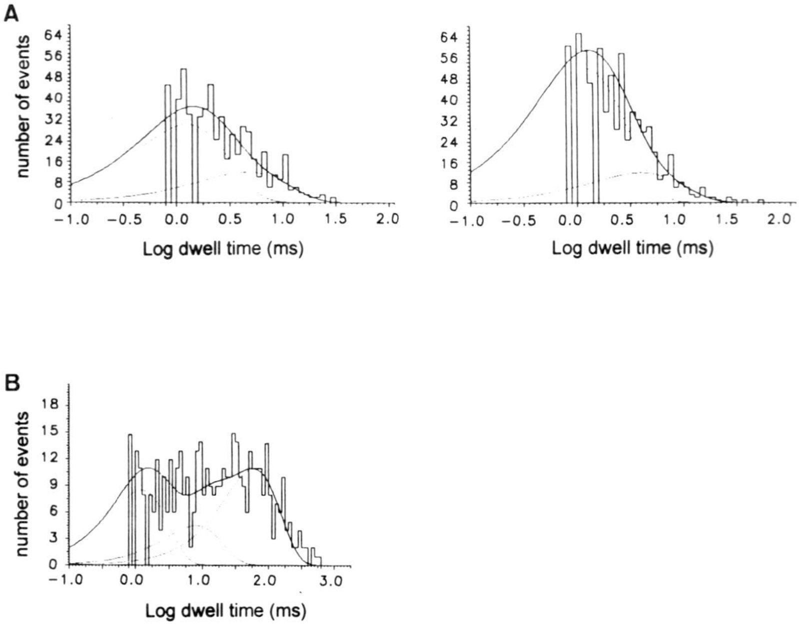

Figure 2. Open and closed dwell times distribution. **A**. Left panel: open dwell times distribution in an excised patch, under control conditions. Right panel: open dwell times distribution (same patch), in the presence of 80 μM cAMP. **B**. Closed times distribution (same patch), in the presence of 80 μM cAMP

time constants nor the relative amplitudes of the two components of open times were affected by the agonist (Table I). In other words, although the probability of channel opening increased significantly as a function of [cAMP], open times durations were independent of [cAMP] and the relative frequency of each component of open dwell times remained constant. The analysis of experimental records allowed also to assign a minimun number of components to the distribution of closed times. As shown in Fig. 2B, closed durations could be fitted by three exponentials, with time constants of 1.2, 7.4 and 60 ms.

The fact that channel open times were independent of agonist concentration discards the notion that the increase in channel open probability induced by cAMP involves agonist binding to open channel conformations, as would be the case for the model shown in scheme 1.

$$\text{ncAMP} \qquad C \rightleftarrows O_f \rightleftarrows O_s \qquad (1)$$

In this model, agonist binding to the O_f open state would cause the fast component of the distribution of open times to depend on [cAMP]. In addition, this model demands also that the relative amplitudes of open dwell times be a function of [cAMP].

On the other hand, the results derived from the analysis of open dwell times are compatible with kinetic models in which agonist binds to closed channel states (Lecar, 1986). Two examples, which are variants of Del Castillo and Katz's (1957) model for the activation of acetylcholine receptors, are :

$$\text{ncAMP} \qquad C_1 \rightleftarrows C_2 \rightleftarrows C_3 \rightleftarrows O_f \rightleftarrows O_s \; (2)$$

$$\begin{array}{ccccc} \text{ncAMP} & & & & \\ C_1 & \rightleftarrows & C_2 & \rightleftarrows & C_3 \\ & & \downarrow\uparrow & & \downarrow\uparrow \\ & & O_f & & O_s \end{array} \qquad (3)$$

Both schemes predict the occurrence of two open states with constant relative amplitudes, and whose probabilitites increase with [cAMP]. Since, as documented in Fig. 1B, channel opening occurs in the absence of agonist, it cannot be argued that cAMP binding makes acessible a conformation which is untenable in its absence. Therefore, the assumption must be made that cAMP binding stabilizes the C_2 conformation from which C_3 and both open states can be reached, via thermal transitions.

Table I. Open dwell time analysis of cAMP-activated, K^+-selective channel. Single-channel currents were recorded in excised, inside-out patches. A_1 and τ_1 are average amplitudes and time constants of fast component of open dwell times; A2 and $\tau 2$ are average amplitudes and time constants of slow component of open dwell times. (Modified from Delgado et al., 1995).

	A_1	τ_1	A_2	τ_2
Control	0.71 ± 0.1	1.1 ± 0.1	0.29 ± 0.1	4.9 ± 1.3
40 μM cAMP	0.75 ± 0.06	1.0 ± 0.2	0.25 ± 0.06	4.9 ± 1.0
80 μM cAMP	0.73 ± 0.08	1.4 ± 0.6	0.27 ± 0.08	3.8 ± 0.4

Open time correlation analysis. Additional information on the gating properties of agonist-gated ion channels can be obtained by resorting to correlation analysis (Jackson et al, 1983; Labarca et al., 1985; Colquhoun and Hawkes, 1995). As shown in Fig. 1B, in the presence of cAMP channel activity tends to occur in bursts, a feature common to many agonist-gated ion channels (Colquhoun and Hawkes, 1995). This pattern of activity makes it tempting to investigate whether or not successive openings are correlated in duration, as reported for the nicotinic receptor (Jackson et al., 1983; Labarca et al., 1985). In fact, in a model like that depicted in scheme (3) one would expect, in principle, to observe correlation in open times, in the sense that openings to O_f could tend to be followed by openings to this same open state. Likewise, visits to O_s would tend to be followed by reopenings to this conformation. Thus, if the cAMP-activated channel under scrutiny here gates through a mechanism similar to the one in scheme (3), one would expect to observe that successive openings are correlated in open durations (Jackson et al., 1983; Labarca et al., 1985; Colqhoun and Hawkes, 1995). In contrast, scheme (2) predicts, rigurously, a lack of correlation (Labarca et al., 1985; Colqhoun and Hawkes, 1995).

The presence of a correlation in open durations can be tested most simply by resorting to a binomial analysis. In such an analysis, the occurrence of a random, uncorrelated pattern of openings, (expected for scheme (2)), would be revealed by the fact that the frequency of occurrence of successive O_f-O_f (fast-fast), O_M (mixed) and O_s-O_s (slow-slow) pairs experimentally observed would obey the binomial distribution.

To carry on the analysis it becomes necessary to establish a criteria to distinguish openings of the type O_f from those of the O_s type. A simple criteria is to choose a time t' at which both open dwell time distributions are equal (Jackson et al., 1983):

$$A_f/t_f \exp\text{-}t'/t_f = A_s/t_s \exp\text{-}t'/t_s \quad (4)$$

At t' the overlap between both components of the distribution of open dwell times is at a minimun, and any event shorter than t' has a better chance of belonging to the fast component of the distribution of open dwell times.

Table II exhibits the results obtained from the analysis of pair of openings, separated by a brief closing gap, in two different patches. The left column (observed) indicates the frequency of O_f-O_f, mixed and O_s-O_s pairs of successive openings obtained directly from the analysis of experimental records. The right column displays the frequencies predicted by the binomial formula. In both examples, the observed frequency of alike pairs is above the binomial values and, as seen, there are too few mixed pairs. The deviation from the predicted values is significant, leading to conclude that there is a correlation in open durations in successive openings. This result allow us to discard class (2) kinetic models, but is compatible with mechanisms similar to that depicted in scheme (3).

<u>Activation by cAMP is transient</u>. As shown in Fig. 1B, channel activation, obtained by chamber perfusion with cAMP, is robust. Nevertheless, the maximal probability of opening obtained in the steady state, many seconds after exposing the patch to the agonist, amounts to only 6-10%. On the other hand, as shown in Fig. 3, fast application of the agonist, using a picospitzer, reveals that channel activation is much higher soon after cAMP exposure than under steady state. In Fig. 3 the arrow signals the moment at which the release of agonist was begun. At the peak of activity, channel open probability can be as high as 40%, some 40 fold above basal levels, as compared to the 6-10% value obtained under steady

Table II. Open time correlation in cAMP-activated, K^+-selective channels. Succesive openings, separated by brief closing gaps (<2msec) were analyzed. Fast and slow openings were classified as explained in text. Expected frequencies were estimated from the binomial distribution. (Modified from Delgado et al., 1995).

	Observed	Expected
Expt. 1		
OF-0F	34	29
OM	15	26
OS-OS	11	6
X^2= 9.7 (0.01<P)		
Expt. 2		
OF-0F	37	32
OM	13	23
OS-OS	9	4
X^2= 11.4 (0.01<P)		

state, many seconds after stimulation. The pattern of activation shown in Fig. 3 is worth some comments. First, it is noticeable that the peak of channel activity is delayed up to a second with respect to the application of the stimulus, revealing that activation might not be diffusion limited. Indeed, diffusion of cAMP to the membrane patch, from a pipette placed 20-40 mm away, takes of the order of one milisecond.

A second feature of channel behavior is evidenced also in Fig. 3. As seen, channel activity reaches a maximun and then decays. In the steady state, several seconds after exposure to the agonist, the activity is much lower than at the peak of activation. Thus, channel activation by cAMP is transient. This behavior, a reproducible feature of this cAMP-modulated K^+-selective channel, ressembles the desensitization of neurotransmitter receptors (Katz and Thesleff, 1957). In contrast, cGMP-gated K^+-selective channels in hyperpolarizing photoreceptors do not exhibit this desensitization-like phenomena (Gomez and Nassi, 1995).

III. SELECTIVITY PROPERTIES.

According to Bruggemann et al. (1993), one unusual feature of the only K^+-selective channel displaying direct modulation by cyclic nucleotide which has been cloned to date, the *eag* channel, is its ability to be permeated by Ca^{2+} (PCa/PK ~ 0.1. In addition, the *eag* channel was reported to follow the selectivity sequence $P_K > P_{Rb} > P_{Cs} > P_{NH4} > P_{Na}$ for monovalent cations. The high permeability to Ca^{2+} ions and the selectivity sequence for monovalent cations reported by Bruggeman and colaborators (1993) for the *eag* channel provides, in the absence of other information, a fingerprint to compare it with other K^+-selective channels modulated directly by cyclic nucleotides.

Figure 4A displays records of cAMP-activated channels from *Drosophila* larval muscle in excised patches at different applied voltages under biionic conditions with K^+ in the pipette and K+, Rb+ or Na+ in the external medium. The inset to Figure 4A offers current-voltage relations, over a range of applied

Figure 3. Pattern of channel activity in an inside-out patch under fast stimulation with cAMP (100 μM) by means of a picospitzer (Agonist-containing pipette 30-50 μm from patch). (Modified from Delgado et al., 1995).

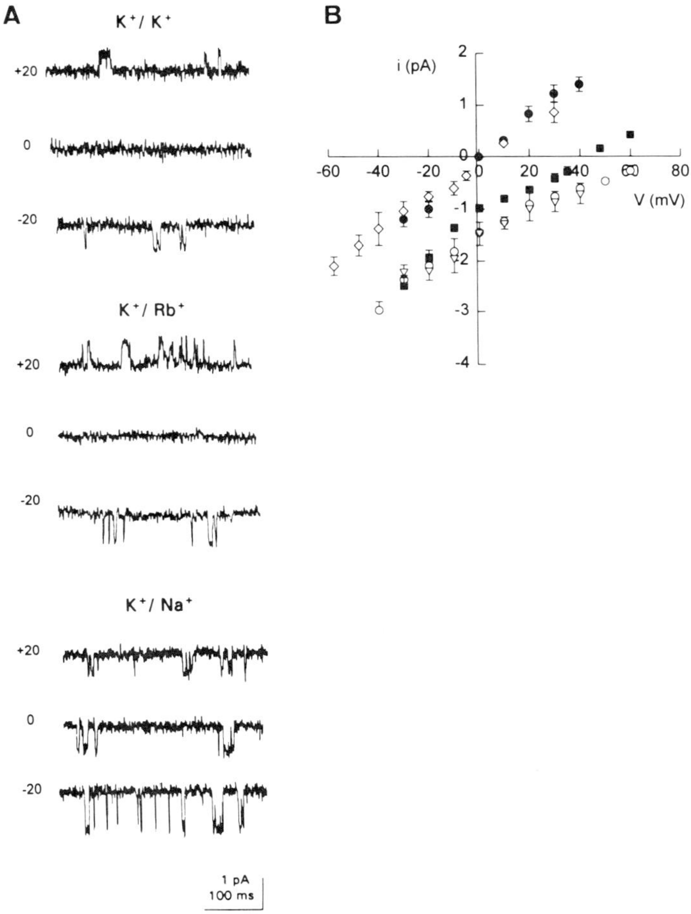

Figure 4. Ion selectivity of cAMP-activated, K^+-selective channel from *Drosophila* larval muscle estimated from permeability ratios. Single channel currents were obtained in excised, inside/out patches, under the indicated cationic conditions, with 60 μM cAMP in external solution. **A**. Single channel currents under biionic conditions, with 115 mM K^+ in pipette/115 mM X^+ in external solution. **B**. Current-voltage relation curves under biionic conditions. Symbols as follows: filled circles: K^+/K^+; diamonds: K^+/Rb^+ ; filled squares: $K^+/NH4^+$; open circles: K^+/Na^+; inverted triangles: K^+/Cs^+. (Modified from Delgado et al., 1995).

voltages, for different monovalent cation pairs. From reversal potentials, the selectivity sequence $P_K > P_{Rb} > P_{NH4} >> P_{Cs}$, P_{Na} was obtained, which differs from that reported for the *eag* channel. On the other hand, this selectivity sequence is similar to that reported for a cAMP-activated channel from sea-urchin sperm (Labarca et al., 1996).

The channel from *Drosophila* larval muscle was found to be essentially impermeable to Ca^{2+} ($P_{Ca}/P_K < 0.02$; Delgado et al., 1995). Other K^+-selective channels activated directly by cyclic nucleotides have been found also to be impermeable to divalent cations or to be blocked by them. For example, cGMP-gated K^+ channels in ciliated photoreceptors of the bay scallop were found to be blocked by Ca^{2+} (Gomez and Nassi, 1995). A K^+ channel activated by cAMP, from sea urchin sperm, was reported to be impermeable to Ca^{2+} and blocked by Ba^{2+} in a voltage-dependent fashion (Labarca et al., 1995).

IV. CONCLUSIONS

Evidence that animals posses K^+-selective channels directly modulated by cyclic nucleotides can be considered conclusive. This evidence invites much needed efforts addressed to achieve the cloning of these channels, and their subsequent expression in heterologous systems. The variety of cells in which these channels have been monitored and the fact that they would seem to be present in invertebrates and vertebrates posses a variety of questions that can be answered only by acceding to the necessary molecular resolution, which currently is unavailable.

With the exception of a channel derived from toad olfactory cilia (Jorquera et al., 1995), most studies of K^+-selective channels directly modulated by cyclic nucleotides have been performed in invertebrates (Delgado et al., 1991; Michel and Ache, 1992; Gomez and Nassi, 1995; Labarca et al., 1996). Among such channels, and to the best of our knowledge, the only one to be cloned and expressed in an heterologous system is the *eag* channel from *Drosophila* (Bruggemann et al., 1993; Robertson et al., 1993). There is evidence that *eag* homologs are present in mouse and human (Warmke and Ganetzky, 1993). The *eag* channel belongs to the S4 superfamily of channels. It contains domains exhibiting a high degree of homology to cyclic nucleotide-binding proteins (Guy et al., 1991). Currently, we lack structural information on most cyclic nucleotide-modulated K^+-selective channels. Thus, at this point, it is not possible to discuss structure-function correlates among this new class of chemically-gated channels, nor to compare them with cation-selective cyclic-nucleotide-gated channels. Indeed, this last channel family has been found to exhibit structural similarities to voltage-gated K^+ channels (Kaupp, 1991; Guy et al., 1991). At this point, the evidence available allows only to speculate that, perhaps, K^+-selective, cyclic

Table III. Properties of cyclic nucleotide, K^+ selective channels.

Channel origin	Agonist	$K_{dissociation}$	Voltage dependence	Selectivity sequence	Ca^{2+}-Permeable	Blockers	Reference
Drosophila larval muscle	cAMP	50 μM	no	$K^+>Rb^+>NH_4^+>Cs^+,Na^+$	no	TEA	Delgado et. al.,1991
Lobster olfactory cells	cAMP, cGMP	(activated by μM cAMP or cGMP)	-	-	-	-	Michel and Ache, 1992
eag channel from *Drosophila*	cAMP	(activated by μM cAMP)	yes	$K^+>Rb^+>Cs^+>NH_4^+>Na^+$	$P_{ca}/P_K= 0.1$	-	Bruggemann et. al., 1992
bay scallop hyperpolarizing fotoreceptors	cGMP	(activated at 50 μM cGMP)	no	-	no	Ca^{2+},Mg^{2+},TEA ,4-AP	Gomez and Nasi,1995
toad olfactory neurons	cAMP	-	no	-	-	-	Jorquera et. al.,1995
sea urchin sperm	cAMP	200 μM	no	$K^+>Rb^+>Na^+>Cs^+$	no	TEA,Br^{2+},Ca^{2+}	Labarca et. al., 1996

nucleotide-modulated channels represent, evolutionarily, the predecessor of cation-selective, cyclic nucleotide-gated ones found in a variety of animal cells.

On the other hand, electrophysiological approaches have provided some information regarding the gating, selectivity and pharmacological properties of K^+-selective cyclic-nucleotide-gated channels. They are summarized in Table III.

Excepting *eag* channels expressed in *Xenopus laevis* oocytes (Bruggemann et al., 1993), activation of K^+ channels by cyclic nucleotides occurs at micromolar concentrations. Also, with the exception of *eag*, other cyclic nucleotide-gated K^+ channels seem to lack voltage-dependent gating.

The mechanisms by which cyclic nucleotides activate directly ion channels in animal cells have been inquired. In cAMP-activated K^+-selective channels from *Drosophila* larval muscle, and in sea urchin sperm, channel open times were found to be independent of [cAMP] (Delgado et al, 1995; Labarca et al., 1996). This observation is in agreement with the idea that agonist binding precedes a conformational change which leads to channel opening (Delgado et al, 1995; Labarca et al., 1996). Gordon and Zagotta (1995) investigated cAMP activation mechanisms in cation-selective channels from photoreceptors and olfactory neurons. Their experimental results were consistent also with the notion that cyclic nucleotides bind to close conformations, channel opening resulting from a conformational transition, after agonist binding. In contrast, the studies of Goulding et al. (1994), carried out in these same channels, led to propose that channel opening is coupled to a conformational change in the binding site which enhances agonist binding. Cyclic nucleotides would activate the channel by binding more tightly to the open state, stabilizing it. The mechanisms of cyclic nucleotide activation of cation-selective channels mentioned above were investigated using macroscopic current measurements. Since the conclusions reached diverge, it would be of interest to perform similar studies, but at the single-channel current level. They would allow, in principle, to discern between these two different activation mechanisms that have been proposed. For example, Gordon and Zagotta's model (1995) predicts that channel open times should be independent of agonist concentration. Goulding et al.'s proposal(1994) demands that channel open times depend on cyclic nucleotide concentration.

The cAMP-activated K^+ channel in *Drosophila* larval muscle was found to be utterly impermeable to divalents (Delgado et al., 1995). Cyclic nucleotide-activated K^+ channels have been reported to undergo blockade by divalents in hyperpolarizing photoreceptors (Gomez and Nassi, 1995) and in sea urchin sperm (Labarca et al., 1996). In contrast, and according to the report by Bruggemann et al. (1993), *eag* channels would be exceptional among cyclic nucleotide-modulated K^+-selective channels, owed to their unusual permeability to Ca^{2+}.

ACKNOWLEDGMENTS. This work was supported by FONDECYT 1950457, and a group of Chilean private companies (COPEC, CMPC, CGEI, CAP, ENERSIS, IBM, La Escondida and XEROX Chile. P. Labarca. was the holder of a Catedra Presidencial en Ciencias and a John Simon Guggenheim Fellowship.

V. REFERENCES

Biel, M., Zong, X., Distler, M., Bosse, E., Klugbauer, N., Murakami, M., Flockerzi, V., and Hoffman, F., 1994, Another member of the cyclic nucleotide-gated channel family, expressed in testis, kidney, and heart. *Proc. Natl. Acad. Sci. USA.* 91: 3505-3509.

Bruggemann, A., Pardo, L. A., Stuhmer, W. And Pongs, O., 1993,. *Ether-a-go-go* encodes a voltage-gated channel permeable a K^+ and Ca^{2+} and moduladted by cAMP. *Nature* 365, 445-448

Colquhoun, D., and Hawkes, A.G., 1995, The principles of the stochastic interpretation of ion channel mechanisms. In: Single channel recording, B. Sakmann and E. Neher, editors, Plenum Press, N.York and London, ch. 18, pp. 397-479.

Del Castillo, J., and Katz, B., 1957, The interaction at end-plate receptors between different choline derivatives. *Proc. R. Soc. Lond. (Biol.)* 146: 369-381.

Delgado, R., Hidalgo, P., Díaz, F., Latorre, R., and Labarca, P., 1991, A cyclic AMP-activated K^+-selective channel in *Drosophila* larval muscle is persistently activated in *dunce*. *Proc. Natl. Acad. Sci. (USA)* 88: 557-560.

Delgado, R., Latorre, R., and Labarca, P., 1995, Selectivity and gating properties of a cAMP-modulated, K^+-selective channel from *Drosophila* larval muscle. *FEBS letts.* 370: 113-117.

Gomez, M., and Nassi, E., 1994. The light-sensitive conductance of hyperpolarizing invertebrate photoreceptors: a patch clamp study. *J. Gen. Physiol.* 103: 939-956.

Gomez,M., and Nassi,E., 1994, Blockage of the light-sensitive conductance in hyperpolarizing photoreceptors of the scallop. *J. Physiol.* 104:487-505.

Gomez, M., and Nasi, E., 1995, Activation of light dependent K^+ channels in ciliary invertebrate photoreceptors involves cGMP but not the IP_3/Ca^{2+} cascade. *Neuron* 15: 607-618.

Gordon, S. E., and Zagotta, W., 1995, Localization of regions affecting an allosteric transition in cyclic nucleotide activated channels. *Neuron* 14: 857-864.

Gotow, T., Nishi, T., and Kijima, H., 1994, Single K^+ channels closed by light and opened by cyclic GMP in molluscan extraocular photoreceptor cells. *Brain Res.* 662: 268-272.

Goulding, E.H., Tibbs, G.R., and Siegelbaum, S.A., 1994, Molecular mechanism of cyclic nucleotide gated channel activation. *Nature* 372: 369-374.

Guy, H.R., Durell, S.R., Warmke, J., Drysdale, R., and Ganetzky, B., 1991, Similarities in aminoacid sequences of *Drosophila eag* and cyclic nucleotide-gated channels. *Science* 254: 730

Jackson, M.B., Wong, B.S., Morris, C.E., Lecar, H., and Christian, C.N., 1983, Sucessive openings of the same acetylcholine receptor channel are correlated in open time. *Biophys. J.* 42: 109-114.

Jorquera, O., Latorre, R., and Labarca, P., 1995, Ion channel classes in purified olfactory cilia membranes: planar lipid bilayer studies. *Am. J. Physiol.* 269: C1235-C1244.

Katz, B., and Thesleff, S., 1957, A study of the "desensitization" produced by acetylcholine at the motor end-plate. *J. Physiol. (Lond.)* 138: 63-80.

Labarca, P., Rice, J., Fredkin, D.R., and Montal, M., 1985, Kinetic analysis of channel gating. Application to the cholinergic receptor and the chloride channel. *Biophys. J.* 47: 469-478.

Labarca, P., Zapata, O., Beltrán, C., and Darszon, A., 1996, A cAMP-modulated K^+-selective channel from sea urchin sperm plasma membrane. *Dev. Biol.* 174: 271-280.

Lecar, H., 1986, Single channel currents and postsynaptic drug actions. In: Ionic channels in cells and model systems, R. Latorre, editor, Plenum Press, N.York and London, ch. 2, pp 17-36.

Michael, W.C., and Ache, B.W., 1992, Cyclic nucleotides mediate an odor-evoked potassium conductance in lobster olfactory receptor cells. *J. Neurosci.* 12: 3979-3984.

Robertson, G., Warmke, J.W., and Ganetzky, B., 1993, Functional expression of the Drosophila *eag* K^+ channel gene. *Biophys. J.* 64: A340.

Warmke, J.W., and Ganetzky, B., 1993, A novel potassium channel gene family: *eag* homologs in *Drosophila*, mouse and human. Biophys. J. 64: A340.

Weyand, Y., Godde, M., Frings, S., Melner, J., Muller, F., Altenhofen, W., Hatt, H., and Kaupp, U.B., 1994, Cloning and functional expression of a cyclic nucleotide-gated channel from mammalian sperm. *Nature* 368: 859-863.

Yau, K-W., 1994, Cyclic nucleotide-gated channels: an expanding new family of ion channels. *Proc. Natl. Acad. Sci. USA* 91: 3481-3483.

REGULATION OF Ca AND Na CHANNELS IN GH3 CELLS BY EPIDERMAL GROWTH FACTOR

Gabriel Cota, Ulises Meza, and Eduardo Monjaraz

Department of Physiology, Biophysics, and Neuroscience
Center for Research and Advanced Studies
National Polytechnic Institute
Mexico City, D.F. 07000, Mexico
Correspondence to: Dr. Gabriel Cota
Department of Physiology, Biophysics, and Neuroscience
Cinvestav-IPN
A.P. 14-740
Mexico City, D.F. 07000, Mexico
Phone number: (525) 747-7001, extension 5131
Fax: (525) 747-7105
Running title: Ion Channel Regulation by EGF

INTRODUCTION

GH3 cells are a clone of rat pituitary tumor cells secreting growth hormone and prolactin.[1] Chronic exposure of these cells and related GH cell lines to epidermal growth factor (EGF) greatly stimulates prolactin synthesis and release.[2,3] In this action, EGF interacts synergistically with calcium to increase the intracellular levels of prolactin mRNA.[4] We show here that the effects of EGF on GH3 cells include regulation of plasma membrane ion channels.

Our research was motivated by three sets of previous findings. First, GH3 cells are able to fire spontaneous action potentials[5] and express several classes of voltage-dependent ion channels in the plasma membrane, including tetrodotoxin-

From Ion Channels to Cell-to-Cell Conversations
Edited by Latorre and Sáez, Plenum Press, New York, 1997

sensitive Na channels[6,7] and low- and high-threshold Ca channels.[8] The high-threshold Ca channels mostly correspond to the L-type.[9,10]

Second, action potential generation in GH cells depends on both Na channels[11] and L-type Ca channels,[12] so each spike leads to a transient rise in cytosolic free Ca^{2+} concentration[12, 13] that is sufficient to trigger prolactin release.[14,15] Calcium influx through L-type channels also sustains prolactin synthesis[16] by enhancing prolactin gene transcription.[17]

Finally, GH3 cultures normally contain bihormonal cells that secrete prolactin and growth hormone, but are practically devoid of classical lactotropes, which produce only prolactin.[18,19] In this regard, GH3 cultures resemble pituitary cell cultures derived from immature rats; by contrast, classical lactotropes account for most prolactin cells in adult rat pituitaries.[18,20] Classical lactotropes are further characterized by their ability to secrete large amounts of prolactin under basal conditions,[19, 20] a feature that depends on a relatively high level of Ca^{2+} current expression[20-22] combined with a large Na channel activity.[23] This specialized subtype of prolactin cell is induced in GH3 cultures after chronic treatment with EGF.[19]

Taken into account these observations, we hypothesized that the stimulatory effects of EGF on prolactin production and lactotrope differentiation could be mediated, at least in part, by changes in Ca^{2+} and Na^{+} current expression. To investigate this possibility, we have carried out patch clamp experiments on GH3 cells grown in the absence or presence of EGF. Our results will be presented after some methodological considerations.

METHODS

GH3 cells were voltage-clamped in the whole-cell recording mode using an Axopatch 200A amplifier (Axon Instruments, Foster City, CA) as previously described.[24] In brief, GH3 cells from the American Type Culture Collection (Rockville, MD) were grown in monolayer culture at 37 C in a $C0_2$ incubator. Cells from a single donor culture were plated on 10 x 2.5 mm glass coverslips placed in 35-mm plastic dishes (150 X 10^3 cells/dish), and then cultured in standard medium (RPMI 1640 medium containing 5% fetal bovine serum and 2 mM L-glutamine) or standard medium supplemented with 5 nM EGF. At this concentration, EGF exerts a maximal stimulatory effect on prolactin production.[2,3] The culture medium was replenished every day, starting on day 3 after plating. On day 5 or 6, the cells were removed from the incubator, placed in a 0.2 ml experimental chamber, and then subjected to patch clamp recording at 21 C.

The composition of the recording solutions is given in the figure legends. The bath solution was continually perfused through the chamber at a rate of 0.3 ml/min. Patch electrodes exhibited resistances of 1.5-2.5 M when filled with internal solution. The series resistance was typically in the range of 2.5 to 3.5 M .

Depolarizing voltage steps were applied from a holding potential of -80 mV at a frequency of 0.5 Hz. Currents were low-pass filtered at 10 kHz, digitally sampled at intervals of 20 μs and stored on disk using a Labmaster-TL1 interface and pCLAMP software (Axon Instruments). Capacitive transients were electronically minimized and the remaining linear components in the current signals were digitally subtracted on-line using the scaled current response to hyperpolarizing pulses. In each cell investigated, measurements of voltage-dependent currents at a given membrane potential were made in triplicate and averaged. Current recording was completed within 3 min after establishing the whole-cell mode.

EGF PROMOTES Ca^{2+} CURRENT EXPRESSION

Figure 1 illustrates the effect of EGF treatment on the Ca channel activity of GH3 cells. The electrophysiological recording in these experiments was carried out under ionic conditions that eliminate K^+ currents. In addition, Na channels were blocked by using tetrodotoxin (TTX). As shown in Figs. 1A and B, culturing the cells during 5 to 6 days in the presence of EGF enhances the inward Ca^{2+} current elicited by pulses to +20 mV, but has little effect on the Ca^{2+} current recorded at -30 mV. The preferential stimulation of Ca channel activity during large depolarizations is better observed in Fig. 1C, which compares the current-voltage (*I-V*) relation for inward Ca^{2+} current in control cells with that obtained in cells treated with the growth factor. In these plots current amplitude has been divided by cell capacitance (C_m) to eliminate membrane area as a variable (C_m was measured as described in Ref. 24). The chronic presence of EGF results in a clear increase in current density at -20 mV or more positive voltages, with no significant change in the density of current for membrane potentials between -50 and -30 mV.

The results shown in Fig. 1 suggest that EGF is a selective regulator of high-threshold Ca^{2+} current expression. Further supporting evidence for this view was derived from tail current measurements as well as from the analysis of Ca^{2+} current behavior during long depolarizations.[(24)] In addition, neither the deactivation rates of the Ca channels nor the voltage dependence of Ca channel activation were significantly modified after chronic treatment with EGF.[(24)]

EGF ALSO STIMULATES Na^+ CURRENT DENSITY

EGF-treated GH3 cells not only generate high-threshold Ca^{2+} currents larger than normal, but also exhibit an enhanced level of Na^+ current expression. Evidence for this additional change in ion channel activity was first obtained by recording Na^+ currents in the absence of Ca channel blockers.[(24)] To study the effects of EGF in more detail, we have now recorded isolated Na^+ currents after blocking the Ca channels with external cadmium.

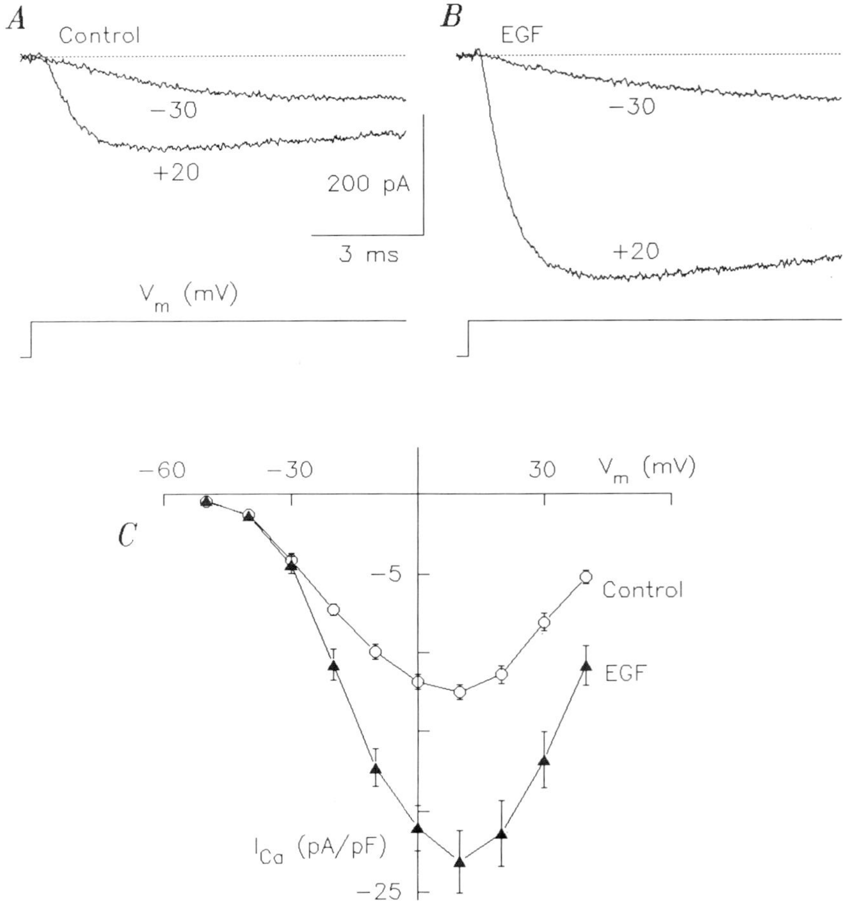

Figure 1. Ca channel activity of GH3 cells and its stimulation after chronic treatment with EGF. (A and B) Ca^{2+} currents generated by two different cells in response to 10-ms voltage steps. The membrane potential (V_m) during the steps is indicated near each current trace. The cells were cultured for 5 or 6 days in standard culture medium (A) or culture medium supplemented with 5 nM EGF (B), then subjected to whole-cell patch clamp within 15-60 min using EGF-free recording solutions (a similar procedure was used to explore the long-term effects of EGF on Na channel activity in the experiments reported in Figs. 2 to 5). As in all subsequent figures, the holding potential was -80 mV. (C) Voltage dependence of Ca^{2+} current density (peak current amplitude during 10-ms pulses normalized to cell capacitance, C_m) for control cells (open symbols) and cells that had been chronically exposed to EGF (filled symbols). Data are presented as means ± SE (n = 8 in each case). The cells were bathed in 133 mM NaCl, 10 mM TEA-Cl, 10 mM $CaCl_2$, 1 μM TTX, 5 mM glucose, and 10 mM Hepes (buffered to pH 7.3 with NaOH), and internally dialyzed by patch pipettes containing 130 mM CsCl, 2 mM $MgCl_2$, 1 mM $CaCl_2$, 10 mM EGTA, 2 mM Na_2ATP, 50 μM GTP, 5 mM glucose, and 10 mM Hepes (adjusted to pH 7.3 with CsOH).

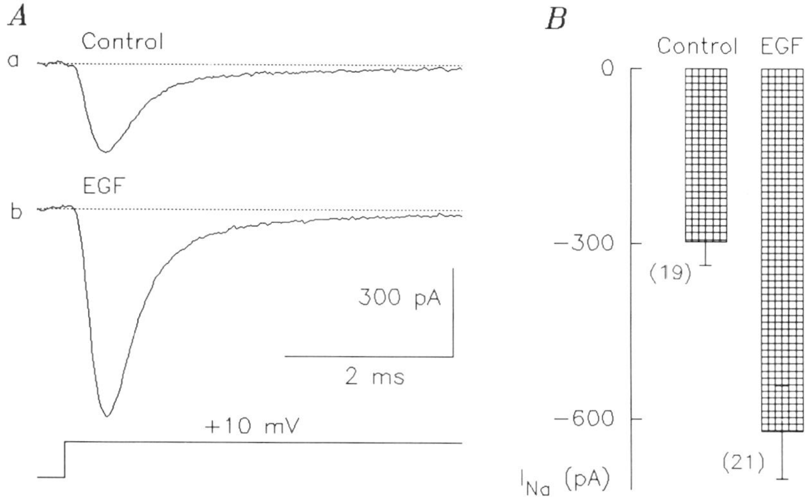

C

Cell group	C_m (pF)
Control	12.4 ± 0.6
EGF	13.8 ± 0.6

Figure 2. EGF treatment also enhances the Na channel activity of GH3 cells. (A) Na^+ currents recorded from a control cell (trace *a*) and a cell treated with EGF (trace *b*). The pulse protocol is shown below. (B and C) Average values for peak Na^+ current at +10 mV (B) and C_m (C) in control cells and cells grown in the presence of EGF. Values represent means ± SE of the results obtained from the number of cells indicated in parentheses in B. In these experiments, the external recording solution was composed of 154 mM NaCl, 2 mM $CaCl_2$, 0.5 mM $CdCl_2$, 5 mM glucose, and 10 mM Hepes (pH 7.3), whereas the composition of the pipette solution was 30 NaCl, 100 mM CsCl, 2 mM $MgCl_2$, 1 mM $CaCl_2$, 10 mM EGTA, 2 mM Na_2ATP, 50 μM GTP, 5 mM glucose, and 10 mM Hepes (pH 7.3). These recording solutions were also used during the current measurements shown in Figs. 3 to 5.

Figure 2A shows representative Na^+ currents induced by brief voltage steps to +10 mV in a control cell and a cell chronically exposed to EGF. The current undergoes a marked enlargement under the influence of the growth factor. This result is confirmed by Fig. 2B, which summarizes data obtained from many cells. Peak Na^+ current at +10 mV increases from about -300 pA in control cells to approximately -620 pA in cells grown in the presence of EGF. In these experiments, the surface membrane area for control cells was close to 1240 μm^2, as estimated from C_m measurements. The mean value of C_m also increased after EGF treatment, but only by a factor of 1.1 (Fig. 2C). It is thus clear that EGF treatment stimulates Na^+ current density.

It is important to note that the Na channel activity of GH3 cells depends on the batch of cells. The reason for this variability is unclear at present. Nevertheless, chronic exposure to EGF consistently increased Na current density by 70-110% relative to its control value.

EGF INCREASES THE MAXIMUM Na^+ CONDUCTANCE

The enhanced Na^+ channel activity of EGF-treated cells reported in Fig. 2 could result from a change in the voltage dependence of Na channel activation. This possibility was explored in six control cells and an equal number of treated cells by applying a series of voltage steps from -50 to +60 mV in 10 mV increments. Figures 3A and B show examples of the resultant families of Na^+ currents, whereas Fig. 3C compares the *I-V* relations for peak Na^+ current obtained from the two groups of cells. EGF treatment amplifies Na^+ current amplitude at every tested membrane potential (V_m), suggesting an increased maximum Na^+ conductance (G_{max}) with no change in voltage dependence of channel activation.

This impression is confirmed in Fig. 3D, which plots the mean value of Na^+ conductance (G_{Na}) as a function of V_m. G_{Na} was computed by dividing peak Na^+ current by the corresponding driving force V_m-V_{rev}, where V_{rev} is the voltage at which current reversed direction from inward to outward. V_{rev} was +41.9 ± 0.6 mV in control cells and +41.2 ± 1.4 mV in cells exposed to EGF. In all cases, G_{Na} was a sigmoidal function of voltage and tended to saturate above +20 mV. Moreover, each data set could be well fitted by the simple Boltzmann function given in the legend of Fig. 3D. The average values for the parameters of the fits, half-maximal activation level, slope factor and G_{max}, were respectively -5.8 ± 1.1 mV, 8.89 ± 0.45 mV and 10.5 ± 1.3 nS in control cells, and -8.1 ± 1.6 mV, 8.67 ± 0.49 mV and 20.9 ± 3.2 nS in EGF-treated cells. In conclusion, chronic exposure to EGF markedly increases the maximum Na^+ conductance, but does not significantly modify the shape of the G_{Na}-V_m curve or its position along the voltage axis.

Na CHANNEL INACTIVATION IS UNAFFECTED BY EGF

The current records in Fig. 2A superimpose almost perfectly after appropriate scaling (Fig. 4A), which suggests that EGF treatment has no significant effect on the activation and inactivation kinetics of Na channels. In fact, as illustrated in Fig. 4B, a single exponential provided a good description of the decaying phase of the Na^+ current in all cells investigated. The average value of the corresponding time constant at +10 mV was close to 0.49 ms in both control cells and cells exposed to EGF (Fig. 4C). Thus, chronic treatment with the growth factor increases Na^+ current density without affecting the macroscopic rate of Na channel inactivation.

Another relevant question is whether the voltage dependence of the steady-state fraction of non-inactivated Na channels (*h*) changes after EGF treatment. To examine this point, the *h*-V_m relation was determined in same cells as in Figs. 3C

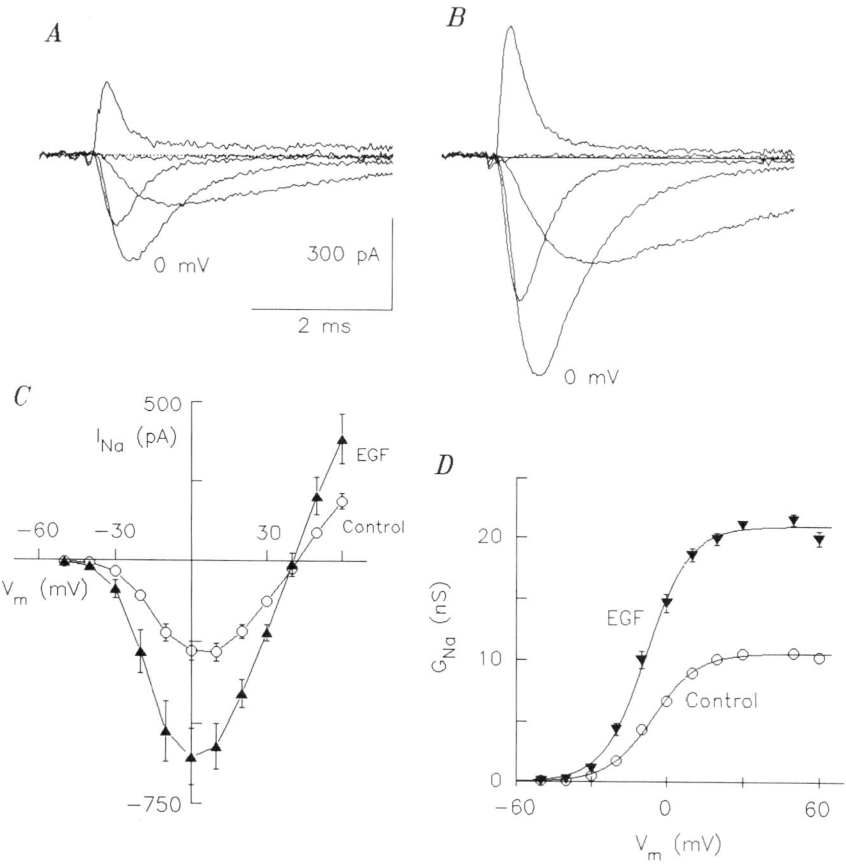

Figure 3. Comparison of the voltage dependence of Na channel activation between control GH3 cells and cells treated with EGF. (A and B) Na^+ currents induced by step depolarizations in a control cell (A) and a cell cultured in the presence of EGF (B). The cells were held at -80 mV and depolarized to test voltages of -20, 0, +20, +40, and +60 mV. (C and D) *I-V* relations for peak Na^+ current (C) and corresponding *G-V* curves (D) obtained from control cells (open symbols) and cells exposed to EGF (filled symbols). Na^+ conductance (G_{Na}) was calculated as described in the text. Data are expressed as means ± SE (n = 6 for both conditions). In each cell, G_{Na} vs. V_m data were fitted to the following function: $G_{Na} = G_{max}/\{1 + \exp[-(V_m - V_{1/2})/s]\}$, where G_{max} is the saturating value of G_{Na} at positive membrane potentials, $V_{1/2}$ is the voltage for half-maximal activation, and s is the slope factor. The smooth lines in D are calculated *G-V* curves using the mean values for G_{max}, $V_{1/2}$, and s reported in the text.

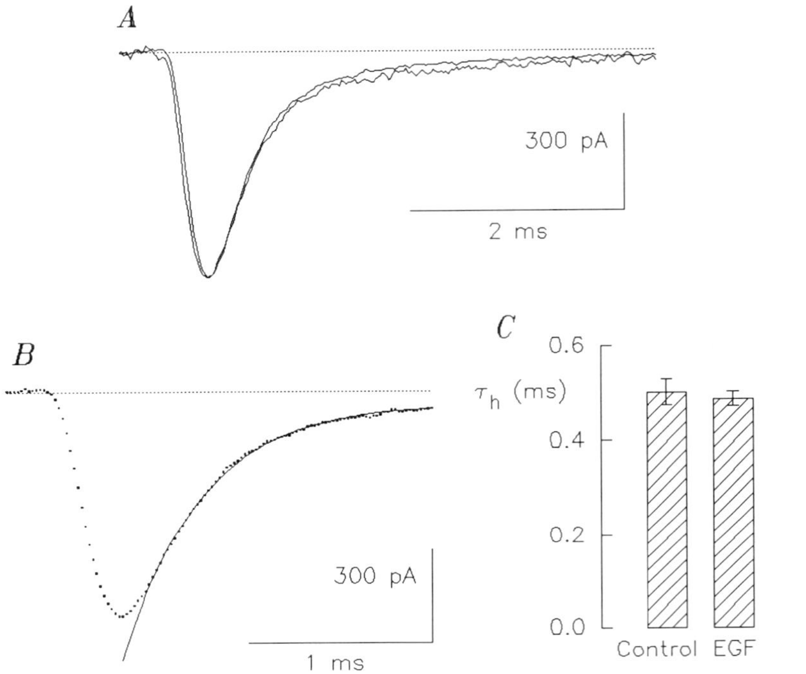

Figure 4. Chronic treatment with EGF does not affect the time course of Na^+ current. (A) Superposition of current traces at +10 mV from Fig. 2A. The Na^+ current generated by the control GH3 cell has been scaled up by a factor of 2.35 to match the peak amplitude of that obtained from the EGF-treated cell. (B) The Na^+ current recorded from the cell treated with EGF has been plotted in an expanded time scale. Current decay is well fit by a single exponential (continuous trace) with a time constant of 0.43 ms. (C) Average values (means ± SE) of the time constant of Na^+ current inactivatio ($_h$) at +10 mV for same cells as in Fig. 2B.

and D using a conventional two-pulse protocol (Fig. 5A). These current measurements are summarized in Fig. 5B. Again, experimental data were fitted with smooth lines according to Boltzmann functions. The midpoint of the h-V_m curve was -50.6 ± 1.8 mV for control cells and -51.1 ± 1.55 mV for cells grown in the presence of EGF, whereas the corresponding slope factor was 10.92 ± 0.20 and 11.06 ± 0.51 mV, respectively. We conclude that the stimulatory effect of EGF on Na^+ current density is not accompanied by any measurable change in the voltage dependence of Na channel inactivation.

TIME COURSE OF EGF ACTIONS

We have previously reported that the increase in Ca^{2+} current density caused by EGF gradually develops over the course of several days, and that it reverses

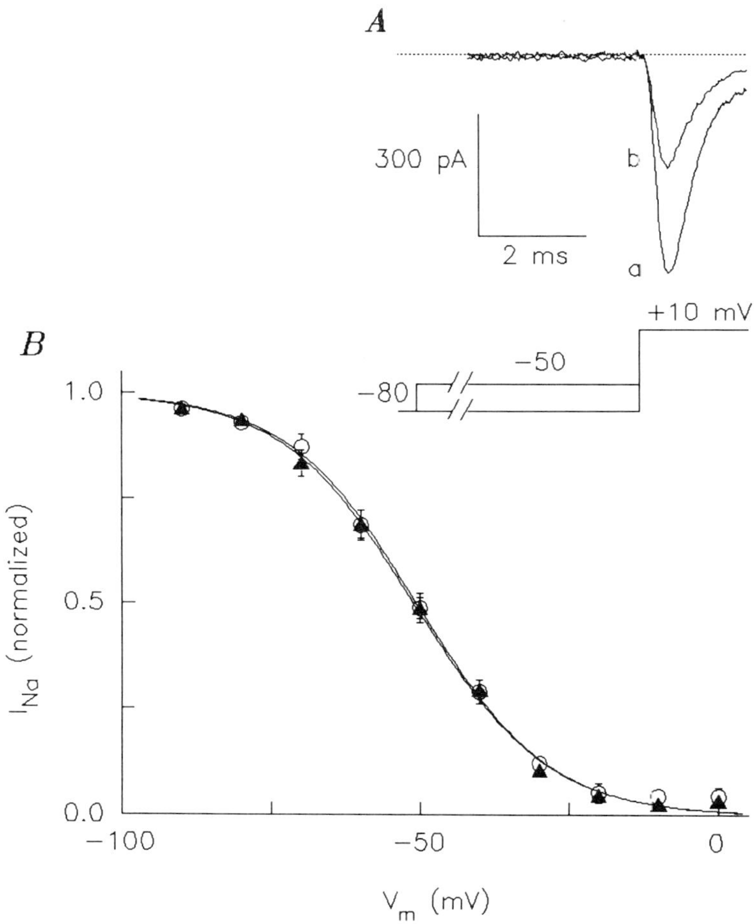

Figure 6. Time-dependent stimulation of Na^+ current density by EGF (A) and its reversibility after growth factor removal (B). Current density is peak Na^+ current during 6-ms steps to +10 mV divided by C_m, and has been converted to a percentage of its average value in control cells (open symbols). Data are presented as means ± SE; the number of cells examined is indicated next to each data point. Ca^{2+} current measurements carried out in these same cells are reported in Ref. 24. (A) Onset of EGF effect. Before the patch clamp experiments, cells were exposed to 5 nM EGF for 0.8, 3 or 5 to 6 days. (B) Recovery from EGF effect. Cells were first grown in the presence of 5 nM EGF for 6 days, then placed in standard culture medium during 0.04, 1, 3 or 6 days, and finally subjected to electrophysiological recording. The externalinternal recording solution contained (in mM): 145 NaCl, 5 $CaCl_2$, 5 glucose, and 10 Hepes30 NaCl, 80 Cs-aspartate, 20 CsF, 1 $CaCl_2$, 10 EGTA, 5 glucose, and 10 Hepes (pH 7.3 for both).

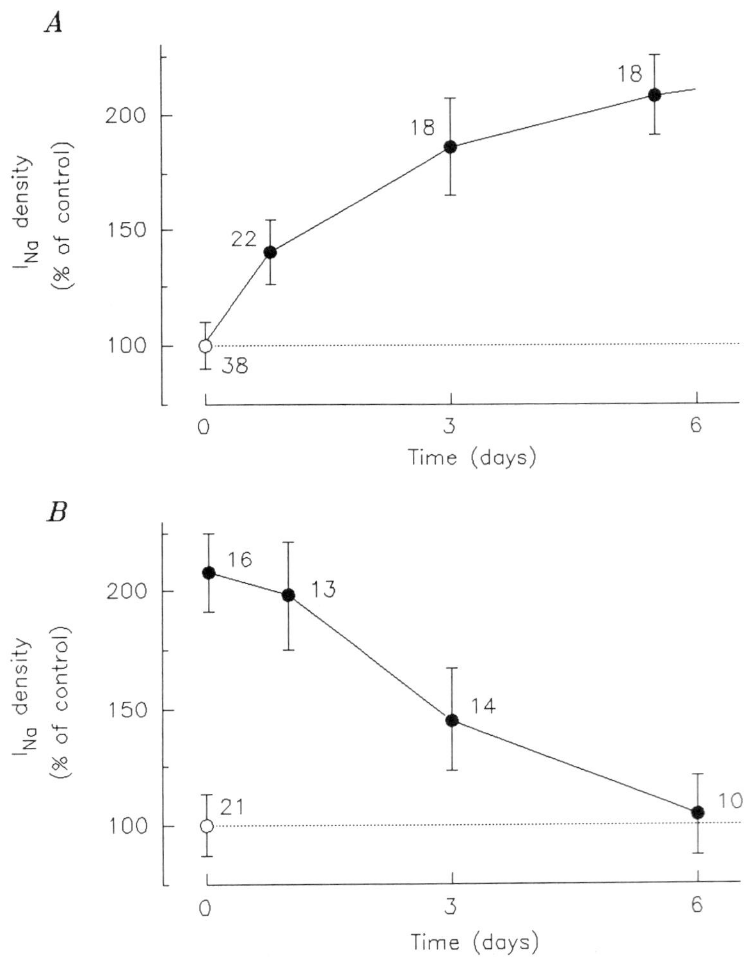

Figure 5. Lack of effect of EGF treatment on the voltage dependence of Na channel inactivation. (A) Na^+ currents generated by an EGF-treated GH3 cell in response to test pulses to +10 mV applied from the holding potential (trace *a*) or after a 30-ms prepulse to -50 mV (trace *b*). (B) Normalized peak Na^+ current during the test pulse plotted as a function of prepulse voltage (test pulse amplitude and prepulse duration as shown in A). Data represent means ± SE of values obtained from same cells as in Figs. 3C and D. The parameters of the fitted Boltzmann functions (smooth lines) are given in the text.

also slowly after growth factor withdrawal.[24] Fig. 6 shows that the increase in Na channel activity follows a similar time course. As can be seen in Fig. 6A, the stimulation of Na^+ current density is already significant in cells exposed to EGF for 19 h. However, a maximal response takes >5 days to develop. In like manner, the effect of EGF is fully reversible, but Na^+ current density is still above its control value 3 days after removing the growth factor from the culture medium (Fig. 5B).

CONCLUDING REMARKS

We have addressed, for the first time, the effects of a growth factor on ion current expression in cultured pituitary cells using whole-cell patch clamp recording. In particular, our studies demonstrate that chronic exposure of clonal GH3 cells to EGF stimulates the density of macroscopic currents flowing through high-threshold Ca channels and Na channels. These effects are not associated with obvious changes in the time- and voltage-dependent properties of the currents and, moreover, both are slowly reversible. In other words, the changes in current density we have observed are not due to short-term modulatory actions on preexisting channels. Instead, they probably result from a long-term increase in either the total number of Ca and Na channels or the fraction of functional channels in the plasma membrane. Even though the molecular bases for the stimulation of Ca^{2+} and Na^+ current expression remain to be fully determined, the present findings improve our understanding about the mechanisms by which EGF enhances prolactin production.

ACKNOWLEDGMENTS

This work was supported by an International Research Scholars award from the Howard Hughes Medical Institute. We thank Araceli Navarrete, Angel Marin and Raul Gonzalez for valuable technical assistance.

REFERENCES

1. Tashjian, Jr., A. H., 1979, Clonal strains of hormone-producing pituitary cells, *Methods Enzymol.* **58:**527-535.
2. Schonbrunn, A., Krasnoff, M., Westendorf, J. M., and Tashjian Jr. A. H., 1980, Epidermal growth factor and thyrotropin-releasing hormone act similarly on a clonal pituitary cell strain, *J. Cell Biol.* **85:**786-797.
3. Aanestad, M., Røtnes, J. S., Torjesen, P. A., Haug, E., Sand, O., and Bjøro, T., 1993, Epidermal growth factor stimulates the prolactin synthesis and secretion in rat pituitary cells in culture (GH_4C_1 cells) by increasing the intracellular concentration of free calcium, *Acta Endocrinol.* **128:**361-366.

4. White, B. A., and Bancroft, F. C., 1983, Epidermal growth factor and thyrotropin-releasing hormone interact synergistically with calcium to regulate prolactin mRNA levels, *J. Biol. Chem.* **258:**4618-4622.
5. Kidokoro, Y., 1975, Spontaneous calcium action potentials in a clonal pituitary cell line and their relationship to prolactin secretion, *Nature* **258:**741-742.
6. Dubinsky J. M., and Oxford, G. S., 1984, Ionic currents in two strains of rat anterior pituitary tumor cells, *J. Gen. Physiol.* **83:**309-339.
7. Matteson, D. R., and Armstrong, C. M., 1984, Na and Ca channels in a transformed line of anterior pituitary cells, *J. Gen. Physiol.* **83:**371-394.
8. Armstrong, C. M., and Matteson, D. R., 1985, Two distinct populations of calcium channels in a clonal line of pituitary cells, *Science* **227:**65-67.
9. Simasko, S. M., Weiland, G. A., and Oswald, R. E., 1988, Pharmacological characterization of two calcium currents in GH_3 cells, *Am. J. Physiol.* **254:**E328-E336.
10. Knaus, H. G., Moshammer, T., Friedrich, K., Kang, H. C., Haugland, R. P., and Glossmann, H., 1992, *In vivo* labeling of L-type Ca^{2+} channels by fluorescent dihydropyridines: evidence for a functional, extracellular heparin-binding site, *Proc. Natl. Acad. Sci. U.S.A.* **89:**3586-3590
11. Biales, B., Dichter, M. A., and Tischler, A., 1977, Sodium and calcium action potential in pituitary cells, *Nature* **267:**172-174.
12. Schlegel, W., Winiger, B. P., Mollard, P., Vacher, P., Wuarin F., Zahnd, G. R., Wollheim, C. B., and Dufy, B., 1987, Oscillations of cytosolic Ca^{2+} in pituitary cells due to action potentials, *Nature* **329:**719-721.
13. Mollard, P., Guerineau, N., Audin, J., and Dufy, B., 1989, Measurement of Ca^{2+} transients using simultaneous dual-emission microspectrofluorimetry and electrophysiology in individual pituitary cells, *Biochem. Biophys. Res. Commun.* **164:**1045-1052.
14. Ronning, S. A., and Martin, T. F. J., 1986, Characterization of Ca^{2+}-stimulatedsecretion in permeable GH_3 pituitary cells, *J. Biol. Chem.* **261:**7834-7839.
15. Zorec, R., Sikdar, S. K., and Mason, W. T., 1991, Increased cytosolic calcium stimulates exocytosis in bovine lactotrophs, *J. Gen. Physiol.* **97:**473-497.
16. Enyeart, J. J., Sheu, S. S., and Hinkle, P. M., 1987, Dihydropyridine modulators of voltage-sensitive Ca^{2+} channels specifically regulate prolactin production by GH_4C_1 pituitary tumor cells, *J. Biol. Chem.* **262:**3154-3159.
17. Day, R. N., and Maurer, R. A., 1990, Pituitary calcium channel modulation and regulation of prolactin gene expression, *Mol. Endocrinol.* **4:**736-742.
18. Frawley, L. S., and Boockfor F. R., 1991, Mammosomatotropes: presence and functions in normal and neoplastic pituitary tissue, *Endocr. Rev.* **12:**337-355.
19. Felix, R., Meza, U., and Cota, G., 1995, Induction of classical lactotropes by epidermal growth factor in rat pituitary cell cultures, *Endocrinology* **136:**939-946.

20. Felix, R., Horta, J., and Cota, G., 1993, Comparison of lactotrope subtypes of neonatal and adult male rats: plaque assays and patch-clamp studies, *Am. J. Physiol.* **265:**E121-E127.
21. Cota, G., Hiriart, M., Horta, J., and Torres-Escalante, J. L., 1990, Calcium channels and basal prolactin secretion in single male rat lactotropes, *Am. J. Physiol.* **259:**C949-C959.
22. Horta, J., and Cota, G., 1993, Lactotrope subtypes are differentially responsive to calcium channel blockers, *Mol. Cell. Endocrinol.* **92:**189-193.
23. Horta, J., Hiriart, M., and Cota, G., 1991, Differential expression of Na channels in functional subpopulations of rat lactotropes, *Am. J. Physiol.* **261:**C865-C871.
24. Meza, U., Avila, G., Felix, R., Gomora, J. C., and Cota, G., 1994, Long-term regulation of calcium channels in clonal pituitary cells by epidermal growth factor, insulin, and glucocorticoids, *J. Gen. Physiol.* **104:**1019-1038.

PHYSICAL STATE OF THE NICOTINIC ACETYLCHOLINE RECEPTOR MEMBRANE AND MODULATION OF THE RECEPTOR CHANNEL BY THE LIPID ENVIRONMENT

F.J. Barrantes

Instituto de Investigaciones Bioquímicas,
Universidad Nac. del Sur, 8000 Bahía Blanca, Argentina.
fax No. (054) 91 860664; e-mail: rtfjb1@criba.edu.ar

INTRODUCTION

Since the discovery of the immobilized lipid surrounding the AChR (Marsh and Barrantes, 1978; see reviews in Barrantes, 1993a,b), the lipid belt region around this protein has become the focus of a variety of biophysical studies aimed at defining the properties of this membrane region. Other groups have studied bulk physical properties of the AChR-rich membrane. Fong and McNamee (1986), for example, have shown that reconstituted AChR is sensitive -within a narrow range of values- to the bulk fluidity of the membrane, as measured by ion flux experiments *in vitro*. Sunshine and McNamee (1994) have varied the fluidity of the host lipid vesicle, as measured by steady-state fluorescence anisotropy of the probe diphenylhexatriene (DPH) at a fixed temperature, and shown that ion permeation through the reconstituted AChR occurs either in low or high fluidity environments.

We have recently characterized the thermotropic behavior of the so-called Generalized Polarization (GP) of the fluorescent probe Laurdan (6-dodecanoyl-2-dimethylamino naphthalene) in AChR-rich membranes from *Torpedo* (Antollini et al., 1996). A still unexploited property of Laurdan, namely its

From Ion Channels to Cell-to-Cell Conversations
Edited by Latorre and Sáez, Plenum Press, New York, 1997

ability to act as a fluorescence energy transfer acceptor of tryptophan emission, has been introduced to measure properties of the protein-vicinal lipid. Energy transfer between AChR-rich membrane protein and Laurdan was observed upon excitation at 290 nm. A minimal donor-acceptor distance r of 14±1 Å could be calculated considering a distance $0 < H < 10$ Å for the separation of the planes containing donor and acceptor molecules, respectively. This value of r corresponds closely to the diameter of the first-shell protein-associated lipid. Only one, ordered (from the point of view of molecular axis rotation)-liquid (from the point of view of lateral diffusion) lipid phase is present in the native AChR membrane, exhibiting a decreasing polarity and an increased solvent dipolar relaxation at the hydrophilic/hydrophobic interface upon increasing the temperature. AChR-vicinal lipid differs structurally and dynamically from the bulk bilayer lipid in terms of polarity and molecular motion. The protein-vicinal lipid region is also characterized by a higher rigidity and a lower degree of water penetration than the bulk bilayer lipid.

More recently, we have extended our initial *in vitro* studies using Laurdan GP to a more physiological situation, namely the cellular environment. We compared the effect of dissimilar cell membrane environments on the activity of the mouse muscle AChR of the embryonic and adult type by studying the thermal sensitivity of its single-channel properties using the patch-clamp technique (Zanello et al., 1996).

We also investigated the effect of the physical state of the cell membrane in situ on the activity of the AChR in clonal cell lines transfected with the cDNAs of embryonic or adult AChR by measuring single-channel properties and the so-called Generalized Polarization (GP) of the lipophilic probe Laurdan as a function of temperature. Unitary conductance and channel closing rate, α, had Q10 values of ≈1.2 and 2.2, respectively. Using Eyring's transition state theory, it was calculated that both embryonic and adult type AChR had relatively low thermal sensitivity of ionic conductance and activation energy (Ea of 3.0-5.0 kcal.mol-1 at 20°C), indicating that once the AChR channel opens, ion movement is dominated by diffusional processes. Channel closure exhibited higher energy requirements, with Ea values of about 13 kcal.mol-1. This process appears to be more endothermic (higher ΔHa values) than that of ion permeation, and it is plausible that the energy acquired by the system can be used in the maintenance of its degree of order, as revealed by the ΔSa ~ 0 calculated for channel closure. The influence of the membrane environment on AChR function is reinforced by the observation that the conductance of the same, embryonic-type AChR protein, expressed in qualitatively different cellular lipid environments, appeared to have different energetic requirements.

A correlation between the electrophysiological and thermodynamic parameters of the AChR and physicochemical properties of the membrane bilayer in which the protein is embedded could be established using measurements of Laurdan GP. Both embryonic and adult AChR exhibited a higher GP and a higher sensitivity to temperature-dependent changes in GP when heterologously expressed in stable form in CHO-derived cells than did the native embryonic AChR in BC3H-1 cells, indicating that these two properties are determined by the host membrane and are not an inherent property of the AChR type. In addition, the differences in the macroscopic physical state of the lipids and membrane-associated solvent (water) dipolar relaxation between BC3H-1 and CHO-derived cells reported by the spectroscopic properties of Laurdan suggest that both lipid and associated water may influence the microscopic activity of individual AChR molecules embedded in the lipid bilayer. Finally, the different dependence of AChR channel conductance and mean open time as a function of GP observed between the different AChR subtypes in clonal cell lines suggests the importance of specific lipid-protein interactions in addition to bulk membrane properties.

RESULTS AND DISCUSSION

Learning about the physical state of the AChR-lipid interface by means of Laurdan General Polarization (GP) of fluorescence.

The intrinsic fluorescence emission spectrum of the AChR-rich membrane protein, centered at 335-340 nm (Barrantes, 1978), is dominated by tryptophan emission. Spectral overlap between the emission spectrum of the intrinsic protein fluorescence, acting as donor, and the excitation spectrum of Laurdan, as acceptor, is an essential requirement for establishing fluorescence resonance energy transfer (FRET) conditions. Excitation under FRET conditions using the tryptophan residues of the AChR-rich membrane as donors was used to obtain information on the properties of the lipids in the immediate vicinity of the AChR in comparison to the average, bulk lipid in the rest of the bilayer. From the spectral overlap integral for the AChR membrane-Laurdan pair we calculated R_O, the Förster critical distance, to be 29 Å (Antollini et al., 1996).

We modeled the AChR as a cylinder of about 80 Å diameter with donor Trp residues lying in a ring within the perimeter of its transmembrane portion. This region of the protein was assumed to have a radius of 32.5 Å (cf. Unwin, 1993). The height of the plane of AChR tryptophan residues was set with respect to the plane of acceptors by using the parameter H, the distance between donor-acceptor planes normal to the membrane surface, which was allowed to vary between 0 and 50 Å in view of the long-axis dimensions of the AChR molecule and the width of the lipid bilayer (Fig. 1). As expected for a constant molar ratio of the donor AChR protein in the membrane, E varied as a function of Laurdan acceptor surface density. The efficiency of FRET in the *Torpedo* AChR-rich membrane reached values of about 50% for molar acceptor:total lipid ratios of 0.03 or higher.

Theoretical curves for different values of minimal donor-acceptor distance, r, and different separation between the planes of donor and acceptor respectively, H,

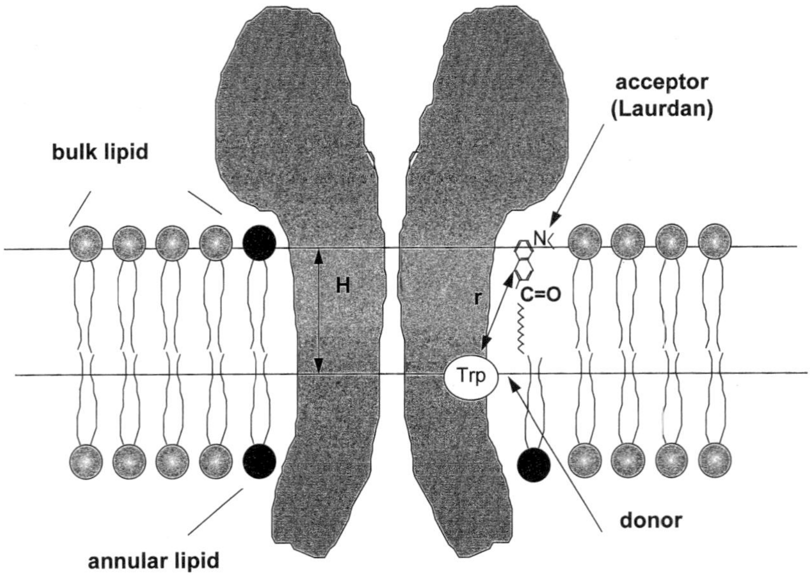

Fig. 1. Topographical relationship between the membrane-bound AChR, surrounding lipid molecules and Laurdan in cross-sectional representation. The low anisotropy of the AChR intrinsic fluorescence indicates that it is highly likely that Trp residues are arranged in a network, with extensive FRET occurring among individual members of such an array. Only one Trp residue (Tr) is represented in the AChR protein for the sake of clarity. This Trp acts as the donor to a Laurdan molecule acceptor in the protein-vicinal lipid annulus (shaded circles indicate the polar head region of protein-vicinal phospholipid in the AChR lipid belt region). The orientation of the acceptor molecule and its depth in the membrane bilayer follows the data of Chong (1988, 1990) and Parasassi et al. (1991). H is the distance between the plane of the donor and that of the acceptor; r is the distance of closest approximation between the donor and acceptor molecules. From Antollini et al. (1996).

were generated. Fig. 2 shows the sets of curves for H = 0 and 10 Å respectively, together with the experimental data. An average value of r = 14 ± 1 Å resulted from these two sets of data. This value matches the thickness of a single layer of phospholipid molecules of about 0.75 nm^2 surface area (Rand, 1981) and is in agreement with the presumptive location of Laurdan in the bilayer (Chong, 1988, 1990; Parasassi et al., 1991) and with the average distance of 10 Å for the depth of AChR tryptophan residues(s) from the center of the bilayer calculated by Chattopadhyay and McNamee (1991).

The probability (α) of occupancy of sites at the lipid belt region by the fluorescent lipid Laurdan (L_l) relative to that of unlabeled lipids has been calculated from its dissociation constant relative to the average dissociation constant of unlabeled lipids in the bilayer (L_2). Thus, using the equilibria:

$P + L_l \approx PL_l$, with K_l as dissociation constant (1)

$P + L_2 \approx PL_2$, with K_2 as dissociation constant (2)

and expressing the concentrations of L_l and L_2 as molar fractions, x_l and x_2 respectively, it follows that:

$$\alpha = (x_l / K_l) / (x_2 / K_2) = (x_l / x_2) \cdot K_r^{-1} \quad (3)$$

where $x_l + x_2 = 1$ and where K_r is the apparent dissociation constant of Laurdan for the lipid belt region. By comparing the experimental values with theoretical curves with different values of K_r between 0.01 and 100 a K_r value $\cong$1 was obtained, indicating that the distribution of Laurdan in the lipid bilayer was random. This result is in agreement with the values of energy transfer efficiency; had Laurdan had affinity for the lipid annulus, these values would be higher than 50% at a low Laurdan:AChR ratio.

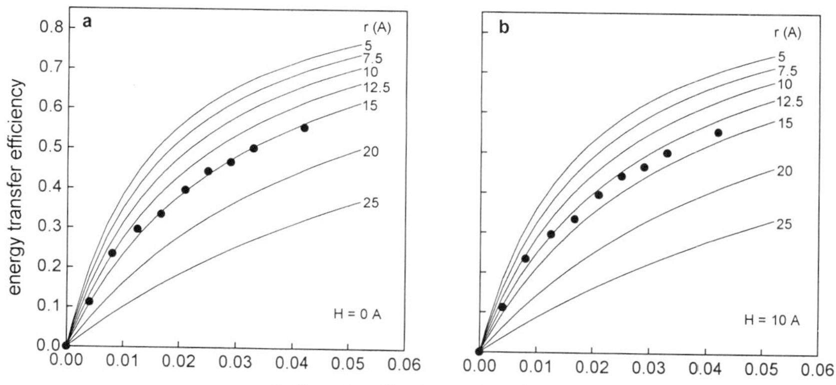

Fig. 2. Efficiency of resonance energy transfer and calculated distances between donor-acceptor pairs (protein intrinsic fluorescence-Laurdan) in the native AChR-rich membrane as a function of the surface density of energy transfer acceptors. A constant concentration of the fluorescent donor (the AChR protein) was titrated with increasing concentrations of the fluorescent acceptor. The curves (continuous lines) correspond to the theoretical fits to the donor-acceptor minimal distances, and the symbols (•) correspond to the experimental values for different *r*, with H = 0 (A) and 10 Å (B) and R_0 = 29 Å (see Fig. 1). From Antollini et al. (1996).

Our results are also in agreement with the more general observation that Laurdan fluorescence is independent of the type of polar head group in the membrane phospholipids (Parasassi et al., 1991).

Considering that the efficiency of the energy transfer process, E, depends on the sixth power of the donor-acceptor distance, and the experimentally determined R_0 of 29 Å, Laurdan molecules excited under FRET conditions are a fraction of the total population localized within a reduced volume around the AChR protein, at a maximum radius slightly larger than the magnitude of R_0. If direct excitation (360 nm) of Laurdan reports on the bulk lipid physical state, and excitation of Laurdan under FRET conditions (290 nm) is used as a probe of the lipid-AChR interface region, from comparison of the two sets of GP values the *larger* values found for the set obtained under FRET conditions (Fig. 3) indicate that the Laurdan molecules within energy transfer distance, i.e. at the AChR-lipid interface, differ from those in the bulk lipid.

When Laurdan was excited under FRET conditions in *T. marmorata* AChR-rich membranes, the GP values were significantly higher than those obtained by direct excitation at 360 nm, especially in the high temperature range (7% higher at 5°C and 67% higher at 50°C). This supports the assumption that Laurdan molecules within FRET distance, and by inference at the lipid-AChR interface region, exhibit lower mobility than that of the bulk lipid, thus allowing for a lesser degree of water penetration.

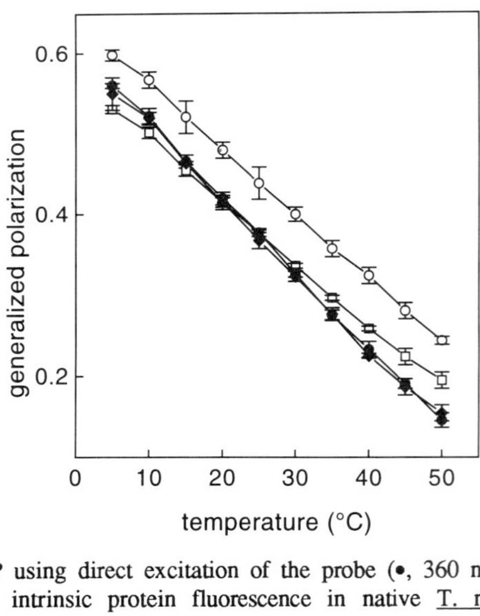

Fig. 3. Laurdan GP using direct excitation of the probe (•, 360 nm) or energy transfer conditions from the intrinsic protein fluorescence in native T. marmorata AChR-rich membranes (o, 290 nm) and from liposomes prepared from their extracted lipids (□, 290 nm; ♦, 360 nm). From Antollini et al. (1996).

We also studied GP in native *T. marmorata* AChR membranes and in liposomes prepared from their lipid extracts as a function of excitation (320-410 nm) and emission (420-500 nm) wavelengths. The spectra exhibited decreasing values of the excitation GP and increasing values of the emission GP with increasing wavelengths, indicating that dipolar relaxation occurs and that there is no selective excitation of subpopulations of Laurdan molecules (Antollini et al., 1996). This in turn suggests that the environment of Laurdan molecules in the native AChR-rich membrane displays the characteristics of the ordered-liquid phase (Parasassi et al., 1993a).

Sunshine and McNamee (1994) have speculated on the possibility that lateral phase separation occurs in the interaction of lipids with the AChR, such that the protein preferentially partitions into the more fluid domains. In our work with Laurdan the probe appeared to lie in a single phase with some of the characteristics of the liquid phase (presence of a dipolar relaxation process) and other properties of the gel phase (high GP values). Thus, the AChR-lipid interface region exhibits a) a higher lipid rigidity in the immediate vicinity of the AChR than in the rest of the membrane, and b) a single phase with intermediate properties, with the characteristics of the so-called liquid-ordered phase. Narayanaswami and McNamee (1993) suggested a more fluid lipid in the immediate vicinity of the AChR. The data on which their hypothesis is based were obtained, however, using reconstituted AChR, with lipids exclusively in the gel phase, and below the phase transition temperature. The *native* AChR-rich membrane, as used in the present study, possesses a high cholesterol:phospholipid ratio and also a high protein:lipid ratio (see review in Barrantes, 1989). Optimal fit of the energy transfer efficiency data indicated a value of 14±1 Å for the minimal donor-acceptor distance, which corresponds roughly to the diameter of the first-shell protein-associated lipid. Combining this piece of information with the data stemming from the GP experiments leads to the interpretation that Laurdan molecules within energy transfer distance, i.e. predominantly first-shell, protein-associated lipid, lie in a relatively *more rigid* environment than the bulk bilayer lipid region. Since Laurdan essentially measures the degree of water penetration in the bilayer (Parasassi et al., 1991) this implies that the lipid mobility in the vicinity of the protein is such that it allows for much lower penetration than the bulk, bilayer lipid. It is highly likely that the rigidifying effect exerted by the transmembrane portion of the AChR protein on boundary lipid is partly counterbalanced by the fluidizing effect of cholesterol located in this region, even though cholesterol in the non-vicinal, bulk lipid domain has the opposite, i.e. rigidifying effect. The *net* effect would thus be an *averaging* one, diminishing abrupt transitions between the two lipid domains.

Single-channel behavior of the AChR homologously and heterologously expressed in clonal cell lines.

The clonal cell lines BC3H-1 (Schubert et al., 1974; Boulter and Patrick, 1977) and CHO-AR42 (Forsayeth et al., 1990) were initially used to compare channel properties of the same γ-type, embryonic mouse muscle AChR (α, β, γ, δ), one of the best characterized AChR channels (Sine and Steinbach, 1984, 1986, 1988) in two potentially different membrane environments. The clone CHO-AR42, a non-muscle cell line in which cDNAs encoding α, β, γ and δ subunits from mouse muscle have been transfected in a stable form (Forsayeth et al., 1990) and which expresses an embryonic-type AChR with single-channel currents showing marked similarities with the endogenous AChR of BC3H-1 cells, was the ideal choice for this comparison. To widen the basis of our observations, we added the CHO-K1/A5 cell line, a new clone that in a stable form expresses the adult, ε-type AChR in the plasmalemma (Roccamo et al., 1993) and which has been obtained in our laboratory by cotransfection of cDNAs of the adult α, β, ε and δ AChR subunits into the CHO-K1 fibroblast cell line.

When ACh was present in the patch pipette at a concentration of 2 μM, activation of the AChR channel was manifested by single-channel currents appearing in isolated short-duration openings and bursts of long-duration openings interrupted by brief intra-burst closures (Sine and Steinbach, 1986, 1988). The increment in temperature caused in all cases a noticeable diminution in the duration of the single-channel events in both the native and the heterologous cellular systems. Furthermore, an increment in the single-channel unitary current was apparent in all cases as temperature increased, as analyzed below in energetic terms.

Thermodynamic analysis of single-channel data.

Eyring's transition state theory (Eyring, 1935) was applied for the calculation of the free energy (ΔG_a), enthalpy (ΔH_a) and entropy (ΔS_a) of activation of the channel's conductance and closure processes. This theory relates a kinetic process, which is a phenomenon that evolves over time, with the energy fluxes associated with the state changes relating the fundamental and the "activated" state of the channel, by means of the following equation:

$$\ln k = \ln (k_B T / h) - \Delta H_a / RT + \Delta S_a / R \qquad (4)$$

where k_B and h are the Boltzmann and Planck constants, respectively. ΔG_a as applied to the processes of channel conductance and gating was calculated from equation (4) as follows:

$$\Delta G_a = - RT \ln k + RT \ln k_B T / h \qquad (5)$$

where k has the same meaning as in equation (4). Values for ΔH_a and ΔS_a associated with the processes were calculated from E_a essentially as described in Zanello and Barrantes (1994), using the following relationships:

$$\Delta H_a = E_a - RT \quad (6)$$

$$\Delta S_a = -(\Delta G_a - \Delta H_a) / T \quad (7)$$

As a measure of the temperature dependence of ion conduction through the pore and on channel kinetics, Q_{10} values were determined from the ratios of the current and the kinetic constants at two temperatures (T_1 and T_2) differing by 10°C, according to the following equation:

$$Q_{10} = \exp[-E_a R^{-1} (T_2^{-1} - T_1^{-1})] \quad (8)$$

According to Eyring's theory, a system that undergoes a state change must overcome an initial energy barrier to reach the activated state. This would correspond to the activation energy empirically found by Arrhenius (1889). The transition state theory conceives this activated complex as a molecular entity having defined thermodynamic properties (see equations 4 to 8). Mean unitary main conductances of the AChR channels were found to increase linearly with increasing temperatures in the three clones studied. From Arrhenius plots constructed for the single-channel unitary conductance of AChR channels in BC3H-1, CHO-AR42 and CHO-K1/A5 cells the energy values associated with the ion conduction process through the AChR could be calculated (Table 1). The E_a value obtained for the γ-type AChR heterologously expressed in CHO-AR42 cells was slightly higher (4.7±0.3 kcal.mol^{-1}) than that obtained for the native embryonic-type AChR in BC3H-1 cells (3.6±0.3 kcal.mol^{-1}) (see also Dilger et al., 1991). In all cases the data were compatible with the existence of low energetic barriers for the passage of ions through the pore of the AChR channel.

For the thermodynamic analysis of the effect of temperature on AChR channel kinetics, the simplest sequential linear model relating open and closed states was considered. The channel closing rate, α, was calculated as the reciprocal of the long-duration mean open lifetimes ($\alpha = \tau_2^{-1}$) on the basis of the classical linear four-state scheme for AChR activation (see review in Lingle et al., 1992) which has proved applicable to the endogenous AChR in BC3H-1 cells (Sine and Steinbach, 1986, 1988):

$$2A + R \underset{k_{-1}}{\overset{k_1}{\rightleftharpoons}} A + AR \underset{k_{-2}}{\overset{k_2}{\rightleftharpoons}} A_2R \underset{\alpha}{\overset{\beta}{\rightleftharpoons}} A_2R^* \quad (9)$$

where two agonist molecules (A) bind to the AChR (R) with association rate constants k_1 and k_2 and dissociation rate constants $k_{-1}$and k_{-2}. The closed

TABLE 1. Thermodynamic parameters for the conductance and closure processes in the native and the heterologously expressed (γ-) and (ε-)AChR type of channel in different clonal cell lines.

Process	Channel conductance and closure rate α	E_a	Q_{10}	Thermodynamic parameters		
				ΔH_a	ΔG_a	ΔS_a
BC3H-1:						
Unitary conductance	46.0±2.8 pS	3.6	1.24	3.0	14.9	-40.6
Channel closure	399±174 s^{-1}	12.9	2.18	12.3	13.7	-4.6
CHO-AR42:						
Unitary conductance	47.2±2.6 pS	4.7	1.33	4.1	14.9	-36.8
Channel closure	426±95 s^{-1}	14.1	2.34	13.5	13.6	-
CHO-K1/A5:						
Unitary conductance	62.8±6.4 pS	3.6	1.24	3.0	14.7	-40.1
Channel closure	1,704±405 s^{-1}	13.15	2.21	12.6	12.8	-

The channel closing rate, α, is defined according to kinetic scheme (9), and corresponds to a membrane potential of -70 mV. All values were calculated from recordings obtained at 20°C. E_a, ΔH_a and ΔG_a are in kcal.mol^{-1}. ΔS_a is in cal.mol^{-1}.°K^{-1}. (-) corresponds to a calculated ΔS_a near zero.

biliganded channel, A_2R, moves to the open configuration (A_2R^*) with an opening rate constant, β, and closes with a closing rate constant α. Values for E_a and the thermodynamic parameters associated with channel closure are listed in Table 1. Q_{10} values (1.7) similar to the ones reported by us (Zanello et al., 1996) were obtained by Sachs and Lecar (1977), whereas Nelson and Sachs (1979) reported lower figures (1.0 and 1.3, respectively). The latter work on the temperature dependence of AChR single-channel conductance and gating kinetics using the patch-clamp technique reported no evidence of discontinuous behavior between 17 and 37°C. Dreyer et al. (1976) had previously reported an abrupt two-fold increase in channel conductance at about 25°C using macroscopic techniques; similarly, Lass and Fishbach (1976) and Fishbach and Lass (1978) had found a nonlinear behavior in the Arrhenius plot of muscle AChR channel conductance, also using macroscopic recordings.

Bulk lipid physical state in living AChR-expressing cells studied using Laurdan GP.

The spectroscopic properties of Laurdan have most conveniently been explained using the concept of GP introduced by Gratton and coworkers (Parasassi et al. 1990, 1991). Characteristic GP values in the pure gel and in the pure liquid-crystalline phospholipid phase have been determined. We have taken advantage of these properties of Laurdan to learn about the physical characteristics of the lipid in the plasmalemma of living cells where the AChR protein is inserted and its function measured in parallel experiments.

GP was found to decrease linearly with increasing temperature, without apparent discontinuities, i.e. without evidence of thermotropic phase transitions, in the three clones (Fig. 4).

GP values for BC3H-1 cells were lower than those obtained for the CHO clones at a given temperature, indicating a difference in the physical state of their cell membranes. The GP value of BC3H-1 cells at 5°C, for instance, was close to that found in CHO-AR42 cells at 12.5°C, and the corresponding value at 22.5°C was approximately that found in the CHO clones at the maximal explored temperature of 35°C. Furthermore, GP varied in a more pronounced manner with heating in BC3H-1 cells than in the other two clones of the CHO family. Q_{10} values of 1.3 and 1.2 were obtained for BC3H-1 and CHO cell lines, respectively (Table 1). This can be related to the ability of the probe to sense the diffusional movement of water molecules in the lipid bilayer (Parasassi et al., 1994). Laurdan excited state dipole is several Debyes larger than its ground state dipole, and can align solvent dipoles having molecular dynamics of the same order of magnitude as the excited state lifetime (Weber and Farris, 1979). The energy required for the solvent reorientation is reflected in the red shift of Laurdan fluorescence emission (Weber and Farris, 1979; Parasassi et al., 1990). It has been hypothesized that the origin of the dipolar relaxation lies in the presence of a few water dipoles at the hydrophobic-hydrophilic interface of the bilayer (Parasassi et al., 1994). The relaxation time of these water molecules is of the same order of magnitude as Laurdan excited state lifetime if the membrane is in the liquid-crystalline state. The rate of Laurdan dipolar relaxation is about $2.5 \times 10^9\ s^{-1}$ in the liquid-crystalline phase and almost negligible -about $4 \times 10^7\ s^{-1}$ - in the gel phase, in which water rotation is much slower and hence the dipoles cannot reorient themselves during the fluorescence lifetime of the probe (Parasassi et al., 1990, 1994). Ion permeation through the AChR channel has been analyzed in terms of an Eyring's model with one free-energy well of a single energy barrier site (Hoffmann and Dionne, 1983). Cations would first bind to this site before passing through the energetically less favorable region of the channel lumen, after acquiring sufficient kinetic energy to leave the site, either crossing the barrier with a virtually intact hydration shell and penetrating the membrane, or being rejected back to the extracellular milieu (Hoffmann and Dionne, 1983). Within this framework, Laurdan may serve as a useful reporter group for the molecular dynamics of water molecules at the hydrophobic-hydrophilic interface in the membrane and for the phase state of the lipid in which the channel protein is immersed. The fluorescence properties of Laurdan do not depend on selective

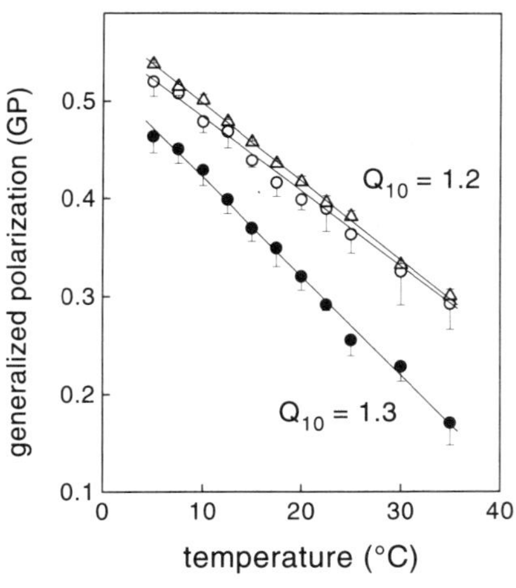

Fig. 4. Generalized polarization (GP) values of the fluorescent probe Laurdan obtained from BC3H-1, CHO-AR42 and CHO-K1/A5 cells as a function of temperature. Symbols are as in previous figures. From Zanello et al. (1996).

partitioning in cholesterol-rich or -poor areas (Parasassi et al., 1994) nor does Laurdan exhibit preferential affinity for the lipid region in the vicinity of the AChR in the *Torpedo* AChR-rich membrane (Antollini et al., 1996).

Correlation between single-channel and fluorescence GP data.

When mean unitary conductances obtained for the AChR channels expressed in each cell clone were plotted versus GP values at the corresponding temperature, a linear relationship was apparent between the two parameters (Fig. 5). The ionic conductance through the embryonic AChR channel in the native BC3H-1 cell system exhibited the lowest sensitivity to changes in GP, as can be seen from the slope of the curves (see legend to Fig. 5). Despite differences in absolute conductance values, the thermal sensitivity of the channel conductance exhibited by embryonic and adult AChRs expressed in the heterologous CHO cell lines were found to be similar. The physical state of CHO-AR42 and CHO-K1/A5 cell membranes sensed by Laurdan GP were similar, and also exhibited temperature dependence, albeit of slightly lower magnitude that that of the BC3H-1 cell (cf. Fig. 5).

A further correlation between the physical state of the lipid bilayer and the conductance process through the AChR channel could be inferred from the present results; the Arrhenius plot indicates that the conductance process through the γ-type AChR heterologously expressed in the clone CHO-AR42 displays the highest temperature dependence, in agreement with the results of Fig. 4. It can be seen that the GP values of BC3H-1 cells were smaller but varied more ostensibly between 5° and 35°C than those of CHO cells (see Fig. 4). The concept of the influence of the membrane environment on AChR function is reinforced by the observation that the conductance of the *same* γ-type AChR protein, expressed in two qualitatively *different* lipid environments (the BC3H-1 and the CHO-AR42 clones), appeared to have different energetic requirements (cf. Table 1). An important conclusion can be drawn from this observation: some functional properties of the AChR revealed by its single-channel behavior do not depend on its molecular constitution but on the lipid microenvironment in which it is inserted.

The above concept is further reinforced by another observation: the unitary conductance of the ε-type AChR channel in CHO-K1/A5 cells varied within a narrower range of GP values than those observed in BC3H-1 cells, and fell within the range of the γ-type AChR in CHO-AR42 cells (Fig. 5). Furthermore, an E_a value nearly equal to that found for the γ-type AChR in the native BC3H-1 cells was found for the ε-type AChR heterologously expressed in the CHO-K1/A5 clone. The latter exhibited a higher dependence of the conductance process on GP -and by inference on membrane fluidity- than the BC3H-1 cell, similar to that exhibited by the γ-type AChR in the CHO-AR42 clone.

The GP values as a function of temperature were correlated with single-channel electrophysiological data in an attempt to evaluate the possible direct modulation of the AChR single-channel activity by the physical state of the lipid environment (Fig. 6). In doing so it became apparent that GP was more discriminant than temperature, revealing differences in the thermal behavior of single-channels that passed inadverted with the latter parameter (Fig. 4). In fact, comparison of Figs. 5 and 6 indicates that it is not temperature itself, but rather a complex property dependent on temperature and sensed by Laurdan that determines functional variations in AChR behavior. That is, the functional properties of the AChR are influenced by changes in the *physical state* of the membrane which are in turn influenced by temperature; differences were more conspicuous at higher temperature values. For instance, the native γ-type AChR channel in BC3H-1 cells exhibited the lowest GP values and the highest thermal sensitivity of GP among the three clones. The same γ-type AChR protein, heterologously expressed in CHO cells, showed a higher sensitivity to small changes in GP than that of the native AChR expressed endogenously in the BC3H-1 cell (Fig. 6). On the other hand, even when the CHO-AR42 and CHO-K1/A5 clones (expressing heterologous embryonic and adult AChR respectively) had similar GP values (cf. Fig. 4), the kinetic processes that determine the rate of ion passage through the pore and lead to channel closure showed a higher sensitivity to the physical state of the membrane reported by GP than that of the homologous BC3H-1 cell system (Fig. 5).

One conclusion from the Laurdan GP work is that there is a single liquid-crystalline lipid phase in the AChR-rich membrane, whose rigidity increases towards the immediate vicinity of the AChR protein. From the combined single-channel/Laurdan fluorescence work on different cell lines expressing embryonic

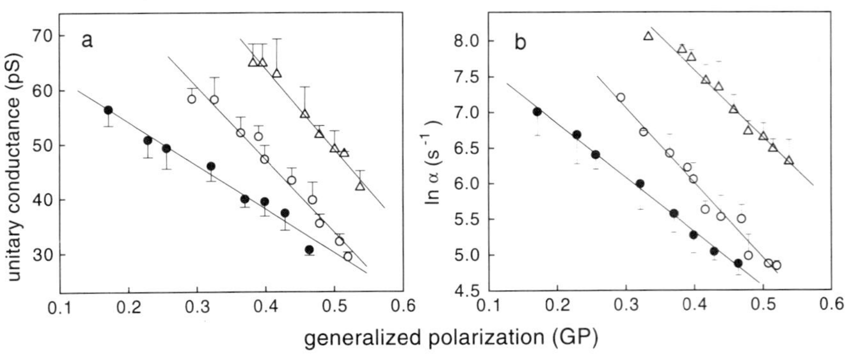

Fig. 5. Relationship between AChR single-channel conductance (a) or channel closing rate α (b) and Laurdan GP values for BC3H-1, CHO-AR42, and CHO-K1/A5 cells respectively. Symbols are as in previous figures. From Zanello et al. (1996).

or adult AChR we can also conclude that the same channel protein can exhibit different functional properties depending on the membrane environment in which it is inserted. That is, ion permeation and channel kinetics do not appear to rely exclusively on AChR subunit composition but are clearly dependent on the lipid and membrane-associated water microenvironment of the protein, which may subtly influence the microscopic activity of individual AChR molecules embedded in the bilayer. This reinforces the importance of specific lipid-protein interactions in addition to bulk membrane properties.

ACKNOWLEDGEMENTS

Thanks are due to Silvia Antollini for her help in the preparation of the illustrations. This work was supported by grants from the Argentinian Scientific Research Council (CONICET), Universidad Nacional del Sur, the European Union No. CI1*-CT94-0127, and the Volkswagen Stiftung to F.J.B.

REFERENCES

Antollini, S. S., Soto, M. A., Bonini de Romanelli, I., Gutiérrez- Merino, C., Sotomayor, P., and Barrantes, F. J., 1996, Physical state of bulk and protein-associated lipid in nicotinic acetylcholine receptor-rich membrane studied by Laurdan generalized polarization and fluorescence energy transfer, *Biophys. J.* **70:**1275-1284.

Arrhenius, S. A., 1889, Uber die Reaktiongeschwindigkeit bei der Inversion von Rohrzucker durch Sauren, *Z. Phys. Chem.* **4**:226-248.

Barrantes, F. J., 1978, Agonist mediated changes of acetylcholine receptor in its membrane environment, *J. Mol. Biol.* **124**:1-26.

Barrantes, F. J., 1989, The lipid environment of the nicotinic acetylcholine receptor in native and reconstituted membranes, *Critic. Rev. Biochem. Molec. Biol.* **24**:437-478.

Barrantes, F. J., 1993a, Structural-functional correlates of a rapid ligand-gated channel and its membrane environment: the nicotinic acetylcholine receptor, *FASEB J.* **7**:1460-1467.

Barrantes, F. J., 1993b, The lipid annulus of the nicotinic acetylcholine receptor as a locus of structural-functional interactions, in: *Protein-Lipid Interactions. New_Comprehensive Biochemistry* (A. Watts, ed.), Elsevier, Amsterdam, vol. 26, pp. 231-257.

Boulter, J., and Patrick, J., 1977, Purification of an acetylcholine receptor from a non-fusing muscle cell line, *Biochemistry* **16**:4900-4908.

Chattopadhyay, A., and McNamee, M. G., 1991, Average membrane penetration depth of tryptophan residues of the nicotinic acetylcholine receptor by the parallax method, *Biochemistry* **30**:7159-7164.

Chong, P.L.-G., 1988, Effect of hydrostatic pressure on the location of PRODAN in lipid bilayers and cellular membranes, *Biochemistry* **27**:399-404.

Chong, P. L.-G., 1990, Interactions of PRODAN and Laurdan with membranes at high pressure, *High Press. Res.* **5**:761-763.

Dilger, J. P., Brett, R. S., Poppers, D. M., and Liu, Y., 1991, The temperature dependence of some kinetic and conductance properties receptor channels, *Biochim. Biophys. Acta* **1063**:253-258.

Fishbach, G. D., and Lass, Y., 1978, A transition temperature for acetylcholine channel conductance in chick myoballs, *J. Physiol.* **280**:527-536.

Fong, T. M., and McNamee, M. G., 1986, Correlation between acetylcholine receptor function and structural properties of membranes, *Biochemistry* **25**:830-840.

Eyring, H., 1935, The activated complex in chemical reactions, *J. Chem. Physics.* **3**:107-115.

Forsayeth, J. R., Franco, A., Jr., Rossi, A. B., Lansmann, J. B., and Hall, Z. W., 1990, Expression of functional mouse muscle acetylcholine receptors in Chinese Hamster Ovary cells, *J. Neurosci.* **10**:2771-2779.

Gutiérrez-Merino, C., 1981, Quantitation of the Förster energy transfer for two-dimensional systems. II. Protein distribution and aggregation state in biological membranes, *Biophys. Chem.* **14:**259-266.

Gutiérrez-Merino, C., Munkonge, F., Mata, A. M., East, J. M., Levinson, B. L., Napier, R. M., and Lee, A. G., 1987, The position of the ATP binding site on the (Ca^{2+} --Mg^{2+})-ATPase, *Biochim. Biophys. Acta* **897**:207-216.

Hoffmann, H. M., and Dionne, V. E., 1983, Temperature dependence of ion channel permeation at the endplate channel, *J. Gen. Physiol.* **81**:687-703.

Lass, Y., and Fishbach, G. D., 1976, A discontinuous relationship between the acetylcholine-activated channel conductance and temperature, *Nature* **263**:150-151.

Magleby, K. L., and Stevens, C.F., 1972, A quantitative description of end-plate currents, *J. Physiol.* **223**:151-171.

Lingle, C. J., Maconochie, D., and Steinbach, J. H., 1992, Activation of skeletal muscle nicotinic acetylcholine receptors, *J. Membr. Biol.* **126**:195-217.

Marsh, D., and Barrantes, F. J., 1978, Immobilized lipid in acetylcholine receptor-rich membranes from *Torpedo marmorata*, *Proc. Natl. Acad. Sci. U.S.A.* **75**:4329-4333.

Narayanaswami, V., and McNamee, M. G., 1993, Protein-lipid interactions and *Torpedo californica* nicotinic acetylcholine receptor function. 2. Membrane fluidity and ligand-mediated alteration in the accessibility of gamma subunit cysteine residues to cholesterol, *Biochemistry* **32**:12420-12427.

Nelson, D. J., and Sachs, F., 1979, Single ionic channels observed in tissue-cultured muscle, *Nature* **282**:861-864.

Parasassi, T., De Stasio, G., d'Ubaldo, A., and Gratton, E., 1990, Phase fluctuation in phospholipid membranes revealed by Laurdan fluorescence, *Biophys. J.* **57**:1179-1186.

Parasassi, T., De Stasio, G., Ravagnan, G. Rusch, R., and Gratton, E., 1991, Quantitation of lipid phases in phospholipid vesicles by the GP of Laurdan fluorescence, *Biophys. J.* **60**:179-189.

Parasassi, T., Ravagnan, R., Rusch M., and Gratton, E., 1993, Modulation and dynamics of phase properties in phospholipid mixtures detected by Laurdan fluorescence. *Photochem. Photobiol.* **57**:403 -410.

Parasassi, T., Di Stefano, M., Loeiro, M., Ravagnan, G., and Gratton, E., 1994, Influence of cholesterol on phospholipid bilayers phase domains as detected by Laurdan fluorescence, *Biophys. J.* **66**:120-132.

Roccamo, A. M., Pediconi, M. F., Aztiria, E., Zanello, L., Politi, L. E., de los Santos, B., Wolstenholme, A., Werkman, I., and Barrantes, F. J., 1993, Stable heterologous expression of nicotinic acetylcholine receptor in temperature-sensitive CHO cells deficient in sphyngomyelin metabolism, *Proc. Arg. Soc. Neurochem.*, p. 44 (abstr.).

Rand, R. T., 1981, Interacting phospholipid bilayer: measured forces and induced structural changes, *Annual Rev. Biophys. Bioenerg.* **10**:277-314.

Sachs, F., and Lecar, H., 1977, Acetylcholine-induced current fluctuations in tissue-cultured muscle cells under voltage clamp, *Biophys. J.* **17**:129-143.

Schubert, D., Harris, A. J., Devine, C. E., and Heinemann, S., 1974, Characterization of a unique muscle cell, *J. Cell Biol.* **61**:398-413.

Sine, S. M., and Steinbach, J. H., 1984, Activation of a nicotinic acetylcholine receptor. *Biophys. J.* **45**:175-185.

Sine, S. M., and Steinbach, J. H., 1986, Acetylcholine receptor activation by a site-selective ligand: Nature of brief open and closed states in BC3H-1 cells, *J.Physiol.* **370**:357-379.

Sine, S. M., and Steinbach, J. H., 1988, Function of mammalian nicotinic acetylcholine receptors: Agonist concentration dependence of single channel current kinetics, in *Current Topics in Membranes and Transport.* **33**:133-145.

Sunshine, C., and McNamee, M. G., 1994, Lipid modulation of nicotinic acetylcholine receptor function: the role of membrane lipid composition and fluidity, *Biochim. Biophys. Acta* **1191**:59-64.

Unwin, N., 1993, The nicotinic acetylcholine receptor at 9 Å resolution, *J. Mol. Biol.* **229**: 1101-1124.

Weber, G., and Farris, F. J., 1979, Synthesis and spectral properties of a hydrophobic fluorescent probe: 6-propionyl-2-dimethylamino-naphtalene, *Biochemistry* **18**:3075-3078.

Zanello, L. P., and Barrantes, F. J., 1994, The effect of temperature on the K+ channel of Chara. A thermodynamic analysis, *Plant Cell Physiol.* **35**:243-255.

Zanello, L. P., Aztiria, E., Antollini, S. and and Barrantes, F. J., 1996, Nicotinic acetylcholine receptor channels are influenced by the physical state of their membrane environment. *Biophys. J.* **70**: 2155-2164.

AMILORIDE-SENSITIVE Na^+ CHANNELS AND HUMAN HYPERTENSION

Dale J. Benos[1], Bakhram K. Berdiev[1], Anne Lynn Bradford[1], James K. Bubien[2], Catherine M. Fuller[1], and Iskander I. Ismailov[1]

[1]Department of Physiology and Biophysics
The University of Alabama at Birmingham
Birmingham, Alabama 35294

[2]Department of Medicine
Division of Nephrology
The University of Alabama at Birmingham

INTRODUCTION

Amiloride-sensitive Na^+ channels play a critical role in Na^+ homeostasis in various tissues and cells throughout the body. With the recent cloning of an epithelial Na^+ channel (ENaC) by two independent groups in 1993[1,2] great strides in understanding the molecular function of these important ion channels have been made. Previously, it was thought that amiloride-sensitive Na^+ channels were confined only to epithelial tissues. However, recently work by Van Renterghem and Lazdunski[3], Kupitz and Atlas[4], Bubien and Warnock[5,6], and Lingueglia *et al.*[7] clearly demonstrate the existence of a regulated amiloride-sensitive Na^+ channel in non-epithelial cells such as vascular smooth muscle, oocytes, human and rat lymphocytes, and neurons, respectively. The distribution of amiloride-sensitive Na^+ channels is essentially ubiquitous. They are present in both the proximal and distal regions of the mammalian lung as well as in a number of sensory organs such as inner ear hair cells and gustatory epithelia[8].

From Ion Channels to Cell-to-Cell Conversations
Edited by Latorre and Sáez, Plenum Press, New York, 1997

Pharmacological, biophysical, and now molecular, evidence indicates that there is not a unique class of amiloride-sensitive Na^+ channel, but rather a family of these channels[9-11]. Amiloride-sensitive Na^+ channels can be differentiated on the basis of several criteria, namely, conductance[11], gating kinetics[11], or amiloride-sensitivity[9]. In addition, these channels appear to have a highly variable Na^+ to K^+ permselectivity. A recent report demonstrates that at least one class of amiloride-sensitive Na^+ channels, namely, the cyclic nucleotide-gated channels found in the kidney, are derived from a gene family different from the so-called ENaCs[12]. However, evidence from our laboratory indicates that some of the observed biophysical diversity may represent post-translational variations of the same gene product[13].

The cDNAs encoding the polypeptides that comprise the ENaC amiloride-sensitive Na^+ channel have been cloned from low salt or dexamethasone-treated rat distal colon[1,2,14]. This channel is a hetero-oligomeric complex consisting of at least three homologous subunits that have predicted molecular masses of 79, 72, and 75 kDa. When expressed in oocytes, this channel has a single channel conductance of approximately 5 pS, and exhibits ion selectivity, gating kinetics, and an amiloride pharmacological profile similar to the low conductance, highly selective Na^+ channel expressed in native Na^+ absorbing epithelial such as renal cortical collecting tubules[15,16]. At present, the relationships between this cloned channel and amiloride-sensitive channels biochemically purified from any native tissue source is unclear. However, it is conceivable that the biochemically purified Na^+ channels are either different isoforms of ENaC or products of a different Na^+ channel gene. Another likely possibility is that the cloned ENaCs represent only a portion of an *in situ* hetero-oligomeric complex. Recent experiments from our laboratory indicate that αrENaC antibodies recognize at least two polypeptides, namely, the 90 and 70 kDa associated proteins, of the renal Na^+ channel complex. Moreover, subsequent to disrupting intermolecular sulfhydryl bonds with low concentrations of DTT, immunopurified renal Na^+ channels display characteristics identical to ENaC, consistent with the hypothesis that the core conduction unit of this native channel is ENaC[17].

We have cloned a bovine renal papillary homolog of α-ENaC, namely, αbENaC. αbENaC by itself produces an amiloride-sensitive conductance when expressed in oocytes, and a mainstate 39 pS highly Na^+ selective channel in lipid bilayers. Single channels of comparable conductance were observed in bilayers using αrENaC. However, the simultaneous expression of α, β, γ-rENaC proteins in bilayers results in a channel of primarily 12-13 pS conductance, although the 39 pS state was still evident. Interestingly, all of these channels were activated by stretch[18]. Our current hypothesis is that the α-ENaC subunit forms the basic conduction unit, and that β and γ act as intrinsic regulatory elements of the ENaC channel. The availability of functionally competent purified channel protein as well as expressible channel cDNA for the first time enables the development of appropriate experiments to elucidate the nature of chemical block by amiloride and structurally related compounds. Moreover, these new molecular tools will permit a detailed analysis of mutational dysfunctions that may impinge upon channel regulatory mechanisms to produce specific diseases such as Liddle's Syndrome. In this article we will summarize our recent experiments concerning the biochemical isolation of an amiloride-sensitive Na^+ channel from human lymphocytes from patients affected with this form of genetic hypertension. We will also discuss a new hypothesis concerning the mechanism of constituitive activation of amiloride-sensitive Na^+ channels in this disease.

Hypertension and Liddle's Syndrome/Disease

Hypertension affects almost 60 million individuals in the United States alone, and produces severe cardiovascular and renal disease. Although anti-hypertensive therapy delays the onset of end-stage renal disease (ESRD) due to hypertension[19], hypertension-induced ESRD nonetheless accounts for at least 25% of all ESRD in the United States[20]. Over 90% of all cases of hypertension are due to unknown causes. Liddle's Syndrome is a rare form of autosomal dominant, low renin, volume-expanded, salt-sensitive hypertension. The basic clinical features of this syndrome (pseudohyperaldosteronism) were first described in an Alabama family in 1963[21]. The original patient developed renal failure in 1989 and underwent successful renal transplantation at UAB. The subsequent resolution of the patient's hypertension[22] supported Grant Liddle's original contention that this disorder results from a greatly enhanced ability of the kidneys to reabsorb salt. Further support for this idea comes from the clinical observations that the hypokalemia and hypertension associated with Liddle's syndrome can be alleviated with triamterene or amiloride therapy[23]. The Nephrology group at UAB has extended the early studies of a Liddle's syndrome patient's red blood cell Na^+-fluxes[24] and has shown, in affected B lymphocytes obtained from members of this pedigree, that there is a constituitive activation of amiloride-sensitive Na^+ channels[25]. The size of the original pedigree coupled with careful clinical evaluation of family members at risk permitted linkage analysis studies to be performed. These genetic studies identified the causal mutation of this hypertensive disease to exist in the β subunit of the recently cloned amiloride-sensitive Na^+ channel, ENaC[26]. Subsequently, mutations in the γ subunit of this channel have been observed; these mutations also result in a constituitive activation of these channels[27], and thus indicate genetic heterogeneity of this disease. The incidence of Liddle's syndrome may be much greater than initially appreciated[23].

All of the aforementioned mutations introduce premature stop codons or frameshift mutations that delete the last 45-76 amino acids of the cytoplasmically-located C-termini of either β or γ ENaC. Expression of β or γ subunits in *Xenopus* oocytes with normal α-ENaC produced a markedly enhanced amiloride-sensitive

Na^+ current as compared to oocytes injected with normal α, β, γ r-ENaC[28,29]. A recent paper from Lifton's group has identified a new mutation in a kindred diagnosed with Liddle's disease in which there is a single amino acid substitution (leucine for proline) at residue 616 in the β subunit of ENaC[30]. Again, expression of ENaC containing this missense mutation in oocytes resulted in a nearly 9-fold enhancement of amiloride-sensitive Na^+ current. All of these results demonstrate that a) abberrations in the primary structure of the cytoplasmic C-terminal regions of the β and γ subunits of ENaC result in constituitive activation of ENaC; b) β and γ subunits of ENaC are important in the modulation of ENaC activity; and c) Liddle's syndrome is a disease of amiloride-sensitive Na^+ channels.

Immunopurification and reconstitution of lymphocyte Na^+ channels

We have used the whole-cell and cell-attached patch clamp technique to characterize an amiloride blockable Na^+ conductance in rat and human lymphocytes. Whole-cell measurements from the lymphocytes (with a high K^+ concentration in the pipette) revealed small outward currents and ragged inward currents when the cells were voltage-clamped in 20 mV increments from -160 mV to +40 mV, from a holding potential of -60 mV (Fig. 1). The inward currents increased significantly upon addition of 40 μM 8-CPT-cAMP to the bathing solution, and this activation was completely reversed by 2 μM amiloride. The ion selectivity of this amiloride-sensitive conductance showed that it was equally

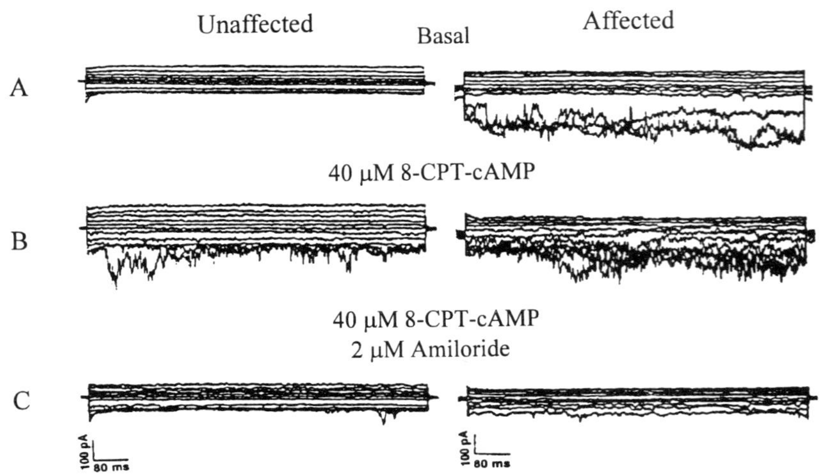

Fig. 1 Whole-cell currents in lymphocytes obtained from unaffected or a Liddle's Disease patient. A. Basal currents; B. Currents after cAMP; C. Effect of amiloride. From Ref. 34, with permission.

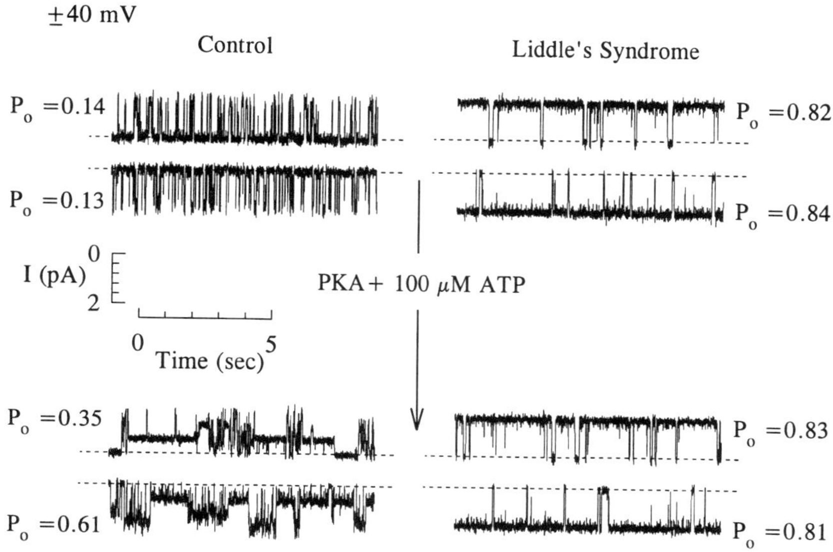

Fig. 2. Amiloride-sensitive Na^+ channels immunopurified from EBV-transformed human lymphocytes from normal (control) of Liddle's-affected individuals, and incorporated into planar lipid bilayers. From Ref. 33, with permission.

permeable to Na^+ and Li^+ with each ion having a 2-fold greater permeability than K^+. In fact, these same characteristics could be observed in isolated rat renal principal cells, i.e., amiloride-sensitive inward currents were stimulated by vasopressin[31]. For comparison, similar whole-cell data obtained from fresh lymphocytes isolated from a patient with Liddle's disease are shown (Fig. 1). Constituitive activation of inward amiloride-sensitive Na^+ current is apparent; this current could not be further activated with 8-cpt-cAMP.

Our laboratory has immunopurified and reconstituted functional amiloride-sensitive Na^+ channels from both rat and human lymphocytes[32,33]. This is a significant advance because it permits us to assess directly the molecular consequences of potential Na^+ channel dysfunction in various hypertensive groups. The existence of regulated amiloride-sensitive Na^+ channels in human lymphocytes and the preservation of the phenotype (i.e., constituitively activated Na^+ channels) in cells from Liddle's affected individuals make the lymphocyte an excellent model system with which to study this human disease. It is difficult to assess Na^+ channel function in human renal tissue for two reasons: first, the risk-to-benefit ratio precludes routine renal biopsy for ion channel experimentation, and second, the conditions required to disperse adult human renal cells from excised biopsy samples are severe because of excessive extracellular matrix proteins, and hence, adequate cell viability is difficult to achieve.

We have performed experiments in which we have reconstituted into planar lipid bilayers Na^+ channels immunopurified from Epstein-Barr virus (EBV)-transformed lymphocytes obtained from either unaffected or affected members of Liddle's kindred. The results are shown in Figure 2. The left half of the figure shows human Na^+ channels immunopurified from lymphocytes obtained from unaffected individuals and reconstituted into planar bilayers. The P_O of these channels under these conditions averaged 0.14, with a conductance of 35 pS. Upon application of protein kinase A plus ATP to one side of the bilayer (i.e., the side opposite from which amiloride was effective) the P_O increased to 0.45. Although not shown, PKA-mediated phosphorylation also induced an inward rectification to channel gating, i.e., channel P_O was significantly greater in the inward (i.e., negative) rather than outward (i.e., positive) direction. Similar observations have been observed in native rat or human lymphocyte whole-cell currents after elevation of intracellular cAMP, and for PKA-phosphorylated renal Na^+ channels[5,34]. Most interestingly, the Na^+ channel protein immunopurified from the Liddle's Syndrome lymphocytes showed a constituitive activation, i.e., a P_O of 0.81. The addition of protein kinase A plus ATP resulted in no further increase in P_O.

An important issue concerns the relationship of the Na^+ channel found in lymphocytes to ENaC. Our initial RT-PCR observations showed that while message for the β ENaC subunit could be detected in rat lymphocyte, RNA, the message for the α subunit could not, at least not at the predicted size (see Fig. 3). However, there was a faint, lower bp product present. Subsequent work in human lymphocytes using an entirely different set of primers has detected an α-like product[35]. Thus, we hypothesize that there is an isoform of α-rENaC present in lymphocytes. As a precedent, Waldmann *et al.*[36] have cloned another isoform of ENaC, termed δ-ENaC , that has characteristics most closely resembling α. Moreover, and pertinent to the Liddle's problem, Oh, Bubien, and Warnock[35] have established by RT-PCR/sequence analysis that the originally described Liddle's point mutation (CGA to TGA at Arg 564) does exist in the β subunit of ENaC, does exist in mRNA isolated from either fresh or Epstein-Barr virus (EBV)-transformed human B lymphocytes obtained from the original Liddle's proband. These observations support the contention that B lymphocytes are a useful model in which to study the dysfunctionality of amiloride-sensitive Na^+ channels in the human. Further, although PCR products of expected sizes for the α subunit were amplified from the lymphocyte mRNA, and these PCR products hybridized with α subunit probes on Southern blots, they showed different restriction mapping patterns (as compared to human or rat epithelial α-ENaC), suggesting that a novel human α-ENaC subunit may be expressed by lymphocytes. Moreover, in preliminary Western blot analysis, we have been able to detect both α and β ENaC subunits in immunopurified channel protein from human lymphocytes.

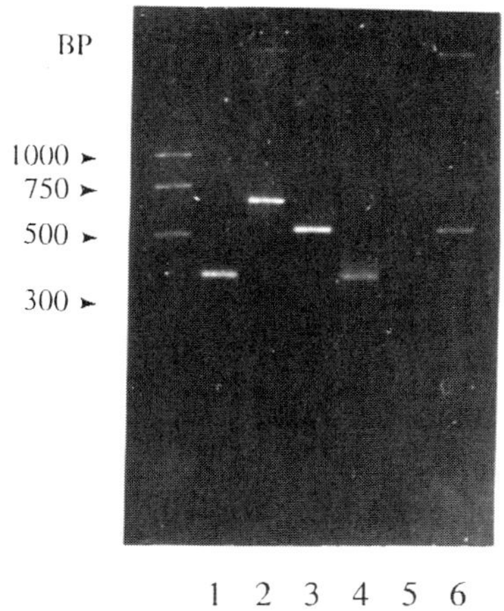

Fig. 3 Reverse transcription (RT)-polymerase chain reaction (PCR) analysis of α-rENaC, β-rENaC, and β-actin in RNA from rat lymphocytes and dexamethasone-treated colon. Primers for an actin product of 380 bp, for a α-rENaC product of 666 bp, and for a β-rENaC product of 525 bp were used. Lanes 1 and 4: β-actin in rat colon and lymphocytes, respectively. Lanes 2 and 3: α-rENaC and β-rENaC in colonic RNA. Lane 5: α-rENaC in rat lymphocytes RNA. Lane 6: β-rENaC in rat lymphocyte RNA. From Ref. 32, with permission.

As indicated earlier, our laboratory has also been successful in reconstituting into planar lipid bilayers *in vitro* translated protein products of α, β, and γ ENaCs[37]. Single channel activity with three unique conductance states, 13 pS, 26 pS, and 39 pS, was observed (Fig. 4). The smallest 13 pS conductance state was most frequently observed, reflecting very well the situation seen in patch-clamp experiments performed on ENaC-expressing oocytes[27]. A triple-barrel model of gating, accounting for these kinetic characteristics of ENaC, has been developed and tested[37]. PKA-induced phosphorylation had no effect on channel activity of either αrENaC alone, or α, β, γrENaC together (Fig. 4). Interestingly, all of the native Na^+ channels studied in bilayers, including those of the lymphocytes, display triple-barrel behavior, similar to the ENaCs.

Does the C-terminal cytoplasmic tail of β and γ ENaC act as an inactivation gate?

As described above, the defining characteristic of Liddle's hypertension is the constitutive activation of an amiloride-sensitive Na^+ channel, ENaC. Expression of the Liddle's phenotype has been linked to mutations in the C terminal cytoplasmic tail of the β subunit of ENaC, although mutations in the C terminal of the γ subunit also give rise to a Liddle's-like hypertensive state. The originally identified mutation (insertion of a premature stop codon at Arg 564) results in a channel with a dramatically increased open probability and/or more channels that

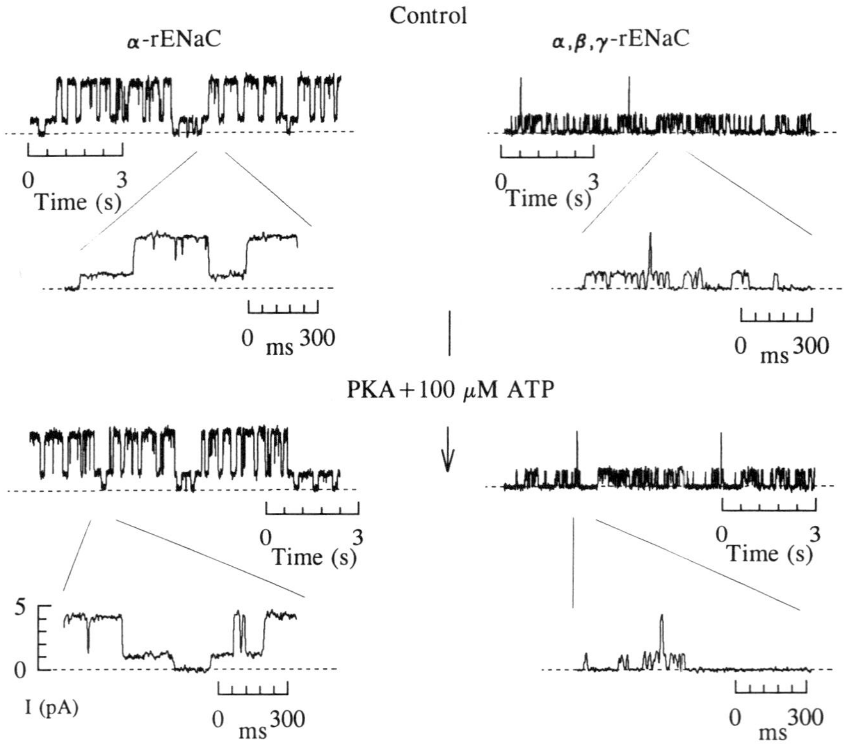

Fig. 4 α-rENaC and α,β,γ-rENaC in planar lipid bilayers. Top half of the figures shows single channel activity, while the bottom half shows that addition of the catalytic subunit of protein kinase A plus ATP had no effect on channel activity. From Ref. 33, with permission.

are trafficked to the cell surface[26,29]. Other β subunit mutations resulting in C terminal truncations also lead to a hypertensive phenotype, as does a recently identified point mutation in the C terminus, P616L[30]. Truncation of the last 76 amino acids of the γ subunit similarly results in a Liddle's type hypertension[27]. We hypothesize that the C-termini of β and γ ENaC, alone or in concert, may act as intrinsic conduction gates for the channel itself. Consistent with this hypothesis has been the observation that synthetic peptides comprising the last ten amino acids of either β or γ hENaC or βrENaC, or the last 30 amino acids of βhENaC, can block the amiloride-sensitive Na^+ channel immunopurified from the lymphocytes of Liddle's patients (Fig. 5). Examination of the peptide sequences of the β and γ blocking peptides reveals that these peptides contain several negatively charged

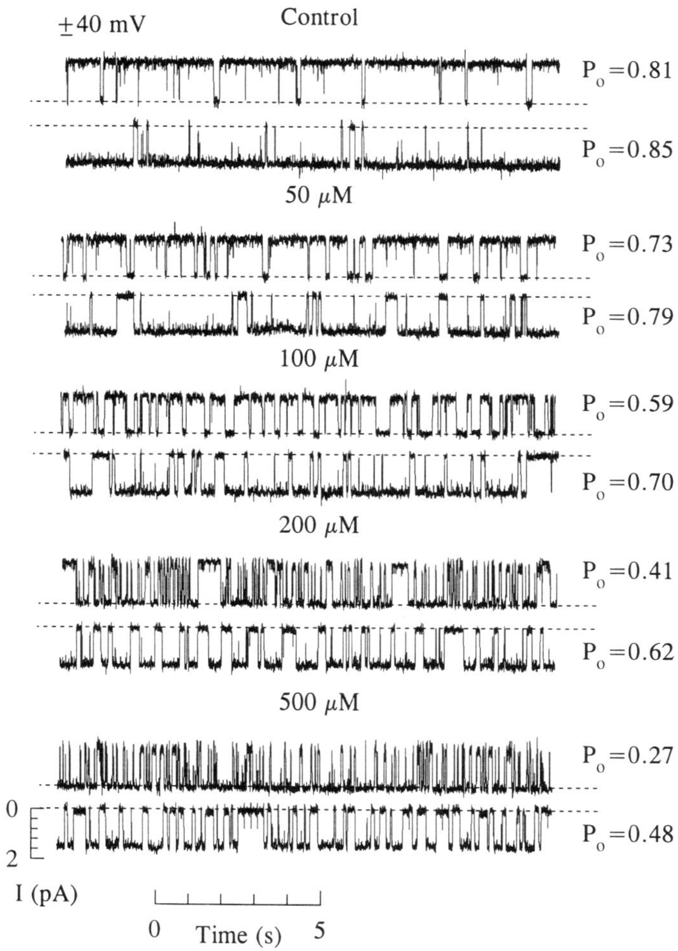

Fig. 5 Peptide block of constitutively activated, amiloride-sensitive Na^+ channel immunopurified from lymphocytes obtained from a Liddle's patient, and incorporated into a planar lipid bilayer. The peptide, NH_2-MESDSEVEAI-COOH is identical in sequence to the last 10 C-terminus amino acids of β-rENaC. From Ref.33, with permission.

residues interspersed by neutral hydrophobic amino acids. In contrast, synthetic peptides that contained either negative residues in a different sequence, or where the negative residues were changed to glycines, were ineffective as channel blockers. Moreover, the terminal ten residues of αhENaC did not block the channel[33]. We also found that the β-peptide blocked inward Na^+ currents in Liddle's lymphocytes[34], and in oocytes expressing truncated β-rENaC[38]. If the C-terminal regions of the β and γ subunits are indeed involved in gating of ENaC, exposure of the channels' internal surface to a proteolytic enzyme such as trypsin should increase single channel open probability (P_O). Preliminary experiments in bilayers reveal this prediction to be accurate. Addition of 0.5 mg/ml of trypsin to the solution bathing the presumptive 'cytoplasmic' surface increased α, β, γ-ENaC P_O from 0.45 ± 0.04 to 0.81 ± 0.08 (N=3). Moreover, addition of 200 μM β-peptide to this same surface restores normal ENaC gating. Cytoplasmic trypsinization also produced activation (i.e., an increase in P_O) of normal lymphocyte Na^+ channels in bilayers[33]. Thus, the C terminal tails of both β and γ ENaC subunits, may act as intrinsic gates for ENaC, analogous to the behavior of a 20 amino acid region at the NH_2 terminus of the Shaker K^+ channel ShB of *Drosophila*[39,40]. To date, potassium channels have been more extensively studied in terms of their structure/function relationships than the ENaCs. Consequently, there currently exists a more detailed understanding of the intrinsic structure-function relations for the K^+ channels. In the case of ShB, (and some other K^+ channels), positively charged residues seem to play a crucial role in the block of the channel, presumably interacting with the negatively charged channel mouth[41]. Similarly, in a second family of K^+ channel blockers, the peptide scorpion toxins, positively charged residues also are involved in the block of the channel, in this case situated on the exposed side of a β sheet[42,43]. We have performed preliminary circular dichroism (CD) studies and these have revealed that the ENaC blocking peptides also adopt a β sheet conformation in solution. However, the residues involved in the gating of the ENaC channel are unknown. Interestingly, only the β and γENaC C-termini are predicted to form β sheets. As mutation of the C termini of either the β or γ subunits can give rise to clinical hypertension, another hypothesis is that the cytoplasmic C-termini of β and γ ENaC associate in a single β-sheet and, in concert, regulate channel gating under physiological conditions. However, at present it is unclear whether (unlike the voltage-activated K^+ channels), it is the negatively charged residues that are important for interaction with the channel pore, or whether mutation of these residues simply causes sufficient structural disruption to prevent efficient β sheet formation, reducing the probability of closure. In order to address this question further, we are generating several different site-directed mutants to identify the region of the primary channel sequence that may form the mouth of the pore and/or interact with the putative inactivation particles, in addition to further substitutions in the sequence of the blocking peptides. We hypothesize that it is the α subunit that forms the conduction region of ENaC. The manner in which these particles interact with the inner mouth region of the conductive pore in the conductive α subunit of the ENaC channel will yield insights into the regulation of channel gating, precisely that property of channel behavior that is directly responsible for Liddle's hypertension.

Summary and Significance

Our observations suggest that amiloride-sensitive Na^+ channels are present in human lymphocytes, and that they operate in a manner comparable to those channels found in renal collecting tubules, including biochemical regulatory modulation. Moreover, these lymphocyte Na^+ channels have at least one gating mechanism similar to that utilized by voltage-sensitive Na^+ and K^+ channels. Our results not only provide new insights into how amiloride-sensitive Na^+ channels work, but provide a new understanding of how alterations in protein-protein interactions and/or channel regulatory pathways may be translated in overt manifestations of hypertensive disease.

AKCNOWLEDGEMENTS

This work was supported by NIH Grant DK37206. We thank Mrs. Charlae Starr for superb secretarial assistance.

REFERENCES

1. C.M. Canessa, J.-D. Horisberger, and B.C. Rossier. Epithelial sodium channel related to proteins involved in neurodegeneration. *Nature* **361**:467-470 (1993).

2. E. Lingueglia, N. Voilley, R. Waldmann, M. Lazdunski, and P. Barbry Expression cloning of an epithelial amiloride-sensitive Na^+ channel. A new channel type with homologies to *Caenorhabditis elegans* degenerins. *FEBS Lett.* **318**:95-99 (1993).

3. C. Van Renterghem, and M. Lazdunski. A new non voltage-dependent epithelial-like Na^+ channel in vascular smooth muscle cells. *Pflugers Arch.* **419**:401-408 (1991).

4. Y. Kupitz, and D. Atlas. A putative ATP-activated Na^+ channel involved in sperm-induced fertilization. *Science* **261**:484-486 (1993).

5. J.K. Bubien, and D.G. Warnock. Amiloride-sensitive sodium conductance in human B lymphoid cells. *Am.J. Physiol.* **265**:C1175-C1183 (1993).

6. D.G. Warnock and J.K. Bubien. Liddle Syndrome: Clinical and Cellular abnormalities. *Hosp. Practice,* **15**:95-105 (1994).

7. E. Lingueglia, G. Champigny, M. Lazdunski, and P. Barbry. Cloning of the amiloride-sensitive FMRFamide peptide-gated sodium channel. *Nature* **378**:730-733 (1995).

8. D.J. Benos, S.A. Cunningham, R.R. Baker, K.B. Beason, Y. Oh, and P.R. Smith. Molecular characteristics of amiloride-sensitive sodium channels. In: *Reviews of Physiology, Biochemistry, and Pharmacology*, Vol. 120, Springer-Verlag, pp.31-113 (1992).

9. P.R. Smith and D.J. Benos. Epithelial Na^+ channels. *Annu. Rev. Physiol.* **53**:509-530 (1991).

10. H. Garty. Molecular properties of epithelial amiloride-blockable Na^+ channels. *FASEB J.* **8**:522-528 (1994).

11. L.G. Palmer. Epithelial Na Channels: function and diversity. *Annu. Rev. Physiol.* **54**:51-66 (1992).

12. D. Vandorpe, C. Sandoe, K. Karlson, and B. Stanton. Cloning and characterization of the cGMP-gated cation channel from mouse kidney inner medullary collecting duct (IMCD). *J. Am. Soc. Nephrol.* **6**:354 (1995).

13. I.I. Ismailov, B.K. Berdiev, and D.J. Benos. Biochemical status of renal epithelial Na^+ channels determines apparent single channel conductance, ion selectivity, and amiloride-sensitivity. *Biophys. J.* **69**:1789-1800 (1995).

14. C. M. Canessa, L. Schild, G. Buell, B. Thoreus, I. Gantschl, J.-D. Horisberger, and B.C. Rossier. Amiloride-sensitive epithelial Na^+ channel is made of three homologous subunits. *Nature* **367**:463-467 (1994).

15. L.G. Palmer and G. Frindt. Amiloride sensitive Na^+ channels from the apical membrane of the rat cortical collecting tubules. *Proc. Natl. Acad. Sci. USA* **83**:2767-2770 (1986).

16. B.N. Ling, C.F. Hinton, and D.C. Eaton. Amiloride-sensitive sodium channels in rabbit cortical collecting tubule primary cultures. *Am. J. Physiol.* **261**:F933-F944 (1991).

17. I.I. Ismailov, B.K. Berdiev, A.L. Bradford, M.S. Awayda, C.M. Fuller, and

D.J. Benos. Associated proteins and renal epithelial Na^+channel function. *J. Membr. Biol.*, **149**:123-132 (1996).

18. M.S. Awayda, I.I. Ismailov, B.K. Berdiev, and D.J. Benos. A cloned renal epithelial Na^+ channel protein displays stretch activation in planar lipid bilayers. *Am. J. Physiol.:Cell* **268**:C1450-C1459 (1995).

19. R. Qualheim, S.G. Rostand, K.A. Kirk, E.A. Rutsky and R.G. Luke. Changing patterns of end-stage renal disease due to hypertension. *Am. J. Kid. Dis.* **23**:336-343 (1991).

20. U.S. Renal Data System: USRDS Annual Report. National Institutes of Health, p. 43-54, Bethesda, MD (1994).

21. G.W. Liddle, T. Bledsoe, and W.S. Coppage. A familial renal disorder simulating primary aldosteronism but with negligible aldosterone secretion. *Trans. Am. Assoc. Phys.* **76**:199-213 (1963).

22. M. Botero-Veley, J.J. Curtis, and D.G. Warnock. Brief report: Liddle's syndrome revisited. *New Engl. J. Med.* **330**:178-181 (1994).

23. M.F. Gadallah, K. Abreo, and J. Work. Liddle's syndrome, an underrecognized entity: a report of four cases, including the first report in black individuals. *Am. J. Kid. Dis.* 25:829-835 (1995).

24. J.D. Gardner, A.Lapey, A.P. Simopoulous, and E.L. Bravo. Abnormal membrane sodium transport in Liddle's syndrome. *J. Clin. Invest.* **50**:2253-2258 (1971).

25. D.G. Warnock, and J.K. Bubien. Cellular physiology of Liddle's syndrome. *Hosp. Pract.* **29**:95-105 (1994).

26. R.A. Shimkets, D.G. Warnock, C.M. Bositis, C. Nelson-Williams, J.H. Hansson, M. Schambelan, J.R. Gil, S. Ulick, R.V. Milora, J.W. Findling, C.M. Canessa, B.C. Rossier, and R.P. Lifton. Liddle's Syndrome: Heritable human hypertension caused by mutations in the beta subunit of the epithelial sodium channel. *Cell* 79:407-414 (1994).

27. J.H. Hansson, C. Nelson-Williams, H. Suzuki, L. Schild, R. Shimkets, Y. Lu, C.M. Canessa, T. Iwasaki, B. Rossier, and R.P. Lifton. Hypertension caused by a truncated epithelial sodium channel γ subunit: genetic heterogeneity of Liddle's syndrome. *Nature Genetics* **11**:76-82 (1995).

28. L. Schild, C. Canessa, R.A. Shimkets, I. Gautschi, R.P. Lifton, and B.C. Rossier. A mutation in the epithelial sodium channel causing Liddle disease increases channel activity in the *Xenopus laevis* oocyte expression system. *Proc. Natl. Acad. Sci. USA* **92**:5699-5703 (1995).

29. P.M. Snyder, M.P. Price, F.J. McDonald, C.M. Adams, K.A. Volk, B.G. Zeiher, J.B. Stokes, and M.J. Welsh. Mechanism by which Liddle's syndrome mutations increase activity of a human epithelial Na^+ channel. *Cell* **83**:969-978 (1995).

30. J.H. Hansson, L. Schild, Y. Lu, T.A. Wilson, I. Gautschi, R. Shimkets, R. Nelson-Williams, B.C. Rossier, and R.P. Lifton. A *de novo* missense mutation of the β subunit of the epithelial sodium channel causes hypertension and Liddle syndrome, identifying a proline-rich segment critical for regulation of channel activity. *Proc. Natl. Acad. Sci. USA* **92**:11495-11499 (1995).

31. J.K. Bubien. Whole cell sodium conductance for principal cells freshly isolated from rat cortical collecting duct. *Am. J. Physiol.* **269**:C791-C796 (1995).

32. A.L. Bradford, I.I. Ismailov, J.-M. Achard, D.G. Warnock, J.K. Bubien, and D.J. Benos. Immunopurification and functional reconstitution of an amiloride-sensitive Na^+ channel protein from rat lymphocytes. *Am. J. Physiol.:Cell* **269**:C601-C611 (1995).

33. I.I. Ismailov, B.K. Berdiev, C.M. Fuller, A.L. Bradford, R.P. Lifton, D.G. Warnock, J.K. Bubien, and D.J. Benos. Peptide block of constituitively activated Na^+ channels in Liddle's Disease. *Am. J. Physiol.:Cell* **270**:C214-C223 (1996).

34. J.K. Bubien, I.I. Ismailov, B.K. Berdiev, T. Cornwell, C.M. Fuller, R.P. Lifton, D.J. Benos, and D.G. Warnock. Liddle's Disease: abnormal regulation of amiloride-sensitive Na^+ channels by β-subunit mutation. *Am. J. Physiol.:Cell*, 270:C208-C213 (1996).

35. Y. Oh, J.K. Bubien, and D.G. Warnock. RT-PCR analysis of amiloride-sensitive sodium channel (ASSC) subunit mRNAs from human B cells. *J. Am. Soc. Neprhol.* **6**:348 (1995).

36. R. Waldmann, G. Champigny, F. Bassilana, N. Voilley, and M. Lazdunski. Molecular cloning and functional expression of a novel amiloride-sensitive Na^+ channel. *J. Biol. Chem.* **270**:27411-28414 (1995).

37. I.I. Ismailov, M.S. Awayda, B.K. Berdiev, J.K. Bubien, J.E. Lucas, C.M. Fuller, and D.J. Benos. Triple-barrel organization of ENaC, a cloned epithelial Na^+ channel. *J. Biol. Chem.* **271**:807-816 (1996).

38. Y. Oh, M.S. Awayda, J.K. Bubien, D.J. Benos, and D.G. Warnock. Expression studies of truncated amiloride-sensitive sodium channel β subunits in *Xenopus* oocytes. *Kidney Intl.*, submitted (1996).

39. T. Hoshi, W.N. Zagotta, and R.W. Aldrich. Biophysical and molecular mechanisms of Shaker potassium channel inactivation. *Science* **250**:533-538 (1990).

40. W.N. Zagotta, T. Hoshi, and R.W. Aldrich. Restoration of inactivation of mutants of shaker potassium channels by a peptide derived from ShB. *Science* **250**:568-571 (1990).

41. M. Kukuljan, P. Labarca, and R. Latorre. Molecular determinants of ion conduction and inactivation in K^+ channels. *Am. J. Physiol.* **268**:C535-C556 (1995).

42. F. Bontems, B. Gilquin, C. Roumestand, A. Menez, and A. Toma. Analysis of side-chain organization on a refined model of charybdotoxin: structural and functional implications. *Biochemistry* **31**:7756-7764 (1992).

43. C.-S. Park and C. Miller. Mapping function to structure in a channel-blocking peptide: electrostatic mutants of charybdotoxin. *Biochemistry* **31**:7749-7755 (1992).

44. M.S. Awayda, I. I. Ismailov, B.K. Berdiev, C.M. Fuller, and D.J. Benos. Protein kinase regulation of a cloned epithelial Na^+ channel. *J. Gen. Physiol.*, in press (1996).

ANION CHANNELS INVOLVED IN VOLUME REGULATION: A COMMON PATHWAY FOR CHLORIDE AND ORGANIC OSMOLYTE PERMEATION?

Andrés Stutzin[1], Ana Luisa Eguiguren[1], L. Pablo Cid[1], and Francisco V. Sepúlveda[1,2].

[1]Departamento de Medicina Experimental, Facultad de Medicina Universidad de Chile. Casilla 70058 Correo 7, Santiago, Chile
[2]Centro de Estudios Científicos de Santiago, Casilla 16443 Correo 9, Santiago, Chile

INTRODUCTION

Volume regulation is a property common to many animal cell types (for references, *see* reviews[1,2]) and when exposed to anisotonic media they modify their intracellular concentration of osmolytes by activation of permeability pathways, leading to osmotically-driven water shifts. In response to swelling induced by hypotonicity, cells lose KCl and organic osmolytes to recover their normal volume, a process known as regulatory volume decrease (RVD).

Although RVD has been extensively studied in a variety of cells, including epithelial cells[3,4], the underlying mechanism is poorly understood. In fact, there is agreement on the ionic species released upon RVD activation, however the sensing, transducing and regulation mechanisms remain essentially unknown.

In this chapter we will address the issue of the anion permeability pathways involved in RVD using the epithelial HeLa cell line as a model. Volume-activated chloride channels in HeLa cells have been previously characterised[5]. Organic osmolyte efflux induced by hypotonicity has also been demonstrated in HeLa cells[6]. These osmolytes are found in high concentrations (10-100 mM) in the cytoplasm of many cells[2]. These molecules play a key role in osmoregulation[7] and may be involved in cellular protective functions[8]. Organic osmolytes are accumulated via energy-dependent transport processes or synthesised from metabolic precursors. Upon cell swelling, osmolyte efflux occurs in a rather fast (seconds), passive and non-saturable manner[9,10]. Interestingly, structurally dissimilar organic osmolytes apparently share a common efflux pathway[11]. This has been supported by the

From Ion Channels to Cell-to-Cell Conversations
Edited by Latorre and Sáez, Plenum Press, New York, 1997

similarities observed for the amino acid, polyol and methylamine loss in diverse cell types and by its inhibition by anion transport blockers[12,13]. The latter observation poses an important question relative to the nature of the exit pathways for organic osmolytes. The transport characteristics for these solutes suggest that their loss is mediated by an anion channel. This hypothesis had first been proposed for swelling-induced efflux of taurine and other amino acids from MDCK cells[4] and a strong argument in favour was presented recently by Roy and Banderali[14]. These authors demonstrated the existence of a swelling-activated, outwardly rectifying anion channel with a relatively high permeability for taurine, glutamate and aspartate in MDCK cells. They postulate this channel, therefore, as a candidate to mediate efflux of both organic osmolytes and chloride ions. The work of Kirk and Kirk in a human lung cancer cell line, in flounder erythrocytes and indeed in HeLa cells[6,15,16] suggests, based mainly upon pharmacological evidence, that taurine and Cl^- move through the same Cl^- permeable channel activated by hypotonicity.

This question, however, is not solved for every cell type. Lambert and Hoffmann[17] in an elegant piece of work reported a differential activation and blockade by some unsaturated fatty acids of swelling-activated taurine and Cl^- efflux in Ehrlich mouse ascites tumour cells. This observation coupled to the differential effects of membrane potential strongly favour the presence of two distinct types of channels or permeability pathways mediating cell swelling-regulated taurine and Cl^- efflux.

As described in this chapter, our results indicate that in HeLa cells hypoosmotic conditions trigger a complex anion-exit response modulated by external Cl^- ions. Thus it appears that external chloride is needed for a Cl^--preferring conductance activated upon cell swelling. The halide, however, seems to inhibit a swelling-induced, taurine-preferring permeability pathway. These observations, in addition to pointing to a heterogeneity in swelling-induced organic osmolyte/anions exit pathways, also suggest a central role for Cl^- in the modulation of these channels or carriers.

METHODS

Cells. The experiments to be described were performed in HeLa cells. The cells were grown to about 50% confluency at 37°C in a 5%/95% : CO_2/air atmosphere in Dulbecco´s modified Eagle´s medium supplemented with 5% foetal calf serum, 80000 I.U/l penicillin and 80 g/l streptomycin.

Electrophysiological experiments. For electrophysiological experiments, cells were seeded onto 35 mm diameter cell culture Petri dishes, which could be directly mounted on the stage of an inverted microscope. Standard whole-cell patch-clamp recordings were performed as described elsewhere[18] using a EPC-7 amplifier (List Medical, Germany). Cells were continuously superfused by gravity with a Cl^--rich, isotonic (315 mosmol/l) solution and a peristaltic pump was used to remove fluid so that the level in the chamber was kept constant. Local changes in composition of the medium around the cell under study were achieved by using a U-tube arrangement as described in[19]. Patch-pipettes were fabricated from thin-walled borosilicate (hard) glass capillary tubing of outside diameter 1.5 mm (Clark Electromedical, Reading, U.K.) and had resistances of 2-4 MΩ when filled with standard intracellular solutions. Pipette solutions always contained 2 mM Na_2ATP with an osmolarity of 287 mosmol/l, the composition of solutions is given in the figure legends. Voltage and current signals from the amplifier were recorded on a digital tape recorder (DTR-1204, Biologic, France) and digitised using a 386-based computer equipped with a Labmaster (Scientific Solutions, Inc., USA) AD/DA interface at a sampling rate of 500 Hz. The voltage pulse generator and voltage-clamp analysis program were written by J. Dempster (University of Strathclyde, Glasgow, Scotland).

Efflux experiments. ^{3}H-taurine and ^{125}I efflux from HeLa cells was measured simultaneously. Cells were grown to 80% confluency (usually 48-72 hours after plating) in 10 mm-diameter, six-well plates. They were loaded with ^{3}H-taurine (1 μCi/ml) and ^{125}I (5 μCi/ml) by incubation for 2-2.5 hours at 37°C in culture medium without foetal calf serum. Immediately before beginning each efflux experiment, the loading solution was removed and the cells washed (3 times at room temperature) with isoosmotic medium. The medium contained in each well (0.75 ml) was replaced at timed intervals and saved for counting. Cells were allowed to equilibrate for 4 minutes and then challenged with the hypoosmotic solutions (see figure legends). The radioactivity remaining in the cells at the end of the experiment was determined after disruption with 0.75 ml 0.25 M NaOH. Samples were first counted in a gamma counter and then in a beta-scintillation counter. Samples containing pure ^{3}H and ^{125}I were counted simultaneously to measure the contribution of ^{125}I radioactivity to beta counting. The amount of ^{3}H-taurine and ^{125}I remaining in the cells as a function of time decreased following a single exponential function. The data were expressed as efflux rate constants. All the experiments were done at room temperature.

VOLUME-ACTIVATED ANION CURRENTS ARE MODULATED BY EXTRACELLULAR CHLORIDE

The effect of lowering the osmolarity of the bathing medium on anion currents in a HeLa cell is depicted in Fig. 1. The composition of the intracellular and extracellular solutions was designed to exclude currents other than anionic. Currents were monitored by stepping the voltage to 120 and -80 mV for 600 ms, from a holding potential of 0 mV. Small currents, linearly related to voltage were observed in isoosmotic conditions. Upon switching to a Cl^--rich hypoosmotic extracellular solution, a slow-developing outwardly-rectifying current was recorded. Under these conditions the zero-current potential was close to 0 mV, consistent with anion permeation. Concomitantly with the development of the currents cell swelling was observed under the phase-contrast microscope. The currents showed a marked outward rectification which was greater at the beginning of the hypotonic challenge. The role of Cl^- as the charge carrier was tested by progressive replacement of the anion with sucrose. Lowering the Cl^- concentration in the bathing solution decreased both, the outward and inward currents and shifted the reversal potential towards more positive values. As described in[5], the current decays during the pulse at depolarised potentials (120 mV). It can be observed that this decay was favoured at low Cl^- concentrations.

The observation that inward currents decrease when the chloride concentration is lowered in the extracellular solution was unexpected. In fact, under the experimental conditions (high N-methyl-D-glucamine-Cl^- in the pipette) defined for

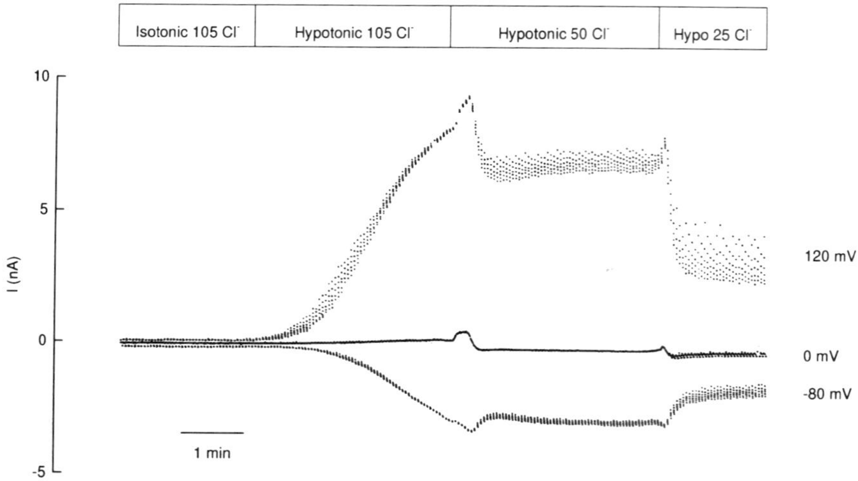

Fig. 1. Volume-activated anion currents as a function of the external chloride concentration. Anion currents were recorded using the patch-clamp technique under the standard whole-cell conditions with the following solutions (in mM): *pipette*: N-methyl-D-glucamine-Cl 105, $MgCl_2$ 1.2, mannitol 70, Tris-HEPES 10, EGTA 1, Na_2-ATP 2, pH 7.4, 287 mosmol/l; *isotonic* Cl^-: NaCl 105, $MgCl_2$ 0.5, $CaCl_2$ 1.3, mannitol 100, Tris-HEPES 10, pH 7.4, 315 mosmol/l; *hypotonic 105* Cl^-: NaCl 105, $MgCl_2$ 0.5, $CaCl_2$ 1.3, Tris-HEPES 10, pH 7.4, 230 mosmol/l; *hypotonic 50* Cl^-: NaCl 50, $MgCl_2$ 0.5, $CaCl_2$ 1.3, sucrose 110, Tris-HEPES 10, pH 7.4, 231 mosmol/l; *hypotonic 25* Cl^-: NaCl 25, $MgCl_2$ 0.5, $CaCl_2$ 1.3, sucrose 160, Tris-HEPES 10, pH 7.4, 237 mosmol/l. Upon switching to hypotonic 105 Cl^- solution, sizeable inward and outward rectifying currents developed. Decreasing the extracellular chloride concentration and keeping the osmolality constant resulted in progressive decrease of both, outward and inward currents.

this experiment, the inward currents (chloride leaving the cell) are expected to increase as the driving force for chloride exit increases. To test whether the anomalous effect on inward current was due specifically to a decrease in Cl^- concentration, rather than the decrease in ionic strength caused by sucrose replacement of NaCl, Cl^- was also replaced by glutamate. Similar results were observed suggesting that the effect was due to Cl^- removal rather than decrease in salt concentration. This non-Goldman-Hodgkin and Katz (GHK) behaviour can be readily seen from Fig. 2 where the peak current at -80 mV is plotted as a function of the extracellular anion (Cl^- or glutamate) concentration. Using the relative permeability coefficient ($P_{glutamate}/P_{Cl}$) measured at the highest extracellular Cl^- concentration it is possible to calculate the expected current for a process following GHK as a function of the anion concentrations. The observed values departed considerably from the behaviour predicted by GHK formalism.

A possible explanation for the anomalous effect of extracellular chloride on chloride currents is that exposure of the cells to low chloride concentration causes a depletion of the intracellular anion, thus causing a decrease in inward anion current. In order to test for this, experiments were performed at low Cl^-, high glutamate intracellular solution and the effect of replacing all but 5 mM extracellular Cl^- with

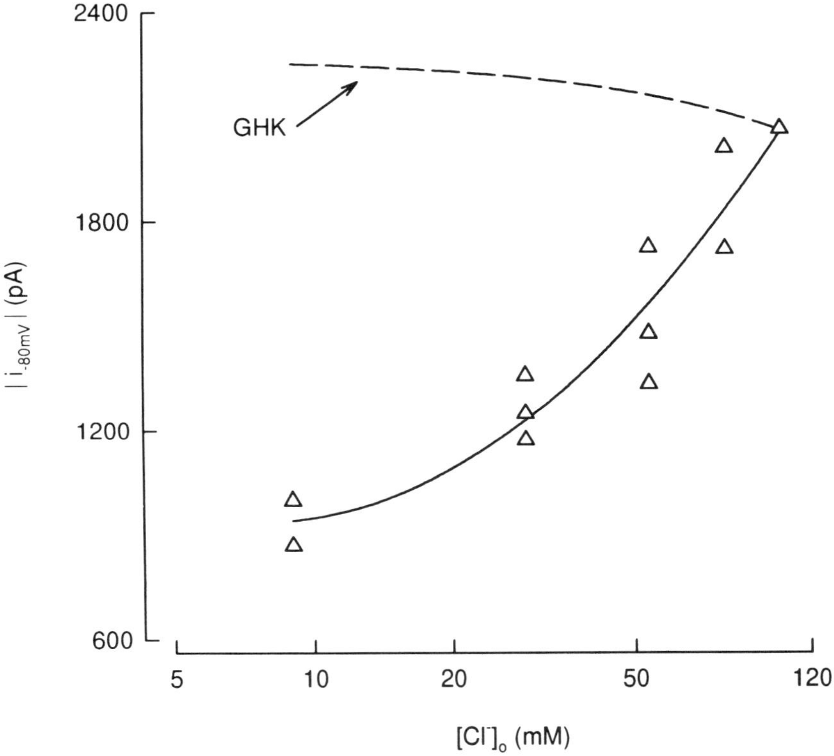

Fig. 2. Inward currents deviate from GHK formalism. Plot of peak inward currents recorded at -80 mV (n=3) as a function of the external anion (chloride or glutamate) concentration. The solid line depicts the expected current for a process following a GHK formalism and was calculated from the GHK current equation using the apparent permeability coefficients ($P_{glutamate}/P_{Cl}$)measured at the highest Cl^- concentration.

sucrose or glutamate explored. As shown in Fig. 3, replacement of Cl^- with an hypotonic-sucrose (H_S) solution or with an hypotonic-glutamate (H_G) solution had similar effects on the inward currents.

The time course of the transients in low extracellular Cl^- are examined in more detail in Fig. 4. The peak outward and inward currents at -80 and 120 mV are plotted against time from the onset of the replacement (time zero) until reaching an apparent steady state condition when either Cl^- or glutamate was the main extracellular anion. As seen in the figure, the data could be fitted to single exponential decays. No difference could be seen between the time constants in H_S or H_G at equal voltages. As during sucrose, but not glutamate, replacement the driving force is for anion exit, the similarity in efflux time course effectively rules out

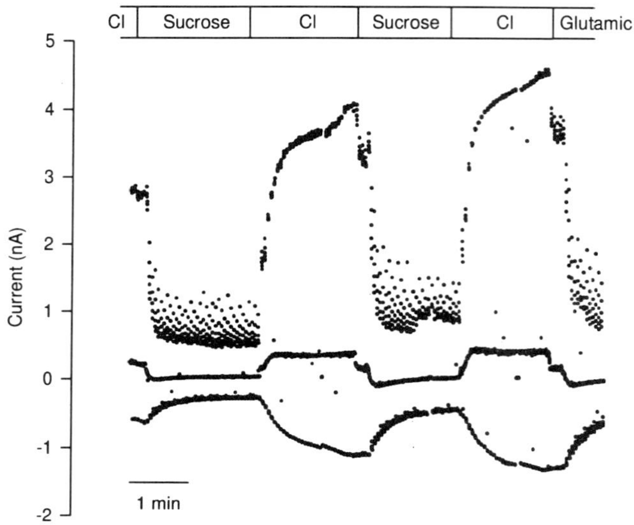

Fig. 3. Glutamate and sucrose share the same effect. Whole-cell volume-activated current in a Cl^--rich extracellular hypotonic solution. In this experiment, the pipette contained (in mM): N-methyl-D-glucamine-Cl 5, N- methyl-D-glucamine-glutamate 100, $MgCl_2$ 1.2, mannitol 70, Tris-HEPES 10, EGTA 1, Na_2-ATP 2, pH 7.4, 275 mosmol/l After changing to low-chloride high-sucrose hypotonic solution (NaCl 5, $MgCl_2$ 0.5, $CaCl_2$ 1.3, Tris-HEPES 10, sucrose 200, pH 7.4, 245 mosmol/l), the inward and outward currents decreased in a comparable size to the reduction observed in low-chloride high-glutamate hypotonic solution (NaCl 5, Na-glutamate 100, $MgCl_2$ 0.5, $CaCl_2$ 1.3, Tris-HEPES 10, pH 7.4, 213 mosmol/l). The effect of Cl^- removal and replacement by sucrose was fully reversible.

depletion as a possible mechanism of inward current decrease. Although no differences between replacements can be measured, there is a 3-4 fold difference between the time constants at -80 and 120 mV in either H_S or H_G. The time constants obtained for the decrease in outward currents are relatively fast (0.08 and 0.1 min) and probably reflect simply the washout of chloride ions from the solution bathing the cell. On the other hand, the time constants at -80 mV were, as stated above, 3-4 times slower (0.08 *vs* 0.32 and 0.1 *vs* 0.31 min) and cannot be solely attributed to the washout of chloride from the vicinity of the membrane. Instead, if it is assumed that chloride is necessary at a specific site in the channel, it could be speculated that this time course reflects the dissociation of the anion from such a putative site. It is suggested that such a binding site is accessible from the extracellular aspect of the membrane. If it were assumed that such a regulatory site exists, it would show a selectivity sequence similar to that of the permeation pathway itself[5], as Br^- and SCN^- are capable of supporting inward currents when used as replacements of Cl^-.

Taken together, these results rule out the role of ionic strength and suggest that this phenomenon is linked rather specifically to the interaction of Cl^- with some site on the external aspect of the cell membrane. In other words, it may be possible that Cl^- permeation through volume-activated anion channels in HeLa cells is specifically modulated by external Cl^-.

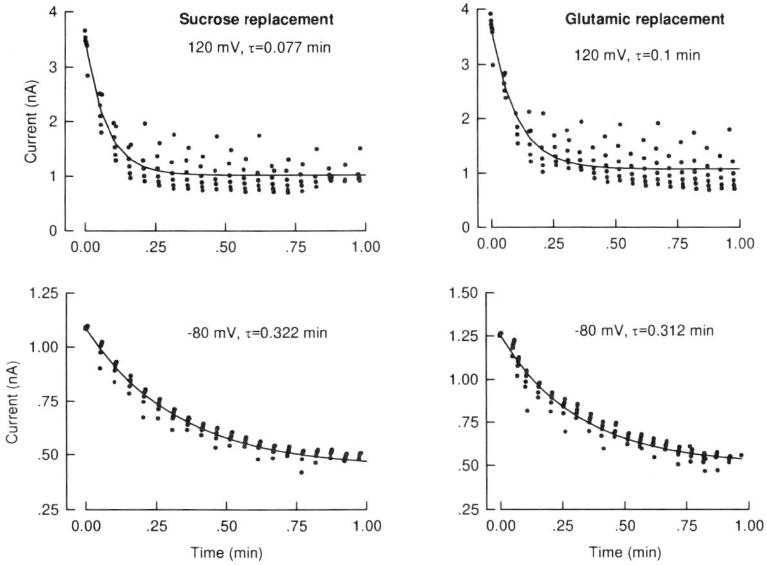

Fig. 4. The decrease in inward current is not due to anion depletion. The peak currents at -80 and 120 mV in low-chloride high-sucrose hypotonic solution and in low-chloride high-glutamate hypotonic solution (see Fig. 2) were fitted to exponential decay functions. No differences were observed between solutions at equal voltages, however, a marked difference was observed for depolarising and hyperpolarising pulses for each condition, ruling out an anion (Cl^-) depletion mechanism.

^{125}I AND TAURINE EFFLUX FROM HeLa CELLS ARE DIFFERENTLY AFFECTED BY EXTRACELLULAR CHLORIDE.

As discussed above, it has been postulated that organic osmolytes share a common efflux pathway with chloride ions during RVD. The rather specific "regulatory" effect of extracellular anion provides a way to test this hypothesis as Cl^- as it would be predicted that replacement of the halide in the extracellular space should lead to a reduction in organic osmolyte efflux. In order to study if inorganic and organic osmolytes share the same efflux pathway upon cell swelling, double tracer experiments were conducted. The cells were loaded with ^{125}I, a halide with permeability properties similar to chloride ions and which has been extensively used as surrogate for Cl^- in radiotracer flux experiments. In many cells the most important organic osmolyte accumulated intracellularly and lost during RVD is the sulphonic amino acid taurine. Therefore ^{3}H-labelled taurine and ^{125}I efflux was studied in HeLa cells challenged with hypotonic stimulus. Under control, isotonic conditions HeLa cells lost previously loaded ^{125}I and ^{3}H-taurine with efflux being well described by single exponential functions. The respective rate constants were around 0.1 min^{-1} for ^{125}I and 0.02 min^{-1} for ^{3}H-taurine. They remained unaltered for up to 15 min and were not affected by replacement of extracellular Cl^- by glutamate. However, challenging the cells with hypotonic solution markedly increased the rate of efflux for both compounds. The left-hand panel of Fig. 5 shows the increase in rate constant for efflux (hypotonic efflux rate constant - isotonic efflux rate constant) of

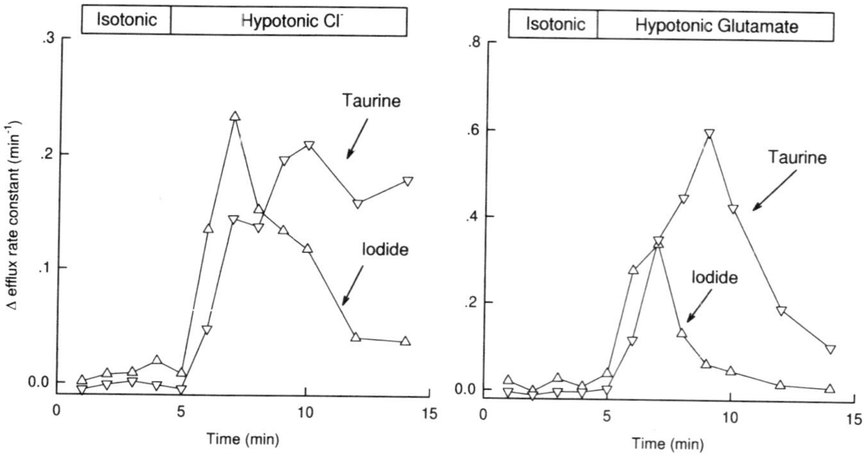

Fig. 5. Hypotonically-induced ^{125}I and ^{3}H-taurine efflux. The efflux rate constants were calculated with the formula $k=cpm_i/\Sigma(cpm_{i=0}..cpm_{i=n})-cpm_i/2$. The data (± SEM) are from n=3 experiments. Left panel: ^{3}H-taurine and ^{125}I efflux under hypotonic Cl^--rich solution (in mM: NaCl 105, KCl 5, $MgCl_2$ 0.5, $CaCl_2$ 1.3, Tris-HEPES 10, pH 7.4, 202 mosmol/l). The change to the hypotonic solution was done after 4 min in isotonic conditions. Iodide efflux preceeds the taurine efflux and is more transient. The peak rate constants are similar. Right panel: ^{3}H-taurine and ^{125}I efflux under hypotonic glutamate-rich solution (in mM: NaCl 5, KCl 5, Na-glutamate 95, $MgCl_2$ 0.5, $CaCl_2$ 1.3, Tris-HEPES 10, pH 7.4, 198 mosmol/l). Taurine efflux follows a more transient temporal course as observed in high-Cl^- conditions, presumably because to a faster regulatory volume decrease.

iodide and taurine in an experiment where a high (105 mM) Cl^- concentration was maintained throughout. The increased in efflux rate was transient for iodide whilst it appeared sustained for the aminoacid within the time frame of the experiments. When the hypotonic challenge experiment was performed in a low Cl^-, glutamate-rich hypotonic solution (right-hand panel in Fig. 5) a slightly larger increase in efflux rate constant for iodide was observed. The increase in taurine efflux, however, was three-fold larger than in the Cl^--rich hypotonic solution. For both compounds the increase was transient, the fall in iodide efflux being faster than in high Cl^-. In both hypotonic conditions, ^{125}I efflux preceded that of ^{3}H-taurine efflux. The peak efflux rate constants for ^{125}I and the cumulative efflux were similar under both conditions. ^{3}H-taurine efflux, on the other hand, was three-fold higher in the absence of external Cl^- (ie hypotonic glutamate solution), cumulative amino acid efflux being about 2.2-fold higher.

The observation that iodide and taurine follow different time courses, in addition to the differential effects of external chloride ions on their efflux rate suggests that there must be two different anion pathways with differing selectivities. Under quasi-physiological (ie high extracellular chloride concentration) conditions, both pathways are activated. However, in the absence of external Cl^-, a permeabilty preferring taurine predominates over a halide preferring one.

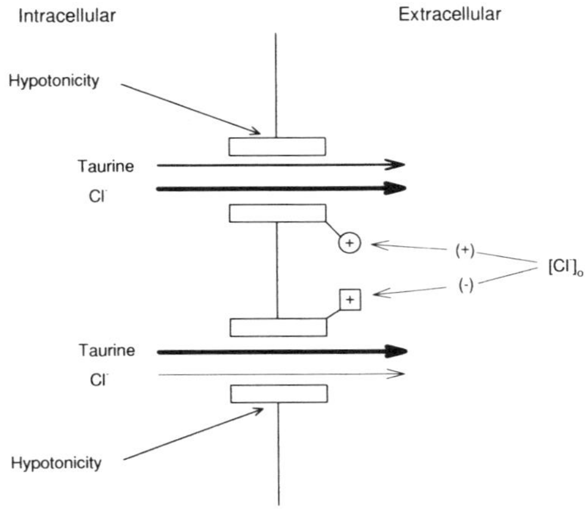

Fig. 6. Model for anion efflux regulation in RVD by external Cl^-. Cell swelling is induced after exposure to hypotonic conditions. To regain a normal volume, cells lose osmolytes. Anion osmolytes exit the cell via carriers or channels. The model depicted considers two separate transport pathways with different selectivity for halides and organic osmolytes. These two pathways are differentially modulated by external Cl^- ions, which bind to a putative site in the outer aspect of the cell membrane. In normal conditions (ie high extracellular Cl^-), hypotonicity activates both pathways, however in the absence of external Cl^- the taurine pathway predominates, suggesting that Cl^- plays a role in its own intracellular homeostasis.

CONCLUSIONS

Volume-activated anion pathways are a requisite for normal cell volume regulation. In view of the complex dynamic response observed here, it seems feasible that more than one anion pathway becomes activated upon cell swelling. Cl^- itself, a precious intracellular anion, seems to be playing a key role in modulating the differential exit of intracellular Cl^- and organic osmolytes available for RVD. A working model for this differential modulation by extracellular Cl^- is shown in Fig. 6. Such effects of Cl^- in modulating Cl^- channel activity are not unprecedented. Recent work from the group of Jentsch[20] have demostrated such effect to reside upon a single lysine residue in a ClC-family Cl^- channel. More work will be neededed, however, to substantiate the hypothesis summarised in the final figure of this paper.

Acknowledgements. The authors wish to thank to the Departamento de Fisiología y Biofísica, Facultad de Medicina, Universidad de Chile for using the beta scintillation counter and the Sección de Medicina Nuclear, Departamento de Medicina, Hospital Clínico José Joaquín Aguirre, Universidad de Chile for the gamma counting facility. This work was supported by Fondecyt grants 1931011 and 1940408.

REFERENCES

1. E.K. Hoffmann and L.O. Simonsen, Membrane mechanisms in volume and pH regulation in vertebrate cells, Physiol. Rev. 69:315 (1989).

2. M.E. Chamberlin and K. Strange, Anisosmotic cell volume regulation: A comparative view, Am. J. Physiol. 257:C159 (1989).

3. U. Banderali and G. Roy, Activation of K^+ and Cl^- channels in MDCK cells during volume regulation in hypotonic media, J. Membr. Biol. 126:219 (1992).

4. G. Roy and C. Malo, Activation of amino acid diffusion by a volume increase in cultured kidney (MDCK) cells. J. Membr. Biol. 130:83 (1992).

5. M. Díaz, M.A. Valverde, C.F. Higgins, C. Rucareanu and F.V. Sepúlveda, Volume-activated chloride channels in HeLa cells are blocked by verapamil and dideoxyforskolin, Pflügers Arch. 422:347 (1993).

6. J. Kirk and K.L. Kirk, Inhibition of volume-activated I^- and taurine efflux from HeLa cells by P-glycoprotein blockers correlates with calmodulin inhibition. J. Biol. Chem. 269:29389 (1994).

7. A. García-Pérez and M.B. Burg, Renal medullary organic osmolytes. Physiol. Rev. 71:1081 (1991).

8. C.E. Wright, H.H. Tallan, Y.Y. Lin and G.E. Gaull, Taurine: Biological update. Ann. Rev. Biochem. 55:427 (1986).

9. J.K. Haynes and L. Goldstein, Volume-sensitive amino acid transport in erythrocytes of the little skate, *Raja erinacea*. Am. J. Physiol. 265:R173 (1993).

10. R. Sánchez-Olea, H. Pasantes-Morales, A. Lázaro and M. Cereijido, Osmolarity-sensitive release of free amino acids from cultured kidney cells (MDCK), J. Membr. Biol. 121:1 (1991).

11. K. Strange and P.S. Jackson, Swelling-activated organic osmolyte efflux: A new role for anion channels. Kidney Int. 48:994 (1995).

12. U. Banderali and G. Roy, Anion channels for amino acids in MDCK cells, Am. J. Physiol. 263:C1200 (1992).

13. R. Sánchez-Olea, C. Pena, J. Moran and H. Pasantes-Morales, Inhibition of volume regulation and efflux of osmoregulatory amino acids by blockers of Cl- transport in cultured astrocytes, Neurosci. Lett. 156:141 (1993).

14. G. Roy and U. Banderali, Channels for ions and amino acids in kidney cultured cells (MDCK) during volume regulation, J. Exp. Zool. 268:121 (1994).

15. K. Kirk and J. Kirk, Volume regulatory taurine release from a human lung cancer cell line: Evidence for amino acid transport via a volume-activated chloride channel. FEBS Lett 336:153 (1993).

16. K. Kirk, J.C. Ellory and J.D. Young, Transport of organic substrates via a volume-activated channel. J. Biol. Chem. 267:23475 (1992).

17. I.H. Lambert and E.K. Hoffmann, Regulation of taurine transport in Ehrlich ascites tumor cells. J. Membr. Biol. 131:67 (1993).

18. O.P. Hamill, A. Marty, E. Neher, B. Sakmann and F.J. Sigworth, Improved patch-clamp techniques for high-resolution current recording from cell and cell-free membrane patches. Pflügers Arch. 391:85 (1981).

19. S. Suzuki, M. Tachibana and A. Kaneko, Effects of glycine and GABA on isolated bipolar cells of the mouse retina. J. Physiol. (Lond) 421:645 (1990).

20. M. Pusch, U. Ludewig, A. Rehfeldt and T.J. Jentsch, Gating of the voltage-dependent chloride channel ClC-0 by the permeant anion. Nature 373:527 (1995).

VOLUME REGULATED ANION CHANNELS AND CYTOSKELETAL INTERACTION

Irena Levitan and Sarah S. Garber

Department of Physiology
2900 Queen Lane
Allegheny University of the Health Sciences
Philadelphia, PA 19129.

INTRODUCTION

All cells must be able to regulate their volume when exposed to variations in external osmolarity. Although mammalian cells generally are not exposed to wide swings in osmolarity because of their highly regulated environment, peripheral blood cells are exceptions. Blood cells, including lymphocytes, are routinely exposed to osmotic stress as they pass through the vasa recta of the kidney, where the osmolarity can change by 800-1000 mOsm. Cells in other tissues which are not normally exposed to changes in osmolarity, however, are also able to exhibit volume regulatory responses that may become important in a variety of disease or trauma states (e.g. edema, stroke, diabetes, polydipsia) which may result in a transient or sustained change in osmolarity.[1, 2]

Cells swell in response to an acute exposure to an external solution hypotonic to the cytosol. An initial response to swelling in many cells is to reduce the intracellular concentration of K^+ and Cl^- so that water will flow out of the cell, allowing the volume to return to normal. Recovery to a normal cell volume occurs within minutes and is termed 'regulatory volume decrease' (RVD).[2, 3] Cell swelling results in the outward flow of Cl^- through volume-

From Ion Channels to Cell-to-Cell Conversations
Edited by Latorre and Sáez, Plenum Press, New York, 1997

regulated anion channels (VRAC) which then leads to the opening of voltage-sensitive K^+ channels.[3-6] VRAC are members of the mechanosensitive ion channel family. The signal responsible for transducing the change in volume to the activation of VRAC is not yet clear.

Several possible mechanisms to account for VRAC activation in response to cell swelling have been suggested. One hypothesis suggests that fusion of intracellular vesicles containing VRAC within their membranes and the plasma membrane as a result of osmotic stress. This hypothesis has been discounted by two groups on the basis of direct measurements of membrane capacitance that show no correlation between VRAC activation and an increase in membrane surface area.[7, 8] Another hypothesis is that VRAC is activated directly by tension within the lipid bilayer of the plasma membrane that is created by cell swelling. This mechanism has been suggested for the activation of several members of the mechanosensitive ion channel family.[9-11] A third suggested mechanism of VRAC activation is transduction of tension to the ion channel protein through membrane-cytoskeletal interactions.[12-14]

In this chapter, we address the mechanism of VRAC activation in response to cell swelling by tension, either in the lipid bilayer itself or through cytoskeletal interaction. Our results support the hypothesis that membrane-cytoskeletal tension, rather than tension within the membrane bilayer alone, regulates VRAC activity and is therefore involved in volume regulation in mammalian cells.

METHODS

RPMI 8226 myeloma cells are a continuous cell line that grows in suspension. The shape of these cells can be well approximated by a sphere, allowing less ambiguity in measurements of diameter or calculated volume. Myeloma cells were grown in suspension with RPMI 1640 (Cell Grow of Fisher Scientific/Mediatech, Washington DC) supplemented with 10% fetal bovine serum (Gibco BRL, Grand Island, NY). Cells were fed and split every 2-3 days. In preparation for recording, cells are washed 2x with RPMI 1640 without media, allowed to settle on an uncoated glass cover slip and washed with extracellular recording solution before experiments.

Standard external recording solution contains (in mM): 155 NaCl, 2 $MgCl_2$, 1 $CaCl_2$, 10 HEPES, pH 7.2. Intracellular solutions contain (in mM): 120 Cs^+-Glutamate, 10 HEPES, 4 ATP, pH 7.2 (CsOH) with free $[Ca^{++}]$ ~10 nM (0.1 $CaCl_2$, 1.1 EGTA). Changes in the composition of recording solutions are indicated in figure legends or text. Substitution of Cs^+ by TEA or inclusion of 0.2 mM GTP (in addition to ATP) in intracellular solution had no effect on activation of VRAC. Intracellular solutions contained Cs^+-glutamate to eliminate any potential contamination of Cl^- currents by outward K^+ currents. The composition of these solutions results in an outward current or inward Cl^- flux at voltages more positive than the reversal potential (-55 to -60 mV).

The osmolarity of all solutions was determined immediately before use with a vapor pressure osmometer (Wescor Inc., Logan UT). The osmolarity of the internal solution was adjusted by addition the sucrose. Relative osmolarity of external solutions is expressed as a percentage of the internal osmolarity ([external osmolarity/internal osmolarity]*100). VRAC activation was dependent only on the osmotic gradient between internal and external solutions. Activation was independent of whether the osmolarity of the extracellular solution was decreased with respect to the internal solution or the osmolarity of the internal solution was raised with respect to the external solution. Increasing the intracellular osmotic and ionic strength with CsCl, instead of sucrose, resulted in an equivalent activation of VRAC. Current activation is also independent of the osmolarity of the media in which cells were grown.

Macroscopic ionic currents were measured using the whole cell configuration of the standard patch clamp technique (Hamill, et al., 1981). Pipettes were pulled from N51A glass (Garner Glass, Claremont, CA), coated with Sylgard (Dow Corning Corp. Midland, MI), and fire polished to give a final resistance of 2-6 MΩ, in the above recording solutions. Currents were recorded using an EPC9 amplifier (HEKA Electronik, Lambrecht, Germany), ITC-16 interface (Instrutech, Inc., Great Neck, NY) and accompanying acquisition and analysis software (Pulse & PulseFit, HEKA Electronik, Lambrecht, Germany) running on a Macintosh Quadra 700 or 800. Further analysis employed IGOR (WaveMetrics, Inc. Lake Oswego OR) and Microsoft EXCEL (Microsoft, Redmond, WA). Junction potentials were nulled immediately before seal formation. Pipette and whole cell capacitance were automatically compensated and recorded. Whole cell capacitance and series resistance (R_S) were monitored throughout the recording. Normal cellular current convention is used when referring to the direction of current. In other words, *outward* current refers to *inward* Cl^- ion flow. In some experiments, cells were treated with cytocholation B for 20 min. or phalloidin for 30 min. before recording was initiated in order to insure that F-actin structure was not undergoing any further change. The effect either agent on VRAC activation was consistent for at least 80 min., after which time recording was terminated. All experiments were performed at room temperature (22°C ± 2).

In experiments using cells with large current amplitudes (> 500 pA), series resistance compensation was employed. In these cases, 95% compensation with a 100 μs lag was used. Occasionally, a lower percentage compensation was required to prevent current oscillation. Cells exhibiting small current amplitudes (< 500 pA) did not require series resistance compensation. In these cases, the voltage error involved was < 5%.

Cell diameter was monitored concurrently with whole cell recording using a ScionImage AG-5 capture board and ScionImage software (a version of NIHImage; Wayne Rasband (wayne@helix.NIH.gov), National Institutes of Health, USA). Measurements were made using NIHImage. Distances were calibrated using latex beads of known diameter.

F-actin was specifically labeled with rhodamine-phalloidin (Molecular Probes, Inc. Eugene, OR). G-actin was specifically labeled with FITC-DNAse I (Molecular Probes). Myeloma cells were treated with 100 μM hydroxyurea for 24 hr. Viable cells were isolated by separation on a ficoll gradient, and then washed in RPMI 1640 media in the absence of sera, re-suspended in isosmotic recording solution and allowed to settle on cover slips coated with polylysine. Cover slips were divided into control and experimental groups and experimentally challenged (e.g. exposure to 50% hyposmotic solution for 5 min. and then returned to isosmotic solutions for recovery over 20 min.) before being subsequently fixed in 3.7% formaldehyde in phosphate-buffered saline (PBS, pH 7.0), washed in PBS, 3 times, 2 min. each, permeablized with 0.1% Triton in PBS and incubated with 300 nM R-P in PBS for 20 min. This was followed by incubation with 300 nM FITC-DNAse I in PBS for 20 min. Cells were then washed in PBS (5 times, 3 min. each) and mounted using SlowFade (Molecular Probes, Lake Owsego, OR) to prevent fading. Staining and fixing was performed at room temperature (22°C ± 2).

Visualization of other cytoskeletal proteins required labeling fixed cells with a specific primary antibody followed by a fluorescent secondary antibody. Cells were prepared, fixed and permeabilized as described above, washed with 3% goat sera (GS: Sigma St. Louis, MO) in PBS for 2 min, 30 min. and 2 min. Cells were then incubated for 60 min. in primary antibody prepared in 3% GS, washed 3 times, 2 min. each in 3% GS, incubated in secondary antibody prepared in 3% GS for 45 min, washed and mounted on slides as described above.

The primary antibodies and dilutions used were: gelsolin (1/200: G4896, monoclonal; Sigma,), β-tubulin (1/10: E7, monoclonal; Developmental Studies Hybridoma Bank (DSHB), Iowa City, IA), alpha-actinin (1/200: A5044, monoclonal; Sigma), ankyrin (1/100: MaAnkHum, monoclonal; East Acres Biologicals, Southbridge, MA), filamin (1/100: F1888, monoclonal; Sigma), myosin (EH7, monoclonal; DSHB), fodrin (1/100: MAB1685, monoclonal; Chemicon, Temecula, CA). The secondary antibody was α-mouse IgG labeled with Cy3 (1/400: Jackson ImmunoResearch Laboratories, Inc., West Grove, PA) and was visualized using an RITC filter set.

Confocal microscopy was performed using a BioRad MRC 600 confocal system with a 63X oil immersion lens fitted on a Zeiss inverted microscope. A YMS filter block was used in conjunction with a 3% laser and a direct filter for single scanned images. Fluorescent images were acquired on a Gateway 2000

4DX 66 MHz computer using CoMOS (v. 3.02, BioRad software) and transferred to a Macintosh Quadra 800 for analysis with NIHImage and Adobe Photoshop.

Fluorescent staining was quantified by taking images through a plane at the center of a cell diameter and determining the mean of the number of pixels per cell using NIHImage. The distribution of pixel intensity was determined over cell diameter using NIHImage. Control cells treated only with isosmotic media were included in each experiment. To control for any inherent variations in fluorescent intensity, gain and black level of the photomultiplier were carefully adjusted at the beginning of each experiment, so that pixel intensity did not saturate in control, isosmotic cells. Quantification of pixel intensity using Microsoft EXCEL (Microsoft, Redmond WA) is expressed as a percentage relative to the isosmotic control. Electrophysiological recordings and staining of F-actin with rhodamine-phaloidin were done on the same day, using the same population of cells.

SWELLING INDUCES UNFOLDING OF PLASMA MEMBRANE INVAGINATIONS

The plasma membrane of most cells is invaginated or folded to varying extents. Electromicrographs of T-lymphocytes, for example, show that these cells are covered with microvilli processes.[15-17] As a cell swells in response to a hyposmotic gradient, these invaginations or villi are likely to unfold. Tension within the plasma membrane bilayer will not increase until the membrane is unfolded to a maximum amount, assuming that the membrane is acting as a fluid lipid bilayer. We have measured changes in the surface area using visual and electrical measurements in order to determine the extent of membrane unfolding in response to a hyposmotic gradient in human myeloma cells. The spherical shape of these cells allow the direct measurement of changes in cell diameter during swelling with a minimum of ambiguity.

Figure 1a shows that the average apparent cell surface area, calculated from a visual measurement of cell diameter, increased as the cell diameter increased in the presence of a hyposmotic solution. VRAC activation was correlated with cell swelling. The osmotic gradient was established concurrently with the whole cell recording configuration. In contrast, the surface area of hyposmotically challenged cells did not increase when the membrane capacitance is measured (Figure 1b). Membrane capacitance is directly proportional to membrane area, and the total membrane surface area can be estimated assuming a typical specific capacitance for biological membranes of $1\mu F/cm^2$.[18] The average cell surface area, regardless of measurement technique, did not change significantly if cells were maintained in an isosmotic solution.

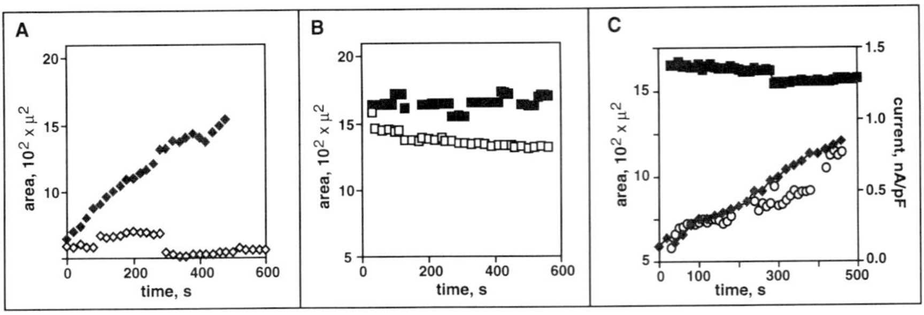

Figure 1. Unfolding of plasma membrane accompanies VRAC activation. (A) The change in average cell surface area over time was determined from visual measurement (diamonds) of the cell radius where $A_c = 4\pi R_c^2$, using light microscopy with phase optics and a video camera. Data were stored directly onto a hard disk for later analysis using NIHImage software. Measurements made in 75% hyposmotic solutions are shown with gray symbols. Isosmotic measurements are shown with open symbols. (B) Average cell capacitance does not change in hyposmotically-challenged (filled squares) or unchallenged cells (open squares). For Panels A and B, each point represents an $n \geq 3$. (C) Simultaneous measurement of VRAC current amplitude (open circles) and surface area as determined from visual (gray diamonds) and capacitance (filled squares) measurements is shown for a representative cell. Activation of VRAC correlates with an increase in surface area determined using visual but not electrical, measurement of cell surface area.

The time course of the change in the cell surface area determined from visual measurements and that determined from capacitance measurements for a typical cell is directly compared in Figure 1c. It can be seen that the area determined from visual measurements approached that determined from capacitance measurements as the cell continued to swell. The initial cell surface area determined from visual measurements averaged 640 ± 50 μm^2 which was 40% of that determined from capacitance measurements (1590 ± 70 μm^2). After 460 s of cell swelling, the surface area determined from visual measurements increased to an average value of 1490 ± 290 μm^2 which was equivalent to the average surface area determined from capacitance measurements after the same period of time (1550 ± 100 μm^2). These results suggest that membrane unfolding accompanies cell swelling and is required for the activation of VRAC.

RE-ARRANGEMENT OF CYTOSKELETAL NETWORK DURING CELL SWELLING

Cytoskeletal architecture is involved in regulating membrane folding and invaginations.[19] F-actin regulates membrane elasticity (ability to stretch) and tension (force within the membrane) at least in part through its associations with membrane proteins. The regulation of membrane elasticity and tension is, by necessity, involved in the regulation of cell swelling. Kleinzeller and Ziyadeh have proposed a model in which linkages between the membrane and cytoskeleton are disrupted as a result of massive swelling.[20] Figure 2 shows a cartoon representing the relationship between a variety of cytoskeletal proteins and how these proteins may be related to VRAC. We have used confocal fluoroscopy to study the distribution of F-actin, gelsolin, an actin binding-protein, and β-tubulin, a major component of microtubules, under isosmotic and hyposmotic conditions.

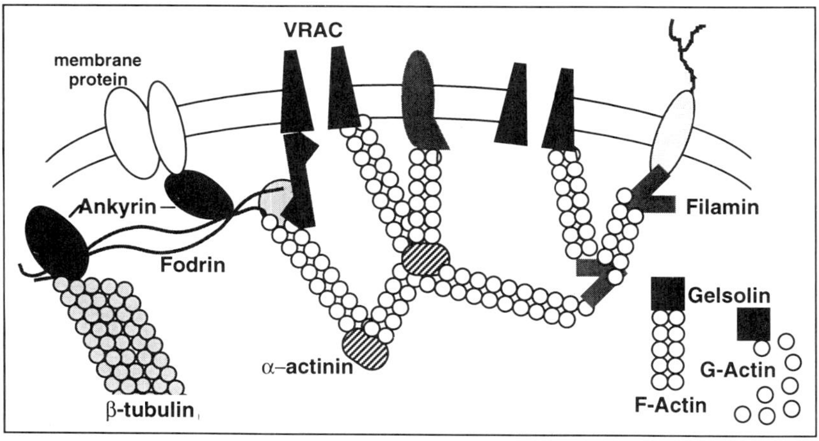

Figure 2. Links between VRAC and cytoskeletal proteins. Cytoskeletal rearrangements are known to effect RVD. The relationships between actin, variety of actin-binding proteins, β-tubulin, and integral membrane proteins including VRAC are indicated. The interaction between cytoskeletal proteins and VRAC are not yet known. Possible interactions between these proteins are suggested by the drawing. Cytoskeletal rearrangement is involved in RVD, but the mechanism of this interaction not known and the role of cytoskeletal rearrangement in VRAC activation is not yet understood.

Actin

F-actin is the filamentous form of actin which is composed of globular, or G-actin, monomers. F-actin is the primary protein of the microfilament cytoskeletal network. The toadstool cyclic oligopeptide, phalloidin, is a specific probe for F-actin and does not bind to G-actin. Rhodamine-labeled phalloidin can be used to visualize the distribution of F-actin within a cell. The enzyme, DNAse I, has a high affinity to the G- actin form. FITC-DNAse I can be used to specifically label G-actin in the same cell population stained for F-actin (for examples see ref. 21, 22). Myeloma cells double stained for F- and G-actin using these probes are shown in Figures 3 and 4.

Figure 3 shows a cross-section of rhodamine-phalloidin stained F-actin in myeloma cells. The pattern of fluorescence shows that F-actin was found in a subcortical ring, just under the plasma membrane of the cells. This F-actin ring was present in cells under isosmotic (non-swelling) conditions (Figure 3a) and was broken up during a 5 min. exposure to a hyposmotic condition (Figure 3b). Figure 3c shows that the total intensity of R-P fluorescence per cell decreased in challenged cells. This is expected if the number of phalloidin binding sites, i.e. F-actin, decreases with cell swelling. Thus, we interpret this result to be a loss of F-actin during cell swelling. The integrity of the cortical F-actin ring recovered within 5 min. if hyposmotic-treated cells are subsequently returned to isosmotic solutions (not shown). The average diameter of the subcortical F-actin ring did not change significantly in cells treated with hyposmotic solutions when compared to isosmotic-treated cells. This suggests that as the cell swells, the diameter of the cell membrane increases but that dimensions of the actin cytoskeleton do not.

The distribution of G-actin is shown in Figure 4 for the same cells as shown in Figure 3. Figure 4 shows that G-actin was distributed through the cytosol, although there was slightly less FITC-DNAse I intensity in the center of the cell over the area of the nucleus. The intensity of G-actin fluorescence increased in hyposmotically treated cells (Figure 4b, c), with slightly more intensity near the membrane. The average G-actin fluorescent intensity increased significantly under these conditions (Figure 4c). The average diameter of G-actin fluorescence was equivalent to that measured for F-actin fluorescence. The increased G-actin fluorescence correlates with the decrease in F-actin distribution in hyposmotically challenged cells.

Gelsolin

Gelsolin is a multi-functional, actin-binding protein known to regulate the equilibrium of F-actin and G-actin by promoting the depolymerization, or severing, of F-actin and sequestration of G-actin in the cytosol. We hypothesized that gelsolin may play a role in mediating the F-actin rearrangement observed in swollen cells. In order to determine the pattern of gelsolin distribution myeloma cells, fixed cells were incubated with a commercially available α-gelsolin monoclonal antibody and visualized with a fluorescent secondary antibody.

Confocal microscopy revealed gelsolin-like immunofluorescence in a subcortical ring below the cell membrane. This ring was intact in cells treated

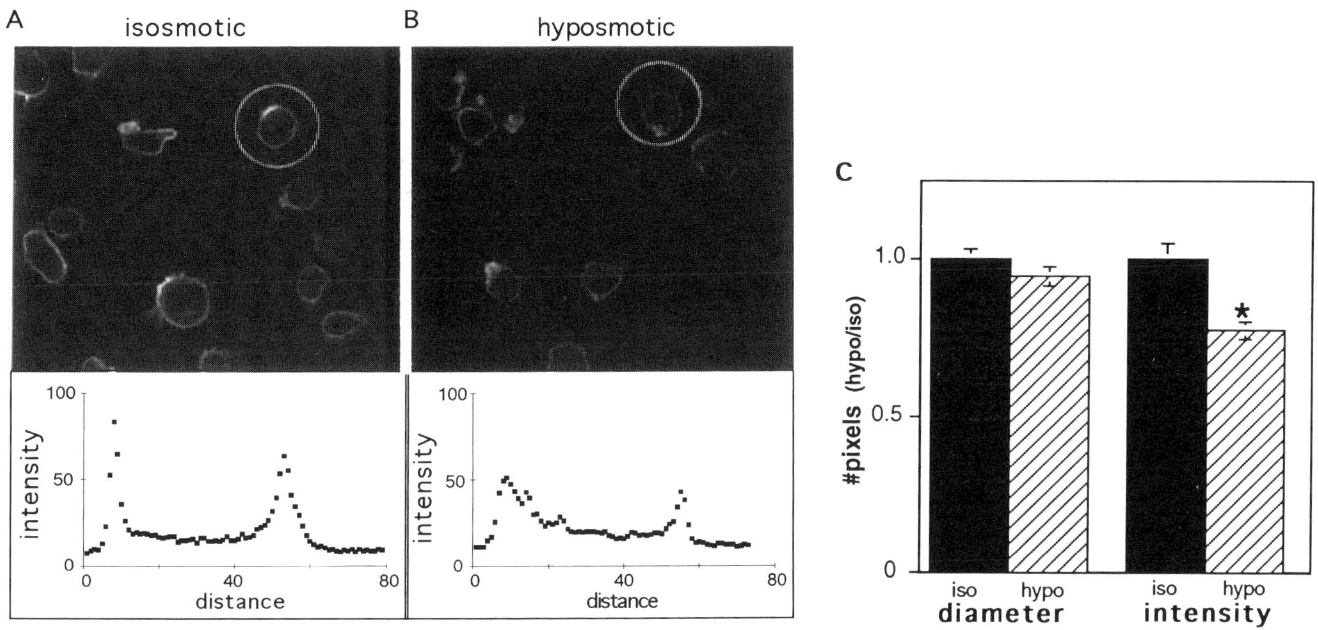

Figure 3. F-actin distribution with cell swelling. Myeloma cells were exposed to an isosmotic (307 ± 5 mOsm) and hyposmotic (165 ± 5 mOsm) gradient are shown in Panels A and B. Rhodamine-phalloidin was used to visualized F-actin with confocal microscopy. The images represent a 2-dimensional cross-section through the approximate center of the cells. Images were recorded using the same gain and threshold values. Broken circles surrounding individual cells indicate cells used to generate histogram of pixel intensity vs. cell diameter, shown below the fluorescent image. Quantification of average diameter of fluorescent ring and intensity of rhodamine-phalloidin fluorescence is shown in Panel C. Pixel intensity was normalized to diameter of fluorescent ring. n = 20 for each experimental condition. Bars indicate SEM and asterisks indicate a significance level of $p < 0.01$ (paired Student's t-test).

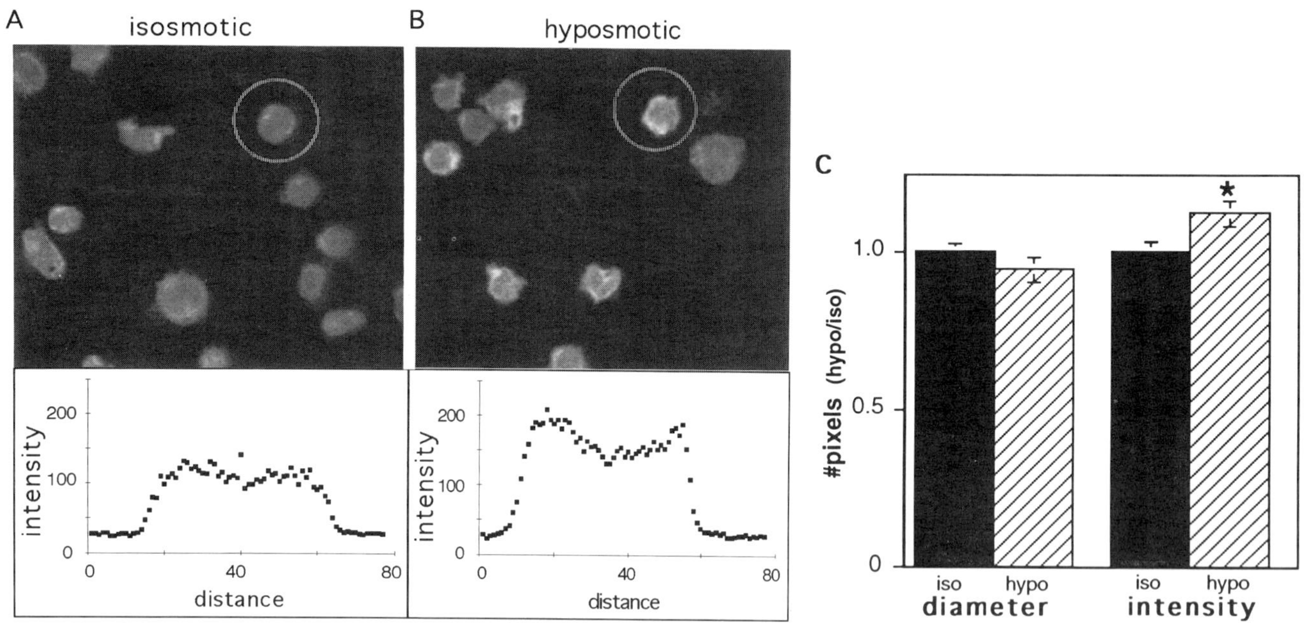

Figure 4. G-actin distribution with cell swelling. Myeloma cells were exposed to an isosmotic (307 ± 5 mOsm) and hyposmotic (165 ± 5 mOsm) gradient are shown in Panels A and B. FITC-DNAse I was used to visualize G-actin with confocal microscopy. Fluorescence images in Figures 3 and 4 show the same cells stained for F- and G-actin using specific fluorescent probes. The images represent a 2-dimensional cross-section through the approximate center of the cells. Images were recorded using the same gain and threshold values. Broken circles surrounding individual cells indicate cells used to generate histogram of pixel intensity vs. cell diameter, shown below fluorescent image. Quantification of average diameter of fluorescent ring and intensity of FITC-DNAse I fluorescence is shown in Panel C. Pixel intensity was normalized to diameter of fluorescent ring. n = 20 for each experimental condition. Bars indicate SEM and asterisks indicate a significance level of $p < 0.01$ (paired Student's t-test).

with iso-osmotic solutions, but broken and with a lower intensity in hyposmotic-treated cells (Figure 5a, b). These differences suggest that the distribution of gelsolin changes during cell swelling. The total fluorescent intensity per cell decreased significantly in challenged cells although the diameter of the fluorescent ring was the same in either condition (Figure 5c). A decrease in the fluorescence intensity may be a reflection of different binding affinities between the α-gelsolin antibody to gelsolin when it is bound to F- or G-actin. Cells incubated in the fluorescent secondary antibody but in the absence of α-gelsolin antibody showed no fluorescent staining (not shown). These results suggest that gelsolin is located near actin, in a subcortical ring and that the distribution of this protein is dependent on cell swelling.

β-Tubulin

β-tubulin is the primary component of microtubules. Microtubule disruption has been implicated in volume regulation in some systems but not others.[23-26] Figure 6a and 6b show that β-tubulin-like immunofluorescence was present in myeloma cells. The distribution of β-tubulin, like that of F-actin and gelsolin, is in a continuous subcortical ring. This ring of β-tubulin-like fluorescence is broken but the width of the subcortical ring is increased in cells that have been treated for 5 min. in a hyposmotic solution (Figure 6a, b). The diameter of the fluorescent ring increased significantly by 18% whereas the fluorescent intensity decreased by 25% when treated cells were compared with untreated cells (Figure 6c). Cells incubated in the fluorescent secondary antibody, but in the absence of α-β-tubulin antibodies, showed no fluorescent staining (not shown). These results suggest that the distribution of β-tubulin, like that of gelsolin, F- and G-actin, is dependent on cell swelling.

Other Actin-binding Proteins

F-actin is known to associate with membrane proteins through direct binding or indirectly through actin associated proteins. We used immunofluorescent techniques as described above for gelsolin and β-tubulin to survey a set of actin associated proteins, alpha-actinin, filamin, fodrin, ankyrin and myosin, to determine the effect of hyposmotic challenge on the distribution of actin-associated proteins (see Figure 2) in myeloma cells. These cytoskeletal proteins tend to be highly conserved and ubiquitous. Antibodies to these proteins are expected to exhibit a high degree of reactivity across different tissue types. We used commercially available mono- or polyclonal antibodies to these proteins. The results of this survey are shown in Table 1. The distribution of alpha-actinin, filamin, fodrin and myosin shifted in hyposmotic-treated cells when compared to unchallenged cells. No specific α-ankyrin fluorescence was observed. Ankyrin is known to have binding sites for several different

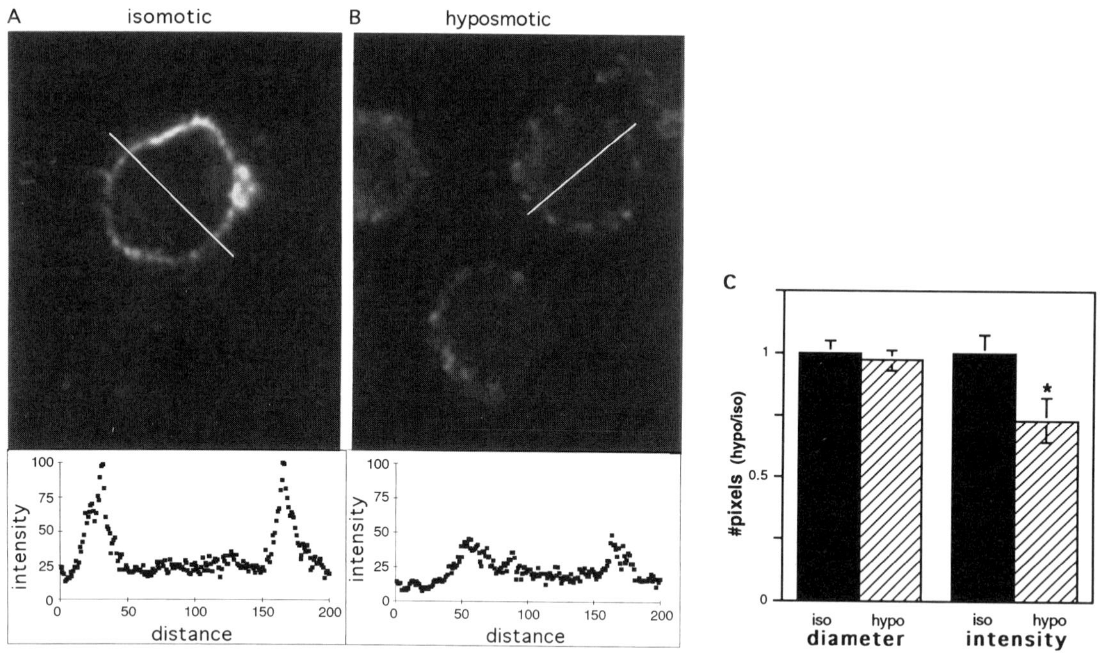

Figure 5. Gelsolin distribution with cell swelling. Myeloma cells were exposed to an isosmotic (307 ± 5 mOsm) and hyposmotic (165 ± 5 mOsm) gradient are shown in Panels A and B. Gelsolin-like immunofluorescence was visualized using confocal microscopy. The images represent a 2-dimensional cross-section through the approximate center of the cells. Images were recorded using the same gain and threshold values. White line across individual cells indicate the position where a histogram of pixel intensity vs. cell diameter, shown below the fluorescent image was determined. Quantification of average diameter of fluorescent ring and intensity of gelsolin-like immunofluorescence is shown in Panel C. Pixel intensity was normalized to diameter of fluorescent ring. n = 20 for each experimental condition. Bars indicate SEM and asterisks indicate a significance level of $p < 0.01$ (paired Student's t-test).

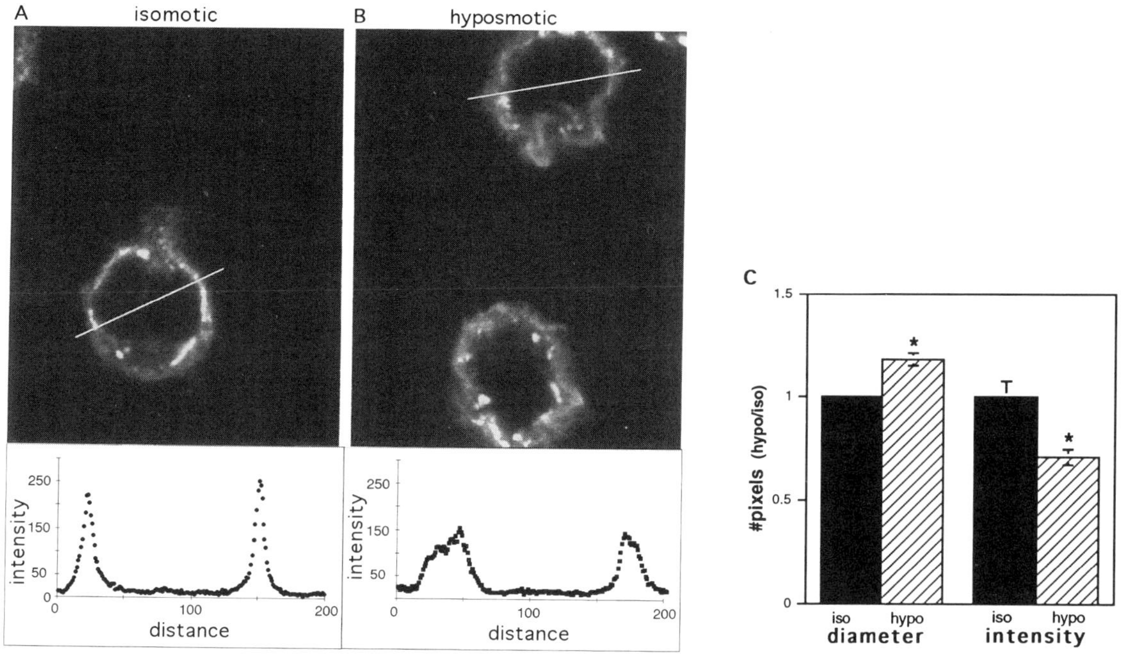

Figure 6. Tubulin distribution with cell swelling. Myeloma cells were exposed to an isosmotic (307 ± 5 mOsm) and hyposmotic (165 ± 5 mOsm) gradient are shown in Panels A and B. Tubulin-like immunofluorescence was visualized using confocal microscopy. The images represent a 2-dimensional cross-section through the approximate center of the cells. Images were recorded using the same gain and threshold values. White line across individual cells indicate the position where a histogram of pixel intensity vs. cell diameter, shown below the fluorescent image was determined. Quantification of average diameter of fluorescent ring and intensity of tubulin-like immunofluorescence is shown in Panel C. Pixel intensity was normalized to diameter of fluorescent ring. n = 20 for each experimental condition. Bars indicate SEM and asterisks indicate a significance level of $p < 0.01$ (paired Student's t-test).

cytoskeletal associated proteins and may be bound to more than one type of protein simultaneously.(27) This high degree of binding may prevent exposure of immunogenic sites on the ankyrin protein such that a negative result is not interpretable as the absence of the protein in this cell line. The results presented in Table 1 indicate that cell swelling has extensive effect on a variety of cytoskeletal elements.

Table 1. Effect of cell swelling on actin-binding proteins.[1]

Protein:	Staining Pattern Iso-osmotic	Staining Pattern Hyposmotic	Change?
α-actinin	ring	cytosol	yes
Filamin	cytosol	ring	yes
Fodrin	ring	none	yes
Ankyrin	none	none	no
Myosin	ring	none	yes

[1]Immunohistochemical techniques (see Methods) were used to visualize the distribution of actin-binding proteins in cells incubated in isosmotic solutions (307 ± 5 mOsm) and hyposmotic (165 ± 5 mOsm) solutions (160 mOsm) for 5 min. Distribution of primary antibodies was visualized by a fluorescent secondary antibody using confocal microscopy. "Ring" refers to immunofluorescence concentrated in the subcortical region. "Cytosol" indicates specific immunofluorescence distributed though out cellular boundaries. "None" indicates that no specific immunofluorescence was observed. Number of experiments in each case was ≥ 3.

MODULATION OF VRAC BY THE F-ACTIN CYTOSKELETON

We addressed the hypothesis that a re-arrangement of the cytoskeleton proteins, specifically, F-actin, during cell swelling is involved in the regulation of volume-regulated Cl^- current. F-actin was disassembled by the incubation of cells with 50 μM cytochalasin B. The degree of rearrangement after incubation with cytochalasin B was similar to that after exposure of the cells to a 75% hyposmotic solution (where the osmotic strength of the extracellular solution is 75% of the intracellular solution). Incubation the cells with cytochalasin B had a strong facilitatory effect on the rate of VRAC activation when cells were exposed to a mild (90%) osmotic gradient (Figure 7a). The rate of VRAC activation increased with the duration of incubation in cytochalasin B (Table 2). There was no effect on the rate of VRAC activation when cells were challenged with a 75% gradient instead of a 90% gradient (Table 2). VRAC was not activated in cells treated with cytochalasin B but maintained in isosmotic solutions, suggesting that rearrangement of F-actin alone was not sufficient to activate VRAC.

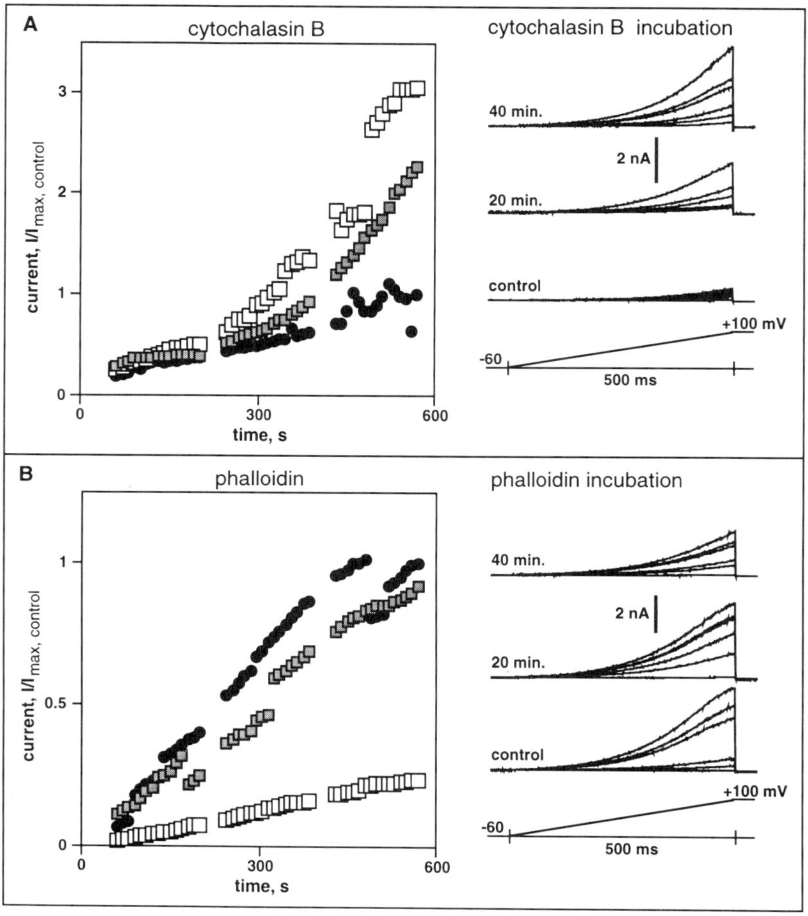

Figure 7. Modulation of VRAC activation by F-actin rearrangement. VRAC activation was elicited in response to a 90% hypo-osmotic gradient. The average initial rate of VRAC activation increases in myeloma cells incubated in (A) externally applied 50 μM cytochalasin B or (B) 15 μM phalloidin for control (filled circles), 20 min. (gray squares) and 40 min. (open squares). VRAC current amplitude increases with longer incubation cytochalasin B. In contrast, the current amplitude decreases with longer incubations of phalloidin. Each point represents n≥ 3. Representative current traces elicited by voltage ramp from -60 mV (holding potential) to +100 mV are shown to the right of each graph. Traces were taken at 75, 225, 300, 450, 525, and 675 s after initiation of whole cell recording under indicated conditions. A schematic of the voltage paradigm is shown below current traces. Under control conditions, extracellular solutions contained 0.05% DMSO or 0.1% MeOH for cytochalasin B or phalloidin solutions, respectively.

Table 2. VRAC activation in the presence and absence of agents which modulate F-actin polymerization.[1]

Gradient	Agent	$Rate_{(agent)}/Rate_{(control)}$
	[cytochalasin B]	
90%	0	1.0
90%	50 μM (<30 min.)	2.1
90%	50 μM (>30 min.)	3.5
75%	0	1.0
75%	50 μM (>30 min.)	1.0
	[phalloidin]	
90%	0	1.0
90%	15 μM (<30 min.)	0.7
90%	15 μM (>30 min.)	0.2
75%	0	1.0
75%	15 μM (>30 min.)	0.8

[1]The rate (pA/(pF-min.) of VRAC activation in response to a 90% hyposmotic gradient increases when cells are incubated in 50 μM cytochalasin B. The rate of activation increases with the duration of the incubation. In contrast, the rate of VRAC activation in response to a 90% hyposmotic gradient in cells treated with phalloidin decreases with increased incubation. Neither agent has an effect on the rate of VRAC activation in cells challenged with a 75% hyposmotic gradient. Recordings were made on paired control and treated cells on the same day. Each measurement is determined from an n ≥ 3. Control conditions contained vehicle used for solublization of cytochalasin B (0.05% DMSO) or phalloidin (1.0% MeOH).

The effect of phalloidin, an F-actin stabilizing agent, on the rate of VRAC activation was opposite of that of cytochalasin B. Incubation of cells with phalloidin significantly depressed the rate of VRAC activation in cells exposed to 90% hyposmotic solutions (Figure 7b). The effect of phalloidin, like that of cytochalasin B, on VRAC activation was also time dependent in cells challenged with a 90% hyposmotic gradient (Table 2.). This effect was lost in cells challenged with a larger (75%) hyposmotic gradient (Table 2.). Inhibition of the rate of VRAC activation by phalloidin confirmed a modulatory role of F-actin on VRAC activation.

DISCUSSION

Membrane Unfolding

In this study, we have shown that an increase in cell volume is likely to be due to an unfolding of invaginations or villi in the plasma membrane. Our results indicate that a major contribution to the increase of apparent cell surface area in myeloma cells during swelling comes from the unfolding of the plasma membrane and not from an increase of membrane surface. This conclusion reqires the assumption that the dielectric constant of the membrane does not change during swelling. Such a change in the dielectric constant has not been previously observed in biological systems. In addition, the width of the membrane is not expected to change significantly due to membrane stretch as a 2-3% increase in membrane area due to a stretching the membrane would be expected to result in rupture. A comparison of measurements made using visual or electrical techniques indicates that plasma membrane invaginations or villi in myeloma cells account for ~60% of the membrane surface area. Membrane unfolding is correlated with the activation of VRAC in myeloma cells. This observation is in agreement with the results of Ross et al. (1994) and Graf et al. (1995) which show that there is no correlation between volume and capacitance increase during cell swelling in T-lymphocytes and hepatocytes respectively.

What are the mechanical forces that are likely to be involved in the activation of VRAC during cell swelling? It has been shown that a mechanosensitive cation channel in *E. coli* is activated by membrane tension and not transmembrane pressure.(9) The mean open time of the bacterial toxin channel, alamethicin, is also dependent on membrane tension.(10) The membrane tension and force of the transmembrane pressure are orthogonal to each other with tension being parallel to the membrane and pressure being perpendicular to the membrane.(28) Tension within a fluid bilayer membrane that is invaginated will not increase until the membrane is fully unfolded. Sokabe, Sachs and Jing showed that in skeletal muscle cells there is a unrestricted flow of membrane lipid around the tip of a patch pipette suggesting that little or no tension develops in the bilayer.(29) These investigators suggested an intracellular cytoskeletal network might underlie the elasticity features of cell membrane and that tension develops in an elastic cytoskeleton rather than in a membrane bilayer. Our observations that cell swelling results in unfolding of a lipid bilayer supports Sokabe, Sachs and Jing's hypothesis.

Rearrangement of Cytoskeletal Proteins

Immunohistochemical techniques were used in this study to visualize changes in the distribution of several of cytoskeletal proteins in when myeloma cells are exposed to a hyposmotic challenge. Both microfilaments, made up of F-actin, and microtubules, of which β-tubulin is a primary component, form a sub-membrane cortical shell. These proteins are known to be either directly, as in the case of F-actin and a lymphocyte-specific protein (LSP1), or indirectly, as in the case of F-actin and Band 3, the erythrocyte Cl^-/bicarbonate transporter, bound to a variety of integral membrane proteins (for a review, see ref. 30). The cytoskeletal structure is also responsible for maintaining membrane invaginations and villi.(19) Therefore, unfolding of the plasma membrane in response to cell swelling is expected to have a profound effect on the sum-membrane cytoskeletal network. Indeed, our results and those of others, show that the distribution of F-actin and β-tubulin are disrupted during cell swelling.[24, 25, 31] Our results show that the disruption of cytoskeletal proteins during cell swelling extends to other actin-associated proteins, such as gelsolin, alpha-actinin, filamin, fodrin and myosin.

The total fluorescence intensity per cell of rhodamine-phalloidin decreased in hyposmotically challenged cells. FITC-DNAse I fluorescence increased in the same population of cells. We interpret this as the loss of F-actin and concurrent increase of G-actin in response to cell swelling. The total fluorescence intensity also changed in assays using immunofluorescent techniques. This result may not be expected as antibodies bind to specific proteins. There are two possible interpretations for the change in total fluorescence intensity during cell swelling when using antibodies to specific cytoskeletal proteins: 1. Cell swelling is accompanied by the destruction of these proteins, and/or 2. Cell swelling is accompanied by a rearrangement in the cytoskeletal structure which results in the change immuno-reactivity of specific cytoskeletal proteins. A change in the immuno-reactivity of these proteins may be due to the change in their association to membrane and other cytoskeletal proteins. We cannot discriminate between these possibilities.

Cytoskeletal Rearrangement and Volume Regulation

Cytoskeletal elements have been reported to participate in the process of RVD in many cell types. The disassembly of F-actin in shark rectal gland recovers gradually as the cell completes RVD. RVD can be inhibited by cytochalasin B, which causes the depolymerization of F-actin.[24, 25, 31] The rate of recovery from swelling is much slower than normal in the presence of cytochalasin B or D. Volume regulation is also blocked in melanoma cells which lack the gene for an F-actin binding protein, non-muscle filamin.[33] RVD in these cells is regulated, at least in part, by a K^+ current. The activation of this current is prevented in the absence of filamin and RVD and the K^+ current can be rescued by expressing filamin in these cells. Our results indicate that the distribution of this actin binding protein is also disrupted by cell swelling.

VRAC Modulation by Cytoskeletal Proteins

VRAC is part of a large functional family of ion channels that respond to membrane stress by activating or increasing macroscopic current amplitude. The activation of the ion channel proteins in this family may be modulated by cytoskeletal proteins. The activity of a large conductance (305 pS) volume-regulated anion channel in renal cortical collecting duct cells and an "intermediate" conductance chloride channel in human skeletal muscle have been shown to be altered by agents which alter F-actin polymerization.[32, 34] On the other hand, Oike and colleagues have shown that alteration of the cytoskeleton in endothelial cells does not regulate a VRAC-type conductance.[35] We have shown that F-actin modulates VRAC activity but does not activate VRAC in myeloma cells.[36, this report] Cytoskeletal modulation of VRAC activity occurs only after the conductance has been activated and only in response to a mild osmotic stress. Modulation of VRAC by F-actin appears to depend critically on the extent of the stimulus, which in this case is the osmotic gradient.

We suggest that the contribution of F-actin rearrangement to VRAC depends on the relative association between the cytoskeleton and integral membrane proteins such an ion channels. Specifically, the transduction of membrane tension developed from cell swelling is due to membrane-cytoskeletal interactions and results in VRAC activation, as is suggested by the cartoon in Figure 8. This hypothesis predicts that the rearrangement of cytoskeletal proteins modulates VRAC activity when the association between the cytoskeletal proteins and the ion channel is still intact. When the membrane-cytoskeletal association is severed, as might be the case with strong osmotic gradient, cytoskeletal rearrangement will not be able to modulate VRAC activity. This interpretation provides an explanation for the apparently

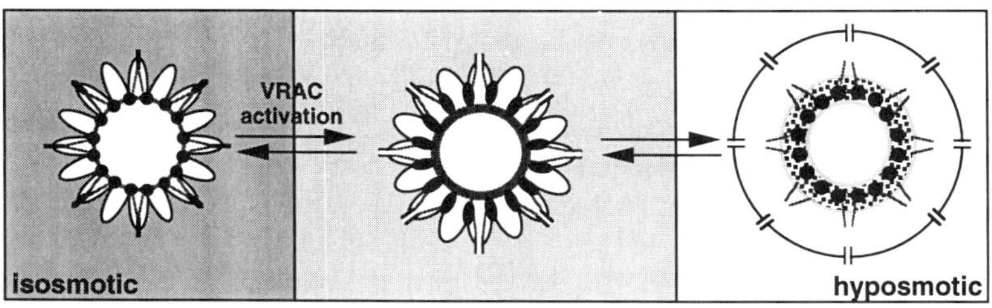

Figure 8. Activation of VRAC due to development of tension through membrane-cytoskeletal interactions. The left panel suggests that proteins in the plasma membrane of cells in an isosmotic environment are bound to cytoskeletal proteins (indicated by dark circles and shaded ring inside cell). This interaction maintains folds, or invaginations, of the plasma membrane. The cytoskeleton also interacts with VRAC, indicated by the solid rectangles in the plasma membrane. A mild hyposmotic stress, as indicated by the middle panel, results in a visual increase in the cell diameter as the plasma membrane begins to unfold. The osmotic stress is not sufficient to break the interactions between the plasma membrane and cytoskeletal proteins, but is sufficient to activate and modulate VRAC activity. VRAC activation is indicated by open channels in the plasma membrane. In the right panel, cell swelling has proceeded to a point where the interaction between the plasma membrane and cytoskeletal proteins has been severed. Under these conditions, VRAC activity can no longer be modulated by cytoskeletal interactions and the distribution of cytoskeletal proteins has been disrupted.

conflicting results found in the literature where the effect of cytoskeletal rearrangement was determined on ion channel activity under different osmotic gradients.

CONCLUSIONS

We have examined changes in cell architecture during cell swelling in human myeloma cells. Cell swelling is due to the unfolding of membrane invaginations rather than addition of the membrane from the intracellular sources. The swelling of human myeloma cells in response to a hyposmotic challenge is accompanied by VRAC activation. Cell swelling is also accompanied by the spatial re-arrangement of a variety of cytoskeletal proteins including major microfilament (F-actin) and microtubule (β-tubulin) components as well as the actin-associated proteins, gelsolin, alpha-actinin, filamin, fodrin, and myosin. Our work shows that the rearrangement of one of at least one of these cytoskeletal proteins, F-actin, is involved in the modulation of VRAC activity. The integrity of the cytoskeleton is dependent on the equilibrium of all cytoskeletal associated proteins, therefore, a disruption of this complex equilibrium would be expected to ultimately result in modulation of volume regulation. Future work will address the role of the cytoskeletal in activation of VRAC and regulation of cell volume.

ACKNOWLEDGMENTS

We thank Mr. Gregory Kowalsky, Ms. Betty Luxama, and Ms. Catherine Sayer for excellent technical assistance. EH7, E7 and A4 antibodies were obtained from the Developmental Studies Hybridoma Bank maintained by the Dept. of Pharmacology and Molec. Sci, Johns Hopkins University, School of Medicine, Baltimore, MD and the Dept. of Biol. Sci., University of Iowa, Iowa City, IA, under contract N01-HD-6-2915 from the NICHD. Dr. Garber is an Established Investigator of the American Heart Association. This work was supported by NIDDK46672, AHA94002340, and NSF BIR-9413528.

REFERENCES

1. M. L. McManus and K. B. Churchwell, Clinical significance of cellular osmoregulation, *in* "Cellular and Molecular Physiology of Cell Volume Regulation", K. Strange, ed., CRC Press, Boca Raton (1994).
2. R. O. Law, Amino acids as volume-regulatory osmolytes in mammalian cells, *Comp. Biochem. Physiol.* 99A:263 (1991).

3. E. K. Hoffman, Cell swelling and volume regulation, *Can. J. Physiol. Pharm.*. 70:S310 (1992).
4. C. Deutsch and S.C. Lee, Cell volume regulation in lymphocytes, *Renal Physiol. Biochem.* 3-5:260 (1988).
5. T.E. DeCoursey, K.G. Chandy, S. Gutpa, and M.D. Cahalan, Voltage-dependent ion channels in T-lymphocytes, *J. Neuroimmun.* 10:71 (1988).
6. S. Grinstein, A. Rothstein, B. Sarkadi, and E.W. Gelfand, Responses of lymphocytes to anisotonic media: Volume regulating behavior, *Am. J. Physiol.* 246:C204 (1988).
7. P.R. Ross, S.S. Garber, and M.D. Cahalan, Membrane chloride conductance and capacitance in Jurkat T lymphocytes during osmotic swelling, *Biophys. J.* 66:169 (1994).
8. J. Graf, M. Rupnik, G. Zupancic, and R. Zorec, Osmotic swelling of hepatocytes increases membrane conductance by not membrane capacitance, *Biophys. J.* 68:1359 (1995).
9. M. C. Gustin, X.-L. Zhou, B. Martinac, and C. Kung, A mechanosensitive ion channel in the yeast plasma membrane, *Science*. 242:762 (1988).
10. L. R. Opsahl and W.W. Webb, Transduction of membrane tension by the ion channel alamethicin, *Biophys. J.* 66:71 (1994).
11. F. Sachs, Stretch-sensitive ion channels, *Sem. Neurosci.* 2:49 (1990).
12. M.D. Cahalan and R.S. Lewis, Role of potassium and chloride channels in volume regulation by T lymphocytes, *in*: "Cell Physiology of Blood", R.B. Gunn and J.C. Parker, eds., Society of General Physioloists Series, Vol. 43, Rockefeller University Press, New York, (1988).
13. S. C. Lee, M. Price, M.B. Prystowsky, and C. Deutsch, Volume response of quiescent and interleukin 2-stimulated T-lymphocytes to hypotonisity, *Am. J. Physiol.*. 254:C286 (1988).
14. R. Motais, H. Guizouarn, and F. Garcia-Romeu, Red cell volume regulation: The pivotal role of ionic strength in controlling swelling dependent transport systems, *Biochim. Biophys. Acta.* 1075:169 (1991).
15. R. K. Cheung, S. Grinstein, H.-M. Dosch, and E.W. Gelfand, Volume regulation by human lymphocytes: Characterization of the ionic basis for regulatory volume decrease, *J. Cell. Physiol.* 112:189 (1982).
16. S. Roath, D. Newell, A. Polliack, E. Alexander, and P.-S. Lin, Scanning electron microscopy and the surface morphology of human lymphocytes, *Nature*. 273:15 (1978).
17. E. Alexander, S. Sanders, and R. Braylan, Purported difference between human T- and B-cell surface morphology is an artifact, *Nature*. 261:239 (1976).
18. B. Hille, "Ionic Channels of Excitable Membranes," Sinauer Associates Inc., Sunderland, MA (1984).

19. A.D. Bershadsky and J.M. Vasiliev, Cellular Organelles, *in*:"Cytoskeleton," 2 edition, P. Siekevitz, ed., Plenum Press, New York (1989).
20. A. Kleinzeller and F.N. Ziyadeh, Cell volume regulation in epithelia- with emphasis on the role of osmolytes and the cytoskeleton, *in*: "Cell Volume Regulation", Beyenbach, ed. Karger, Comparative Physiology, Basel (1990).
21. B. Polzar, E. Nowak, R. Goody, and H. Mannherz, The complex of actin and deoxyribonuclease I as a model system to study the interactions of nucleotides, cations and cytochalasin D with monomeric actin, *Eur. J. Biochem.* 182:267 (1989).
22. Sanger, J., Dabiri, G., Mittal, B., Kowalski, M., Haddad, J. and Sanger, J., 1990, Disruption of microfilament organization in living non-muscle cells by micro-injection of plasma vitamin D-binding protein or DNAase I, *Proc. Natl. Acad. Sci. USA*. 87.
23. S. vom Dahl, B. Stoll, W. Gerok, and D. Haussinger, Inhibition of proteolysis by cell swelling in the liver requires intact microtubular structures, *Biochem. J.*. 308:529 (1995).
24. J.K. Foskett and K.R. Spring, Involvement of calcium and cytoskeleton in gallbladder epithelial cell volume regulation, *Am. J. Physiol.* 248:C27 (1985).
25. M. Cornet, J. Ubl, and H.-A. Kolb, Cytoskeleton and ion movements during volume regulation in cultured PC12 cells, *J. Mem. Biol.* 133:161 (1993).
26. D.A. Baron, R.M. Burch, P.V. Halushka, and S.S. Spicer, Blockade of colchicine-induced inhibition of vasopressin-stimulated osmotic water flow: Failure to influence microtubule formation, *Am. J. Physiol.* 249:F464 (1985).
27. T. Kreis and R. Vale, eds., "Guidebook to the Cytoskeletal and Motor Proteins," Oxford University Press, Oxford (1993).
28. H. Sackin, Stretch-activated ion channels, *in* "Cellular and Molecular Physiology of Cell Volume Regulation", K. Strange, ed., CRC Press, Boca Raton (1994).
29. M. Sokabe, F. Sachs, and Z.Q. Jing, Quantitative video microscopy of patch clamped membranes stress, strain, capacitance, and stretch channel activation, *Biophys. J.* 59:722 (1991).
30. P.A. Janmey and C. Chaponnier, Medical aspects of the actin cytoskeleton, *Curr. Opin. Cell Biol.* 7:111 (1995).
31. R. Gilles, E. Delpire, C. Duchene, M. Cornet, and A. Pequeux, The effect of cytochalasin B on the volume regulation response of isolated axons of the green crab *Carcinus maenas* submitted to hypo-osmotic media, *Comp. Biochem. Physiol.* 85A:523 (1986).

32. E.M. Schwiebert, J.W. Mills and B.A. Stanton, Actin-based cytoskeleton regulates a chloride channel and cell volume in a renal cortical collecting duct cell line, *J. Biol. Chem.* 269:7081 (1994).
33. H.F. Cantiello, A.G. Prat, J.V. Bonventre, C.C. Cunningham, J.H. Hartwig and D.A. Ausiello, Actin-binding protein contributes to cell volume regulatory ion channel activation in melanoma cells, *J. Biol. Chem.*. 268:4596 (1993).
34. U. Häussler, B.M. Rivet, C. Fahlke, D. Müller, E. Zachar, and R. Rüdel, Role of the cytoskeleton in the regulation of Cl- channels in human embryonic skeletal muscle cells, *Pflugers Arch.* 428:323 (1994).
35. M. Oike, G. Schwarz, J. Sehrer, M. Jost, V. Gerke, K. Weber, G. Droogmans and B. Nilius, Cytoskeletal modulation of the response to mechanical stimulation in human vascular endothelial cells, *Pflügers Arch.* 428:569 (1994).
36. I. Levitan and S.S. Garber, Inactivation of volume regulated Cl- current in T84 and Myeloma cells, *Pflügers Arch.*, 431:297 (1995).

INHIBITORY RESPONSES TO ODORANTS IN VERTEBRATE OLFACTORY NEURONS

Juan Bacigalupo+, Bernardo Morales#, Pedro Labarca*+, Gonzalo Ugarte+, Rodolfo Madrid+.

Departamento de Biología, Facultad de Ciencias, Universidad de Chile (+); Centro de Estudios Científicos de Santiago (*); and Departamento de Fisiología, Facultad de Ciencias Biológicas, Pontificia Universidad Católica de Chile (#).

THE EXCITATORY MECHANISM OF VERTEBRATE OLFACTORY NEURONS.

In vertebrates, the primary steps in olfaction take place at the olfactory epithelium. In this neuroepithelium, bipolar sensory olfactory neurons transduce chemical stimuli into electrical signals. Olfactory neurons project from their apical pole, towards the epithelial surface, a dendrite ending in a knob, to which a variable number of olfactory cilia are attached. Unmyelinated axons, originating from the basal pole of olfactory neurons, reach the olfactory bulb to make contact with secondary cells. Typically, olfactory neurons exhibit spontaneous action potential firing. Exposure to odorants gives place to changes in action potential firing rates.

There is convincing evidence that chemotransduction takes place in olfactory cilia (Kurahashi, 1989; Morales et al., 1995). Receptor proteins that bind odorant molecules, ionic channels that initiate the electrical changes triggered by the odorants, and other elements engaged in chemotransduction are either associated to the ciliary membrane or confined to the ciliary lumen (Restrepo et al., 1996).

The ability of olfactory cells to undergo excitatory responses to odorants is well documented (Firestein, 1992). A variety of experimental approaches allowed, in recent years, to expand decisively our understanding of olfactory transduction in vertebrates. One important step forward came out from the availability of biochemically relevant amounts of highly purified olfactory cilia membranes (Anholt et al., 1986). This development paved the way to a preliminary definition of crucial aspects of the molecular machinery engaged in excitatory responses (Ronnet and Snyder, 1992; Breer et al., 1994; Morales and Bacigalupo, 1996). In parallel, methods to dissociate functional olfactory neurons were developed, rendering these small sensory neurons amenable to patch-clamp inquiries (Trotier, 1986; Firestein and Werblin, 1987; Frings and Lindemann, 1988). The electrophysiological avenue permitted the resolution of the conductance mechanism engaged in the development of excitatory receptor potentials (Trotier, 1986; Kurahashi, 1989; Firestein et al., 1990). In turn, use of recombinant DNA approaches allowed the cloning of the components of the excitatory cascade (Buck and Axel, 1991; Jones and Reed, 1989; Bakalyar and Reed, 1990; Dhallan et al., 1990).

More recently, a large family of genes encoding for thousands of proteins pertaining to the superfamily of G-protein coupled receptors was identified by Buck and Axel (1991) in the olfactory tissue. This discovery solved, in all appearance, an old mystery in physiology, namely the nature of the mechanism underlying the ability of olfactory neurons to discriminate among a huge variety of olfactory stimuli. Each olfactory neuron seems to express a large number of a single receptor type or a limited variety of them (Ngai et al., 1993; Reesler et al., 1993; Vassar et al., 1993). It is speculated that each receptor protein might recognize more than one different odorant molecule. It must be said, however, that the ability of these putative receptors to bind odorants or to trigger G protein-coupled cascades remains to be demonstrated (Raming et al., 1993; Morales and Bacigalupo, 1996).

The elements engaged in the development of excitatory responses in vertebrate olfactory cells are known. The events are initiated by binding of odorants to Axel-Buck type membrane receptors. Receptor occupancy activates a G-protein, G_{olf} exclusive of the olfactory epithelia (Jones and Reed, 1989). Activation of G_{olf} in turn stimulates an adenylyl cyclase, rising cAMP levels in olfactory cilia. The elevation in cAMP is followed by the opening of cAMP-activated, cation-selective channels (Zufall et al., 1994) in the ciliary membrane, leading to the development of an excitatory receptor potential and the subsequent increase in action potential firing. Calcium, one of the ions permeating the

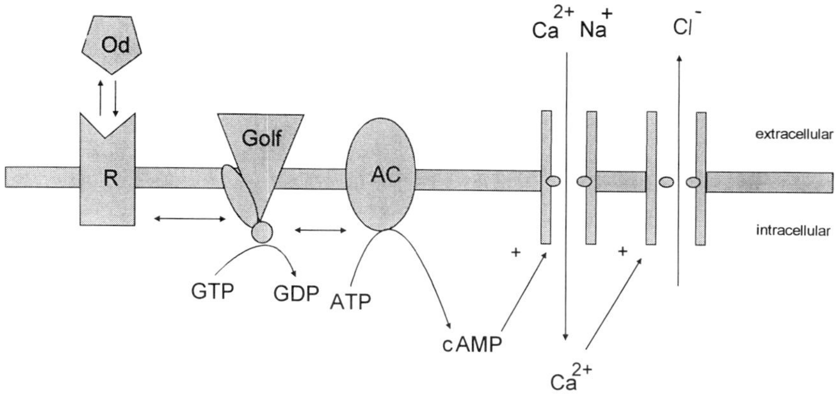

Figure 1. Excitatory pathway in vertebrate olfactory transduction. The interaction of odorants (**Od**) with specific receptor proteins (**R**) localized in ciliary plasma membrane activates an enzyme cascade involving a specific G protein (**G$_{olf}$**) and an adenylyl cyclase (**AC**). By increasing cAMP levels, this cascade leads to the activation of a cyclic nucleotide-gated, cation-selective channel, and subsequently a Ca^{2+}-activated chloride channel. The inward currents through both of these channels generate an excitatory receptor potential, incrementing the firing rate. The + sign denotes activation.

cAMP-gated channel, in turn activates a Cl^- channel. The resulting Cl^- efflux contributes to the receptor potential. This mechanism is schematized in Fig. 1.

INHIBITORY ODORANT RESPONSES IN VERTEBRATE OLFACTORY CELLS.

Excitatory phenomena in vertebrate olfactory cells received much attention and, as summarized in previous lines, led to the identification of the molecular elements underlying excitatory responses (reviewed by Firestein, 1992; Reed, 1992; Breer et al., 1994). For a while, the idea that odorants elicited only excitatory responses in olfactory neurons prevailed. However, it is well known that many odorants fail to activate the excitatory cascade (Sklar et al., 1985; Boekhoff et al., 1990). There is evidence that olfactory cilia posses other transduction mechanisms, in addition to the excitatory cascade (Boekhoff et al., 1990; Restrepo, et al., 1993; Broillet and Firestein, 1996). Furthermore, electrophysiological studies have provided direct evidence for the presence of

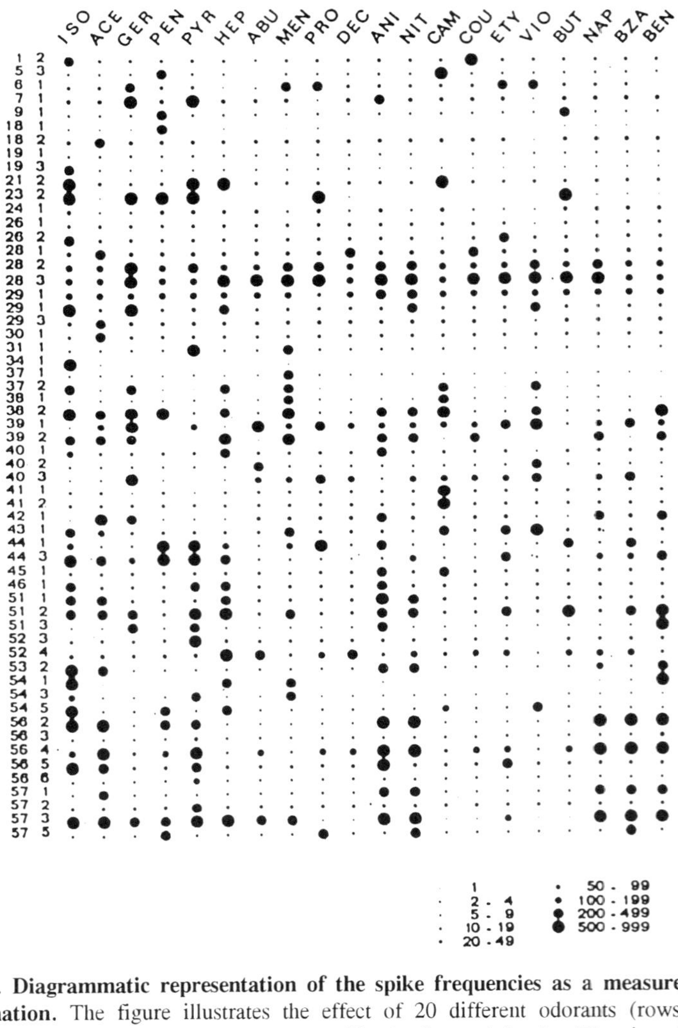

Figure 2. Diagrammatic representation of the spike frequencies as a measure of odor discrimination. The figure illustrates the effect of 20 different odorants (rows) applied separately on 58 individual olfactory neurons (lines) of a toad *in situ*. The size of a dot is proportional to the spike frequency (spikes/min) evoked on a particular cell by the respective odor. The numbers by each dot at the bottom of figure, give the frequency range represented by the size of the dot. The odorants, from left to right, are the following: isoamylacetate, acetophenone, geraniol, pentanol, pyridine, heptanol, butyric acid, menthol, propanol, decanol, anisole, nitrobenzene, camphor, coumarin, ethyl-aceto-acetate, ß-ionone, butanol, naphthalene, benzaldehyde and benzene. (From Duchamp et al., 1974).

inhibitory responses in olfactory neurons. In what follows, this evidence will be analyzed, the possible mechanisms underlying inhibitory odor transduction overviewed and the importance of the inhibitory phenomenon in olfaction will be discussed briefly.

Early evidence for the existence of inhibitory responses in vertebrate olfactory cells was obtained by Gesteland et al. (1965), who recorded extracellularly from single units in frogs and found that olfactory neurons could be excited or inhibited, depending on the stimuli. This observation led them to suggest that inhibition formed part of the repertoire of responses in vertebrate olfactory neurons. In addition, based on their observations, Gesteland et al. forwarded the notion that an olfactory neuron has the ability of undergoing both excitatory and inhibitory responses to odorants, a concept that was to be confirmed later (Morales et al., 1994; Kang and Caprio, 1995).

Duchamp and colleagues (1974) investigated the response of olfactory neurons to odors by applying a number of different stimuli odorants to a frog epithelium *in situ*. For this aim they recorded the response of a large number of single units stimulated separately with 20 different odorants. Single unit recordings revealed that 80% of responses were excitatory. The remaining 20% exhibited inhibitory responses to some odorants (Fig. 2). Inhibitory responses were observed less frequently than excitatory responses perhaps due to difficulties in detecting inhibition in cells with a low spontaneous rate of action potential firing.

Evidence for the presence of inhibitory responses in dissociated vertebrate olfactory cells, using patch-clamp approaches, was reported first by Dionne (1992), in the salamander *Necturus maculosus*. Amino acids, which are odorous stimuli in this animal, were delivered from a puffer pipette containing either mixtures or individual amino acids. When action currents were recorded under cell-attached conditions, excitatory responses to odorant stimulation were revealed by increases in action potential firing rates. Inhibitory responses consisted of a temporal decrease in firing frequency, or transient suppression of firing. In addition to excitatory and inhibitory responses, some cells were found to respond to odorants with changes in firing rates, without detectable potential changes. In such neurons, odorants were effective in changing action potential firing patterns evoked by depolarizing current pulses. Dionne termed these responses 'silent responses' and proposed that they are due to subtle modulation by odorants of membrane conductances by multiple transduction pathways.

Morales and collaborators (1994) reported the presence of inhibitory responses to odorants in isolated olfactory neurons from the Chilean toad *Caudiverbera caudiverbera*. This animal spends much of its life in the water, but sticks its head into the air for prolonged periods of time, thus being extensively exposed to volatile odors. Two mixtures of odorants were used as stimuli by these authors. One of them was made of floral and frutal fragrances known to trigger the excitatory cascade. The other was a mixture of putrid odorants, which are utterly ineffective in activating the excitatory cascade (Morales et al., 1994). Responses were recorded first in the form of action currents, under cell-attached patch-clamp conditions. The results obtained showed that, while odorants known to activate the excitatory cascade caused increases in action potential firing, putrid odorants caused diminishment in firing rates (Fig. 3A). Furthermore, under whole-cell conditions the mixture of putrid odorants induced an outward current and a hyperpolarizing receptor potential (Fig. 3B). Inhibition by putrid odorants was recorded also in *Xenopus* olfactory neurons (Morales et al., 1994).

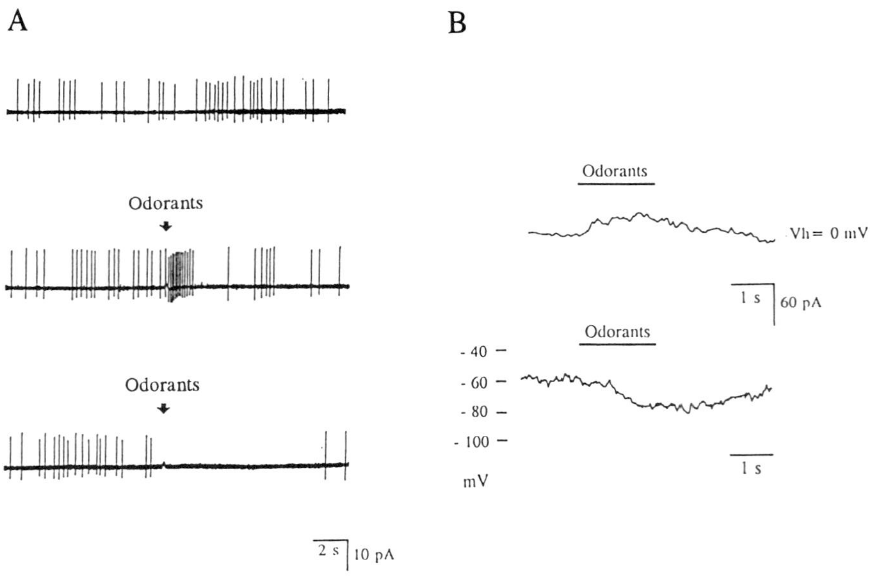

Figure 3. Odorants induce inhibition of electrical activity by triggering a hyperpolarizing outward current. A. Excitatory and inhibitory responses induced by odorants on the same olfactory neuron. Electrical activity of olfactory neurons from the Chilean toad *Caudiverbera caudiverbera*, recorded with the cell-attached configuration of the patch-clamp technique. The upper trace shows the basal (spontaneous) firing activity of this neuron. An excitatory response, consisting on an increase in the spike frequency, was induced by a mixture odors containing 10 μM citralva, citronellal and geraniol (middle trace). The lower trace shows the inhibition of the electrical activity after exposition to 100 μM of an putrid odorants mixture containing isovaleric acid, triethylamine and pyrazine. Odorants were applied focally onto the ciliary region of the neuron for 40 ms. **B.** Outward current and hyperpolarization induced by the mixture of putrid odorants in an olfactory neuron. The current was recorded in voltage-clamp conditions (V_{hold} = 0 mV) and the membrane potential under current-clamp (I_{hold} = 0 pA). The upper trace shows the outward current induced by a puff of 40 μM of a mixture containing isovaleric acid, triethylamine and pyrazine. The lower trace shows the membrane potential change in response an identical stimulus.

In toad, Morales et al (1994) established quite clearly that an olfactory neuron can undergo both excitatory and inhibitory responses, depending on the stimuli (Fig. 3A). Similar observations were made later in olfactory neurons from the channel catfish (Kang and Caprio, 1995). In view of the finding that the same olfactory cell can undergo excitation or inhibition, depending on the stimuli, it is reasonable to postulate that distinct transduction pathways underlie both types of chemosensitive events.

MECHANISM OF INHIBITION.

Currently, the existence of inhibitory responses has been established only in some anurans (Duchamp et al., 1978; Dionne, 1992; Morales et al., 1994) and in

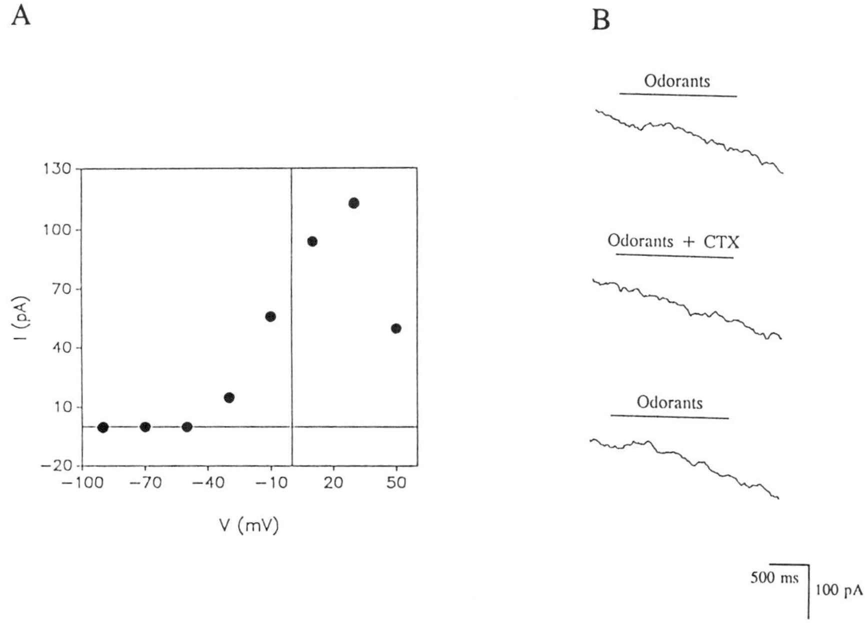

Figure 4. Inhibition is due to the activation of a Ca^{2+}-activated K^+ conductance. A. I-V relation of the odorant-induced conductance. The currents induced by identical odor stimuli (750 ms, 30 μM of the putrid mixture made of isovaleric acid, triethylamine and pyrazine) applied at various voltages stepped from a holding potential of -70 mV, are plotted as a function of voltage. The N-shape of this curve is similar to that of Ca^{2+}-activated K^+ conductances; it has a maximal value at ~+30 mV, decaying at more positive voltages. **B.** Blockage of the odorant-induced outward current by CTX. The upper trace shows the outward current evoked by a puff of a mixture containing the odorants isovaleric acid, triethylamine and pyrazine. An identical stimulus, supplemented with 10 nM CTX, blocked the response (middle trace). The lower trace shows the recovery of the response after removal of the toxin. V_{hold} = 0 mV.

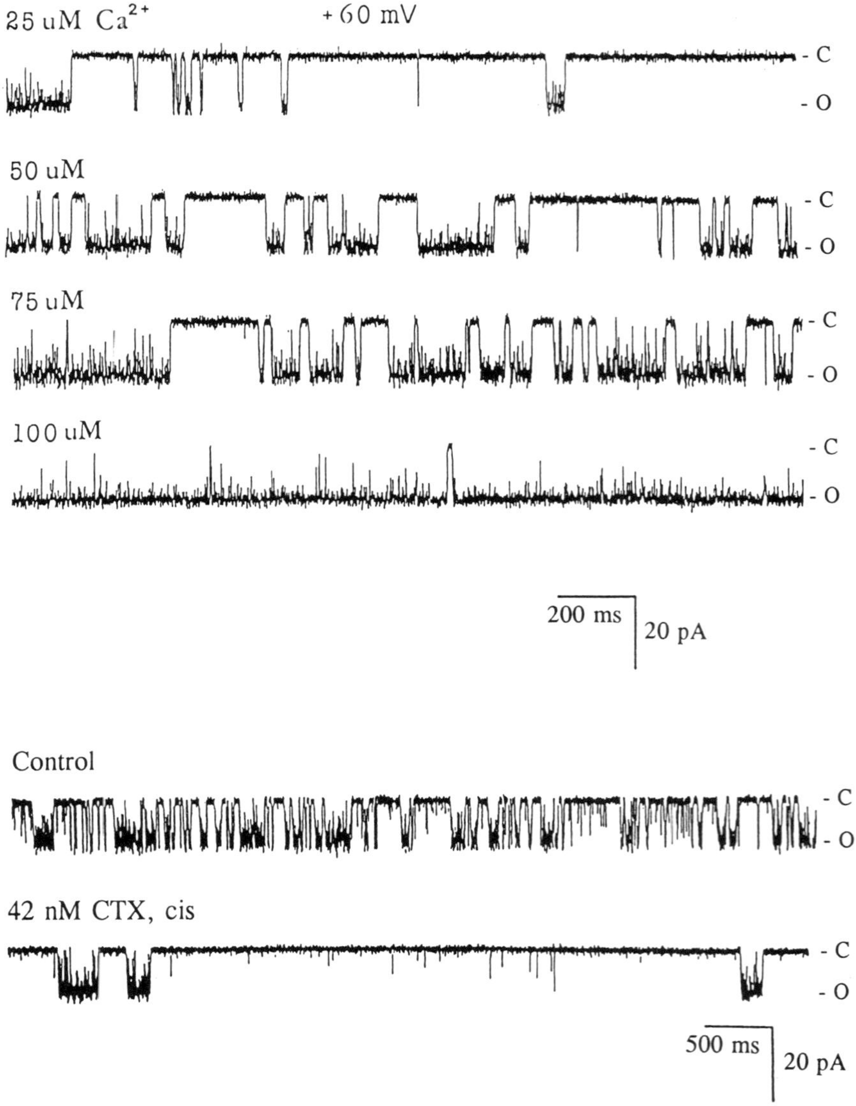

Figure 5. Ca^{2+}-activated K^+ channel from olfactory cilia membranes in planar lipid bilayer. A. Effect of Ca^{2+} on the channel open probability. **c** = zero-current level; **o** = current level of the open channel. The transmembrane voltage is indicated. **B.** Sensitivity of this channel to CTX. Recordings were obtained at -50 mV, under symmetric ionic conditions (200 mM KCl). Upper record: control; lower record: same membrane in the presence of 42 nM CTX.

Table 1. Odorant-induced increase in second messenger concentration in olfactory cilia preparation. The data were obtained using fast kinetics methodology, after application of 1 uM of the different odorants. They are expressed as a percentage of the effect induced by citralva (cAMP) or lyral (IP_3) (From Breer and Boekhoff, 1991).

Odorants	cAMP	IP_3
Floral		
Hedione	108 ± 21	6 ± 7
Geraniol	62 ± 25	6 ± 7
Acetophenone	36 ± 8	4 ± 6
Phenylethyalcohol	16 ± 9	8 ± 7
Lilial	1 ± 2	106 ± 22
Fruity		
Citralva	100	4 ± 7
Citraldimethylacetal	63 ± 10	3 ± 5
Citronellal	61 ± 7	3 ± 5
Citronellylacetate	62 ± 10	1 ± 1
Lyral	0	100
Herbaceous		
Eugenol	70 ± 18	4 ± 4
Isoeugenol	49 ± 13	2 ± 3
Ethylvanillin	0	63 ± 15
Putrid		
Furfurylmercaptan	37 ± 9	8 ± 7
Triethylamine	4 ± 2	141 ± 60
Phenylethylamine	6 ± 2	137 ± 78
Pyrrolidine	0	70 ± 25
Isovaleric acid	0	68 ± 24

the channel catfish (Ivanova and Caprio, 1992; Kang and Caprio, 1995). It remains to be determined whether inhibition occurs in other vertebrates.

The mechanism underlying inhibition has been investigated further in olfactory neurons from the Chilean toad. Under whole-cell conditions it became clear that putrid odorants activate K^+ currents. The N-shaped I-V curve (Fig. 4A) and the fact that these currents were blocked by nanomolar concentrations of the scorpion toxin charybdotoxin (CTX; Morales et al., 1995; Fig. 4B), suggested that a Ca^{2+}-activated K^+ conductance is involved in inhibition (Morales et al., 1995). Removal of extracellular Ca^{2+} abolished K^+ currents induced by stimulation with putrid odorants, suggesting that Ca^{2+} influx was required. More recently, Morales and collaborators (Morales et al, 1996) have gathered evidence that putrid odorants trigger a small Ca^{2+} current (< 1 pA), leading to increases in Ca^{2+} only in the apical region of olfactory neurons. This raise in Ca^{2+}, monitored with Ca^{2+}-sensitive dyes under the confocal microscope, might open Ca^{2+}-activated K^+ channels responsible for the hyperpolarizing receptor potential.

The presence of Ca^{2+}-activated K^+ channels in chemosensory olfactory membranes was scrutinized by Jorquera and colleagues (1995), using planar lipid bilayers reconstitution. Fusion of purified olfactory cilia to planar lipid bilayers led to the insertion of CTX-sensitive, voltage-dependent Ca^{2+}-activated K^+ channels (Fig. 5). These studies also allowed to monitor the cAMP-gated cation-selective channels engaged in excitatory responses (Jorquera et al., 1995).

One crucial question that remains to be answered is the mechanisms by which binding of putrid odorants to ciliary receptors leads to the Ca^{2+} influx which, presumably, activates Ca^{2+}-dependent outward currents. The time delay observed between the onset of the chemical stimulus and the activation of the Ca^{2+} current (> 100 ms) is consistent indeed with a second messenger-mediated process.

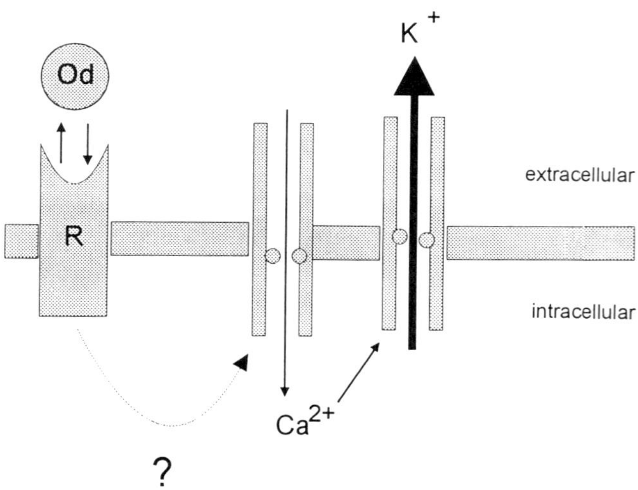

Figure 6. Model for the inhibitory transduction to putrid odorants. After binding of an odorant to a receptor at the ciliary membrane, an unknown mechanism leads to the activation of a Ca^{2+} conductance. The Ca^{2+} influx increases the cytosolic Ca^{2+} concentration, determining the activation of a Ca^{2+}-activated K^+ conductance. This results on the hyperpolarization of receptor cell, inhibiting action potential firing.

DIVERSITY OF TRANSDUCTION PATHWAYS IN VERTEBRATE OLFACTORY NEURONS.

The presence of distinct and opposite electrophysiological responses to odors in olfactory neurons suggest the involvement of multiple transduction pathways associated to each response type. Fast kinetic approaches revealed that in vertebrate olfactory cilia certain odorants induce increases in cAMP, whereas other odorants induce elevation in IP_3. Breer and collaborators (Boekhoff and Breer, 1990; Breer et al., 1990) reported that, in rat, certain odorants, mostly floral, fruity and herbaceous, induce a rapid increase in cAMP. Another group of odorants, mostly putrid, triggered increases in IP_3 levels. Breer and collaborators proposed that odorants affect either one or the other pathway, but not both (Table 1). In contrast, Ronnett et al. (1993) reported that in permeabilized olfactory neurons from rat, all odorants tested induced increases in both cAMP and IP_3. The reason for the difference in the results obtained by these two groups remains to be explained.

The electrophysiological target of the cAMP, excitatory olfactory transduction pathway, is the cAMP-activated channel (Zufall et al., 1994; Fig. 1). However, the target for the IP_3 pathway has not yet been established. Tareilus et al. (1995) reported that activation of both pathways gives place to increases in intracellular Ca^{2+}.

It is known that the increase in Ca^{2+} associated to the cAMP pathway is due to its influx through cAMP-activated channels (Zufall et al., 1994). One effect of this influx is the activation of a Ca^{2+}-activated Cl^- current (Kurahashi and Yau, 1993, see Fig. 1).

The origin of the intracellular increase in Ca^{2+} associated to the activation of the IP_3 pathway is unclear. If this pathway operates in toad, it follows that the IP_3-dependent elevation in ciliary Ca^{2+} must lead to the activation of Ca^{2+}-activated K^+ channels in the cilia and, therefore, to the development of an inhibitory receptor potential (Fig. 6). If so, it would be necessary to postulate that IP_3 is directly involved in activating influx of this divalent in the membrane of olfactory cilia. As discussed in previous lines, the development of inhibitory responses in toad depends on external Ca^{2+}. Moreover, putrid odorant, have been found to activate Ca^{2+} currents in toad olfactory cells.

Alternatively, the Ca^{2+} influx induced by putrid odorants could be triggered by other, as yet unknown transduction mechanism. In such a case, the elevation in IP_3 levels might be secondary, and perhaps causally related to elevations in intracellular Ca^{2+}. IP_3 production could then trigger the release of the divalent ion from intracellular stores. This last possibility would be consistent with Ronnet et al's observation that odorants from all classes cause elevations in IP_3 levels in olfactory neurons. As stated previously, all odorants, inhibitory and excitatory, cause influx of Ca^{2+} in olfactory neurons.

INHIBITION IN INVERTEBRATE OLFACTORY NEURONS.

The presence of odorant-induced inhibition is well established in invertebrates, such as insects (Boeckh, 1967; Kaissling, 1986), squid (Lucero et al., 1992) and crustaceans (Seelinger, 1983; Derby and Harpaz, 1988; McClintock and Ache, 1989; Derby et al., 1988; Michel et al., 1991; Michel and Ache, 1994).

Among invertebrates, the best understood olfactory receptor neuron is that of the lobster *Panulirus argus*. In this species, it has been shown that the olfactory neurons display both, excitatory and inhibitory responses (Michel and Ache, 1994). The chemical stimuli used by Michel and Ache were either a complex commercial food mixture for fish, which excited the neurons, or the amino acid L-proline, which had an inhibitory effect on them. Each type of response is caused by a different mechanism, and trigger increases in membrane conductance. Contrary to vertebrates, in the lobster excitatory responses are associated to an IP_3 pathway (Fadool and Ache, 1992). IP_3 is thought to activate both, a Ca^{2+} channel and a cation-selective cation channel, triggering a depolarizing receptor potential.

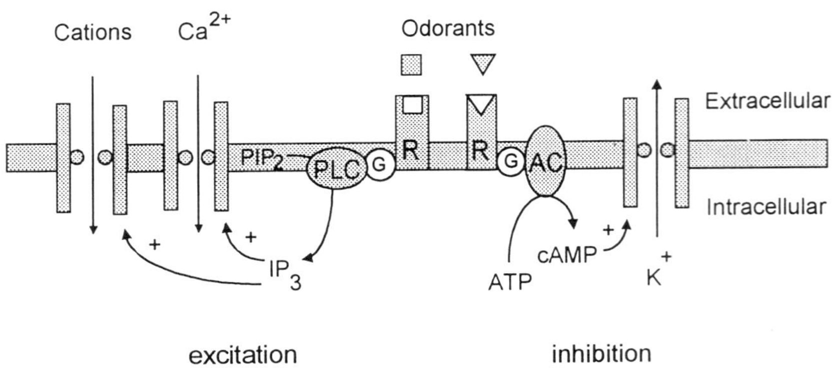

Figure 7. Chemotransduction pathways in lobster olfactory receptor cells. Evidence is consistent with the existence of two pathways. One of them involves an elevation of cAMP levels, determining the activation of a K^+ conductance that inhibits the electrical activity. A separate pathway, triggered by different odorants, involves an increase of IP_3 concentration. This second messenger appears to gate directly both a cation-selective channel and a Ca^{2+}-selective cation channel, producing an excitatory response.

The cAMP pathway is, in this species, associated to the inhibitory response (Michel and Ache, 1992). cAMP would lead to the opening of K^+ channels, which are responsible for an hyperpolarizing receptor potential. The transduction pathways that participates in odorant detection in lobster olfactory neurons are illustrated in Figure 7.

PHYSIOLOGICAL RELEVANCE OF INHIBITION.

Recognition and discrimination of odors involves the transmission of the information as action potentials from the periphery to the olfactory bulb. Until recently, the available evidence supported the notion that this code was formed exclusively by increases in the firing rate generated at the olfactory neurons. The evidence described above demonstrates that inhibition is also part of the physiology of olfactory neurons and suggest that it may play an important role on olfactory coding, by increasing the diversity of neuronal firing patterns, which are the basis for odor discrimination. It is not yet known how not to what extent inhibition participates in this process. It may be hypothesized that inhibition may regulate the magnitude and time course of the excitatory receptor potential in the olfactory neurons, or it may express itself independently in the form of a reduction in the spontaneous firing. A purely excitatory system would be limited to increases in firing. In real life odorants are usually presented as blends, likely including odorants having both excitatory and inhibitory effects.

ACKNOWLEDGEMENTS.

Supported by Fondo Nacional de Ciencia y Tecnología, grant # 1960878 and TWAS, grant # 94-463 RG/BIO/LA. PL held a John Simon Guggenheim Fellowship and a Presidential Chair in Science.

REFERENCES

Anholt, R.R.H., Aebi, U., and Snyder, S, 1986, A partially purified preparation of isolated chemosensory cilia from the olfactory epithelium of the bullfrog, *Rana catesbiana*, *J. Neurosci.* 6:1962.

Bakalyar, H.A., and Reed, R.R., 1990, Identification of a specialized denylate cyclase that may mediate odorant detection, *Science*. 250:1403.

Boeckh, J., 1967, Inhibition and excitation of single insect olfactory receptors, and their role as a primary sensory code, in: "Olfaction and Taste", Y. Zotterman, ed., Vol II. Pergamon Press, New York, p. 721.

Boekhoff, I., and Breer, H., 1990, Differential stimulation of second messenger pathways by distinct classes of odorants, *Neurochem. Intl.* 7(4):553.

Boekhoff, I., Tareilus, E., Strotmann, J., and Breer, H., 1990, Rapid activation of alternative second messenger pathways in olfactory cilia from rats by different odorants, *EMBO J.* 9:2453.

Breer, H., Boekhoff, I., and Tareilus, E., 1990, Rapid kinetics of second messenger formation in olfactory transduction, *Nature.* 345:65.

Breer, H., Raming, K., and Krieger, J., 1994, Signal recognition and transduction in olfactory neurons, *Biochem. Biophys. Acta.* 1224:277

Broillet, M.C., and Firestein, S, 1996, Gaseous second messengers in vertebrate olfaction, *J Neurobiol.* 30:49.

Buck, L., and Axel, R., 1991, A novel multigene family may encode odorant receptor: a molecular basis for odorant recognition, *Cell.* 65:175.

Derby, C.D., and Harpaz, S., 1988, Physiology of Chemoreceptor cells in the legs of the freshwater prawn, *Macrobrachium rosenbergii*, *Comp. Biochem. Physiol.* A 90:85.

Derby, C.D., Girardot, M.N., and Harpaz, S, 1988, Inhibition in the periphery: occurrence in olfaction and gustatory receptor cells of aquatic crustaceans, correlation with mixture suppression and effect on quality coding, *Chem. Senses.* 13:683.

Dhallan, R.S., Yau, K.W., Schrader, K.A., and Reed, R.R., 1990, Primary structure and functional expression of a cyclic nucleotide-activated channel from olfactory neurons, *Nature.* 347:184.

Dionne, V.E., 1992, Chemosensory responses in isolated olfactory receptor neurons from *Necturus maculosus*, *J. Gen. Physiol.* 99:415.

Duchamp, A., Revial, M.F., Holley, A., and Mac Leod, P., 1974, Odor discrimination by frog olfactory receptors, *Chemi. Senses Flavor.* 1:213.

Fadool, D.A., and Ache, B.W., 1992, Plasma membrane inositol 1,4,5-trisphosphate-activated channels mediate signal transduction in lobster olfactory receptor neurons, *Neuron.* 9:907.

Firestein, S., 1992, Electrical signals in olfactory transduction, *Curr. Opinion Neurobiol.* 2:444.

Firestein, S., Shepherd, G.M., and Werblin, F.S., 1990, Time course of the membrane current underlying sensory transduction in salamander olfactory receptor neurons, *J. Physiol.* (London) 430:135.

Firestein, S., and Werblin, F.S., 1987, Gated currents in isolated olfactory receptor neurons of the larval tiger salamander, *Proc. Natl. Acad. Sci. USA.* 84:6292.

Frings, S., and Lindemann, B., 1988. Odorant response of isolated olfactory receptor cells is blocked by amiloride, *J. Memb. Biol.* 105:233.

Gesteland, R.C., Lettvin, J.Y., and Pitts, W.H., 1965, Chemical transmission in the nose of the frog, *J. Physiol.* (London) 181:525.

Ivanova, T., and Caprio, J., 1993, Odorant receptor activated by amino acids in sensory neurons of the channel catfish Ictalurus punctatus, *J. Gen. Physiol.* 102:1085.

Jones, D.T., and Reed, R.R., 1989, G_{olf}: An olfactory neuron specific G-protein involved in odorant signal transduction, *Science*. 244:790.

Jorquera, O., Latorre, R., and Labarca, P., 1995, Ion channel classes in the ciliary membrane of olfactory neurons: Planar lipid bilayer studies, *Am. J. Physiol.* 38:C1235.

Kaissling, K.E., 1986, Chemo-electrical transduction in insect olfactory receptors, *Ann. Rev. Neurosci.* 9:1221.

Kang, J., and Caprio, J., 1995, In vivo responses of single olfactory receptor neurons in the channel catfish, *Ictalurus punctatus, J. Neurophysiol.* 73:172.

Kurahashi, T., 1989, Activation by odorants of cation-selective conductance in the olfactory receptor cells isolated from the newt, *J. Physiol.* 419:177.

Kurahashi, T., and Yau, K.W., 1993, Co-existence of cationic and chloride components in odorant-induced current of vertebrate olfactory receptor cells, *Nature*. 363:71.

Lucero, M.T., Horrigan, F.T., and Gilly, W.F., 1992, Electrical responses to chemical stimulation of squid olfactory receptor cells, *J. Exp. Biol.* 162:231.

McClintock, T.S., and Ache, B.W., 1989, Hyperpolarizing receptor potentials in lobster olfactory receptor cells: implications for transduction and mixture suppression, *Chem. Senses*. 14:637.

Michel, W.C., and Ache, B.W., 1992, Cyclic nucleotides mediate an odor-evoked potassium conductance in lobster olfactory receptor cells, *J. Neurosci.* 12:3979.

Michel, W.C., and Ache, B.W., 1994, Odor-evoked inhibition in primary olfactory receptor neurons, *Chem. Senses*. 19:11.

Michel, W.C., McClintock, T.S., and Ache, B.W., 1991, Inhibition of lobster olfactory receptor cells by an odorant-activated potassium conductance, *J. Neurophysiol.* 65:446.

Morales, B., Labarca, P., and Bacigalupo, J., 1995, A ciliary conductance sensitive to charybdotoxin underlies inhibitory responses in toad olfactory receptor neurons, *FEBS Lett.* 359:41.

Morales, B., Ugarte, G., Labarca, P., and Bacigalupo, J., 1994, Inhibitory K^+ currents activated by odorants in toad olfactory neurons, *Proc. R. Soc. Lond. B.* 257:235.

Morales, B., and Bacigalupo, J., 1996, Chemical reception in vertebrate olfaction: evidence for multiple transduction pathways, *Biol. Res.* 201: In press.

Morales, B., Madrid, R., and Bacigalupo, J., 1996, Influx of Ca^{2+} mediates the activation of the odorant-induced inhibitory current in vertebrate olfactory receptors neurons, *Chem. Senses*. abstract. In press.

Ngai, J., Chess, A., Dowling, M., Necles, N., Macagno, E., and Axel, R., 1993, Coding of olfactory information: topography of odorant receptor expression in the catfish olfactory epithelium, *Cell*. 72:667.

Raming, K., Krieger, J., Strotmann, J., Boekhoff, I., Kubick, S., Baumstark, C., and Breer, H., 1993, Cloning and expression of odorant receptor, *Nature*. 361:353.

Reed, R.R., 1992, Signaling pathways in odorant detection. *Neuron*. 8:205.

Ressler, R., Sullivan, S., and Buck, L., 1993, A zonal organization of odorant receptor gene expression in the olfactory epithelium, *Cell*. 73:597.

Restrepo, D., Boekhoff, I. and Breer, H. 1993. Rapid kinetic measurements of second messenger formation in olfactory cilia from channel catfish. Am. J. Physiol. 264: C906-C911

Restrepo, D., Teeter, J.H., and Schild, D., 1996, Second messenger signaling in olfactory transduction, *J. Neurobiol*. 30:37.

Ronnett, G.V., Cho, H., Hester, L.D., Wood, S.F. and Snyder, S.H., 1993, Odorants differentially enhance phosphoinositide turnover and adenylyl cyclase in olfactory receptor neuronal cultures, *J. Neurosci*. 13:1751.

Ronnett, G.V. and Snyder, S.H., 1992, Molecular messengers of olfaction, *TINS*. 15:508.

Sklar, P.B., Anholt, R.R.H., and Snyder, S.H., 1986, The odorant-sensitive adenylate cyclase of olfactory receptor cells: differential stimulation by distinct classes of odorants, *J. Biol. Chem*. 261:15538.

Seelinger, G., 1983, Response characteristics and specificity of chemoreceptors in *Hemilepistus reaumuri* (Crustacea, Isopoda), *J. Comp. Physiol*. A. 152:219.

Trotier, D., 1996, A patch clamp analysis of membrane currents in salamander olfactory cells, *Pflügers Archiv*. 407:589.

Tareilus, E., Noe, J., and Breer, H., 1995, Calcium signals in olfactory neurons, *Biochem. Biophys. Acta*. 1269:129.

Vassar, R., Ngai, J., and Axel, R., 1993. Spacial segregation of odorant receptor expression in the mammalian olfactory epithelium, *Cell*. 74:309.

Zufall, F., Firestein, S., and Shepherd, G.M., 1994, Cyclic nucleotide-gated ion channels and sensory transduction in olfactory receptor neurons, *Annu. Rev. Biophys. Biomol. Struct*. 23:577.

MULTIPLE PATHWAYS IN INVERTEBRATE VISUAL TRANSDUCTION

Peter M. O'Day[1], Juan Bacigalupo[2], Cecilia Vergara[2], Gonzalo Ugarte[2], Diana M. Bautista[1,3], and Joan E. Haab[1]

[1] Institute of Neuroscience & Department of Biology, University of Oregon, Eugene OR USA
[2] Department of Biology, University of Chile, Santiago Chile
[3] Present address: Department of Molecular and Cellular Physiology, Stanford University, Stanford CA USA

Intracellular information transfer is basic to virtually every biological process, therefore it is of great value to understand and appreciate the complexities and the diversity of modes of signalling. Studies of biochemical events underlying phototransduction, the process in photoreceptor cells whereby light generates a nerve signal, have set the stage for discoveries of several general themes in cellular signalling. Invertebrate visual transduction is particularly interesting because, in spite of its apparent complexity, it bears similarity to other signalling systems. Phototransduction may involve activation of and interaction between two common biochemical pathways. Activation of multiple interacting pathways may reflect common themes in cell signalling.

Phototransduction in invertebrates begins with photon absorption by a receptor molecule capable of triggering intracellular biochemical cascades. It ends with the gating of plasma membrane ion channels. Additionally, regulatory processes at intermediate steps control the gain (measured as the

From Ion Channels to Cell-to-Cell Conversations
Edited by Latorre and Sáez, Plenum Press, New York, 1997

number of ion channels opened per incident photon). Negative and positive feedback processes make photoreceptors non-linear transducers, providing adaptation that enables cells to respond effectively to illumination varying in intensity over an enormous range.

In most known invertebrate photoreceptors, the physiological response to illumination is a depolarizing receptor potential caused by ion channel opening, although there are interesting notable exceptions. In these cases, phosphoinositide signalling cascades appear to be at the heart of invertebrate phototransduction. There is much evidence also for the involvement of the cGMP pathway. Despite a growing body of experimental evidence, however, there is no complete picture of phototransduction in any invertebrate photoreceptor.

A comprehensive understanding of the biochemical orchestration of phototransduction will require identification and characterization of each of the key participating enzymes and messengers, the factors affecting each of them, their interactions, and their spatio-temporal distributions. This includes details of enzyme activation and deactivation and, in some cases, inactivation, modulation, and translocation of each participant in the cascades. It is important to realize that all of these processes proceed in innumerable microenvironments in the same cell at the same time. Spatio-temporal characterization of the synthesis, degradation, and diffusion of each cellular messenger will be required. A full reconciliation of the biochemical and molecular information with physiological measurements will aid in the evaluation of phototransduction models. Such a description has not yet been achieved in any invertebrate system, but a great deal of progress has been made on many fronts. There are several recent reviews relating to these topics.[1-4]

In this article, we outline basic features of invertebrate phototransduction and we discuss some current controversy. We focus primarily on the widely studied *Limulus* and *Drosophila* photoreceptors, but there is much to be learned from the differences and similarities among these and other invertebrates.

RECEPTOR ACTIVATION AND REGULATION

Phototransduction is begun by the absorption of an incident photon by the visual pigment protein, rhodopsin.[5-7] This induces conformational changes of the rhodopsin, enabling it to bind and activate heterotrimeric G-proteins, continuing the phototransduction cascade.[8-11] (review: Ref. 1). G-proteins (GTP-binding proteins) relay information from the receptor to effector molecules in excitation cascades. In addition to momentary binding to activated rhodopsin, G-protein activation requires exchange of bound GDP for GTP and liberation of the catalytic α-subunit (G_α) from the heterotrimer.[12]

Following activation, components are inactivated in several steps: rhodopsin phosphorylation by rhodopsin kinase,[13-16] phosphorylation of another regulatory protein, arrestin,[17-24] by CaM kinase, and binding of phosphorylated arrestin to rhodopsin,[12,22,25] preventing further G_α binding. G-protein inactivation is accomplished by its autocatalytic GTPase activity. In vertebrate photoreceptors, inactivation of the heterotrimeric G-protein appears to be aided by a GTPase accelerating protein;[26,27] however, GTPase accelerating proteins have not been examined in invertebrate photoreceptors. Resetting the components of receptor activation involves reisomerization of the pigment triggered by absorption of a second photon, and dephosphorylation of both rhodopsin and arrestin.[22,28-30] The G-protein cycle is completed by reassociation of the G_α and $G_{\beta,\gamma}$ subunits. While the components of such an intricate regulatory scheme have not been fully characterized in any invertebrate preparation, it appears likely that this general "strategy" is similar among the invertebrate rhabdomeral photoreceptors studied (review: Ref 1).

In *Drosophila*, the G-protein is of the class G_q, which activate the polyphosphoinositide (PI) cascades.[31-34] This is presumably true of all species with PI-pathway dependent phototransduction. In *pecten* photoreceptors, in which excitation appears to be independent of PI-mediated signalling, light-sensitive channels are activated by cGMP;[35] however, G-protein involvement has not yet been examined.

PHOSPHOINOSITIDE (PI)-PATHWAY

The next steps in phototransduction involve activation of the PI-pathway by activated $G_{q\alpha}$-proteins. $G_{q\alpha}$ activates phospholipase C (PLC),[36-41] which cleaves membrane phospholipids, 4,5-phosphatidylinositol bisphosphate PIP_2, generating two intracellular messengers: cytosolic messenger, inositol-(1,4,5)-trisphosphate (IP_3), and membrane lipid soluble messenger, diacylglycerol (DAG). These two messengers localize to different intracellular environments and initiate separate, parallel signaling events before being recycled *via* phospholipid metabolic cycles. IP_3 binds to the IP_3-Receptor (IP_3-R), a Ca^{2+}-channel on the membranes of intracellular Ca^{2+}-stores (ER, or endoplasmic reticulum)[0]. Binding of IP_3 to IP_3-R causes the release of Ca^{2+} from the ER to the cytosol and transient elevation of intracellular Ca^{2+} concentration ($[Ca_i^{2+}]$). This $[Ca_i^{2+}]$ elevation has multiple effects in phototransduction, since many components are Ca^{2+}-dependent. The DAG generated in the PLC reaction activates protein kinase C (PKC), a Ca^{2+}-dependent enzyme.[42-44] A *Drosophila* eye-specific PKC appears to be involved in response termination and in adaptation, the regulation of photoreceptor sensitivity.[44,45] Mutants lacking PKC exhibit normal light-induced activation properties but dramatically slowed response deactivation. Possible PKC substrates have been reported,[46,47] however the specific mechanisms underlying the physiological effects are not understood.

The "final" step in light-induced excitation is the opening of plasma membrane ion channels. This step has been an issue of great continuing controversy, both in *Limulus*, where light-sensitive channels can be gated by cGMP, and in *Drosophila* where gating mechanism(s) for light-sensitive channels remain unresolved (see discussions on CHANNELS below). While direct involvement of components of the PI-pathway in gating light-sensitive plasma membrane channels has not been demonstrated, several lines of evidence suggest a role for Ca^{2+}-store operated channels (SOC) in *Drosophila*.

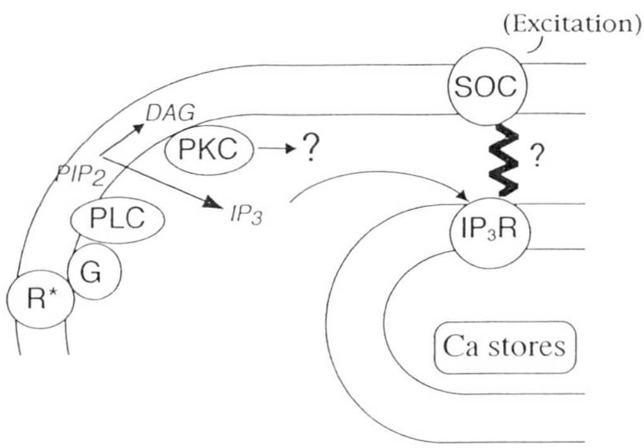

Figure 1. Basic scheme of PI-pathway dependent excitation with Ca^{2+}-store depletion operated channel. **R*** activated Rhodopsin; **G** trimeric G-protein; **PLC** activated Phospholipase C; **PKC** protein kinase C; **SOC** Ca^{2+}-stores operated plasma membrane channel; **IP_3-R** ER membrane IP_3-Receptor Ca^{2+}-channel; **PIP_2** membrane phospholipid, 4,5-phosphatidylinositol; **IP_3** soluble intracellular messenger, inositol-(1,4,5)trisphosphate; **DAG** lipid soluble plasma membrane messenger, diacylglycerol.

SOC are a class of channels activated by depletion of intracellular Ca^{2+}-stores, and they have been described in several preparations.[48-51] SOC are opened following Ca^{2+}-depletion induced by IP_3-R activation, but the specific mechanism of channel gating is unclear. A diagram of a basic scheme of PI-pathway dependent excitation is presented in Figure 1.

PHOSPHOLIPASE C REQUIREMENT

Central to the argument that the PI-pathway is at the heart of *Drosophila* phototransduction are the demonstrations 1) that excitation cannot occur without functional PLC;[39] 2) that excitation can be recovered at permissive conditions in temperature sensitive mutants of PLC;[52] and 3) that light triggers PLC activity.[10,53,54] Oddly, however, exogenous applications of DAG, IP_3, or Ca^{2+} have not mimicked excitation in whole cell recordings from *Drosophila* photoreceptors[3] (Bacigalupo & O'Day unpublished observations), raising questions about the precise role of PLC. Furthermore, it remains to be demonstrated that specifically light-induced PLC activity causes the subsequent steps in excitation to proceed. It is a formal possibility that excitation does not require PLC activity triggered by a specific stimulus for the generation of the response to that stimulus. It may be, for instance, that PLC-dependent PI turnover is required for maintenance of cellular functions and PLC is activated by the effects of light, but it plays no direct role in response generation. This idea is consistent with the need for a functional PI pathway for the maintenance of retinal structural integrity in *Drosophila* (review: Refs. 6,1).

In *Limulus*, introduction of IP_3 is sufficient to induce channel opening unless intracellular Ca^{2+} is buffered,[55] suggesting that Ca^{2+} elevation can trigger phototransduction events downstream of IP_3-R function. Thus, even if there is an excitation pathway separate from PLC in *Limulus*, it would appear that the light-induced activation of PLC should cause channel opening.

CYCLIC GMP (cGMP)-PATHWAY

The best evidence for the involvement of cGMP in PI-pathway dependent invertebrate phototransduction is the direct demonstration that cGMP can induce excitation and serve as a ligand that gates light-sensitive channels of *Limulus* photoreceptors.[56] Supporting this notion are observations that injection of cGMP and non-hydrolyzable cGMP-analog, 8-Br-cGMP, can induce excitation in these cells.[57-59] Furthermore, pharmacological agents that affect cGMP synthesis or cGMP degradation alter phototransduction.[60-63] The strong Ca^{2+}-dependence of this phenomenon lends support to models (*e.g.*, refs. 61,64) in which cytosolic Ca^{2+}, elevated by IP_3-R mediated Ca^{2+}-release from intracellular stores, modulates the activity of guanylate cyclase (GC, the enzyme that synthesizes cGMP) and of cGMP-phosphodiesterase (cG-PDE, the enzyme that degrades cGMP), thereby inducing transient elevation of cGMP and channel opening. A model of cGMP-gated channel functioning in PI-dependent excitation is diagrammed in Figure 2. Consistent with this idea is the report of cGMP phosphodiesterase (PDE) in *Limulus* ventral eyes[65] and the report that introduction of phosphodiesterase inhibits a phase of the light response in *Limulus*.[66] Furthermore, guanylate cyclase has been detected in other invertebrate retinas (*e.g.* squid[67] and *Drosophila*[68-70]), but results of measurements of light-dependent cGMP in *Limulus* and other invertebrates have been equivocal.[62,71,57]

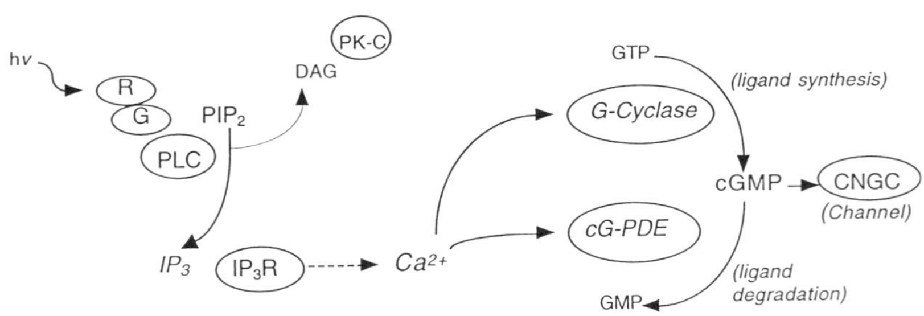

Figure 2. Hypothetical model of Limulus phototransduction. This is designed to reconcile the following features of *Limulus* phototransduction: 1) cGMP-gated channels; 2) PI-pathway dependent excitation; 3) Ca^{2+}-dependent inhibition of excitation by cGMP-pathway inhibitors. In this model, the PI and cGMP pathways would be linked by Ca^{2+} in a 'series' fashion. Several alternative models have been proposed in which independent intracellular pathways operate in parallel to yield excitation.[125,126,103]

In *Drosophila*, there is also evidence for a role for cGMP in phototransduction, but a direct role has not been shown. Excised patch experiments analogous to those in *Limulus*[56] have not been possible because light-sensitive channels have not been found in plasma membrane patch recordings. Figure 3 illustrates that exogenous application of 8-Br-cGMP caused enhancement of light responses, without significantly changing the reversal potential of the responses.[72] Figure 4 shows that 8-Br-cGMP application also induced membrane currents by itself, and this cGMP-induced current showed marked dependence on the intensity of background illumination.[72] We have also found that inhibitors of cGMP metabolism affect light-induced excitation.[73] The results are consistent with direct or indirect roles for cGMP in phototransduction, but specific mechanisms have not been elucidated.

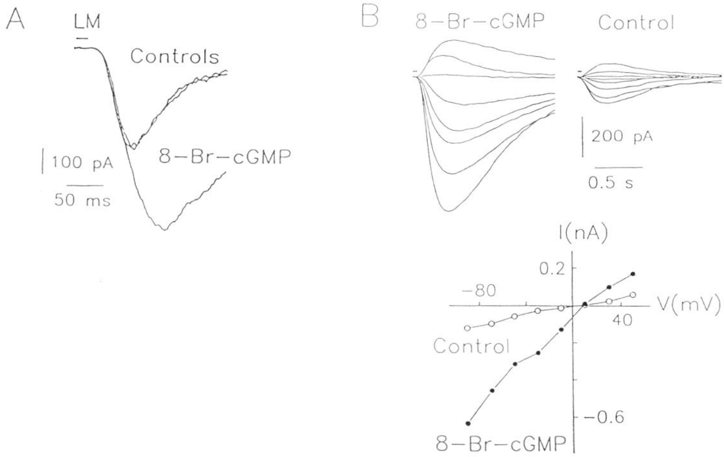

Figure 3. Enhancement of light-induced excitation in Drosophila photoreceptors. Experimental details are in Ref. 72. (A) Enhancement of light responses induced by non-hydrolyzable cGMP analog, 8-Br-cGMP. (B) Voltage-dependence measurements illustrate that reversal voltage of the light responses were substantially affected by cGMP treatment. Adapted from Ref. 72.

Response enhancement by 8-Br-cGMP (Figure 3) and 8-Br-cGMP induced currents (Figure 4) are consistent with a role of cGMP as channel ligand. Illumination may elevate [cGMP] near the channels, and, if gating requires several bound cGMP molecules, the probability of channel opening (*P*) would vary sigmoidally with [cGMP]. In this case, exogenously applied cGMP may elevate [cGMP] to higher than normal levels in the dynamic region of the *P* vs [cGMP] curve, causing light response enhancement.

On the other hand, the results are also consistent with the notion that cGMP plays only a modulatory role in phototransduction. It has been reported that currents normally activated by light can undergo spontaneous activation in whole-cell recordings ("rundown current").[74] It is possible that the

8-Br-cGMP induced currents of Figure 4 reflect modulation (enhancement) of this rundown current rather than direct, *de novo* activation of excitation components. We are currently investigating this possibility.

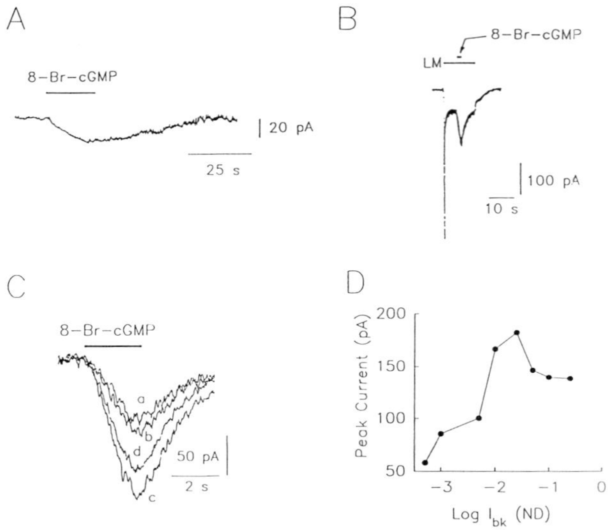

Figure 4. Membrane currents induced directly by cGMP analog. Experimental details are in Ref. 72. (A) 8-Br-cGMP induced currents in the dark. (B) Light and 8-Br-cGMP induced currents in the presence of background illumination. (C) 8-Br-cGMP induced currents from trials as in (B) superimposed for four different light intensities, increasing from a to d. (D) Response-intensity plot for experiments as in (C). Adapted from Ref. 72.

Molecular and biochemical findings suggest a role for cGMP in retinal function in flies. Firstly, a *Drosophila* homolog of the vertebrate light-sensitive cGMP-gated channel (CNGC) gene appears to be expressed in the retina.[75] When this gene is expressed heterologously in *Xenopus* oocytes, a cGMP-activated conductance is found, possessing similar voltage-dependence, Ca^{2+}-selectivity, and Mg^{2+}-dependent inhibition to the light-induced conductance.[1,2,75,124] Secondly, retinal expression of *Drosophila* guanylate cyclases has been reported, including both soluble[68,69] and receptor type guanylate cyclase;[70] although there is some controversy concerning the localization of these gene products in the functioning eye. Thirdly, retinal localization of cyclic nucleotide phosphodiesterase has been reported in the fly *Calliphora*.[76]

cGMP INVOLVEMENT IN PI-MEDIATED PHOTOTRANSDUCTION

The specific involvement of cGMP in phototransduction is a controversial issue, particularly in *Drosophila*. As we have mentioned, it is possible that cGMP is not a ligand for light-activated channels as it appears to be in *Limulus*. We are currently examining the possibility that cGMP acts in a modulatory role mediated by cGMP-dependent protein phosphorylation. Two genes for cGMP-dependent protein kinase (G-kinase) have been cloned from *Drosophila*.(77) Although detailed expression patterns and enzyme substrates have not been reported, G-kinase substrates in other signalling systems include components whose analogs in *Drosophila* retina are central to photoreceptor function, among them: Ca^{2+} pumps, Na/Ca-exchanger, the IP_3-Receptor, and Ca^{2+}-channels.(78-86) Thus, cGMP-dependent phosphorylation could influence light-induced excitation by regulating the activity some key protein(s). One particularly interesting possibility is that cGMP-dependent signalling influences one or more processes regulating intracellular Ca^{2+}, making it crucial for proper PI-dependent phototransduction. A hypothetical model depicting various steps in PI-mediated excitation is presented in Figure 5.

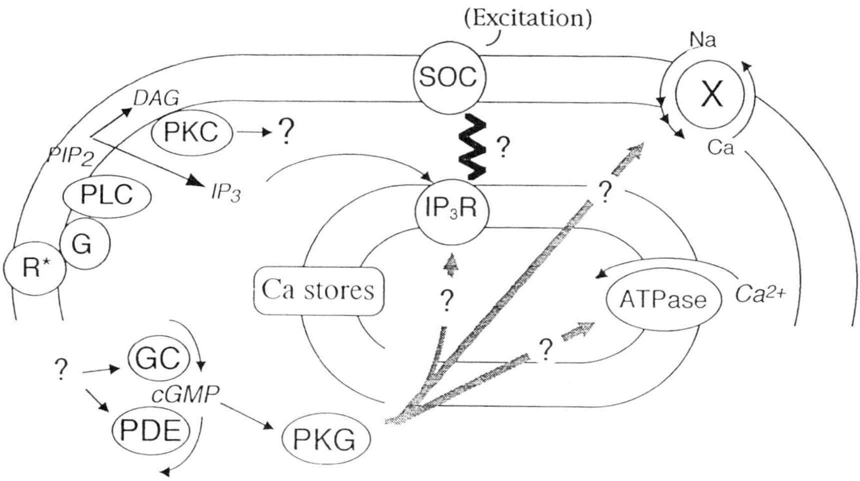

Figure 5. Possible scheme whereby cGMP modulates light-activated PI pathway in phototransduction via phosphorylation of Ca^{2+} regulatory components. **GC** guanylate cyclase; **PDE** cGMP-phosphodiesterase; **PKG** cGMP-dependent protein kinase; **ATPase** Ca^{2+}-pump across ER membrane into Ca^{2+} stores; **X** Na/Ca-exchanger across plasma membrane.

Light-sensitive Ion Channels

The identity and characteristics of invertebrate light-activated channels are currently issues of intense interest, as is the number of classes of channels activated by light. In *Limulus*, physiological evidence clearly indicates that a light-activated channel is gated by cGMP.[56] The *Limulus* cGMP-gated channel differs in ionic specificity from that in *Drosophila*, being principally permeable to Na^+, somewhat less permeable to K^+, and very poorly permeable to Ca^{2+}.[87] The existence of other light-activated channel types in *Limulus* remains unclear. In *Pecten* photoreceptors cGMP gates a light-activated K^+ channel, yielding a hyperpolarizing light response.[88]

In *Drosophila*, there are three genes for ion channel subunits expressed in the retina that are candidates for light-activated channels: *trp*, *trp-l*, and *DmCNGC*.[89,90,75] The DmCNGC channel is cGMP-gated, *trp-l* can be activated indirectly by IP_3, and *trp* reportedly functions as a Ca^{2+}-store operated channel, SOC.[91,92] The direct mode of activation of both *trp* and *trp-l* in phototransduction are unclear. *trp* and *trp-l* localize to the rhabdomeres, the regions where phototransduction takes place, but their precise subcellular localization are unknown, as is that of DmCNGC.

Physiological evidence indicates that *trp* protein is of fundamental importance in *Drosophila* phototransduction. Null mutants of *trp* lack a component of the normal light-induced current.[93] Responses to prolonged bright illumination in normal flies consist of a brief latency period followed by a quickly rising inward current that reaches a transient peak and a decline to a sustained plateau current. Illumination of *trp* mutants elicits only a transient response; the sustained plateau phase is absent. This feature can be mimicked in wild type by a Ca^{2+} channel blocker (review: Ref 3).

The *trp* protein has six potential transmembrane segments, and in several regions it resembles voltage-gated Ca^{2+} channels. However it lacks the positive residues in the S4 region characteristic of voltage-gated channels.[94,90]

Studies of *trp* expressed heterologously in insect (Sf9) cells suggest that the *trp* gene product operates as a SOC channel.[91,95] *trp* fits this category because it is highly Ca^{2+} selective and it can be opened by the introduction of any agent that depletes Ca^{2+}-stores: a) IP_3, the ligand that gates the ER IP_3-R Ca^{2+}-channel; b) thapsigargin, which inhibits Ca^{2+} reuptake into stores; c) ionomycin, which directly empties stores; and d) Ca^{2+} chelators. However *trp* has a high unitary conductance by comparison with other known SOC channels ($\geq$3pS compared with 0.02pS.[3,48]

The *trp-l* channels differ from *trp* in significant ways. *trp-l* channels do not exhibit Ca^{2+}-selectivity. In the Sf9 cell expression system, *trp-l* channels are constitutively open and insensitive to thapsigargin application; however they can be further stimulated by IP_3.[(91)] This suggests that *trp-l* is not a SOC and has a different mechanism of activation from *trp*. Structurally, *trp* has a C-terminal "tail" that is absent in *trp-l*. Experiments with transgenic chimeras of demonstrate that *trp-l* channels augmented with the *trp*-tail region exhibit thapsigargin-dependent activation. Conversely, *trp* channels lacking tails lose thapsigargin-dependent activation,[(95)] suggesting a role for the *trp*-tail region in Ca^{2+}-depletion dependent gating.

These findings, together with the finding that introduction of ionomycin into *Drosophila* photoreceptors induces channel opening,[(3)] suggest that phototransduction involves Ca^{2+}-store depletion. Furthermore, *Drosophila* double mutants of *trp/trp-l* exhibit very reduced light-activated currents,[(96)] suggesting, if there are no pleiotropic effects, that these may be the only important light-sensitive channels. It is not ruled out, however, that the presence of *trp* & *trp-l* is required for activation of other channels.

There are important differences between observations from Sf9 cells and those from the native photoreceptor preparation, however, that cloud the picture. Firstly, mutant photoreceptors lacking *trp* channels, even though they presumably express the *trp-l* channels, differ from *trp-l*-containing Sf9 cells in that their photoreceptors are not constitutively active. Rather, they require light for activation.[(93)] Thus, native *trp-l* channels may possess an additional regulatory feature. Secondly, although ionomycin can induce channel opening in photoreceptors,[(3)] thapsigargin does not substantially affect light-induced excitation of photoreceptors, in spite of measurable Ca^{2+} elevation in its presence.[(97)] These discrepancies could arise from differences in Ca^{2+}-stores, Ca^{2+}-regulation, or coupling between IP_3-R activation and channel opening.

Further questions concerning the nature of the *trp* and *trp-l* channels remain to be addressed. Each polypeptide is presumably one subunit of a tetrameric channel. Do *trp* and *trp-l* channels segregate as subunits of different channels, or are they subunits of the same heterotetramer? One cannot conclude the former from the fact that mutants lacking either of these gene products can form functional subunits. It may be that abnormal, homotetrameric channels can form from the aggregation of identical subunits when only one type of subunit is expressed; whereas normally some specific interaction between *trp* and *trp-l* monomers makes assortment with proper stoichiometry energetically favorable (similar to *Drosophila* K^+ channels[(98,99)]).

A recent immunolocalization study suggests that *trp* segregates to regions of the plasma membrane directly adjacent to the intracellular Ca^{2+}-stores,[100] supporting a phototransduction model involving direct coupling between channels and IP_3-R. It remains to be determined, however, whether such coupling exists and, in fact, whether the *trp* and/or *trp-l* channels are actually plasma membrane channels.

MULTIPLE CHANNEL TYPES

There are several interesting electrophysiological observations supporting the notion that invertebrate photoreceptors possess more than one class of light-sensitive ion channel. A key observation is that light-induced currents measured in voltage-clamp conditions can have very complex waveforms. In particular, light-induced currents may have no unique reversal voltage; rather, they may appear to have different components that reverse sign at different clamped voltages. The stimulus-induced currents appear biphasic and sometimes multiphasic. This phenomenon, which can be termed BIPHASIC REVERSAL, has been observed in *Drosophila*, *Limulus*, and other photoreceptors.[101-103,121]

Biphasic reversal can result when more than one class of ion channel contributes to the response. For instance, if light activates two or more channel types possessing different ionic selectivities and gating kinetics (opening, closing, and/or inactivation), different components of the current may reverse at different voltages.

However, the existing data are insufficient to make the firm conclusion that there are multiple channel types involved in excitation. Biphasic reversal may arise from a variety of sources in cells possessing only one type of ion channel.[104] It is difficult to evaluate whether multiple channel types are responsible for biphasic reversal. One might try to separate the components with ion substitution experiments and examine each separately, to investigate the basis of the separate components. However, light response kinetics are dependent on membrane voltage, stimulus intensity, and light-induced changes in intracellular chemistry, making it difficult to identify and distinguish between individual components even under rigorously controlled experimental conditions.

In Figure 6, we illustrate some simulations of membrane currents predicted for two such sources in *Limulus* photoreceptors. We have included a brief discussion in the figure legend.

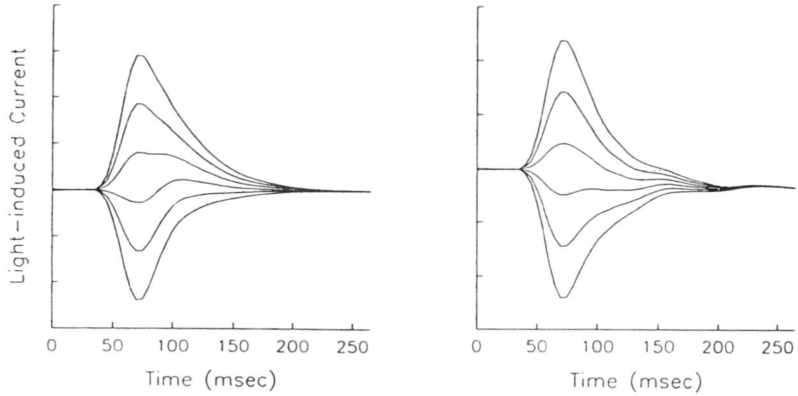

Figure 6. Simulation of BIPHASIC REVERSAL. Biphasic reversal arises from (A) change in reversal potential of the light response (E_{hv}) and (B) activation of Na/Ca-exchange and exchange current, I_X, due to light-induced elevation of Ca^{2+}. In *Drosophila* and in *Limulus*, biphasic reversal is very labile, very intensity dependent, and very Ca^{2+}-dependent.

(A) In a situation where there is only one class of light-activated channel, E_{hv} may change during the response itself as a result of Ca^{2+} elevation (e.g., Ca^{2+}-release from stores and/or inward Ca^{2+} current, I^{hv}). Light evokes inward, followed by outward, currents in a cell clamped near reversal potential. The E_{hv} shift is intensity-dependent, since Ca^{2+}-release and I_{hv} are intensity-dependent. Shown are simulated light-induced currents under voltage-clamp to -70mV, using Ca^{2+}-elevation measured in *Limulus* photoreceptors[113] and a light-activated channel permeable to Ca^{2+} and Na^+.

(B) Light-induced Ca^{2+}-elevation accelerates the plasma membrane electrogenic Na/Ca-exchanger,[111,113-117] increasing I_X. Thus, when V_m is clamped very slightly positive of the E_{hv}, a bright flash evokes a very small outward I_{hv} and a large rise in $[Ca_i^{2+}]$ from intracellular release. If $[Ca_i^{2+}]$ elevation is such that the reversal voltage of the Na/Ca-exchange, E_X, is substantially positive of V_m, the I_X will be inward current while extruding Ca^{2+}. When I_X outweighs I_{hv} a net inward current appears late during the response, yielding an outward-to-inward biphasic response may result. This biphasic reversal is intensity- and Ca^{2+}- dependent. The simulation here is based on I_X as modeled in Ref 114.

(C) *Other possibilities not shown*: (1) Physiological events that cause transient voltage clamp currents could elicit biphasic reversal. *e.g.* A significant Ca^{2+} conductance between intracellular Ca^{2+} stores and the extracellular space is proposed in smooth muscle cells.[118] Light-induced IP_3-R gating in ER membranes would provide a leakage pathway from cytosol to ground via ER lumen, causing momentary clamp failure. This would introduce a stimulus-dependent component of the clamp output, appearing as a current possibly comparable to the I_{hv} near E_{hv}.
(2) Light may induce rapid alteration of the rhabdomeral membranes (microvilli and ER), introducing capacitative currents even under voltage-clamp. Extensive transient reorganization of the rhabdomeral membranes is triggered by bright illumination in photoreceptors of the *Limulus* compound eyes.[119] This may be related to the vesicle fusion models of SOC activation.[119] Vesicle fusion may induce capacitative currents, similar to those induced by secretory vesicles,[120] and these currents may appear as a component causing biphasic reversal.

QUANTUM BUMPS

Finally we raise the question "What is the mechanistic basis of the "quantum bump" the elemental physiological unit of the light response, corresponding to a single photon response? Each light response is a summation of these single photon responses.(105,106) In invertebrates, each quantum bump develops sharply following a somewhat variable latency period, suggestive of a discrete event having a variable probability of initiation. In *Limulus*, quantum bumps result from the opening of 1,000 to 10,000 channels.(107) Quantum bumps often occur spontaneously in the absence of light.(108) Individual spontaneous quantum bumps appear normal in *Drosophila* mutants of PLC and rhodopsin.(109,110) In *Drosophila* PKC mutants, quantum bump termination is slowed, but initiation is unaffected,(42,45) reflecting the role of PKC in light response termination. It has been suggested that the number of IP_3-R activated in a light response is many times smaller than the number of quantum bumps that make up that response.(111) These results suggest that, for models in which the PI-pathway is required and sufficient for excitation, quantum bump initiation occurs downstream of the IP_3-R.

If a Ca^{2+}-store depletion model of excitation is correct, one might imagine that the quantum bump arises from a discrete, local depletion of Ca^{2+}-stores that then activates a local population of membrane channels.

Recently, a vesicle transport and fusion model of SOC channel current activation has been proposed.(112) In this model, SOC channel activation involves their translocation from the Ca^{2+}-stores within vesicle membranes that fuse with the plasma membrane as a result of Ca^{2+}-store depletion. This is based on the observation that vesicular transport inhibitors block activation of SOC channels by Ca^{2+}-store depletion in mammalian cell lines. Along these lines, an alternative is that vesicular release from Ca^{2+}-stores liberates a substance that results in activation of the plasma membrane ion channels. In either case, quantum bump generation might involve a small localized vesicle fusion event that leads to the activation of thousands of channels.

Excitation models involving cGMP (*e.g.* Figure 2) would not appear to place additional constraints on the mechanism of quantum bump generation.

ACKNOWLEDGEMENTS. We are very grateful to Drs. Roger Hardie and Charles Zuker for making material available to us before publication. This work was supported by NIH Grant EY09388, NSF Grant INT9301719)/Fundación Andes-CONICYT, and TWAS 94-456 RG/BIO/LA.

REFERENCES

1. Ranganathan R, Malicki DM, and Zuker CS, 1995, Signal transduction in *Drosophila* photoreceptors. *Ann Rev Neurosci* **18:** 283-317.
2. Hardie RC, and Minke B, 1993, Novel Ca^{2+} channels underlying transduction in *Drosophila* photoreceptors: implications for inositide-mediated Ca^{2+} mobilization. *Trends Neurosci* **16:** 371-376.
3. Hardie RC, and Minke B, 1995, Phosphoinositide-mediated phototransduction in *Drosophila* photoreceptors: the role of Ca^{2+} and *trp*. *Cell Calcium* **18:** 256-274.
4. Richard EA, Sampat P, and Lisman JE, 1995, Distinguishing between roles for calcium in *Limulus* photoreceptor excitation. *Cell Calcium* **18:** 331-341.
5. Hubbard R, and Wald G, 1960, Visual pigment of the horseshoe crab, *Limulus polyphemus*. *Nature* **186:** 212-215.
6. O'Tousa JE, Baehr W, Martin RL, Hirsh J, Pak WL, and Applebury ML, 1985, The *Drosophila ninaE* gene encodes an opsin. *Cell* **40:** 839-850.
7. Zuker CS, Cowman AF, and Rubin GM, 1985, isolation and structure of a rhodopsin gene from *Drosophila melanogaster*. *Cell* **40:** 851-858.
8. Blumenfeld A, Erusalimsky J, Heichal O, Selinger Z, and Minke B, 1985, Light-activated GTPase in *Musca* eye membrane resembles the prolonged depolarizing afterpotential in photoreceptor cells. *Proc Nat Acad Sci* **82:** 7116-7120.
9. Bentrop J, and Paulsen R, 1986, light-modulated ADP-ribosylation, protein phosphorylation and protein binding in isolated fly photoreceptor membranes. *Eur J Biochem* **161:** 61-67.
10. Devary O, Heichal O, Blumenfeld A, Cassel A, Barash A, Rubinstein T, Minke B, and Selinger Z, 1987, Coupling of photoexcited rhodopsin to phosphoinositide hydrolysis in fly photoreceptors. *Proc Natl Acad Sci USA* **84:** 6939-6943.
11. Robinson PR, Wood SF, Szuts EZ, Fein A, Hamm H, and Lisman JE, 1989, GTP-Binding proteins in squid photoreceptor membranes. *Biophys J* **55:** 60a.
12. Dolph PJ, Ranganathan R, Colley NJ, Hardy RW, Socolich M, and Zuker CS, 1993, Arrestin function in inactivation of G-protein coupled receptor rhodopsin in vivo. *Science* **260:** 1910-1916.
13. Paulsen R, and Bentrop J, 1984, Reversible phosphorylation of opsin induced by irradiation of blowfly retinae. *J Comp Physiol A* **155:** 39-45.
14. Vandenberg CA, and Montal M, 1984, Light-regulated biochemical events in invertebrate photoreceptors. 2. Light-regulated phosphorylation of rhodopsin and phosphoinositides in squid photoreceptor membranes. *Biochem* **23:** 2347-2352.
15. Tsuda M, Tsuda T, and Hirata H, 1989, Cyclic nucleotides and GTP analogues stimulate light-induced phosphorylation of octopus rhodopsin. *FEBS lett* **257:**38-40.
16. Doza YN, Minke B, Chorev M, and Selinger Z, 1992, Characterization of fly rhodopsin kinase. *Eur J Biochem* **209:** 1035-1040.
17. Matsumoto H, and Pak WL, 1984, Light-induced phosphorylation of retina-specific polypeptides of *Drosophila in vivo*. *Science* **223:** 184-186.

18. Edwards SC, Wishart AC, Wiebe EM, and Battelle B-A, 1989, Light-regulated proteins in the *Limulus* ventral photoreceptor. *Vis Neurosci* **3:** 95-105.
19. Yamada T, Takeuchi Y, Komori N, Kobayashi H, Sakai Y, Hotta Y, and Matsumoto H, 1990, A 40-kD phosphoprotein in the *Drosophila* photoreceptor is and arrestin homolog. *Science* **248:** 483-486.
20. LeVine H III, Smith DP, Whitney M, Malicki DM, Dolph PJ, Smith GFH, Burkhart W, and Zuker CS, 1991, Isolation of a novel visual-system-specific arrestin: an *in vivo* substrate for light-dependent phosphorylation. *Mech Devel* **33:** 19-26.
21. Calman BG, Andrews AW, and Battelle B-A, 1992, The 46K protein in *Limulus* photoreceptors is phosphorylated by a calcium-calmodulin dependent protein kinase. *Invest Ophthalmol Vis Sci* **33:** 874.
22. Byk T, Bar-Yaacov M, Doza YN, Minke B, and Selinger Z, 1993, Regulatory arrestin cycle secures the fidelity and maintenance of the fly photoreceptor cell. *Proc Natl Acad Sci* **90:** 1907-1911.
23. Matsumoto H, Kurien BT, Takagi Y, Kahn ES, Kinumi T, Komori N, Yamada T, Hayashi F, Isono K, and Pak WL, 1994, Phosrestin I undergoes the earliest light-induced phosphorylation by a calcium/calmodulin-dependent protein kinase in Drosophila photoreceptors. *Neuron* **12:** 997-1010.
24. Smith WC, Greenberg RM, Hendrix MM, and Battelle B-A, 1995, Cloning of arrestin from the horseshoe crab lateral eye: *Limulus* arrestin is a phosphoprotein. *J Neurochem* **64:** 1-13.
25. Richard EA, and Lisman JE, 1992, Rhodopsin inactivation is a modulated process in *Limulus* photoreceptors. *Nature* **356:** 336-338.
26. Angleson JK, and Wensel TG, 1993, A GTPase-accelerating factor for transducin, distinct from its effector cGMP phosphodiesterase, in rod outer segment membranes. *Neuron* **11:** 939-949.
27. Arshavsky VY, Dumke CL, Zhu Y, Artemyev NO, Skiba NP, Hamm HE, and Bownds MD, 1994, Regulation of transducin GTPase activity in bovine rod outer segments. *J Biol Chem* **269:** 19882-19887.
28. Steele FR, Washburn T, Rieger R, and O'Tousa JE, 1992, *Drosophila retinal degeneration C (rdgC)* encodes a novel serine/threonine protein phosphatase. *Cell* **69:** 669-676.
29. Van Dyke TH, Windelspeckt M, Shah M, Edwards SC, and Smith WC, 1994, Protein phosphatase 2A characterization and evidence for *Limulus* and *Drosophila* arrestin dephosphorylation. *Invest Ophthalmol Vis Sci* **35:** 2129.
30. Tsibris AMN, Van Dyke TH, and Edwards SC, 1994, Protein phosphatase type 2A removes arrestin from the membrane of *Limulus* photoreceptors and is inhibited by PKA phosphorylation. *Soc Neurosci Abs* **20:** 131.
31. Lee Y-J, Dobbs MB, Verardi ML, and Hyde DR, 1990, *dgq*: a *Drosophila* gene encoding a visual system specific G_α molecule. *Neuron* **5:** 889-898.
32. Lee Y-J, Shah S, Suzuki E, Zars T, O'Day PM, and Hyde DR, 1994, The Drosophila *dgq* gene encodes a G_α protein that mediates phototransduction. *Neuron* **13:** 1143-1157.
33. Yarfitz S, Niemi GA, McConnell JL, Fitch CL, and Hurley, 1991, A G beta protein in the *Drosophila* compound eye is different from that in the brain.

Neuron **7:** 429-438.

34. Scott K, Becker A, Sun Y, Hardy R, and Zuker CS, 1995, $G_{q\alpha}$ protein function in vivo: genetic dissection of its role in photoreceptor cell physiology. *Neuron* **15:** 919-927.
35. Gomez MP, and Nasi E, 1995, Activation of light-dependent K^+ channels in ciliary invertebrate photoreceptors involves cGMP but not the IP_3/Ca^{2+} cascade. *Neuron* **15:** 607-618.
37. Pak WL, Grossfield J, and Arnold K, 1970, Mutants of the visual pathway of *Drosophila melanogaster. Nature* **227:** 518-520.
38. Yoshioka T, Inoue H, and Hotta Y, 1985, Absence of phosphatidylinositol phosphodiesterase in the head if *Drosophila* visual mutant, *norpA* (no receptor potential A). *J Biochem* **97:** 1251-1254.
39. Bloomquist BT, Shortridge RD, Schneuwly W, Perdew M, Montell C, Steller H, Rubin G, and Pak WL, 1988, Isolation of a putative phospholipase C gene of *Drosophila, norpA*, and its role in phototransduction. *Cell* **54:** 723-733.
40. Mitchell J, Gutierrez J, and Northrup JK, 1995, Purification, characterization, and partial amino acid sequence of a G protein-activated Phospholipase C from squid photoreceptors. *J Biol Chem* **270:** 854-859.
41. Running-Deer JL, Hurley JB, and Yarfitz SL, 1995, G protein control of *Drosophila* photoreceptor phospholipase C. *J Biol Chem* **270:** 12623-12628
42. Ranganathan R, Harris GL, Stevens CF, and Zuker CS, 1991a, A *Drosophila* mutant defective in extracellular calcium-dependent photoreceptor deactivation and rapid desensitization. *Nature* **354:** 230-232.
43. Schaeffer E, Smith D, Mardon G, Quinn W, and Zuker C, 1989, Isolation and characterization of two new *Drosophila* protein kinase C genes, including one specifically expressed in photoreceptor cells. *Cell* **57:** 403-412.
44. Smith DP, Ranganathan R, Hardy RW, Marx J, Tsuchida T, and Zuker CS, 1991, Photoreceptor deactivation and retinal degeneration mediated by a photoreceptor-specific protein kinase C. *Science* **254:** 1478-1484.
45. Hardie RC, Peretz A, Suss-Toby E, Rom-Glas A, Bishop SA, Selinger Z, and Minke B, 1993, Protein kinase C is required for light adaptation in *Drosophila* photoreceptors. *Nature* **363:** 634-637.
46. Paulsen R, Sander P, and Huber A, 1996, Visual transduction of fly photoreceptors may be controlled by PKC-mediated phosphorylation of the INAD protein. *Invest Ophthalmol Vis Sci* **37:** S807.
47. Shieh B-H, and Niemeyer B, 1995, A novel protein encoded by the *inaD* gene regulates recovery of visual transduction in *Drosophila. Neuron* **14:** 201-210.
48. Hoth M, and Penner R, 1992, Depletion of intracellular calcium stores activates a Ca^{2+} current in mast cells. *Nature* **355:** 353-356.
49. Zweifach A, and Lewis RS, 1993, Mitogen-regulated Ca^{2+} current of T lymphocytes is activated by depletion of intracellular Ca^{2+} stores. *Proc Nat Acad Sci USA* **90:** 6295-6299.
50. Putney JWJr, and Bird StJ, 1993, The signal for capacitative calcium entry. *Cell* **75:** 199-201.
51. Bennett DL, Petersen CCL, and Cheek T, 1995, Cracking I_{CRAC} in the eye. *Curr Biol* **5:** 1225-1228.

52. Minke B, 1979, Transduction in photoreceptors with bistable pigments: intermediate processes. *Biophys Struct Mech* **5:** 163-174.
53. Suzuki E, and Hirosawa K, 1992, Immunocytochemical studies on light-induced changes in phosphatidylinositol 4,5-bisphosphate immunoreactivity in the visual systems of normal and *norpA* mutant of *Drosophila. Neurosci Res* **13:** 155-160.
54. Szuts EZ, Wood SF, Reid MS, and Fein A, 1986, Light stimulates the rapid formation if inositol trisphosphate in squid retinas. *Biochem J* **240:** 929-932.
55. Payne R, Corson DW, Fein A, Berridge, and MJ, 1986, Excitation and adaptation of *Limulus* ventral photoreceptors by inositol 1,4,5 trisphosphate result from a raise in intracellular calcium. *J Gen Physiol* **88:** 127-142.
56. Bacigalupo J, Johnson EC, Vergara C, and Lisman JE, 1991, Light-dependent channels from excised patches of *Limulus* ventral photoreceptors are opened by cyclic GMP. *Proc Nat Acad Sci USA* **88:** 7938-7942.
57. Johnson EC, Robinson PR, and Lisman JE, 1986, Cyclic GMP is involved in the excitation of invertebrate photoreceptors. *Nature* **324:** 468-470.
58. Bacigalupo J, Johnson EC, Robinson PR, and Lisman JE, 1990, Second messengers in invertebrate phototransduction, in: *Transduction in Biological Systems,* (Hidalgo C, Bacigalupo J, Jaimovich E, Vergara J, eds.) Plenum Press, New York, pp 27-45.
59. Feng J, Frank TM, and Fein A, 1991, Excitation of *Limulus* photoreceptors by hydrolysis-resistant analogs of cGMP and cAMP. *Brain Res* **552:** 291-294.
60. Brown JE, Kaupp UB, and Malbon CC, 1984a, 3',5'-cyclic adenosine monophosphate and adenylate cyclase in phototransduction by *Limulus* ventral photoreceptors. *J Physiol* **353:** 523-539.
61. O'Day PM, Johnson EC, and Baumgard M, 1991, Effects of lithium, calcium and PDE-inhibitors on excitation in *Limulus* photoreceptors. *Biophys J* **59:** 540a.
62. Inoue M, Ackermann K, and Brown JE, 1992, Cyclic GMP phosphodiesterase in photoreceptor cells in *Limulus* ventral eye, in: *Signal Transduction in Photoreceptor Cells* (Hargrave PA, Hofmann KP, Kaupp UB, eds.) Springer-Verlag, Berlin, pp 271-280.
63. Johnson EC, and O'Day PM, 1995, Inhibitors of cyclic-GMP phosphodiesterase alter excitation of *Limulus* ventral photoreceptors in Ca^{2+}-dependent fashion. *J Neurosci* **15:** 6586-6591.
64. Shin J, Richard EA, and Lisman JE, 1993, Ca^{2+} is an obligatory intermediate in the excitation cascade of *Limulus* photoreceptors. *Neuron* **11:** 845-855.
65. Inoue M, and Brown JE, 1988, Cyclic GMP phosphodiesterase in Limulus ventral eye. *Invest Ophthalmol Vis Sci* **29:** 218.
66. Nagy K, 1993, Cyclic nucleotides and inositol trisphosphate activate different components of the receptor current in *Limulus* ventral nerve photoreceptors. *Neurosci Lett* **152:** 1-4.
67. Robinson PR, and Cote R, 1998, Characterization of guanylate cyclase in squid photoreceptors, *Vis Neurosci* 3: 1-7.
68. Shah S, and Hyde DR, 1995, Two *Drosophila* genes that encode that encode the α and β subunits of the brain soluble guanylyl cyclase. *J Biol Chem* **270:** 15368-15376.
69. Yoshikawa S, Miyamoto I, Aruga J, Furuichi T, Okano H, and Mikoshiba K,

1993, Isolation of a *Drosophila* gene encoding a head-specific guanylyl cyclase. *J Neurochem* **60:** 1570-1573.

70. Liu W, Yoon J, Burg M, Chen L, and Pak WL, 1995, Molecular characterization of two *Drosophila* guanylate cyclases expressed in the nervous system. *J Biol Chem* **270:** 12418-12427.
71. Saibil HR, 1984, A light-stimulated increase in cyclic GMP in squid photoreceptors. *FEBS Letters* **168:** 213-216.
72. Bacigalupo J, Bautista DM, Brink DL, Hetzer JF, and O'Day PM, 1995, Cyclic-GMP enhances light-induced excitation and induces membrane currents in *Drosophila* retinal photoreceptors. *J Neurosci* **15:** 7196-7200.
73. Haab JE, Farjah F, and O'Day PM, 1996, Inhibitors of cGMP synthesis and degradation affect light-induced excitation in *Drosophila* retinal photoreceptors. *Biophysical J* **70:** A137.
74. Hardie RC, and Minke B, 1994, Spontaneous activation of light-sensitive channels in *Drosophila* photoreceptors. *J Gen Physiol* **103:** 389-407.
75. Baumann A, Frings S, Godde M, Seifert R, and Kaupp UB, 1994, Primary structure and functional expression of a *Drosophila* cyclic nucleotide-gated channel present in eyes and antennae. *EMBO J* **13:** 5040-5050.
76. Schraermeyer U, Stieve H, and Rack M, 1993, Cyclic 3'5'-nucleotide phosphodiesterase: cytochemical localization in photoreceptor cells of the fly *Calliphora erythrocephala*. *J Neurocytol* **10:** 845-853.
77. Kalderon D, and Rubin GM, 1989, cGMP-dependent protein kinase genes in *Drosophila*. *J Biol Chem* **264:** 10738-10748.
78. Komalavilas P, and Lincoln TM, 1994, Phosphorylation of the inositol 1,4,5 trisphosphate receptor by cyclic GMP dependent protein kinase. *J Biol Chem* **269:** 8701-8707.
79. Ehrlich BE, Kaftan E, Bezprozvannaya S, and Bezprozvanny I, 1994, The pharmacology of intracellular Ca^{2+}-release channels. *Trends Pharmacol Sci* **15:** 145-149.
80. Alioua A, Huggins JP, and Rousseau E, 1995, PKG-I alpha phosphorylates the alpha-subunit and upregulates reconstituted GKCa channels from tracheal smooth muscle. *Am J Physiol* **268:** L1057-1063.
81. Sperelakis N, 1994, Regulation of calcium slow channels of heart by cyclic nucleotides and effects of ischemia. *Adv Pharmacol* **31:** 1-24.
82. Light DB, Corbin JD, and Stanton BA, 1990, Dual ion-channel regulation by cyclic GMP and cyclic GMP-dependent protein kinase. *Nature* **344:** 336-339.
83. Chik CL, Liu QY, Li B, Karpinski E, and Ho AK, 1995, cGMP inhibits L-type Ca^{2+} channel currents through protein phosphorylation in rat pinealocytes. *J Neurosci* **15:** 3104-3109.
84. Cornwell TL, Pryzwansky KB, Wyatt TA, and Lincoln TM, 1991, Regulation of sarcoplasmic reticulum protein phosphorylation by localized cyclic GMP dependent protein kinase in vascular smooth muscle cells. *Mol Pharmacol* **40:** 923-31.
85. Yoshida Y, Sun HT, Cai JQ, and Imai S, 1991, Cyclic GMP-dependent protein kinase stimulates the plasma membrane Ca^{2+} pump ATPase of vascular smooth muscle via phosphorylation of a 240-kDa protein. *J Biol Chem* **266:**

19819-19825.

86. Mene P, Pugliese F, and Cinotti GA, 1993, Cyclic nucleotides inhibit Na^+/Ca^{2+} exchange in cultured human mesangial cells. *Exp Nephrol* **1:** 245-252.
87. Brown JE, and Mote MI, 1974, Ionic dependence of reversal voltage of the light response in *Limulus* ventral photoreceptors. *J Gen Physiol* **63:** 337-350.
88. Nasi E, and Gomez M, 1992, Light-activated ion channels in solitary photoreceptors of the scallop *Pecten irradians. J Gen Physiol* **99:** 747-769.
90. Phillips AM, Bull A, and Kelly LE, 1992, Identification of a *Drosophila* gene encoding a calmodulin binding protein with homology to the *trp* phototransduction gene. *Neuron* **8:** 631-642.
91. Hu Y, and Schilling WP, 1995, Receptor-mediated activation of recombinant Trpl expressed in Sf9 insect cells. *Biochem J* **305:** 605-611.
92. Petersen C, Berridge M, Borgese MF, and Bennett DL, 1995, Putative capacitative calcium entry channels: expression of *Drosophila trp* and evidence for the existence of vertebrate homologues. *Biochem J* **311:** 41-44.
93. Minke B, 1982, Light-induced reduction in excitation efficiency in the *trp* membrane of *Drosophila. J Gen Physiol* **79:** 361-385.
94. Montell C, and Rubin GM, 1989, Molecular characterization of the *Drosophila trp* locus: a putative integral membrane protein required for phototransduction. *Neuron* **2:** 1313-1323.
95. Sinkins WG, Vaca L, Hu Y, Kunze D, and Schilling WP, 1996, The COOH-terminal domain of *Drosophila* TRP channels confers thapsigargin sensitivity. *J Biol Chem* **271:** 2955-2960.
96. Niemeyer BA, Suzuki E, Scott K, Jalink K, and Zuker CS, 1996, *Drosophila* light-activated conductance is composed of two channels *trp* and *trp-l. Cell* (In press).
97. Ranganathan R, Bacskai BJ, Tsien RY, and Zuker CS, 1994, Cytosolic calcium transients: spatial localization and role in *Drosophila* photoreceptor cell function. *Neuron* **13:** 837-848.
98. Zhong Y, and Wu C-F, 1993, Modulation of different K^+ currents in *Drosophila*: A hypothetical role for the Eag subunit in multimeric K^+ channels. *J Neurosci* **13:** 4669-4679.
99. Papazian DM, Shao XM, Seoh S-A, Mock AF, Huang Y, and Wainstock DH, 1995, Electrostatic interactions of S4 voltage sensor in Shaker K^+ channel. *Neuron* **14:** 1293-1301.
100. Pollock JA, Assaf A, Peretz A, Nichols CD, Mojet MH, Hardie RC, and Minke B, 1995, TRP, a protein essential for inositide-mediated influx is localized adjacent to the calcium stores in *Drosophila* photoreceptors. *J Neurosci* **15:** 3747-3760.
101. Hardie RC, and Minke B, 1992, The *trp* gene is essential for a light-activated Ca^{2+} channel in *Drosophila* photoreceptors. *Neuron* **8:** 643-651.
102. Nagy K, 1991, Biophysical processes in invertebrate photoreceptor: recent progress and a critical overview based on *Limulus* photoreceptors. *Q Rev Biophys* **24:** 165-226.
103. Deckert A, Nagy K, Helrich CS, and Stieve H, 1992, Three components of the light-induced current of the *Limulus* ventral photoreceptor. *J Physiol* **453:** 69-96.

104. O'Day PM, Johnson EC, and Freeman S, 1993, A single class of light-activated channel can yield biphasic light-induced currents at Erev. *Biophys J* **64:** 134a.
105. Wong F, Knight BW, and Dodge FA, 1982, Adapting bump model for ventral photoreceptors of *Limulus. J Gen Physiol* **79:** 1089-1113.
106. Stieve H, 1986, Bumps the elementary responses of invertebrates, in: *The molecular mechanism of photoreception*, (Stieve H, ed.) Springer Verlag, Berlin, pp. 199-230.
107. Bacigalupo J, and Lisman JE, 1983, Single-channel currents activated by light in *Limulus* ventral photoreceptors. *Nature* **304:** 268-270.
108. Yeandle S, and Spiegler JB, 1973, Light-evoked and spontaneous discrete waves in the ventral nerve photoreceptor of *Limulus. J Gen Physiol* **61:** 552-571.
109. Pak WL, Ostroy SE, Deland MC, and Wu C-F, 1976, Photoreceptor mutant of *Drosophila*: Is protein involved in intermediate steps of phototransduction? *Science* **194:** 956-959.
110. Johnson EC, and Pak WL, 1986, Electrophysiological study of *Drosophila* rhodopsin mutants. *J Gen Physiol* **88:** 651-673.
111. Hardie RC, 1996, INDO-1 measurements of absolute resting and light-induced Ca^{2+} concentration in *Drosophila* photoreceptors. *J Neurosci* **16:** (In press).
112. Somasundaram B, Norman JC, and Mahaut-Smith MP, 1995, Primaquine, an inhibitor of vesicular transport, blocks the calcium-release-activated current in rat megakaryocytes. *Biochem J* **309:** 725-729.
113. O'Day PM, and Gray-Keller MP, 1989, Evidence for electrogenic Na^+/Ca^{2+} exchange in *Limulus* ventral photoreceptors. *J Gen Physiol* **93:** 473-492.
114. O'Day PM, 1991, Sodium-Calcium exchange in invertebrate photoreceptors. *Ann NY Acad Sci* **639:** 285-299.
115. Armon E, and Minke B, 1983, Light-activated Na^+/Ca^{2+}-exchange in fly photoreceptors: modulation by Na^+/K^+-pump activity. *Biophys Struct Mech* **9:** 349-357.
116. Hardie RC, 1995a, Photolysis of caged Ca^{2+} facilitates and inactivates but does not directly excite light-sensitive channels in *Drosophila* photoreceptors. *J Neurosci* **15:** 889-902.
117. Hardie RC, 1995b, Effects of intracellular Ca^{2+} chelation on the light response in dependent photoreceptors. *J Comp Physiol A* **177:** 707-721.
118. Rys-Sikora KE, Ghosh TK, and Gill DL, 1994, Modification of GTP-activated calcium translocation by fatty acyl-CoA esters. Evidence for a GTP-induced prefusion event. *J Biol Chem* **269:** 31607-31613.
119. Chamberlain SC, and Barlow RB Jr, 1979, Light and efferent activity control rhabdom turnover in *Limulus* photoreceptors. *Science* **206:** 361-363.
120. Breckenridge LJ, and Almers W, 1987, Currents through the fusion pore that forms during exocytosis of a secretory vesicle. *Nature* **328:**814-817.
121. Nasi E, 1991, Two light-dependent conductances in *Lima* rhabdomeric photoreceptors. *J Gen Physiol* **97:** 55-72.
122. Matsumoto H, O'Tousa JE, and Pak WL, 1982, Light-induced modification of *Drosophila* retinal polypeptides *in vivo. Science* **217:** 839-841.
123. O'Tousa JE, 1990, Genetic analysis of phototransduction in *Drosophila. Semin Neurosci* **2:** 207-215.

124. Hardie RC, and Mojet MH, 1995, Mg^{2+} dependent block of the light-activated and *trp* dependent conductance in *Drosophila* photoreceptors. *J Neurophysiol* **74:** 2590-2599.
125. Frank TM, and Fein A, 1991, The role of the inositol phosphate cascade in visual excitation of invertebrate microvillar photoreceptors. *J Gen Physiol* **97:** 697-723.
126. Payne R, and Fein A, 1986, The initial response of *Limulus* ventral photoreceptors to bright flashes: released calcium as a synergist to excitation. *J Gen Physiol* **87:** 243-269.

Part II

Intercellular Channels

STRUCTURE AND FUNCTION OF CELL-TO-CELL CHANNELS PURIFIED FROM THE LENS AND OF HEMICHANNELS EXPRESSED IN OOCYTES

Guido A. Zampighi, [1] Nicola Konig, [2] and Donald D.F. Loo[3]

[1]Department of Neurobiology UCLA School of Medicine, Los Angeles CA, 90024.
[2] MRC Laboratory of Molecular Biology, Hills Road, CB2 2QH Cambridge, England.
[3]Department of Physiology, UCLA School of Medicine, Los Angeles CA, 90024.

INTRODUCTION

The lens is a spherical organ whose only function is to focus images on the retina. It is composed of a mass of highly elongated cells, called fibers, and of a simple cuboidal epithelium on its anterior surface. The fiber cells arise from differentiation of anterior epithelial cells at the lens equator, a process that continues throughout the life of the animal. To attain transparency, fiber cells accumulate in the cytoplasm high concentrations of soluble proteins, called crystalline, and eliminate cellular organelles, including nuclei and mitochondria, that may scatter light. In spite of the loss of cellular organelles, however, the lens is able to maintain homeostasis (Duncan and Croghan, 1989; Mathias et al., 1981; Mathias and Rae, 1985). It is currently believed that this property is due, at least partially, to the establishment of networks of cell-to-cell channels connecting directly the cytoplasm of lenticular cells (Mathias et al., 1981; Mathias and Rae, 1985). This chapter deals with the structure and some functional properties of the proteins composing the cell-to-cell channels connecting fiber cells.

Cell-to-cell channels are present in many tissues from different animal species where are organized in plaques called "gap junctions". In thin sections, the junctions are characterized by a close apposition of the plasma membranes

From Ion Channels to Cell-to-Cell Conversations
Edited by Latorre and Sáez, Plenum Press, New York, 1997

and a decrease in the width of the extracellular clefts between cells from 10-20 nm to 1-2 nm (Fig. 1B). In the lens, fiber cells have >60% of their surface area covered by structures called "fiber cell junctions" (Fig. 1 A). Early studies proposed that this enormous area of inter-cellular contact was composed of channels mediating cell-to-cell communication (Johnson et al., 1985; Johnson et al., 1988; Peracchia and Girsch, 1985). However, more recent studies demonstrated that fiber cell junctions were composed of junctions exhibiting different structures, chemical composition and functional properties (for review see Zampighi et al., 1992). In the following paragraphs, the data supporting this hypothesis are reviewed briefly.

Studies using electron microscopy and immunocytochemistry methods identified two different types of junctions contained in the structures commonly referred to as "fiber cell junctions". One type of junction was 11-13 nm in overall thickness (Zampighi et al., 1982) composed principally of the "major intrinsic protein" (MIP) (Zampighi et al., 1989). MIP is a 28 kDa protein that composes >50% of the total intrinsic protein of fiber cell plasma membranes (6). In the lens, MIP is assembled in tetramers that crystallize in two-dimensional lattices spaced 6-7 nm center-to-center apart (Costello et al., 1989; Lo and Harding, 1984; Zampighi et al. 1982). MIP-junctions are formed when a tetragonal array in one membrane become abutted against an opposing protein-free membrane (Zampighi et al., 1989).

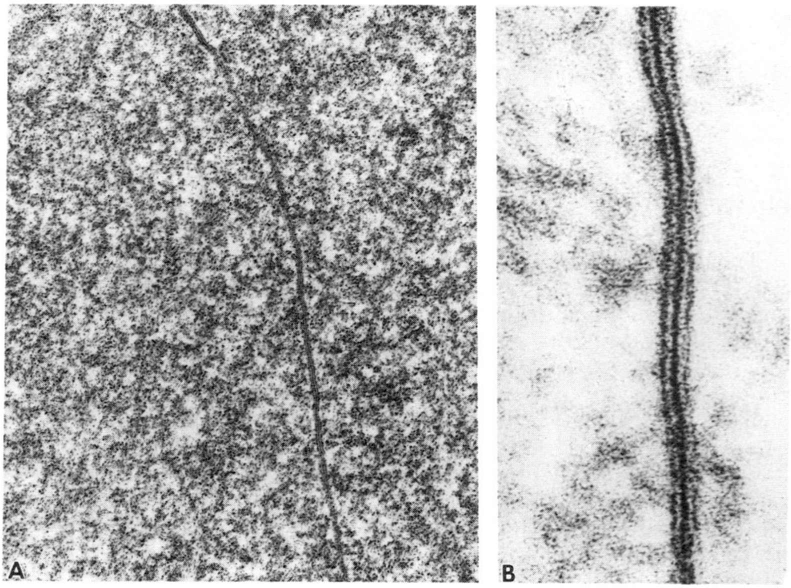

Figure 1. A low magnification view of a section through a calf lens. The sections show two fiber cells separated by an extensive junction, seen here as a dark line extending diagonally across panel A. The fiber cell cytoplasm lacks cellular organelles and is filled with amorphous material that correspond to the crystallines. Magnification: X50,000. Panel B shows a higher magnification view of a junction connecting fiber cells. The junction appears in transverse section as a pentalamellar structure formed by the close apposition of two plasma membranes. Magnification: X250,000.

MIP belongs to an ancient family of proteins expressed in plants, E. coli, yeast, Drosophyla and human red cells (Agre and Preston, 1992; Baker and Saier, 1990; Guerrero et al., 1990; Heller et al. 1980; Johnson et al., 1990; Preston et al., 1992). Proteins of the MIP family share substantial amino acid sequence homology and a common topology composed of six transmembrane domains with amino and carboxy-terminus exposed on the cytoplasmic surface (Gorin et al., 1984). They function in transport of either water or neutral molecules across membranes. MIP also forms water channels when expressed in Xenopus oocytes (Zampighi et al., 1995), however, such a function is yet to be demonstrated in the lens.

The other type of junctions present in the lens fiber cell junctions is 18-20 nm in overall thickness and is composed of proteins of the "connexin" family (Beyer et al., 1990). Connexins form cell-to-cell channels in tissues from many different animal species. In lens fiber cells, the first connexin was MP70, a 55-75 kDa protein located primarily in the 18-20 nm in thick junctions (Gruijters et al., 1987; Kistler et al., 1985; Kistler and Bullivant, 1988). MP70 was identified as a connexin because the amino acid sequence of its N-terminus was homologous to connexin 32 and connexin 43; proteins which form cell-to-cell channels in liver hepatocytes and heart myocytes, respectively (Kistler et al., 1988). The structure of MP70 junctions, however, exhibits both similarities and differences to regular gap junctions. Immunocytochemistry shows that the MP70-junction contains the protein in both membranes of the pair, but freeze-fracture showed that they differ from gap junctions in that the channels are not arranged in hexagonal lattices (Gruijters et al., 1987; Zampighi et al., 1989). Resolution of this puzzle required knowledge of the sequence of the connexins expressed in fiber cells.

Cloning methods have demonstrated that connexin 50 and connexin 44 (connexin 46 in rat) are expressed in fiber cells of calf lenses (Gupta et al., 1994; Paul et al., 1991; White et al., 1992). However, the relationship of connexin 44 and connexin 50 to the "old" MP70 protein has not been completely ascertained. Also, it is presently unknown whether both connexins are expressed in different regions of the lens (i.e., cortex versus nucleus) or if they are assembled as "hybrids" in the same cell-to-cell channel. To answer some of these question, we purified "complete" cell-to-cell channels from calf lenses and characterized them both chemically and structurally. This chapter shows that "complete" (dodecamers) cell-to-cell channels are composed of almost equal amounts of connexin 44 and connexin 50. In addition, we expressed connexin 50 in Xenopus laevis oocytes and studied the hemichannels by electron microscopy and electrophysiological methods. We show that hemichannels in the oolemma can form plaques that resemble those formed by complete channels (i.e., gap junctions). Functionally, however, hemichannels are closed to the diffusion of ions and water.

PURIFICATION OF LENS CELL-TO-CELL CHANNELS

The starting material for the purification of cell-to-cell channels were plasma membranes fractions isolated from calf lenses (Konig and Zampighi, 1995; Zampighi et al., 1992). SDS-PAGE shows that lens plasma membranes were composed of numerous proteins that show up as distinct bands (Fig. 2A). A band migrating at ~26 kDa, named the "major intrinsic protein" (MIP), composed >50% of the total protein of fiber cell plasma membranes (Fig. 2A, lanes 1 and 2). The proteins composing the cell-to-cell channels, on the other hand, appear as a diffuse band of 55-75 kDa apparent molecular weight (Fig. 2A, lanes 4 and 5). Thin sectioning electron microscopy showed that the isolated plasma membrane fractions were composed of both junctions and single membranes (Zampighi et al. 1982). Immunocytochemistry demonstrated that junctions made of MIP and of MP70 were present in the same junction and, thus, isolation of fractions enriched in MP70 was not a viable strategy.

We developed an alternative strategy that consisted in transferring all the intrinsic proteins in fiber cell plasma membranes (i.e., these that spanned the bilayers) into micelles of the nonionic detergent octyl-b-glucopyranoside (OG) (Konig and Zampighi, 1995). The aim of this treatment was to produce stable, monodisperse "mixed" micelles (i.e., a channel surrounded by detergent molecules) suitable for subsequent purification by column chromatography. Comparison of SDS-PAGE gels of plasma membranes before and after treatment (Fig. 3B, lanes 1 and 2) showed that OG effectively solubilized all the intrinsic proteins, including the 55-75 kDa protein band composing the cell-to-cell channels.

The channels were purified from the OG-solubilized membranes by ion and gel filtration column chromatography. Injection of the OG-solubilized material in a cation exchange column (MonoS) eluted the cell-to-cell channels in the "flow-through" fractions (shaded in the first panel of Fig. 3). SDS-PAGE shows that this column removed most of the MIP from the OG-solubilized protein solution (Fig. 2A, lane 3). MP70 become a prominent band of the "flow-

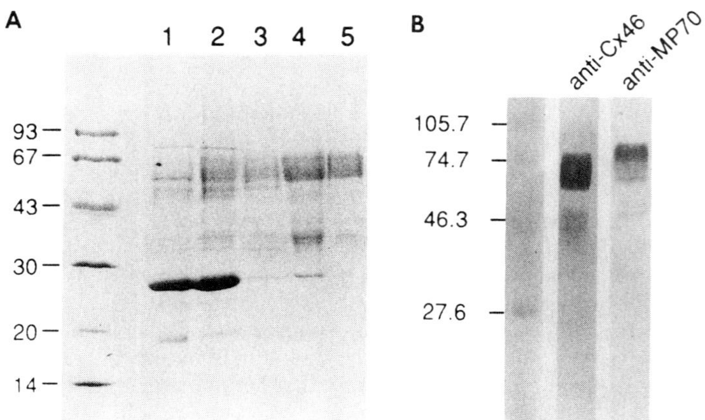

Figure 2. A. SDS-PAGE of the principal fractions of the procedure used to purified cell-to-cell channels from lens fiber cells. Lane 1 shows the protein composition of plasma membranes from calf lenses and lane 2 the OG-solubilized proteins. Lane 3 shows the protein composition of the "flow-through" fractions (shaded in first panel of 3) and lane 4 the proteins in the 0.35-0.45 M NaCl peak from the MonoQ (shaded in the second panel of fig. 3). Lane 5 shows the "purified cell-to-cell channel" (shaded in third panel of fig. 3). B. Western blot analysis of "purified cell-to-cell channels" (lane 5 in A) were blotted with either a polyclonal anti-connexin 46 (for connexin 44) or a monoclonal anti-MP70 (for connexin 50). The diffuse 55-75 kDa protein band contained both connexins. Molecular weight markers are in kDa. (From Konig and Zampighi, 1995)

through" peaks. Further purification required exchanging OG for decylmaltopyranoside, another nonionic detergent, followed by anion exchange column chromatography.

The replacement of OG was necessary because intrinsic proteins aggregated in solutions of this detergent and made further purification of the channels impossi-ble. After trying several nonionic detergents, we found that decylmaltopyranoside (DeM) formed mixed micelles with the solubilized channels that remained stable in solution for over extended periods of time. This allowed further purification of the channels by binding the cell-to-cell channel/detergent micelle to a MonoQ column. A continuous NaCl gradient eluted the channels in a peak at 0.35-0.45 M (Fig. 3, shaded region in second panel). SDS-PAge showed the peak to be composed principally of the diffuse 55-75 kDa band (MP70) and of weaker bands at ~35 kDa and 26 kDa (Fig. 2A, lane 4). Gel filtration chromatography produced fractions from 6.5 ml to 10.5 ml containing MP70 only (Fig. 3, shaded in third panel).

CHARACTERIZATION OF LENS CELL-TO-CELL CHANNELS

Negative staining showed that fractions eluting from 7.5 ml to 9.5 ml (shaded in Fig. 3, third panel) contained annuli 11±0.6 nm (S.D., n=105) diameter and 16±0.8 nm long particles (S.D., n=96) (Fig. 4). The ~16 nm long particles were composed of two domains joined through a 3 nm wide, lightly stained region (arrows Fig. 4). SDS-PAGE showed that these fractions contained the diffuse 55-75 kDa band and only small amount of a ~35 kDa protein which likely correspond to an endogenous proteolytic fragment of MP70

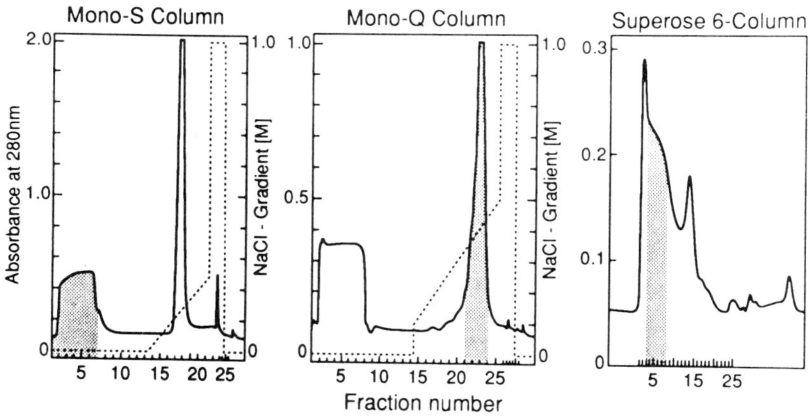

Figure 3. Steps in the purification of cell-to-cell channels starting from the octyl-b-D-glucopyranoside solubilized proteins of calf lenses. The cell-to-cell channels eluted in the "flow-through" fractions of a MonoS cation exchange column (shaded in first panel). The "flow-trough" fractions were injected in a MonoQ anion exchange column and eluted with a NaCl gradient (dotted line in second panel). The channels eluted in a sharp peak at 0.35-0.45 M NaCl (shaded in second panel). Chromatography in a Superose 6 gel filtration column removed contaminants and aggregated channels. Fractions from 7.5 ml to 9.5 ml (shaded in third panel) contained the "purified cell-to-cell channels". (From Konig and Zampighi, 1995).

(Fig. 2, lane 5). These fractions were referred to as "purified cell-to-cell channels".

Fractions eluting from 6.5-7.5 ml contained principally the diffuse 55-75 kDa band. Negative staining showed aggregates composed of 3-6 channels. Fractions eluting from 9.5 ml to 10.5 ml contained single, complete cell-to-cell channels, however, SDS-PAGE showed a substantial increase in the amount of the ~35 kDa protein. Both of these fractions were discarded.

The ~16 nm particles are long enough to span the two membranes forming the junction and the intervening extracellular gap. They also exhibit two distinct domains separated by a lightly stained region that is consistent with the model of a channel composed of two hemichannels joined through the external domains (Unwin and Zampighi, 1980; Unwin and Ennis, 1984). Therefore, this particle must correspond to the side view of a "complete" cell-to-cell channel. The annulus, on the other hand, had a larger diameter than channels forming liver gap junctions (~11 nm instead of ~7 nm). Such a discrepancy is not unexpected since the chan-nels were purified as a complex with DeM molecules, a fact that must increase their diameter but not the length.

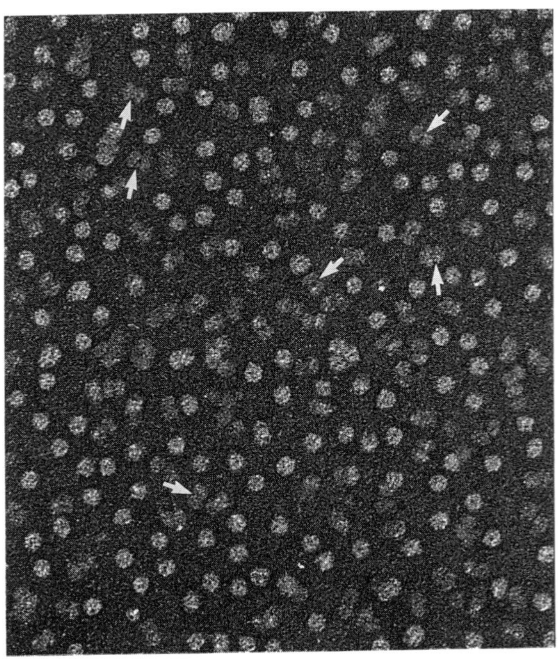

Figure 4. Negative staining of "purified cell-to-cell channel" eluting the Superose 6 gel filtration column. The channel/detergent complexes appear as annuli ~11 nm in diameter and as ~16 nm long particles (arrows). These correspond to top and side views of "complete" cell-to-cell channels. Magnification: X300,000.

Immunoblots showed that the 55-75 kDa band was recognized by two non-crossreacting antibodies. The upper part of the diffuse band was recognized by anti-MP70, a monoclonal that crossreacts with bovine, ovine and mouse connexin 50 (Fig. 3C) (Kistler et al., 1985; White et al., 1992). The lower part of the 55-75 kDa band was recognized by a polyclonal anti-rat connexin 46 that also crossreacts with bovine connexin 44 (Fig. 3C) (11). Both antibodies recognized lower molec-ular weight bands migrating at about 44 kDa and 50 kDa, respectively (Fig. 3C). Thus, the diffuse band previously named MP70 contains connexin 44 and 50.

Whereas the weaker bands detected in the immunoblotting correlated well with the molecular masses predicted from the amino acid sequence of connexin 44 and 50, the stronger, diffuse bands were 15-20 kDa larger. A likely explanation of the increase in molecular mass is that both connexins were posttranslationally modi-fied. Glycosylation and phosphorylation are two modifications that can account for the increase in molecular mass of the connexin 44 and connexin 50. Treatment with endoglycosidase H and neuroaminidase did not change the electrophoretic mobility of the 55-75 kDa band. However, treatment with alkaline phosphatase resolved the diffuse band into two distinct bands, at a ratio close to unity, but did not change the molecular masses of both connexins to that predicted from the sequences.

IS THE LENS CELL-TO-CELL CHANNEL A "HYBRID" CHANNEL?

Connexin 46 and connexin 50 had been located in the same fiber cells, and more importantly, in the same junction (Louis et al., 1993; Paul et al., 1991; TenBroek et al., 1992). Our results show that purified cell-to-cell channels are composed of equal amounts of connexin 44 and 50. These results are consistent with channels assembled of either connexin 46 or 50 (homotypic) co-localized in the same junction or as a channels composed of both connexins (hybrids). A hybrid channels can be assembled of hemichannels formed of the same connexins (heterotypic) or of hemichannels formed of different connexins (heteromeric).

Although not conclusive, our results support the hypothesis that lens cell-to-cell channels are assembled as "hybrids" (i.e., heterotypic or heteromeric). The channels were purified from other intrinsic proteins based on differences of their molecular mass, hydrodynamic radii and charge. Homotypic channels assembled of either connexin 44 or connexin 50 should have been separated by charge because their calculated pIs differs from each other by more than 2 pH units. Thus, the simplest rationalization of results obtained by studying the connexins in fractions composed of purified cell-to-cell channels, in plasma membranes (Paul et al., 1991; TenBroek et al. 1992) and in fiber cells is that lens cell-to-cell channels are "hybrids" composed of connexin 44 and connexin 50.

Expression of multiple connexins and their posttranslational modification may represent a control mechanism of cell-to-cell communication in the lens. Expression of multiple types of connexins may ensure formation of cell-to-cell channels when a particular connexin is not expressed, is expressed at low levels or is vulnerable to endogenous proteolysis. Hybrid channels may also differ in the conductance, gating or voltage-dependence with channels assem-bled homotypical-ly. Presently, it is not possible to determine which, if any, of these possibili-ties can explain the presence of channels made of two connexins in the lens. To under-stand the structure and function of both channels and hemichannels, connexin 46 and 50 had been expressed in Xenopus laevis oocytes (Swenson et al., 1989; Werner et al., 1989) and studied by combined functional and structural methods.

PROPERTIES OF HEMICHANNELS OF CONNEXIN 50 EXPRESSED IN OOCYTES

Injection of cRNA coding for connexin 50 in oocytes resulted in expression of connexin 50, as determined by Western blotting. Expression increased steadily with time after cRNA injection up to day 3. To gather information about whether connexin 50 is inserted in the oolemma as monomers (single polypeptide) or as hemichannels (hexamers), oocytes expressing connexin 50 were studied using a method that measures, in the same oocyte, both the total number of newly-inserted particles, using freeze-fracture electron microscopy, and the current induced by them, using 2-electrodes voltage-clamp (Zampighi et al., 1995). This allows the determination of the functional properties of "single hemichannels".

In oocytes expressing connexin 50, but not in water-injected controls, an inward current was induced by decreasing the external calcium concentration

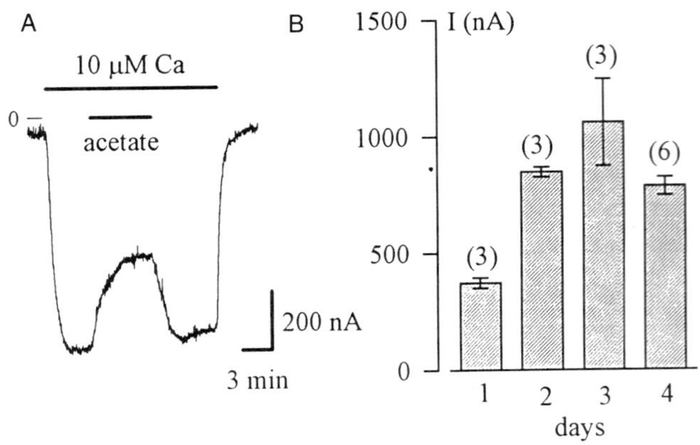

Figure 5. Currents associated with the expression of connexin 50 after reducing the external calcium concentration from 1 mM to 10 μM. Oocytes were bathed in buffer (in mM: 100 NaCl, 2 KCl, 1 CaCl2, 1 MgCl2, 10 HEPES/pH 7.4 at 20° C). A) The membrane potential was held at -50 mV with a two-electrode voltage clamp (Loo et al., 1993). The current was reversibly blocked by exchanging 50% of the NaCl for acetate in the external solution. B). Dependence of the current induced by the low external calcium on the days of incubation post-cRNA injection. Oocytes were maintained at 18° C in Barth media (in mM: 88 NaCl, 0.33 Ca(NO3)2, 0.41 CaCl2, 0.82 MgSO4, 2.4 NaHCO3, 10 HEPES, 0.1 mg/ml gentamycin, pH 7.4) after injection of 50 ng (1μg/μl) of cRNA. The current activated by low external calcium and the inhibito-ry effect of acetate was not observed in control noninjected or H2O-injected oocytes.

from 1 mM to 10 μM (Fig. 5A). The current was proportional to the level of expression of connexin 50: it appears only hours after injection and reached a maximum ~3 days postinjection (Fig. 5B). The current was blocked reversibly by cytoplasmic acidifi-cation induced by exchanging Na-chloride for Na-acetate in the external solution. Because the conductance of complete cell-to-cell channels is also blocked by acidification, this experiment supports the hypothesis that the inward current is due to the expression of connexin 50.

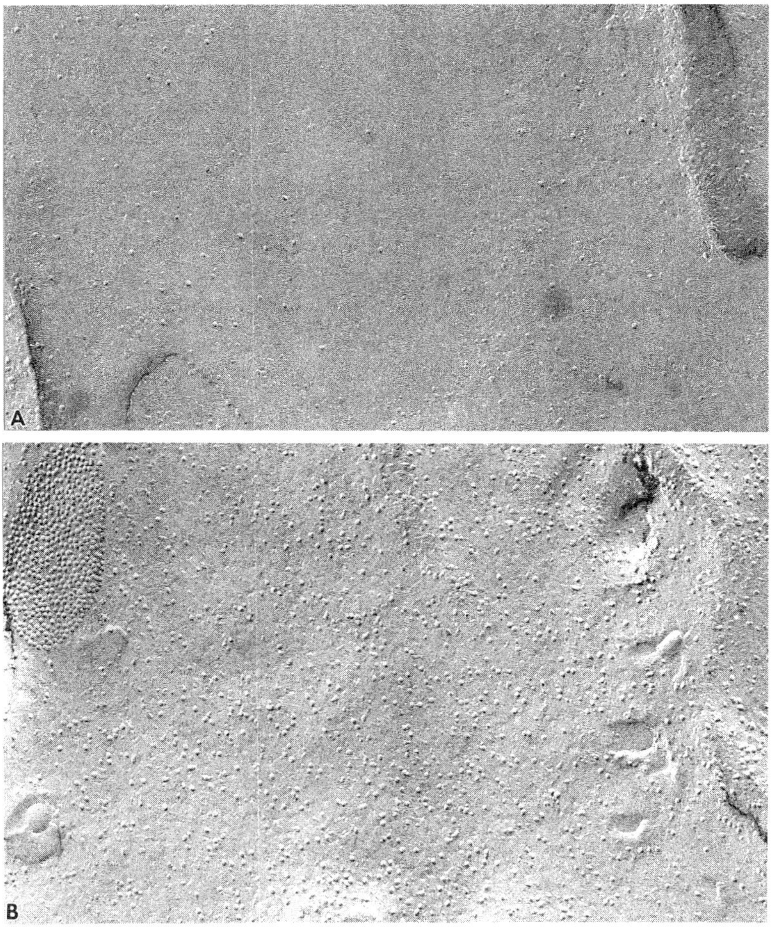

Figure 6. Freeze-fracture replicas showing the P (protoplasmic) fracture face of the plasma membrane of oocytes expressing connexin 50. Panel A shows the P face of the oolemma of an oocyte incubated for 6 hours after cRNA injection. The density of particles is only slightly above that of water-injected controls (~100/mm2). Panel B shows the P face of an oocyte incubated for 3 days after injection. The density of particles is greatly increased and, more importantly, they aggregate in plaques resembling gap junctions. Magnification A and B: X100,000.

The total number of hemichannels in the oolemma of oocytes expressing connexin 50 was determined from the density of newly-inserted particles in the oolemma using freeze-fracture electron microscopy. In freeze-fracture replicas, membranes fracture into two complementary faces; the protoplasmic (P) and the external (E) faces. Intrinsic proteins appear as distinct 8-10 nm diameter particles on either the P or E faces. In oocytes, two unique properties of the oolemma permit to estimate the density of exogeneously expressed intrinsic proteins. First, the oolemma of water-injected, control oocytes contains a very low density of particles (~200/µm2) on the P face (Zampighi et al., 1995).

Most of the endogenous intrinsic proteins appear as particles on the complementary E face. Second, most exoge-neously expressed proteins, such as connexin 50, appear as particles on the P face, the face with the lowest density of endogenous particles (Zampighi et al., 1995). We took advantage of these properties of the oolemma to quantify the insertion of hemichannels made of connexin 50 in the oolemma.

In oocytes expressing connexin 50, increases in the magnitude of the current (Fig. 5A) and in the time of incubation postinjection were correlated to increases of the particle density on the P face of the oolemma (Fig. 6). The newly-inserted particles were ~11 nm in diameter, a dimension consisting with connexin 50 being inserted in the oolemma as pre-formed hemichannels, not as a monomer. By combining measurements of total capacitance, the magnitude of the inward current, the water permeability and the density of hemichannels in the same oocyte, we determined two properties of hemichannels made of connexin 50. First, at -50 mV and 1 mM external calcium, hemichannels do not allow diffusion of ions, or even water, into the cell. Second, at a critical density they aggregate in plaques, similar to gap junctions seen in most tissues. These properties are likely to be present in tissues coupled via gap junctions.

CONCLUSIONS

1. To achieve transparency, the lens eliminated most cellular organelles during development. However, it can still maintain homeostasis probably via networks of cell-to-cell channels which allow the diffusion of ions, nonelectrolytes and water between the cytoplasm of adjacent cells.

2. Lens fiber cell-to-cell channels are aggregated in plaques devoid of crys-talline order. This random arrangement is, most likely, not related to the functional state of the channel (open/closed) but to the chemical composition of the connexins composing the channel.

3. Lens cell-to-cell channels are composed of a protein migrating as a diffuse band of 55-75 kDa apparent molecular weight (MP70). The diffuse band is composed of almost equal amounts of connexin 44 and 50. Both connexins are posttranslationally modified, thereby increasing their molecular weight by 15-20 kDa above what is predicted from the amino acid sequence. Phosphorylation explains part but not all the posttranslational modification of the connexins.

4. The "complete" cell-to-cell channel likely is a "hybrid" assembled from subunits of both connexin 44 and connexin 50.

5. Expression of connexin 50 in Xenopus laevis oocytes results in the insertion of hemichannels (hexamers) in the oolemma. These hemichannels do not conduct ions or even water. They are, however, able to aggregate themselves into plaques when the density in the oolemma reaches a critical concentration.

ACKNOWLEDGMENTS

We thank Mr. M. Kreman for excellent technical help. The work was supported by NIH Grant EY-04110 (GAZ), NS25554 (D.D.F.L.) and MRC Protein Engineering LINK Program (NK).

BIBLIOGRAPHY

Agre, P., and Preston GM., 1992, Isolation of the cDNA for erythrocyte integral membrane protein of 28 kilodaltons: Member of an ancient family, *Proc. Natl. Acad. Sci. USA*. 88:1110.

Baker, M.E., and Saier, M.H., 1990, A common ancestor for bovine lens fiber major intrinsic protein, soybean nodulin-26 protein and E. coli glycerol facilitator, *Cell* 185:185.

Barrio, L.C., Suchyna, T., Bargiello, T., Xu, L.X., Roginski, R.S., Bennet, M.V.L., and Nicholson, B.J.,1991, Gap junctions formed by connexins 26 and 32 alone and in combinations are differently affected by applied voltage, *Proc. Natl. Acad. Sci. USA* 88:8410.

Beyer, E.C., Paul, D.L. and Goodenough, D.A., 1990, Connexin family of gap junction proteins, *J. Membr. Biol.* 116:187.

Costello, M.J., McIntosh, T.J., and Robertson, J.D., 1989, Distribution of gap junctions and square arrays in the mammalian lens, *Invest. Opthalmol. Vis. Sci.* 30:975.

Bloemendal H., Vermorken A.J.M., Kibbelaar, M., Dunia, A., Benedetti, E.L. 1977, Nomenclature for the polypetide chains of lens plasma membranes, *Exp. Eye Res.* 24:413.

Duncan,G., and Croghan, P.C., 1989, Mechanisms for the regulation of cell volume with particular reference to the lens, *Exp. Eye Res.* 8:421.

Gorin, M.B., Yancey, S.B., Cline, J., Revel, J.P., and Horwitz, J., 1984, The major intrinsic protein (MIP) of the bovine lens fiber membrane: characterization and structure based on cDNA cloning, *Cell* 39:49.

Gruijters, W.T.M., Kistler, J., Bullivant, S., and Goodenough, D.A., 1987, Immunolocalization of MP70 in lens fiber 16-17 nm intercellular junctions, *J. Cell Biol.* 104:565.

Guerrero, F.D., Jones, J.T., and Mullet, J.E., 1990, Turgor responsive gene transcription and RNA levels increase rapidly when pea shoots are wilted. Sequence and expression of three inducible genes, *Plant Mol. Biol.* 15:11.

Gupta, V.K., Berthoud, V.M., Atal, N., Jarillo, J.A., Barrio, L.C., and Beyer, E.C., 1994, Bovine connexin44, a lens gap junction protein: molecular cloning, immunologic characterization, and functional expression, *Invest. Opthalmol. Vis. Sci.* 35:3747.

Heller, K.B., Lin, E.C., and Wilson, T.H., 1980, Substrate specificity and transport properties of the glycerol facilitator of Escherichia coli, *J. Bacteriol.* 144:274.

Johnson, R., Frenzel, E., Johnson, K, Klukas, K., Lampe, P, Sas, D., Biegon, R., and Louis, C., 1985, Is lens MP26 a gap junction protein? A phosphoprotein?, in: "Gap Junctions," M.V.L. Bennet, and D.C. Spray, eds., Cold Spring Harbor Laboratory, Cold Spring Harbor, NY.

Johnson, R.G., Klukas, K.A., Tze-Hong, L., and Spray, D.C., 1988, Antibodies to MP28 are localized to lens junctions, alter intercellular permeability and demonstrate increased expression during development, in: "Gap Junctions", Modern Cell Biology, E.L. Hertzberg, and R.G. Johnson, eds., Alan R. Liss, NY.

Johnson, K.D., Hofte, H., and Chrispeels, M.J., 1990, An intrinsic tonoplast protein storage vacuoles in seed is structurally related to a bacterial solute transporter (GlpF), *The Plant Cell* 2:525.

Kistler, J., Kirkland, B., and Bullivant, S., 1985, Identification of a 70,000-D protein in lens membrane junctional domains, *J. Cell Biol.* 101:28.

Kistler, J., and Bullivant, S., 1988, Dissociation of lens fibre gap junctions releases MP70, *J. Cell Sci.* 91:415.

Kistler, J., Christie, D., and Bullivant, S., 1988, Homologies between gap junction proteins in lens, heart and liver, *Nature* 331:721.

Konig, N., and Zampighi, G.A., 1995, Purification of bovine lens cell-to-cell channels composed of connexin44 and connexin50, *J. Cell Sci.* 108:3091.

Lo, W.K., and Harding, C.V., 1984, Square arrays and their role in ridge formation in human lens fibers, *J. Ultrast. Res.* 86:228.

Loo, D.D.F., Hazama, A., Supplisson, S., Turk, E., and Wright, E.M., 1993, Relaxation kinetics of the Na+/glucose cotransporter, *Proc Natl. Acad. Sci. USA* 90:5767.

Louis, C.F., Arneson, M., Jarvis, L., and TenBroek, E.M., 1993, The developmental expression and organization of membran proteins of the mammalian lens, in: "Gap Junctions". Progress in Cell Research, J.E. Hall, G.A. Zampighi, and R.M. Davies, eds., Elsevier, Amsterdam.

Mathias, R.T., Rae, J.L., and Eisenberg R.S., 1981, The lens as a nonuniform spherical syncytium, *Biophys. J.* 34:61.

Mathias, R.T., and Rae, J.L., 1985, Transport properties of the lens, *Amer. J. Physiol.* 249:C181.

Peracchia, C., and Girsch, S.J., 1985, Is the C-terminal arm of lens gap junction channel protein the channel gate, *Biochem. Biophys. Res. Comm.* 133:688.

Preston, G.M., Carroll, T.P., Guggino, W.B., and Agre, P., 1992, Appearance of water channels in Xenopus oocytes expressing red cell CHIP28 protein, *Science* 256:385.

Paul, D.L, Ebihara, L., Takemoto, L.J., Swenson, K.I., and Goodenough, D.A., 1991, Connexin46, a novel lens gap junction protein, induces voltage-gated currents in nonjunctional plasma membrane of Xenopus oocytes, *J. Cell Biol.* 115:1077.

Swenson, K.I., Jordan, J.R., Beyer, E.C., and Paul, D.L., 1989, Formation of gap junctions by expression of connexins in Xenopus oocyte pairs, *Cell* 57:145.

TenBroek, E., Arneson, M., Jarvis, L., and Louis, C.F., 1992, The distribution of the fiber cell intrinsic membrane proteins MP20 and connexin46 in the bovine lens, *J. Cell Sci.* 103:245.

Unwin, P.N.T., and Zampighi, G., 1980, Structure of junctions between communicating cells, *Nature* 283:545.

Unwin, P.N.T., and Ennis, P.D., 1984, Two configurations of a channel forming protein, *Nature* 609:609.

Verselis, V.K., Bargiello, T.A., Rubin, J.R., and Bennett, M.V.L., 1993, Comparison of voltage-dependent properties of gap junctions in hepatocytes and in Xenopus oocytes expressing Cx32 and Cx26, in: "Gap Junctions". Progress in Cell Research, J.E. Hall, G.A. Zampighi, and R.M. Davies, eds., Elsevier, Amsterdam.

Werner, R.E., Levine, E., Rabadan-Diehl, C., and Dahl, G., 1989, Formation of hybrid cell-cell channels, *Proc. Natl Acad. Sci USA* 86:5380.

White, T.W., Bruzzone, R., Goodenough, D.A., and Paul, D.L., 1992, Mouse CX50, a functional member of the connexin family of gap junction proteins, is the lens fiber protein MP70, *Mol. Biol. Cell* 3:711.

Zampighi, G., Simon, S.A., Robertson, J.D., McIntosh, T.J., Costello, M.J., 1982, On the structural organization of isolated bovine lens fiber junctions, *J. Cell Biol.* 93:175.

Zampighi, G.A., Hall, J.E., Ehring, G.R., and Simon, S.A., 1989, The structural organization and protein composition of lens fiber junctions, *J. Cell Biol.* 108:2255.

Zampighi, G., Simon, S.A., and Hall, J.E., 1992, The specialized junctions of the lens, *Inter. Rev Cytol.* 136:185.

Zampighi, G.A., Kreman, M., Boorer, K.J., Loo, D.D.F., Bezanilla, F., Chandy, G., Hall, J.E., and Wright, E.M., 1995, A method for determining the unitary functional capacity of cloned channels and transporters expressed in Xenopus laevis oocytes, *J. Membr. Biol.* 148:65.

STUDIES OF VOLTAGE GATING OF GAP JUNCTIONS AND HEMICHANNELS FORMED BY CONNEXIN PROTEINS

Vytas K. Verselis, E. Brady Trexler, Thaddeus A. Bargiello, and Michael V.L. Bennett

Department of Neuroscience
Albert Einstein College of Medicine
Bronx, NY 10461

INTRODUCTION

The cloning of a cDNA encoding a gap junction forming protein from rat liver rapidly led to identification by low stringency hybridization and PCR of many related proteins, now termed connexins (Bennett *et al.*, 1991). At this point, at least thirteen are known from rodents (rat and mouse) and several homologs have been found in human, cow, dog, chicken, *Xenopus,* and several fishes (Bennett *et al.*, 1994; White *et al.*, 1995). Given their multiplicity, their molecular divergence, and the many tissues in which they are found, one asks how do the properties of connexins differ and why, in the structure-function sense as well as teleologically. A readily measured parameter of gap junctions is their conductance, g_j. As it has turned out, all known vertebrate gap junctions are electrically non-linear; their conductance changes as a function of applied voltage. Both the open probability and conductance of the open state of single junctional channels can change as a function of voltage. In this presentation we demonstrate that polarity of gating differs among junctions formed of different connexins, and we use mutagenesis to identify amino acid residues that are part of the voltage sensor; by changing charge on critical residues we can we increase voltage sensitivity and even reverse the gating polarity. Moreover, several connexins form hemichannels that can be opened by voltage without their being connected to another hemichannel in an apposed

From Ion Channels to Cell-to-Cell Conversations
Edited by Latorre and Sáez, Plenum Press, New York, 1997

membrane. We characterize voltage gating in one of these hemichannels by single channel patch recording; these (hemi)channels exhibit two gating processes, one that corresponds to that in the complete junctional channel and a second, distinct process that probably corresponds to the binding of one hemichannel to another to form a junctional channel.

The most widely used nomenclature abbreviates connexin as Cx and appends its predicted molecular weight in kilodaltons (kD) to two or three significant figures, whatever is necessary to distinguish connexins of nearly the same size. A prefix, hum, Xen, rat, etc., may be used to indicate the animal of origin. A nomenclature based on tissue of origin is not possible, as many tissues express more than one connexin and many connexins are expressed in more than one tissue. The connexins appear to have a conserved domain structure (Fig. 1A), verified in several connexins by antibody and protease studies (see Bennett *et al.*, 1991). There are four membrane spanning regions (M1-M4) and both N and C termini (NT, CT) are cytoplasmic. There are two extracellular loops (E1, E2) and a cytoplasmic loop (CL) connecting M2 and M3. A single gap junction channel is a duodecamer, made up of two connexin hexamers (hemichannels or connexons) in each of the apposed membranes.

Comparison of function when sequences differ permits two classes of inference. If a function is the same, differences in the sequences are not important for that function. Conversely, if a function differs, regions where the sequences differ are candidates for mediators (or modulators) of that function. Evolution of multiple connexins provides an initial mutagenesis study that can be used to guide site-directed mutagenesis for more precise analysis of how structure relates to function. Our initial studies used Cx26 and Cx32, which differ in their voltage gating properties; we transferred domains between them to identify regions responsible for the functional differences.

Because gap junctions connect two cells, they can be subjected to voltages applied between either (or both) cells and the extracellular medium. It is convenient to consider that there are two kinds of voltage, that between the inside and outside of the cells, inside-outside voltage ($V_{i\text{-}o}$), and that between the two cell interiors, transjunctional voltage (V_j). $V_{i\text{-}o}$ can be manipulated independently of V_j with $V_j = 0$ by polarizing each cell simultaneously and equally from the same holding potential. Presumably, most of $V_{i\text{-}o}$ is developed across the channel wall and little across the extracellular medium in the intercellular gap (Fig. 1B), since the gap is wide enough to permit easy access by small ions and the channel wall passes very little current as indicated by coupling coefficients between cells that can approach unity.

V_j can be made non-zero by changing the voltage in either cell with respect to the external medium. Of necessity, a change in V_j produced by a

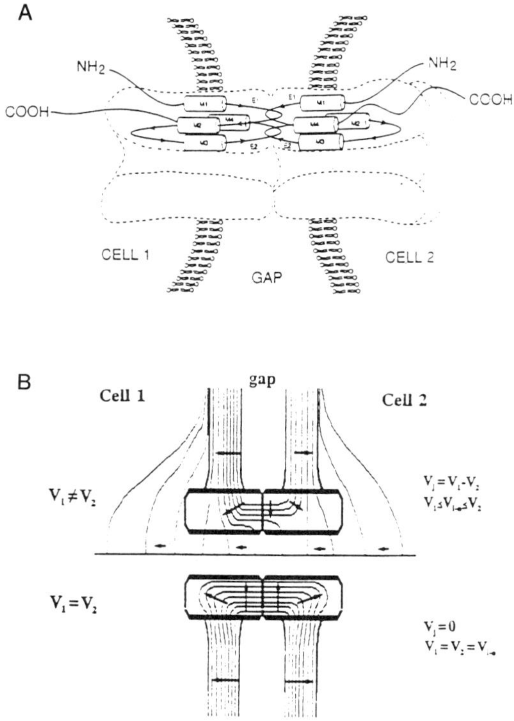

Figure 1. Connexin topology and fields due to applied voltages. **(A)** Representation of a gap junction channel between two cells. The dotted lines indicate the channel boundaries and the solid lines and cylinders the location of one connexin subunit in each hemichannel. **(B)** Diagram of isopotential lines in and around a gap junction channel when $V_j = |V_1 - V_2| > 0$, shown above, and when only a $V_{i\text{-}o}$ is present, where $V_{i\text{-}o} = V_1 = V_2$, shown below. Arrows indicate field direction and their length suggests relative field strength. From Rubin *et al.* (1992b).

voltage applied in one cell changes the $V_{i\text{-}o}$ in that cell and along the channel. However, for many junctions, changes in g_j produced by depolarizations or hyperpolarizations in one cell are identical to the changes produced by equal and opposite polarizations in the other cell; thus, the junctions exhibit V_j sensitivity with no $V_{i\text{-}o}$ sensitivity. For homotypic junctions, i.e., those made of the same connexin in each of the apposed cells, the (steady state) g_j/V_j relation is symmetrical around $V_j = 0$; asymmetry implies $V_{i\text{-}o}$ sensitivity. For heterotypic junctions, i.e. those made of different connexins, the g_j/V_j relation may be asymmetric, but still independent of $V_{i\text{-}o}$. Symmetry can be ascribed to independent action of identical hemichannels joined head to head. Asymmetry arises from joining of dissimilar hemichannels.

Generally, one studies voltage sensitivity of gap junctions by independently voltage clamping each cell of a coupled pair (Spray *et al.*, 1981). Stepping one cell while holding the voltage of the other cell constant results in a step in voltage across the junctions. Current applied to the second cell to maintain its voltage constant is equal and opposite to the junctional current, I_j. The time course of I_j represents the time course of change in g_j in response to the voltage step. (The clamping current in the voltage stepped cell includes both junctional and non-junctional current.) Provided the cytoplasm on the two sides of the junctions is of the same composition, there is no voltage generated in the junctions. When displaying voltage and current as a function of time, the convention is upward for positive going voltage and for outward current through an electrode. Thus, the voltage applied in a cell and the current required to generate it are both displayed in the same direction. This convention leads to an apparent anomaly, when voltage clamping a pair of coupled cells and measuring g_j by applying a voltage step in one cell, while holding the voltage in the second cell constant; in this case the current in the second cell is equal and opposite to the junctional current and is displayed in the direction opposite to that of the current and voltage in the first cell. One further convention is important for this presentation. We often express different connexins in pairs of *Xenopus* oocytes and label these pairs by the two connexins as for example, Cx32/Cx26. When we display the g_j/V_j relation, which in general is asymmetric, we define positive V_j (displayed on the right side of the g_j/V_j relation) as relative positivity in the oocyte expressing the connexin on the right side of the pairing designation, in this case Cx26.

Typical V_j dependence of g_j is exhibited by Cx32 junctions (Fig. 2). During a voltage step, current, and therefore g_j, decays more or less exponentially to a steady state value. The decay is faster and greater for larger voltages of either sign. g_j is maximal for $V_j = 0$ and decreases for V_j of either sign with symmetry around $V_j = 0$. g_j approaches a non-zero minimum value, g_{jmin}, for large voltages. The g_j/V_j relation for one polarity of V_j can be reasonably well fit by a Boltzmann relation of the form:

$$g_j = (g_{jmax} - g_{jmin})/(1 + \exp(A(V_j - V_0))) + g_{jmin}$$

where g_j is the steady state value at the transjunctional voltage, V_j; g_{jmax} is the maximum value of g_j, which may be obtained by extrapolation because a fraction of the hemichannels are closed at $V_j = 0$ and non-zero V_j's that open

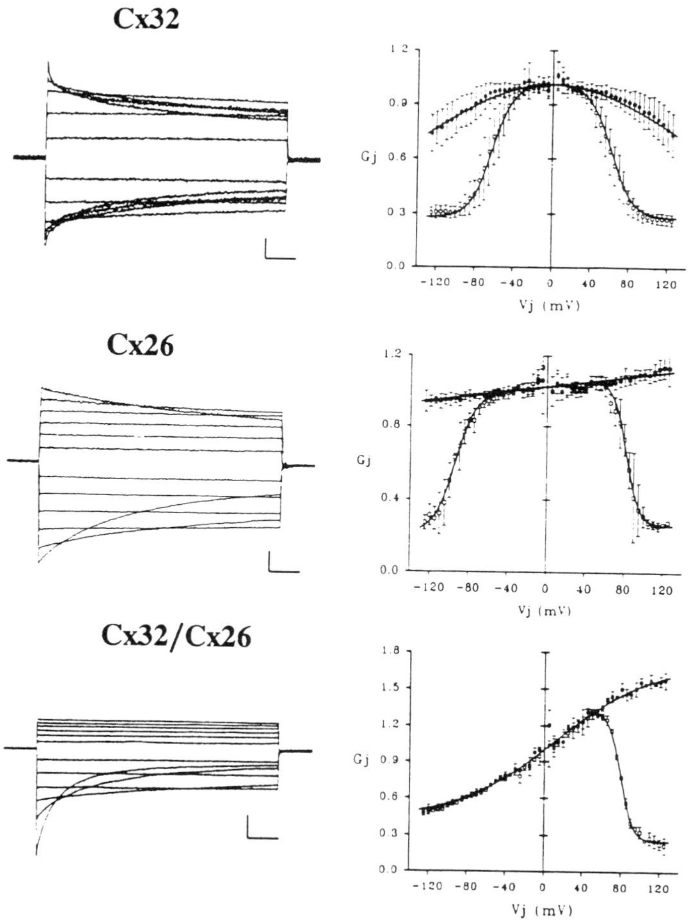

Figure 2. g_j/V_j relations for Cx32 and Cx26 homotypic junctions and for heteromeric Cx32/Cx26 junctions. On the left are junctional currents for V_j steps up to $\pm$ 120 mV in 20 mV increments. For Cx32/Cx26 junctions, positive and negative V_j are relatively positive and negative in the Cx26 expressing cell and elicit downward and upward currents, respectively, in the postjunctional cell. Time calibrations, 1 s. Current calibrations: 20 nA for Cx32 and Cx26 and 50 nA for Cx32/Cx26. On the right graphs show initial and steady state values (closed and open symbols, respectively) of G_j which is g_j normalized to the value of g_j at $V_j = 0$. Cx32 junctions show symmetric g_j/V_j relations for both initial and steady state. Cx26 junctions show $V_{i\text{-}o}$ dependence of initial g_j (a fast process) with g_j increasing for depolarization of either cell (shown as positive V_j) and decreasing for hyperpolarization of either cell (negative V_j). The slow changes in g_j are somewhat faster for depolarization of a given amplitude than for hyperpolarization, and the steady state g_j/V_j relation is a bit asymmetric, indicating an effect of $V_{i\text{-}o}$ on the slow process. The g_j/V_j relation for heterotypic Cx32/Cx26 junctions is markedly asymmetric as described in the text. From Bennett *et al.*, 1993.

closed hemichannels on one side, close open hemichannels on the other side, g_{jmin} is the minimum value of g_j, V_0 is the voltage at which the reduction in V_j is half maximal and A is a constant given by zq/kT where z is the effective number of gating charges that move through the entire applied field, q is the charge on the electron and kT has its usual meaning. Although for some connexins there are additional components of decay, which contradicts the Boltzmann assumptions, the Boltzmann parameters are a convenient measure of the degree of voltage dependence and permit the calculation of an effective gating charge. In the first demonstration of this kind of g_j /V_j relation, it was proposed that each hemichannel had its own gate that reacted to one or the other polarity of V_j across the junction (Fig. 3) (Harris *et al.*, 1981). Because of symmetry, it was uncertain which gate closed with which polarity of V_j.

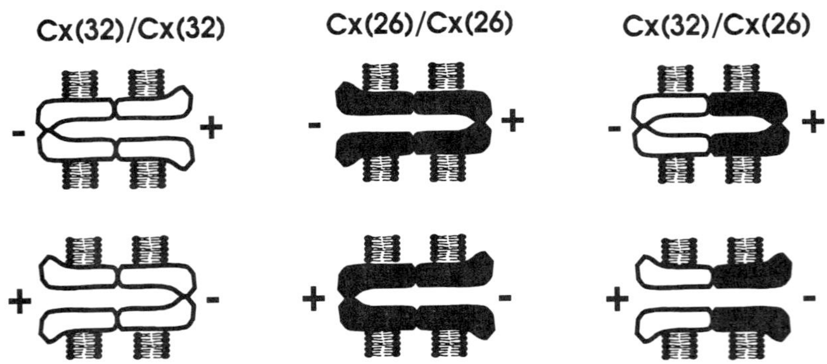

Figure 3. Diagram of gating polarity and how opposite polarity of sensitivity results in asymmetry of the slow process in heterotypic junctions. In Cx32/Cx32 junctions the hemichannel on the relatively negative side closes, but the g_j/V_j relation is symmetric because there are two hemichannels in each channel. In Cx26/Cx26 junctions the hemichannel on the relatively positive side closes, but again, the g_j/V_j relation is symmetric because there are two oppositely oriented hemichannels in each channel. In heterotypic Cx32/Cx26 channels the hemichannels are of opposite sensitivity but are oppositely oriented; both hemichannels tend to close for relative positivity on the Cx26 side and open (or stay open) for relative negativity on the Cx26 side. Double closure as diagrammed may be infrequent because when one hemichannel closes, the voltage across the other hemichannel is likely to be reduced. From Bennett *et al.* (1993)

In addition to the relatively slow changes in g_j, i.e., with time constants of tenths of seconds to seconds, Cx32 exhibits an additional V_j dependence, which was not obvious in our earlier studies. The initial conductance at the start of the clamping pulses is decreased by larger V_j, and the change occurs too rapidly to be resolved by our voltage clamp (about 5 ms for *Xenopus* oocytes). We call this fast V_j dependence mediated by a fast process. It may arise from rectification of the single channel conductance as discussed below.

HEMICHANNELS IN JUNCTIONS DIFFER IN POLARITY OF GATING; SOME CLOSE FOR POSITIVITY ON THE CYTOPLASMIC SIDE, SOME CLOSE FOR NEGATIVITY

When oocytes are injected with a connexin cRNA and paired, gap junctions can form between them, and this system has been widely used to characterize the properties of junctions formed by various cloned connexins. Cx26 junctions show a slow decrease in conductance for large V_j of either sign (Fig. 2) (Barrio *et al.*, 1992). However, there is a slight asymmetry in the both the initial and steady state conductance for depolarization and hyperpolarization of either cell, indicating a small degree of $V_{i\text{-}o}$ dependence. Initial and steady state g_j are both slightly increased for depolarization and decreased for hyperpolarization. Since hepatocytes express both Cx32 and Cx26 it was of interest to determine if oocytes expressing Cx32 would form junctions with oocytes expressing Cx26. Such junctions did form, and the voltage dependence had very unexpected properties. From the more or less symmetric g_j/V_j relations of Cx26 and Cx32 homotypic junctions, we had expected that the g_j/V_j relation of the heterotypic junctions would have been like that of Cx32 junctions on one side and Cx26 junctions on the other and more or less symmetric; it was, however, markedly asymmetric (Fig. 2). There was a pronounced asymmetric fast process that increased initial g_j for relative positivity on the Cx26 side and decreased initial g_j for relative negativity on the Cx26 side. There was an asymmetric slow process that decreased steady state g_j only for relative positivity on the Cx26 side and not for relative negativity. Both fast and slow processes were purely V_j dependent.

The presence of slow closure for only one polarity of V_j was unexpected because we thought that the two kinds of hemichannel would have the same polarity of gating. We had previously shown for Cx38 junctions between *Xenopus* blastomeres that V_j's close the hemichannel on the relatively positive side (Bennett *et al.*, 1988). The demonstration depended on the observation that V_j dependence was altered at reduced cytoplasmic pH. Injection of acid into one cell affected voltage dependence in the same manner as bilateral acidification, but for only one sign of V_j, that for relative positivity on the acidified side. Since the injected hydrogen ions should act more prominently on the nearer hemichannel, we could conclude that Cx38 hemichannels closed for positivity on their cytoplasmic side. Subsequently, pairing of Cx43 and Cx38 expressing cells led to the same conclusion (Swenson *et al.*, 1989). Cx38 junctions are much more voltage sensitive than Cx43 junctions, and the g_j/V_j relation was highly asymmetric; g_j decreased with Cx38 characteristics for positivity on the Cx38 side; the properties were like Cx43 junctions for positivity on the Cx43 side. (Actually, Swenson et al. saw little voltage dependence in Cx43 junctions; it is now agreed that Cx43 junctions are voltage dependent, although much less so that Cx38 junctions.) Thus, Cx32/Cx26 junctions presented us with a problem. Although it had crossed our minds that the two channel types might have opposite gating polarities, which could account for the asymmetry of the slow V_j dependence, we came out rather strongly for the idea that the gating properties of one or both hemichannels were greatly modified by being apposed to different kinds of hemichannel in the heterotypic as compared to the homotypic junctions. We now recognize that this view is for most connexins incorrect; properties of a given kind of hemichannel are usually, but not always, quite well conserved in its heterotypic combinations.

We subsequently found that we could increase voltage sensitivity of g_j by reciprocal substitution of the first two amino acids of E1 in Cx26 and Cx32 (Bennett *et al.*, 1993; Verselis *et al.*, 1994). These residues are ES in Cx32 and KE in Cx26 (Fig. 4). Homotypic junctions of the mutants show steeper

```
          <----------NT----------><--------M1--------><---E1---
POSITIVE GATING POLARITY

          + -     +   -   -       +          +         --   --
ratCx46   MGDWSFLGRL LENAQEHSTV IGKVWLTVLF IFRILVLGAA AEEVWGDEQS

          + -     +   -           +          +         -    --
mouCx43   MGDWSALGKL LDKVQAYSTA GGKVWLSVLF IFRILLLGTA VESAWGDEQS

          + -     -   --  +       +          +         -    --
mouCx40   MGDWSFLGEF LEEVHKHSTV IGKVWLTVLF IFRMLVLGTA AESSWGDEQA

          + -    -    -   -       +          +         -    --
mouCx37   MGDWGFLEKL LDQVQEHSTV VGKIWLTVLF IFRILILGLA GESVWGDEQS

          + -             +       +          +          -   -- -
mouCx30   M.DWGTLHTV IGGVNKHSTS IGKVWITVIF IFRVMILVVA AQEVWGDEQE

          + -             +       +          +         +-   --
mouCx26   M.DWGTLQSI LGGVNKHSTS IGKIWLTVLF IFRIMILVVA AKEVWGDEQA

NEGATIVE GATING POLARITY

          +               +       +          +         -    --
mouCx32   M.NWTGLYTL LSGVNRHSTA IGRVWLSVIF IFRIMVLVVA AESVWGDEKS

          +      +    --          +          +         -    --
mouCx45   M.SWSFLTRL LEEIHNHSTF VGKIWLTVLI VFRIVLTAVG GESIYYDEQS
          <----------NT----------><--------M1--------><---E1---
```

Figure 4. Sequences of connexins from the N terminal domain, NT, through the first membrane spanning region, M1, and the beginning of the first extracellular loop, E1. Charged residues are indicated. Connexins whose hemichannels normally gate on cytoplasmic positivity, including Cx26, are above; those gating on cytoplasmic negativity, including Cx32, are below. Gating polarity is determined by the presence of a negative charge at position 2 or 3, but changing charges at the first two amino acids of E1 can also alter polarity. Additional negative charges at positions 7-16 of Cx37 and Cx40 may account for the greater voltage sensitivity of junctions formed of these connexins, but these positions may "see" less of V_j than charges at the beginning of NT and E1.

voltage dependence of g_j, and the heterotypic junctions with the parent wild type, Cx26*ES/Cx26 and Cx32*KE/Cx32, are asymmetric with greater voltage sensitivity on one side (Fig. 5). For Cx26*ES/Cx26 junctions, the greater sensitivity is for positivity on the side expressing the mutant and the lesser sensitivity is for the positivity on the side expressing the wildtype, indicating that cytoplasmic positivity closes mutant as well as wildtype Cx26 hemichannels, as we expected from comparison with Cx38. However, for Cx32*KE/Cx32 junctions, the greater sensitivity is for negativity on the mutant side and the lesser sensitivity is for negativity on the wildtype side. Thus, Cx32 and Cx32*KE hemichannels close for relative negativity on their cytoplasmic side; their sensitivity is opposite in polarity to that of Cx26. The slow process in Cx32/Cx26 junctions is asymmetric because relative positivity on the Cx26 side tends to close both Cx26 and Cx32 hemichannels and relative negativity on this side tends to keep both hemichannels open (Fig. 3). (If one hemichannel closes, the voltage drop across the other should be greatly reduced; thus, for the most part only one hemichannel may be closed at a time, see (Bennett and Spray, 1987)).

Although these data indicate that gating polarity is conserved in the different heterotypic pairings, there may be differences in hemichannel g_j/V_j relations in different pairings that are ascribable to changes in hemichannel conformation of open, closed and transition states associated with altered interactions of the extracellular loops. The asymmetry in fast V_j dependence of g_j is not so clearly explained. We did not see pure fast V_j dependence in Cx26 homotypic junctions, but it may have been masked by the $V_{i\text{-}o}$ dependence. The fast V_j dependence of the Cx32/Cx26 junctions may arise exclusively in the Cx32 hemichannels (in which case it would have opposite sensitivity to the slow process). Models with asymmetric barriers (associated with charges in the channel lining?) may account for the rectification (in preparation), which occurs at the single channel level (Bukauskas *et al.*, 1995b). Fast $V_{i\text{-}o}$ dependence is not simply a hemichannel property since it is absent in Cx32/Cx26 junctions. The extracellular loops of Cx26 may permit $V_{i\text{-}o}$ dependent conformational changes in homotypic junctions, which are prevented when Cx26 and Cx32 form heterotypic junctions. Fast $V_{i\text{-}o}$ dependence may represent electroconstriction and -dilation of the channel, but slow $V_{i\text{-}o}$ dependence, which is observed in Cx43, Cx45 (unpublished and L. Barrio, personal communication) and some hepatocyte junctions (Verselis *et al.*, 1993) is likely to result from changes in open probability. Similar slow $V_{i\text{-}o}$ dependence in conjunction with V_j dependence is found in insect junctions (Bukauskas and Weingart, 1994; Verselis *et al.*, 1991). However, gap junctions of the nematode, mollusc, arthropod line appear to be formed of a protein family unrelated to the connexins (Barnes, 1994).

SENSITIVITY AND EVEN POLARITY OF V_j GATING CAN BE REVERSED BY CHANGING THE CHARGE OF SPECIFIC RESIDUES, WHICH THUS BECOME PUTATIVE GATING CHARGES

We asked what is the structural basis of the opposite gating polarity of Cx26 and Cx32. An extensive series of domain shifts between Cx26 and Cx32 indicated that putting just the NT domain of Cx32 into Cx26 was sufficient to reverse its gating polarity (Verselis *et al.*, 1994). The amino acid sequences of this domain differ in charge only at position 2 where Cx26 has a negatively charged aspartate (D) and Cx32 has a neutral asparagine (N) (Fig. 4). Changing the charge at this site altered gating in ways that are consistent with its being part of the voltage sensor. For Cx32 the mutations, N2D and N2E, which make this site more negative, reversed the gating polarity from closing on cytoplasmic negativity to closing on cytoplasmic positivity. Reversal of polarity was inferred from Cx32/Cx32(N2D) and Cx32/Cx32(N2E) junctions, whose g_j/V_j relations are very similar to that of Cx32/Cx26 junctions and from Cx26/Cx32(N2D) and Cx26/Cx32(N2E) junctions, whose g_j/V_j relations have slow closure for both polarities of V_j (Fig. 6). The mutations, N2Q and N2A, which do not affect charge, did not change gating polarity. For Cx26 the mutations, D2N and D2Q, which make this site less negative, appeared to change the gating polarity from closing on cytoplasmic positivity to closing on cytoplasmic negativity. However, there were large shifts in V_0, which obscured closure by Cx26 and Cx32 when they were paired with the mutants. The conservative mutation Cx26(D2E) did not change polarity. As noted in the previous section, mutations at the beginning of E1 that effectively exchange charges between Cx32 and Cx26 increased voltage sensitivity. Cx26*ES is more negative at this site than wildtype Cx26, and Cx32*KE is more positive than wildtype Cx32. These differences suggest than voltage sensor includes the beginning of M1 as well as position 2 of NT and that charges in these positions move towards the cytoplasm in the gating process. In agreement the mutation,

Figure 5. Mutations increasing voltage sensitivity permit determination of gating polarity in heterotypic junctions with wild type channels. Display showing currents and g_j/V_j relations as in Fig. 2 except values of initial g_j are omitted. Diagrams of oocytes above indicating the connexin expressed and the voltage stepped cell. **Upper row, left**: Wild type Cx32. **Middle:** A Cx32 mutant, Cx32*KE, shows increased voltage sensitivity. **Right:** In heterotypic Cx32*KE/Cx32 junctions the less sensitive change is for relative negativity on the Cx32 side, indicating that this polarity closes Cx32 hemichannels. **Lower row, left:** Wild type Cx26. **Middle:** A Cx26 mutant, Cx26*ES, shows increased voltage sensitivity. **Right:** In heterotypic Cx26*ES/Cx26 junctions the lesser sensitivity is for relative positivity on the Cx26 side, indicating that this polarity closes the Cx26 hemichannels. From (Verselis *et al.*, 1994).

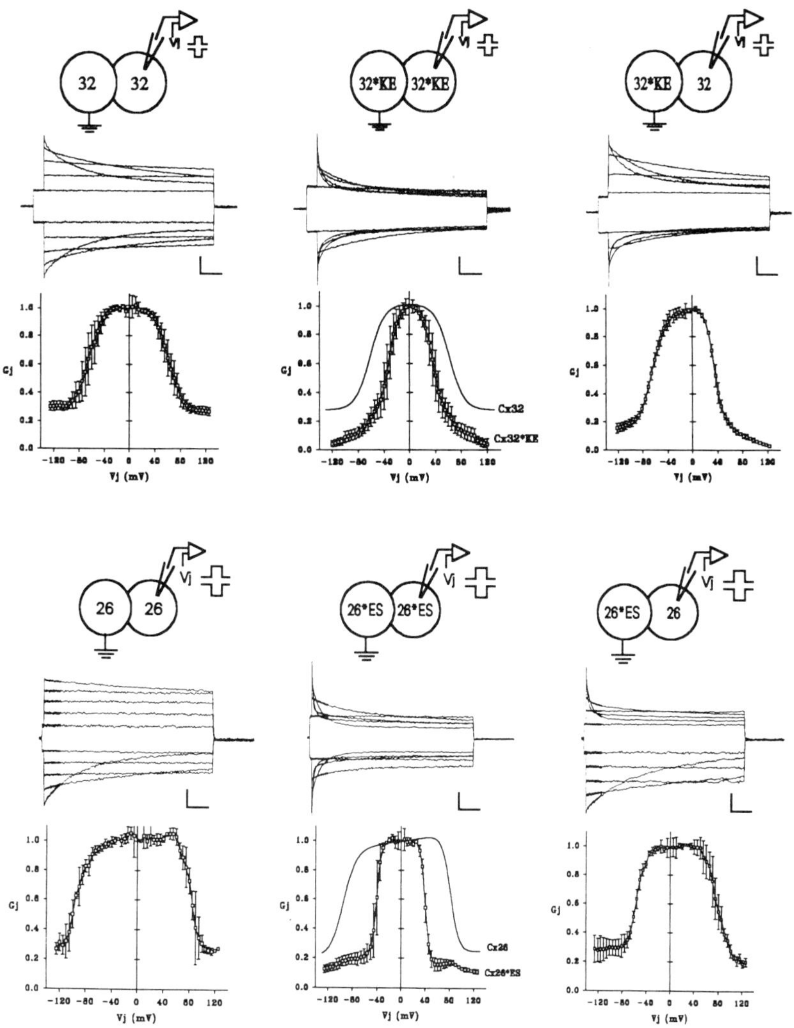

32
32
Vj
32*KE
32*KE
32*KE
32
Cx32
Cx32*KE
26
26
26*ES
26*ES
26*ES
26
Cx26
Cx26*ES
Gj
Vj (mV)

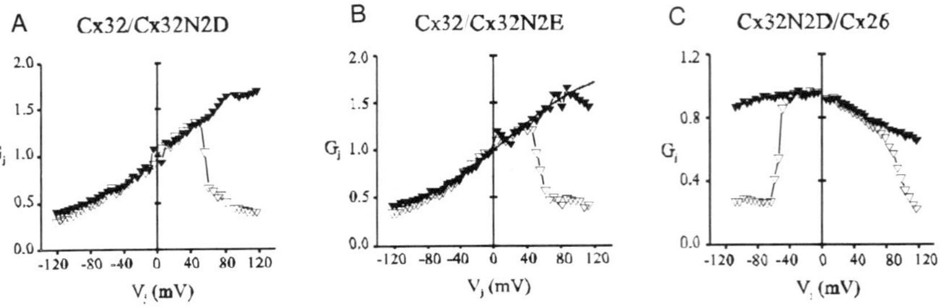

Figure 6. Mutations that reverse gating polarity of Cx32. **(A)** The g_j/V_j relation of Cx32/Cx32(N2D) junctions is like that of Cx32/Cx26 junctions (see Fig. 2), indicating that Cx32(N2D), like Cx26, has the opposite gating polarity from Cx32, i.e., Cx32(N2D) closes on relative positivity at its cytoplasmic end. **(B)** The g_j/V_j relation of Cx32/Cx32(N2E) junctions is also like that of Cx32/Cx26 junctions, indicating that the N2E mutation has the same effect as the N2D mutation. **(C)** The g_j/V_j relation of Cx32(N2E)/Cx26 junctions has slow processes for both polarities of V_j, confirming that Cx32(N2D) has the same gating polarity as Cx26.

Cx32(S42E), which changes charge in the opposite direction from wild type than Cx32*KE, reversed the gating polarity in junctions with Cx26. Moreover, the double mutation Cx32(N2D)*KE did not reverse polarity when paired with Cx26.

The foregoing data strongly indicate that charges at the beginning of NT and at E1 just after M1 are part of the gating charge. Thus, the NT-M1-E1 domains may move together in gating, carrying the gating charges towards the cytoplasm. These data present several problems for the generally proposed structure of the connexin molecule. Electrophysiological reasoning would place the gating charges near the wall of the channel at the cytoplasmic end, where they would be more sensitive to V_j and less sensitive to $V_{i\text{-}o}$. Moreover, independent gating by hemichannels would suggest that the gates are at the cytoplasmic end of the hemichannels, where they are less likely to interfere. It may be that NT is bent back around so that its charged end is near the channel opening. Introducing a negative charge to Cx32 at position 11 [Cx32(S11D)] did not reverse gating polarity (Verselis *et al.*, 1993). Either this position is out of the field or the domains do not move as a unit. Of the four membrane spanning regions, M3 is the most amphipathic and is a candidate to line the channel. The effect of charge changes at the M1-E1 border suggests that M3 may not line the channel, at least in this region, and that the gate is not a simple flap at the cytoplasmic mouth of the channel.

Whatever the conformational changes in connexin gating, the evidence for reversal of gating polarity is strong. The simplest interpretation is that the same conformational change occurs with gating of positive and negative polarity, and that change of a single charge can alter the sign of the net charge on the voltage sensor. Although the actual number of charges that move during gating can exceed the calculated value of gating charge (using the Boltzmann equation), the charges may not move through the entire V_j, however, the net charge moving in each connexin subunit must be small because a single charge change can reverse its sign. The calculated gating charges for Cx32 and Cx26 are about +2 and -4 (the sign is chosen on the assumption that the charge moves towards the cytoplasm). If all six subunits moved together, the charges per subunit are just about right. In the Na, Ca, K family of voltage gated channels, the S4 domain has 3 or 4 negative charges. Although mutations can change the effective gating charge, reversal has not been achieved. To make a functional channel, one might have to change not only the positively charged R's and K's, but also the counter ions that may be necessary to maintain the structural integrity of the molecule. If the connexin gating charges are in the channel lining, they may well be less important in maintaining structure; however, altering gating charges would be likely to affect permeability and we do see large shifts in V_0 with some mutations.

Polarity of V_j gating has been determined for a number of additional connexins (Fig. 4). In general, those with a negative charge at position 2 or 3 close on positivity, while those with neutral residues at this position close on negativity. Corresponding to the mutations in Cx26 noted above, the mutations Cx46(D3N) and Cx43(D3G) reverse polarity in junctions with Cx32 and Cx38 respectively (unpublished). In a few cases gating polarity changes with the pairing combination. For example, Cx38 has positive gating polarity in Cx43/Cx38 junctions but negative polarity in Cx46/Cx38 junctions.

Cx46 HEMICHANNELS CAN OPERATE WITHOUT AN APPOSED HEMICHANNEL.

It has generally been thought, with some experimental evidence, that formation of gap junctions occurs by association of preassembled hemichannels in apposed membranes and that before junction formation the hemichannels are closed. Certainly, closely coupled cells have junctional conductances that are high compared to their nonjunctional conductances, and junction formation is not associated with a large decrease in nonjunctional conductance. A further argument is that open hemichannels would provide a non-selective leak to the extracellular milieu. Thus, it was not surprising that, when expression of Cx46 was found to kill *Xenopus* oocytes, the mechanism appeared to be that it formed open hemichannels, i.e., hemichannels that opened by themselves without contacting another hemichannel in an apposed membrane (Paul *et al.*, 1991). The cells were sufficiently viable at low levels of expression to permit determination of the macroscopic gating properties of the Cx46 hemichannels (Ebihara and Steiner, 1993). The oocytes show large slowly-activating currents upon depolarization from a holding potential of -70 mV. In K^+ salt solutions with no added Ca^{2+}, the currents reverse near 0 mV. With larger inside positive voltages (e.g. +65 mV) and relatively low levels of expression, currents show a peak followed by relaxation to a lower value (Trexler *et al.*, 1996). Thus, increasing inside positivity appears to open and then close the channels. (With higher levels of expression, the currents no longer show relaxation to a lower level at higher voltages, which is attributable to a higher ratio of series access resistance to junctional resistance and reduced voltage drop across the junctions (Wilders and Jongsma, 1992)).

Patch recordings from oocytes expressing Cx46 show large conductance channels (~300 pS, Fig. 7) (Trexler *et al.*, 1996). These channels are not seen in oocytes injected only with DNA antisense to the endogenous oocyte connexin, *Xen*Cx38, or this antisense together with Cx32 RNA. Furthermore, currents from multichannel patches from Cx46 expressing oocytes display properties consistent with those observed macroscopically, i.e., slowly activating currents that reverse near 0 mV and that peak and relax to lower values at higher inside positive voltages.

The single Cx46 hemichannels are gated by voltage (Fig. 7A shows current as a function of time at different voltages, all points histograms are at the right of each record). At large inside negative voltages, the hemichannel remains predominantly in γ_{closed}, with occasional openings to γ_{open}. There are frequent lower amplitude noisy bursts that are not resolvable as discrete events and may be incomplete hemichannel openings. (Another possibility would be fast and unresolved complete openings, but substates between γ_{closed} and γ_{open} are often resolvable, see next paragraph). With depolarization, dwell times in γ_{open} markedly increase. From γ_{open} there are brief sojourns to several substates including γ_{sub}, a conductance state about 1/3 that of γ_{open}. At voltages near zero of either polarity, the hemichannel remains predominantly in γ_{open} with continued brief sojourns to substates and γ_{closed}. At larger inside positive voltages, dwell times in γ_{sub} become strikingly longer.

The complex gating behavior of Cx46 hemichannels clearly evident in I-V curves obtained by applying slow voltage ramps between -70 mV and +70 mV to patches containing a single hemichannel (Fig. 7B). The negative quadrant is dominated by transitions between γ_{closed} and γ_{open}, whereas the positive quadrant exhibits transitions between γ_{open} and γ_{sub}. Residence in γ_{open}

dominates at intermediate voltages. γ_{open} shows marked rectification; the chord conductance changes as a smooth monotonic function over the entire voltage range and in Fig. 7B is 300 pS at -50 mV and 135 pS at +50 mV. The rectification in symmetric salts as shown by Cx46 hemichannels is also observed in Cx32/Cx26 channels (Bukauskas *et al.*, 1995b). Our data indicate that rectification can be a hemichannel property. Thus, opposing identical the hemichannels in series to form homotypic channels gives rise to symmetry about $V_j=0$, and possibly linearity in the single channel I/V relation, whereas opposing different hemichannels to form heterotypic channels gives rise to various degrees of rectification depending on the intrinsic properties of the hemichannels (in preparation).

Gating at inside positive and negative voltages is not only to different conductance levels, i.e. γ_{closed} and γ_{sub}, respectively, but the transitions to those levels have markedly different kinetics (Fig. 7C). Transitions between γ_{open} and γ_{closed} are slow and usually involve several transient subconductance states distinct from γ_{sub} (Fig. 7C, epoch b). In some cases, these transitions appear to be continuous, lasting 10-20 ms (Fig. 7C, epoch e). The apparently slow transitions between γ_{closed} and γ_{open} are likely to result from unresolved rapid gating among the identified transient substates and, perhaps, other unresolved substates. Thus, Cx46 gating may actually be like that in most other ion channels, where transition times are unresolved. Transitions between γ_{open} and γ_{sub} are rapid and do not appear to involve any other substates (Fig. 7A, +35mV).

We attribute the voltage-dependent behavior of Cx46 hemichannels to the actions of two distinct voltage gates as diagrammed in Fig. 8. As noted above hemichannels in general are believed to be closed when not apposed to another hemichannel, which would prevent loss of cytoplasmic solutes and entry of extracellular ions. After contact between hemichannels, i.e., docking, the two hemichannels open to create a continuous channel connecting the cytoplasms of the two apposed cells. This process can be viewed as comparable to ligand-receptor binding, with hemichannels acting in both roles. When formed, the cell-cell channel can gate in response to V_j by closing either hemichannel depending on the polarity of V_j. In the case of Cx46, moderate depolarization opens hemichannels that are not contacting an apposed hemichannel. The openings are characterized by transitions between γ_{closed} and γ_{open} involving multiple, short-lived substates. We ascribe this gating mechanism to the extracellular loops of the connexin subunits and term it "loop gating" (Fig. 8, inset). The transitions of loop gating resemble "docking" currents observed during formation of cell-cell channels (Bukauskas *et al.*, 1995a). Activation by depolarization is consistent with the effect of moderate depolarization on macroscopic currents in *Xenopus* oocytes expressing Cx46. At large inside positive voltages, when the loop gate is open, the hemichannel closes by a gating mechanism characterized by transitions between γ_{open} and γ_{sub} (Fig. 7A, +35 mV); γ_{sub} accounts for the residual conductance, g_{min}. V_j gating to a non-zero substate has been observed in other connexins as well (Bukauskas *et al.*, 1995a; Moreno *et al.*, 1994; Perez-Armendariz *et al.*, 1994); however, there is at least one connexin whose junctions have a voltage induced state with γ not significantly

Figure 7. Cell-attached patch recordings from single Cx46 hemichannels in *Xenopus* oocytes. **(A)** Representative 5 sec segments of 60 sec recordings of single channel activity at positive and negative voltages. The channel assumes three principal conductance states: γ_{open}, the fully open state; γ_{sub}, a subconductance state approximately 1/3 the magnitude of γ_{open}; and γ_{closed}, a fully closed state. All-points amplitude histograms of the 60 sec recordings are shown to the right of each trace. Arrowheads indicate the histogram peak assigned to γ_{open}. At -55 mV the hemichannel spends most of the time in γ_{closed}, with occasional openings to γ_{open}. With depolarization to -35 mV, openings to γ_{open} increase in frequency and duration. At smaller voltages of either polarity, e.g., −15 mV, the hemichannel remains predominantly in γ_{open}, with occasional closures to either γ_{closed} or γ_{sub}. At more positive voltages the channel exhibits longer residence times in γ_{sub}, and at +35mV, closures to γ_{closed} are rare. **(B)** Currents obtained with 16 sec voltage ramps from -70 mV to +70 mV applied to an on-cell patch containing a single hemichannel from a different cell than in **(A)** clearly illustrate the differential distribution of conductance states with voltage. Shown are three ramps applied in succession. At large negative and positive voltages the hemichannel closes to γ_{closed} and to γ_{sub}, respectively, and remains predominantly in γ_{open} at intermediate voltages of either polarity. The conductance level of γ_{closed} corresponds to the leak conductance obtained by applying voltage ramps to the same patch after channel activity spontaneously disappeared (current through the leak conductance is superimposed on the three ramps). Leak conductances of patches containing no active hemichannels were of similar magnitude. **(C)** Examples of the complex gating transitions between γ_{open} and γ_{closed}. On the left are shown three consecutive 1 sec traces of an on-cell patch recording (V = -45 mV). Epochs in γ_{open} are designated by solid bars labeled a, b and c. At the onset and end of open epochs, the hemichannel transitions between γ_{closed} and γ_{open} often involve at least two identifiable transient substates distinct from γ_{sub}. Epoch "b" shown in an expanded time scale exhibits an initial opening to the first identifiable transient substate (S1) lasting approximately 15 msec, followed by an opening to the second transient substate (S2), before finally reaching γ_{open}. Within this open epoch a sojourn to γ_{sub} is evident, along with brief sojourns to S2 (asterisks). At the end of this epoch, closure from γ_{open} occurs rapidly to S1, then slowly decays to γ_{closed} over approximately 20 ms. Periods in γ_{closed} are interrupted by a number of short noisy bursts (arrows) that may represent incomplete openings. Illustrated on the right side is another epoch in γ_{open} from the same patch at the same voltage. Segments representing the initial opening (d) and final closing (e) are shown below at an expanded time scale. Numbered arrows mark specific events. An incomplete opening [*1*] falls back to S1, then transits to S2 [*2*] before fully opening [*3*]. Closure from γ_{open} appears slow and continuous [*4*] and may be due to fast unresolved transitions among the identified transient substates and/or other unresolved transient substates; with a -3dB cutoff at 1kHz, the transitions would have to be <400 μsec to go unresolved. A short, noisy burst [*5*] yields to γ_{closed}. From Trexler *et al.*(1996).

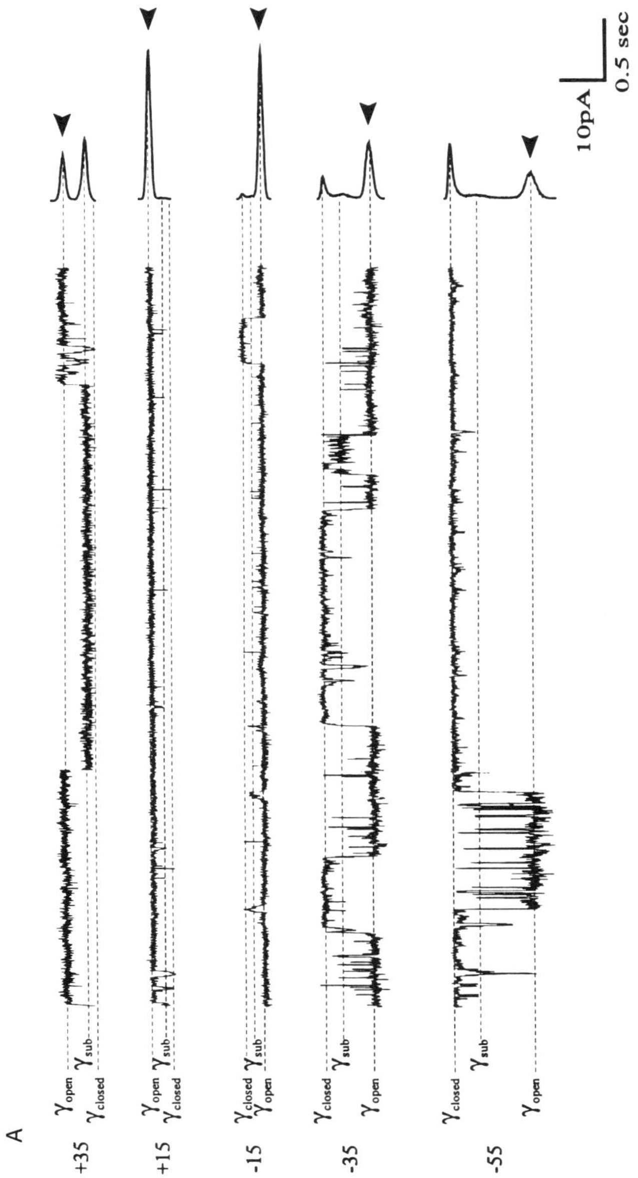
A
+35
+15
-15
-35
-55
γ_{open}
γ_{sub}
γ_{closed}
10pA
0.5 sec

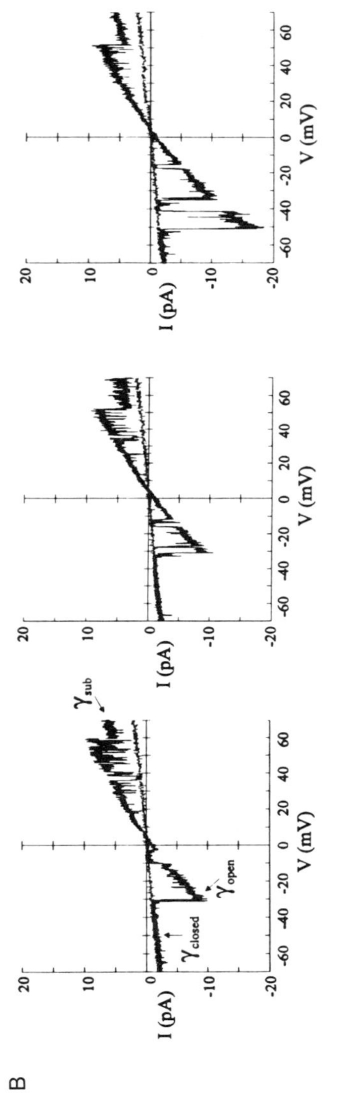

Figure 7. Continued.

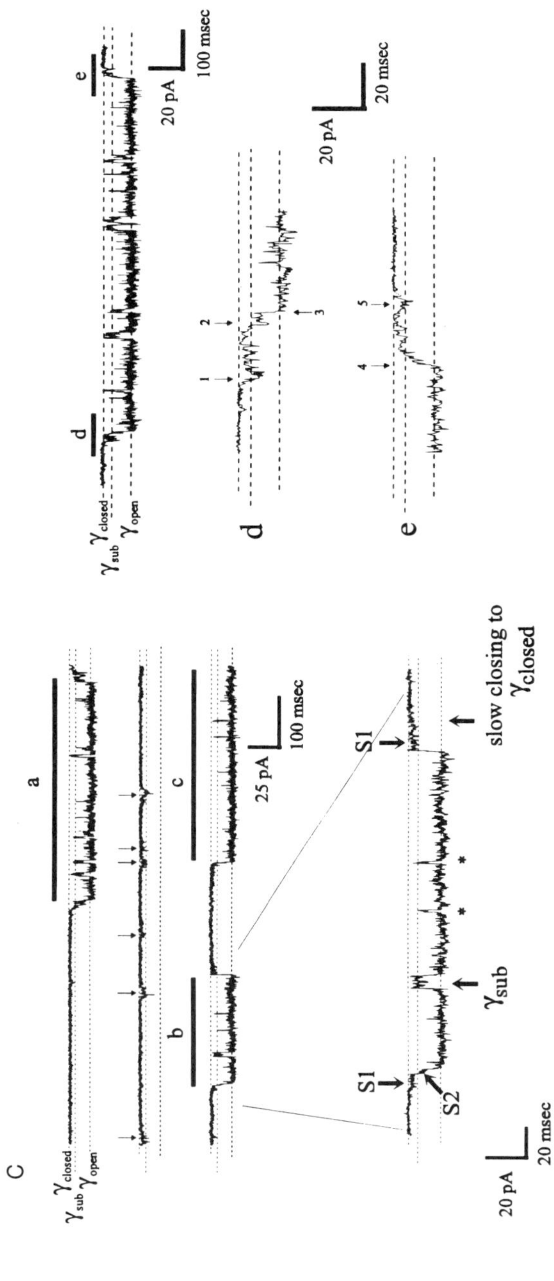

Figure 7. Continued.

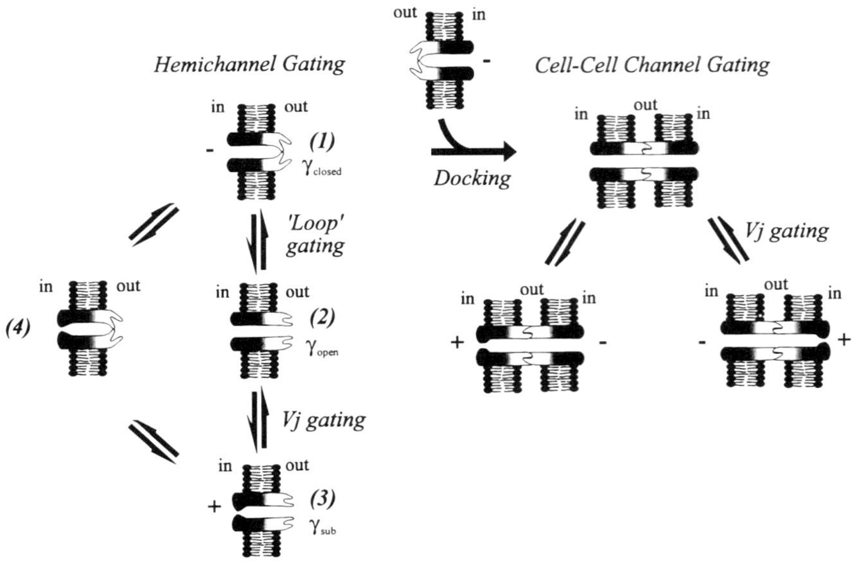

Figure 8. Schematic diagram of voltage gating of hemichannels when unapposed and when incorporated into cell-cell channels. Hemichannels in the surface membrane of cells are generally closed and require docking with another hemichannel to open. In cell-cell channels each hemichannel contains a separate V_j sensor/gate that can close the channel. Cx46 hemichannels are closed by V_j that is relatively positive on their cytoplasmic side (+ and - in the diagram indicate voltage of one cell relative to the other). Since V_j gating shows little sensitivity to $V_{i\text{-}o}$ and amino acid residues located at the beginning of the NT of Cx26 and Cx32 are integral components of the V_j sensor, the V_j gates is diagrammed at the cytoplasmic end of each hemichannel. For Cx46 hemichannels unapposed by another hemichannel, the closed state is favored at inside negative voltages, state *(1)*, but moderate depolarization tends to open the hemichannels, state *(2)*. We ascribe this form of voltage gating to the extracellular loops. Further polarization to inside positive voltages closes the V_j gate of the hemichannel, state *(3)*. Thus, the loop and V_j gates are in series and have opposite polarities of closure, inside negative and positive, respectively. The large difference in the voltage at which these two mechanisms operate, together with long dwell times in the open state near 0 mV (Fig. 7), indicates that simultaneous closure of both gates would seldom occur, state *(4)*. From Trexler *et al.* (1996).

different from zero (Chanson *et al.*, 1993). Closure of the Cx46 hemichannel to γ_{sub} is consistent with the presence of a g_{min} in V_j gating of Cx46/Cx46 channels (although the values are not quantitatively in agreement with simple series connection of the hemichannels). Closing by cytoplasmic positivity is indicated by the g_j/V_j relations of Cx46/Cx32 and Cx46/Cx26 heterotypic junctions (White *et al.*, 1994); the former is asymmetric like that of Cx26/Cx32 heterotypic junctions; the latter is more nearly symmetric like that of Cx26/Cx26*ES junctions (Fig. 5). These data indicate that the polarity of V_j

gating is the same in isolated hemichannels and junctional channels. White et al. (1994) suggested that Cx46 had opposite gating polarity in junctional channels and isolated hemichannels; our data indicate that the opposite polarity in single Cx46 expressing *Xenopus* oocytes is not a change in the V_j gate but a result of the second, loop gating mechanism.

Cx46 hemichannels can be activated by moderate depolarization, which may be important physiologically. Thus far, Cx46 has been shown to be expressed in the lens (Jiang *et al.*, 1993) and, at the mRNA level, in proliferating Schwann cells (Chandross *et al.*, 1995), but there are no data on the conductance properties of either junctions or isolated hemichannels. Cx56, an avian homolog of Cx46, also forms hemichannels that function in isolation (Ebihara *et al.*, 1995). Teleost and elasmobranch retinal horizontal cells appear to have gap junction-like hemichannels that can be opened in low Ca^{2+} solution by depolarization or in normal saline by pharmacological agents, although the putative connexin(s) involved are unidentified (DeVries and Schwartz, 1992; Malchow *et al.*, 1993).

The biophysical study of gap junctions has been hampered by the difficulty of patch clamping from the inside of cells. Considerable improvement was obtained when single channel activity was recorded by patch clamping of small, low conductance cells (Neyton and Trautmann, 1985), but frequency response, noise level and ease of solution exchange remain suboptimal. The applicability of hemichannel recording to connexins other than Cx46 and Cx56 remains to be determined. Replacing the extracellular domains of other connexins with those of Cx46 (Rubin *et al.*, 1992a) may convert them into functional hemichannels, which could greatly facilitate determination of the characteristics of the unmodified part of the molecule. Although it is necessary to confirm that characteristics of V_j gating in isolated hemichannels are the same in junctional channels, our experience with Cx46 suggests that the properties are likely to be very similar in the two situations.

PHYSIOLOGICAL SIGNIFICANCE OF VOLTAGE DEPENDENCE

Fast rectifying synapses have an obvious function in escape systems, and are found in vertebrates as well as arthropods (Auerbach and Bennett, 1969; Furshpan and Potter, 1959). We can now suggest that these rectifying synapses are asymmetric in their component subunits, and exhibit fast rectification similar to that seen in Cx32/Cx26 junctions. Symmetrical V_j dependence and slow asymmmetrical V_j dependence have yet to be shown to have any physiological role. The large V_j values required to modulate g_j in a number of junctions suggest that there is no physiological closure, which could be important in such tissues as heart where long lasting action potentials cross the junctions. Cx38 in amphibian embryos is much more voltage dependent, and moderate differences in resting potential could lead to formation of boundaries between communication compartments (Harris *et al.*, 1983). Cx37 is also quite voltage sensitive (Reed *et al.*, 1993) and is found in neural progenitor cells (R. Rozental, personal communication) as well as skin (Goliger and Paul, 1994). This sensitivity might give rise to functional modulation of transmission at electrical synapses. Furthermore, heterotypic junctions with more and less voltage sensitive hemichannels provide a form of rectification in addition to that of hemichannels with opposite gating polarity.

An additional consideration is connexin compatibility. Some connexins are compatible in that they form functional heterotypic junctions, which may be quite symmetrical or very rectifying. Other connexins are incompatible and will not form functional junctions and may even interfere with junction formation by otherwise compatible connexins (Chang *et al.*, 1996; (White *et al.*, 1995b). To be sure, formation of junctions depends on cell adhesion as well as the connexins involved (Mege *et al.*, 1988). Junctional properties are quite variable and are usually, but not always, predictable from the properties of their constitutive connexins in other combinations. Single channel recording is illuminating aspects of the gating mechanism; many but not all of these aspects would have been predicted from macroscopic properties of the junctions.

REFERENCES

Auerbach, A. A., and Bennett, M. V. (1969). A rectifying electrotonic synapse in the central nervous system of a vertebrate. *J. Gen. Physiol.* **53,** 211-37.

Barnes, T. M. (1994). OPUS: a growing family of gap junction proteins? [letter]. *Trends Genet.* **10,** 303-305.

Barrio, L. C., Suchyna, T., Bargiello, T., Xu, L. X., Roginski, R. S., Bennett, M. V., and Nicholson, B. J. (1992). Gap junctions formed by connexins 26 and 32 alone and in combination are differently affected by applied voltage. *Proc.Nat.Acad.Sci.USA* **89,** 4220.

Bennett, M. V. L., Barrio, L. C., Bargiello, T. A., Spray, D. C., Hertzberg, E., and Saez, J. C. (1991). Gap junctions: new tools, new answers, new questions. *Neuron* **6,** 305-20.

Bennett, M. V. L., Rubin, J. B., Bargiello, T. A., and Verselis, V. K. (1993). Structure-function studies of voltage sensitivity of connexins, the family of gap junction forming proteins. *Jap. J. Physiol.* **43 Suppl 1,** S301-S310.

Bennett, M. V., Zheng, X., and Sogin, M. L. (1994). The connexins and their family tree. *Soc. Gen.Physiol. Series* **49,** 223-33.

Bennett, M. V. L., and Spray, D. C. (1987). Intercellular communication mediated by gap junctions can be controlled in many ways. *In* "Synaptic Function" (G. M. Edelman, W. E. Gall, and W. M. Cowan, Eds.), pp. 109-135. John Wiley, New York.

Bennett, M. V. L., Verselis, V., White, R. L., and Spray, D. C. (1988). Gap junctional conductance: Gating. *In* "Gap Junctions" (E. L. Hertzberg and R. G. Johnson, Eds.), pp. 287-304. Alan R. Liss,Inc., New York.

Bukauskas, F. F., Elfgang, C., Willecke, K., and Weingart, R. (1995a). Biophysical properties of gap junction channels formed by mouse connexin40 in induced pairs of transfected human HeLa cells. *Biophys. J.* **68,** 2289-2298.

Bukauskas, F. F., Elfgang, C., Willecke, K., and Weingart, R. (1995b). Heterotypic gap junction channels (connexin26-connexin32) violate the paradigm of unitary conductance. *Pflugers Arch.* **429,** 870-872.

Bukauskas, F. F., and Weingart, R. (1994). Voltage-dependent gating of single gap junction channels in an insect cell line. *Biophys. J.* **67,** 613-625.

Chandross, K. J., Dermietzel, R., Simnburger, E., Spray, D. C., and Kessler, J. A. (1995). Changes in gap junction expression after peripheral nerve injury. *Soc. Neurosci. Abstr.* **21,** 814.

Chang, M., Werner, R., and Dahl, G. (1996). A role for an inhibitory connexin in testis? *Devel. Biol.* **175,** 50-56.

Chanson, M., Chandross, K. J., Rook, M. B., Kessler, J. A., and Spray, D. C. (1993). Gating characteristics of a steeply voltage-dependent gap junction channel in rat Schwann cells. *J. Gen. Physiol.* **102,** 925-946.

DeVries, S. H., and Schwartz, E. A. (1992). Hemi-gap-junction channels in solitary horizontal cells of the catfish retina. *J. Physiol.* **445,** 201-230.

Ebihara, L., Berthoud, V. M., and Beyer, E. C. (1995). Distinct behavior of connexin56 and connexin46 gap junctional channels can be predicted from the behavior of their hemi-gap- junctional channels. *Biophys. J.* **68,** 1796-1803.

Ebihara, L., and Steiner, E. (1993). Properties of a nonjunctional current expressed from a rat connexin46 cDNA in Xenopus oocytes. *J. Gen. Physiol.* **102,** 59-74.

Furshpan, E. J., and Potter, D. D. (1959). Transmission at the giant motor synapses of the crayfish. *J. Physiol.* **145,** 289-325.

Goliger, J. A., and Paul, D. L. (1994). Expression of gap junction proteins Cx26, Cx31.1, Cx37, and Cx43 in developing and mature rat epidermis. *Devel. Dynamics* **200,** 1-13.

Harris, A. L., Spray, D. C., and Bennett, M. V. (1981). Kinetic properties of a voltage-dependent junctional conductance. *J. Gen. Physiol.* **77,** 95-117.

Harris, A. L., Spray, D. C., and Bennett, M. V. (1983). Control of intercellular communication by voltage dependence of gap junctional conductance. *J. Neurosci.* **3,** 79-100.

Jiang, J. X., Paul, D. L., and Goodenough, D. A. (1993). Posttranslational phosphorylation of lens fiber connexin46: a slow occurrence. *Invest. Ophthalmol. Vis. Sci.* **34,** 3558-3565.

Malchow, R. P., Qian, H., and Ripps, H. (1993). Evidence for hemi-gap junctional channels in isolated horizontal cells of the skate retina. *J. Neurosci. Res.* **35,** 237-245.

Mege, R. M., Matsuzaki, F., Gallin, W. J., Goldberg, J. I., Cunningham, B. A., and Edelman, G. M. (1988). Construction of epithelioid sheets by transfection of mouse sarcoma cells with cDNAs for chicken cell adhesion molecules. *Proc. Nat. Acad. Sci. USA* **85,** 7274-7278.

Moreno, A. P., Rook, M. B., Fishman, G. I., and Spray, D. C. (1994). Gap junction channels: distinct voltage-sensitive and - insensitive conductance states. *Biophys. J.* **67,** 113-119.

Neyton, J., and Trautmann, A. (1985). Single-channel currents of an intercellular junction. *Nature* **317,** 331-335.

Paul, D. L., Ebihara, L., Takemoto, L. J., Swenson, K. I., and Goodenough, D. A. (1991). Connexin46, a novel lens gap junction protein, induces voltage-gated currents in nonjunctional plasma membrane of Xenopus oocytes. *J. Cell Biol.* **115,** 1077-1089.

Perez-Armendariz, E. M., Romano, M. C., Luna, J., Miranda, C., Bennett, M. V., and Moreno, A. P. (1994). Characterization of gap junctions between pairs of Leydig cells from mouse testis. *Amer. J. Physiol.* **267,** C570-80.

Reed, K. E., Westphale, E. M., Larson, D. M., Wang, H. Z., Veenstra, R. D., and Beyer, E. C. (1993). Molecular cloning and functional expression of human connexin37, an endothelial cell gap junction protein. *J. Clin. Invest.* **91,** 997-1004.

Rubin, J. B., Verselis, V. K., Bennett, M. V., and Bargiello, T. A. (1992a). A domain substitution procedure and its use to analyze voltage dependence of homotypic gap junctions formed by connexins 26 and 32. *Proc. Nat. Acad. Sci. USA* **89,** 3820-4.

Rubin, J. B., Verselis, V. K., Bennett, M. V., and Bargiello, T. A. (1992b). Molecular analysis of voltage dependence of heterotypic gap junctions formed by connexins 26 and 32. *Biophys. J.* **62,** 183-93; discussion 193-5.

Spray, D. C., Harris, A. L., and Bennett, M. V. (1981). Equilibrium properties of a voltage-dependent junctional conductance. *J. Gen. Physiol.* **77,** 77-93.

Swenson, K. I., Jordan, J. R., Beyer, E. C., and Paul, D. L. (1989). Formation of gap junctions by expression of connexins in Xenopus oocyte pairs. *Cell* **57,** 145-155.

Trexler, E. B., Bennett, M. V., Bargiello, T. A., and Verselis, V. K. (1996). Voltage gating and permeation in a gap junction hemichannel. *Proc. Nat. Acad. Sci. USA* **93,** 5836-41.

Verselis, V. K., Bargiello, T. A., Rubin, J. B., and Bennett, M. V. L. (93). Comparison of voltage dependent properties of gap junctions in hepatocytes and in *Xenopus* oocytes expressing Cx32 and Cx26. *In* "Progress in Cell Research" (J. E. Hall, G. A. Zampighi, and R. M. Davies, Eds.), Vol. 3, pp. 105-112. Elsevier Science.

Verselis, V. K., Bennett, M. V., and Bargiello, T. A. (1991). A voltage-dependent gap junction in *Drosophila melanogaster. Biophys. J.* **59,** 114-26.

Verselis, V. K., Ginter, C. S., and Bargiello, T. A. (1994). Opposite voltage gating polarities of two closely related connexins. *Nature* **368,** 348-351.

White, T. W., Bruzzone, R., and Paul, D. L. (1995). The connexin family of intercellular channel forming proteins. *Kidney Int.* **48,** 1148-1157.

White, T. W., Bruzzone, R., Wolfram, S., Paul, D. L., and Goodenough, D. A. (1994). Selective interactions among the multiple connexin proteins expressed in the vertebrate lens: the second extracellular domain is a determinant of compatibility between connexins. *J. Cell Biol.* **125,** 879-892.

White, T. W., Paul, D. L., Goodenough, D. A., and Bruzzone, R. (1995b). Functional analysis of selective interactions among rodent connexins. *Mol. Biol. Cell* **6,** 459-470.

Wilders, R., and Jongsma, H. J. (1992). Limitations of the dual voltage clamp method in assaying conductance and kinetics of gap junction channels. *Biophys. J.* **63,** 942-953.

GAP JUNCTIONS BETWEEN LEUKOCYTE AND ENDOTHELIUM: EXPRESSION OF CONNEXIN 43 IN ADHERENT OR ACTIVATED LEUKOCYTES

Mauricio P. Boric, Alejandro Roth, Patricio Jara,
and Juan C. Sáez
Departamento de Ciencias Fisiológicas,
Facultad de Ciencias Biológicas
Pontificia Universidad Católica de Chile
Casilla 114-D, Santiago, Chile

INTRODUCTION

In this chapter we will discuss the possibility that intercellular communication mediated by gap junctions takes place between adhering leukocytes and endothelial cells during the inflammatory process.

Leukocyte activation occurs at sites of acute inflammation to provide a successful defense response after tissue injury or against pathogenic agents. However, since activated leukocytes can be lethal for pathogens as well as for healthy tissue, the activation process is carefully regulated by the organism. Our knowledge of the molecular signals that regulate the interaction between leukocytes and endothelial cells has growth enormously in the last decade. Leukocyte margination, adhesion, and its posterior migration through the vascular wall are sequential processes mediated by specific surface molecules that act as cellular receptors. These molecules are expressed and/or exposed in the endothelium and leukocytes at the right place and moment.

Leukocyte Rolling.

Under normal conditions leukocytes have little contact with the endothelium because of hydrodynamic forces that tend to keep them flowing. Due to

From Ion Channels to Cell-to-Cell Conversations
Edited by Latorre and Sáez, Plenum Press, New York, 1997

trauma, or inflammatory mediators, leukocytes begin to roll over the internal surface of postcapillary and larger venules. This phenomenon has been characterized by intravital microscopic observations in a variety of tissues[1-3]. Leukocyte preference for venules may be favored by the low shear rate and low shear stress conditions present in this vascular segment; however the major factor that determines leukocyte rolling is the expression of pro-adherent molecules in the endothelial cells as well as in the surface of leukocytes since at equal shear rates leukocytes do not roll nor adhere to arteriolar walls[4,5]. Leukocyte rolling is mediated by carbohydrate-binding glycoproteins of the selectin family[6]. There are three types of selectins: L-selectin, present in leukocytes, formerly known as LAM-1 (leukocyte adhesion molecule; or LECCAM-1 Lectin, Epidermal growth factor, Complement regulatory unit, Cell Adhesion Molecule; or LHR, Lymphocyte Homing Receptor); E-selectin from endothelium, also known as ELAM-1 (Endothelium Leukocyte Adhesion Molecule), or LECCAM-2; and P-selectin, from platelets, also known as PADGEM, LECCAM-3 or GMP-140. The selectin macromolecules have a lectin domain that allows them to bind other cell surface glycoproteins, and several complement regulatory sequence repeats. While L-selectin is normally expressed by leukocytes, the E and P forms are expressed by activated endothelial cells. Moreover, expression of L-selectin by leukocytes increases transiently during the inflammatory process, in concomitance with increased leukocyte rolling.

Leukocyte Adhesion.

Leukocyte attachment to the endothelial surface (cellular adhesion, or sticking) depends on the translocation of integrins across the leukocyte plasma membrane. Integrins are transmembrane glycoproteins and in leukocytes the most important are the CD11/CD18 complexes[5,7]. They are heterodimers of a β_2 (CD18) subunit and one of three α subunits that are expressed by myeloid and lymphoid cells. CD11b/CD18 or $\alpha_L\beta_2$, known as lymphocyte function antigen (LFA-1), is found in all kinds of immune cells; CD11a/CD18 or $\alpha_M\beta_2$ was first identified in macrophages (MAC-1) and is found mostly in neutrophils and myeloid cells, but is also expressed in some lymphocytes; the same distribution is found for CD11c/CD18 or $\alpha_X\beta_2$ (p150,95)[8]. These β_2 integrins are stored in intracellular granules and rapidly exposed on the cell surface in response to chemical attractants[9]. *In vitro*, leukocyte up-regulation of integrins is time associated with the proteolytic shedding of L-selectin from the cell surface[10]. In this way, leukocytes change their surface from a "rolling" to a "sticking" phenotype within minutes. Endothelial cells, on the other hand, expose proteins that belong to the immunoglobulin gene family, known as

Inter Cellular Adhesion Molecules, or ICAMs, which act as counter receptors for CD11/CD18 integrins[11,12]. ICAMs are translocated from endothelial granules (Waibel-Palade bodies) during inflammation.

Functional studies using specific antibodies have helped to understand the role of each adhesion molecule. Leukocytes sticking to venular endothelium are released to free-flowing motion shortly after intravenous injection of anti CD18/CD11 antibodies[7], whereas leukocyte rolling is not affected by these antibodies. In contrast, anti-selectin antibodies dramatically reduce the number of rolling cells[2,13]. Based on these and other evidence, a current two-step model for leukocyte adhesion has been proposed[13-15] (Figure 1).

In the first stage, leukocyte rolling is achieved by interactions between L and E selectins and their respective ligands. These interacting forces act as a tether to reduce the speed of white cells and to allow leukocytes to get in contact with other signal molecules presented by the endothelium. The second stage involves the expression/exposure of integrins by leukocytes, accompanied by L-selectin shedding, and exposure of endothelial integrin receptors. This process is triggered by chemoattractant molecules "presented" by the endothelium to the rolling leukocyte at sites of inflammation[16]. Subsequent binding between complementary proteins, i.e., the leukocyte integrins to endothelium adhesion molecules (ICAM-1 or ELAM-1), produce leukocyte adherence that becomes progressively stronger and irreversible (sticking).

The proposed model can answer some of the questions that puzzled investigators for years, such as: how are leukocytes activated during their fast passage through the inflamed zone?, and why leukocytes do not adhere to other segments of the vasculature once activated? The earlier concept of a "chemoattractant gradient" leading the leukocytes towards the affected area by amoeboid motion is well-suited for a culture plate, but not in the context of fast flowing capillaries or venules. In contrast, the new idea considers the chemoattractant as fixed molecular signals that, after being released by the injured tissue (including endothelium), are trapped by proteoglycans in the surface of neighboring endothelial cells, where they can interact with the rolling leukocytes. Among the molecules that play a pivotal role for leukocyte activation and adhesion are cytokines, such as interleukins (IL-1, IL-2 and IL-8) and the tumoral necrosis factor (TNF)[17]. These molecules are able to initiate up-regulation of leukocyte integrins and, in addition, they are released in great amounts by activated leukocytes, providing with a mechanism for a rapidly expanding activation signal that can recruit more granulocytes or other defense cells to the inflamed area. In this scenario, the platelet activating factor (PAF) is recognized as a juxtacrine mediator that is retained and presented to rolling leukocytes by the endothelium[18,19].

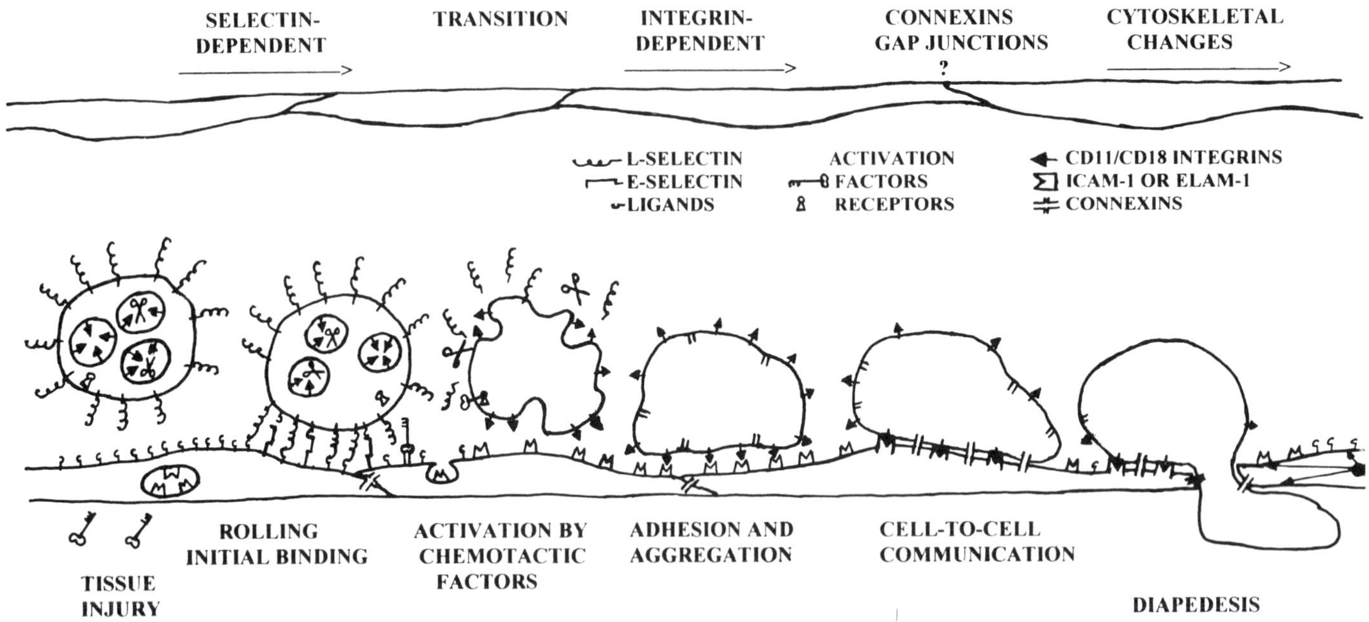

Figure 1. Proposed model for Leukocyte-Endothelium Interactions showing cell contact proteins involved in each step. (Modified from Argenbright, Ref. 15). See text for details.

Leukocyte Transmigration.

The initial adherence of leukocytes comprises the first 5 to 30 minutes of the inflammation process. Adhesion is usually followed by diapedesis, i. e., leukocyte crossing of the venular wall towards the interstitial space. Ultrastructural studies have shown that leukocytes cross through intercellular endothelial junctions squeezing between adjacent endothelial cell processes[20,21]. Furthermore, during diapedesis a tight junction contact is maintained between the leukocyte and the endothelium, thus preventing discontinuity of the endothelial lining. As diapedesis is completed, the endothelial cells extend their lateral processes, resealing the barrier. Thereafter, leukocytes leave the vascular wall digesting the basement membrane by action of proteolytic enzymes, at a site distant from the crossing point, a phenomenon that also helps to maintain endothelial integrity. In fact, functional studies have shown that leukocyte transmigration through endothelial cell monolayers in culture, or venules *in vivo*, does not alter endothelial permeability to macromolecules[20-22]. Thus, the typical increase in vascular permeability to macromolecules observed in inflammation is not due to leukocyte diapedesis. Increased permeability is a well studied phenomenon involving receptor-mediated endothelial cell contraction that produces small disruptions of inter-endothelial junctions in post-capillary venules with a time course faster than leukocyte diapedesis[23-25]. Inter-endothelial gaps are formed transiently, in response to inflammatory mediators liberated by the damaged tissue, such as bradykinin, histamine, or thrombin. These mediators generate a rise in cytosolic Ca^{++} which promotes actin depolymerization and cell contraction, allowing the filtration of plasma into the interstitium. *In vivo* it is often difficult to differentiate between leukocyte activation and permeability changes since both are part of an orchestrated cascade of events that may be triggered from different starting points. For instance, activated leukocytes contribute to the increased permeability by releasing pro-inflammatory agents[21,26,27].

Diapedesis involves important changes in cytoskeletal organization of both cell types; the endothelium contracts to allow for the leukocyte passage, and the white cell elongates and squeezes through the inter-endothelial opening. These changes are coordinated and require cell-to-cell communication. The group lead by D. Shepro has studied how the permeability of cultured endothelial cell monolayers is correlated, or associated with the arrangement of actin filaments within endothelial cells[28-31]. These authors postulate that chemical signals that activate endothelial adenylyl cyclase (increasing cAMP) induce a state of cytoskeletal stability, associated with an enhanced barrier function (lower permeability). This state is characterized by stress fibers attached to cell adhesion points to the matrix substrate or to other cells, and organized in a dense peripheral band. Conversely, agents that mobilize intracellular Ca^{++}, through activation of PLC (and IP_3 and DAG) produce cytoskeleton instability through PKC-mediated phosphorylation of myosin light chain and subsequent increased actin-myosin interaction leading to higher cell motility, cell-cell detachment and/or cell contraction and diapedesis.

Intercellular Signaling.

Integrins serve as extracellular anchorage points to the substrate by binding to interstitial matrix proteins such as collagen, elastin and fibronectin. In addition, integrins act as intracellular anchor points for cytoskeletal structures (actin microfilaments). Therefore, it is likely that the coordinated shape changes occurring in endothelium and leukocytes during diapedesis may be triggered or mediated by the binding of matching adhesion molecules in both cell types. Nevertheless, in addition to the above mentioned hypothesis, leukocytes and endothelium may communicate to each other by direct cell-to-cell cytoplasmatic contacts by forming gap junctions.

Formation of most cell junctions, including gap junctions, tight junctions and those involved in cell adhesion, such as spot desmosomes and adherent junctions, depends on an intimate plasma membrane interaction between adjacent cells mediated by cell adhesion molecules[32-35]. As explained above, extensive leukocyte-leukocyte and leukocyte-endothelial cell contact occurs during the inflammatory response. A sticking leukocyte preparing to cross the endothelial layer is in perfect position for cell-to-cell communication through gap junctions. However, to our knowledge, gap junctions which are present between most cell types[36,37], have not been found at leukocyte-leukocyte or leukocyte-endothelial cell contacts *in vivo*. Gap junctions appear to occur *in vitro* between some activated white blood cells. For example, functional and morphological evidence of gap junction communication between phytohemagglutinin-activated lymphocytes have been reported[38-40]. In addition, dye transfer has been observed between lymphocytes and endothelial cells, suggesting that functional coupling occurs between these cell types[41].

The mRNA for connexin 43 (Cx43), a gap junction protein, is expressed by cultured vascular endothelial cells[42], a mouse macrophage cell line[43] and the macrophage foam cells of atherosclerotic lesions[44]. These findings suggest that leukocytes could form homotypic gap junctions with each other and heterotypic gap junctions with endothelial cells at the site of inflammation provided that Cx43 is expressed during the acute response. On the other hand, Cx43 and other connexins (Cx40, Cx37) have been found in cultured and arteriolar endothelial cells[45-47]. Furthermore, electrical coupling and dye transfer has been demonstrated to exist between smooth muscle and endothelium of the arterioles of the hamster cheek pouch microcirculation[48]. In the arteriolar segment, cell-to-cell communication via gap junctions has been proposed as a mechanism to explain propagated vasodilation, a phenomenon were distal-to-proximal vasodilatory waves are observed experimentally[49].

Hereafter, we will provide evidence obtained in our laboratory supporting the idea that leukocytes form gap junctions with endothelial cells during the activation-adhesion process.

EXPRESSION OF CONNEXIN 43 IN ACTIVATED LEUKOCYTES

Connexin 43 in Leukocytes Activated *in vitro*.

We reported that freshly isolated hamster leukocytes do not express significant levels of Cx43, as determined by indirect immunofluorescence[50]. Similarly, leukocytes incubated 1-3 h in control Hanks' solution were negative for Cx43. In contrast, Cx43 immunostaining was prominent in smears of isolated leukocytes incubated for 1-3 h with E. coli lipopolysaccharide (LPS, 1 μg/ml) (Figure 2). Approximately 25% of LPS treated leukocytes were immunoreactive for Cx43 (Table 1). An analysis whereby white cells were briefly stained with diamidino phenylindole to visualize the nuclei before labeling for Cx43, revealed that Cx43 positive cells corresponded to polymorphonuclear and monocytes[50]. We were unable to recognize LPS-treated lymphocytes that were positive to Cx43.

Interestingly, the expression of Cx43 was associated with cluster formation among white cells (Table 1). Clusters with three or more cells per field (20X objective) were significantly more numerous in LPS-treated than in control leukocytes. Cluster size increased, although moderately, with this activation maneuver. Cx43 labeling was diffuse in the cytoplasm whereas a strong localized labeling was detected at membrane appositions of LPS-treated cells.

Table 1. Agglutination and Cx43 expression in hamsters leukocytes.

	Time 0	3 h Hanks	1 h LPS	3 h LPS
Clusters per field	0.86 ± 0.49	1.96 ± 0.16	2.57 ± 0.32	3.21 ± 0.53
P vs 3 h Hanks *	<0.001		<0.001	<0.0001
Cells per cluster	2.51 ± 0.07	2.59 ± 0.09	3.65 ± 0.10	3.7 ± 0.10
P vs 3 h Hanks *	*n.s.*		<0.001	<0.001
Cx43 Positive (%)	n.d.	n.d.	23.2 ± 2.5	24.8 ± 4.1

* Student t test; n.s. = not significant; n.d.= not detected

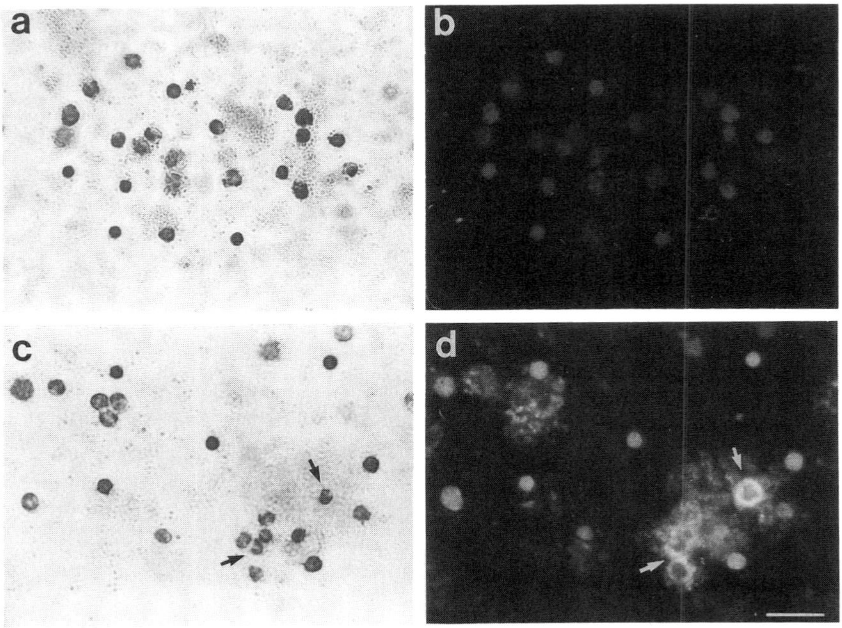

Figure 2. Immunocytochemical detection of Cx43 in circulating hamster leukocytes. White cells were incubated 1 h with Hanks' solution (a, b) or 1 μg/ml LPS (c, d) and mounted in glass smears for Cx43 fluorescence staining. Control cells show only background fluorescence (b), while Cx43 labeling was clear in LPS treated cells (d). (a) and (c) show phase-contrast views of the same areas. Cx43 labeling was found in cytoplasm as well as in membrane appositions of clustered cells (arrows).

We obtained similar results (both qualitatively and quantitatively) with rat leukocytes activated *in vitro* with LPS. Moreover, human granulocytes, isolated by standard techniques of differential centrifugation in discontinuous ficoll gradients, also expressed Cx43 when incubated 1, 3 or 6 h with LPS. After treatment, up to 80-90% of cells of the homogeneous human granulocyte population showed positive staining for Cx43. In contrast, human lymphocytes were not reactive to Cx43. These findings indicate that Cx43 expression by freshly isolated circulating neutrophils and monocytes after *in vitro* exposure to LPS is a common feature in all three species studied. The *in vitro* induction of Cx43 by this stimulus is apparently restricted to myeloid cells, and could have been the direct consequence of the interaction between these cells and LPS, or to an autocrine effect of LPS-induced release of leukocyte cytokines[51,52].

In contrast to the results obtained with LPS, either hamster or human leukocytes incubated with PAF (1-10 nM) did not express significant levels of Cx43. These negative results obtained with PAF *in vitro*, are quite different from those observed *in vivo* (see below).

Cx43 in Leukocytes Activated *in vivo*.

Peritoneal Lavage. We studied whether leukocytes also expressed Cx43 after *in vivo* activation. Hamsters were injected intraperitoneally with 1 μg LPS or the vehicle (1 ml Hanks' solution). A week after, the peritoneal cavity was washed with 5 ml saline and drained under ether anesthesia. Cells contained in the peritoneal lavage were washed in Hanks' solution and prepared for immunocytochemistry or Western analysis for Cx43. As seen in Table 2, slightly more cells were recovered from the peritoneum of LPS-injected than from control animals, but the proportion of Cx43-immunoreactive cells was twice that of controls. Positive cells from LPS-treated animals frequently formed clusters and showed strong Cx43 immunoreactivity. As observed after *in vitro* activation, in some clusters there was a prominent Cx43 labeling at membrane appositions[50]. In control peritoneal lavage, few cells (<5%) showed mainly a diffuse cytoplasmatic labeling whereas in most Cx43 positive cells the protein was detected in perinuclear as well as cell contact regions.

Table 2: Cx43 expression in cells obtained by peritoneal lavage.

	Control	LPS treated	P vs Control
Cells / ml (x 10^6)	1.39 ± 0.20	1.76 ± 0.25	< 0.01
Cx43 positive (%)	23.2 ± 2.5	45 ±3.1	< 0.0001

Immunoblots. Cell pellets were lysed by a brief sonication in the presence of standard protease and phosphatase inhibitors. Equivalent amounts of protein (80 μg) from cells of LPS and control treated hamsters were separated in SDS-PAGE and transferred to nitrocellulose for Western blot analysis. In agreement with immunocytochemical observations, a barely detectable band was observed in control treated cells, corresponding to the unphosphorylated form of Cx43. In contrast, clearly detectable levels of the unphosphorylated form of Cx43 was detected in LPS-treated animals. The phosphorylated forms of Cx43 (P1, P2, P3 and HP)[53] were clearly detected in heart homogenate samples used as positive controls for Cx43 but were almost undetectable in control peritoneal white cells[50]. Studies on leukocyte kinetics during inflammation have shown that monocytes accumulate at sites of tissue injury more slowly than neutrophils, but monocytes predominate at extravasation sites after 24 hours[54]. In agreement with this notion, the majority of the cells that we obtained from the peritoneum several days after endotoxin-induced inflammation probably were monocytes since they did not show a polylobulated nucleus characteristic of granulocytes. Certainly, all these cells have crossed the vascular wall long before the observation period, and under this condition, Cx43 was mostly unphosphorylated and presumably corresponded to the diffuse intracellular labeling. Only a small Cx43 fraction was phosphorylated and presumably corresponded to the protein located at membrane appositions. Consistent with this interpretation, it has been previously found in other cell types that unphosphorylated Cx43 is mainly located in a cytoplasmic compartment and not forming functional gap junctions[34].

EXPRESSION OF Cx43 IN THE HAMSTER CHEEK POUCH MICROCIRCULATION *IN VIVO*.

Adhesion Induced by PAF.

The hamster cheek pouch was prepared for intravital microscopy as described[55-57]. Briefly, a lucite plate and a fiber optic bundle were inserted through the mouth and a hollow lucite chamber was placed through a skin incision on the subepithelial surface of the pouch, exposing 2.25 cm^2 tissue area. The chamber was superfused with 1 ml/min bicarbonate buffer at 37^oC, pH 7.35, equilibrated with 95% N_2-5% CO_2. The animal was placed under a Nikon Optiphot microscope connected to a TV camera, video display and recording system. After 30 min of equilibration, the pouch was superfused with 1 or 10 nM PAF during 5 min, followed by drug-free superfusion for periods of 15, 30, 60 or 90 min. Control animals received just the PAF vehicle (superfusion buffer with traces of DMSO). Thereafter, the animal was killed and the tissue was washed by intraarterial perfusion in order to analyze only leukocytes that had adhered to the endothelium. The tissue was fixed *in situ* with Bouin fixative for immunohistochemical analysis as described[50]. Parafin-embedded 5 μm thick sections were incubated with anti-Cx43 antibody, followed by incubation with a FITC-conjugated secondary antibody, washed with PBS, and mounted. Cx43 labeling was evaluated under a Nikon epifluorescent microscope equipped with a xenon arc lamp. Specificity of the immu-

Table 3: Leukocyte - endothelium interactions in postcapillary venules before (C) and at different times after stimulation with 10 nM PAF.

	C	10 min	20 min	30 min	40 min
Rollers (cells/min)	32.1 ± 4.7	20.3 ± 4.8 *	21.5 ± 7.4	24.4 ± 8.5	29.8 ± 11.6
Stickers (cells / 100 μm)	1.5 ± 0.8	9.4 ± 3.6 *	14.5 ± 3.1 *	14.5 ± 4.4 *	17.3 ± 6.5 *

* $p<0.05$ vs Control (C), student t test.

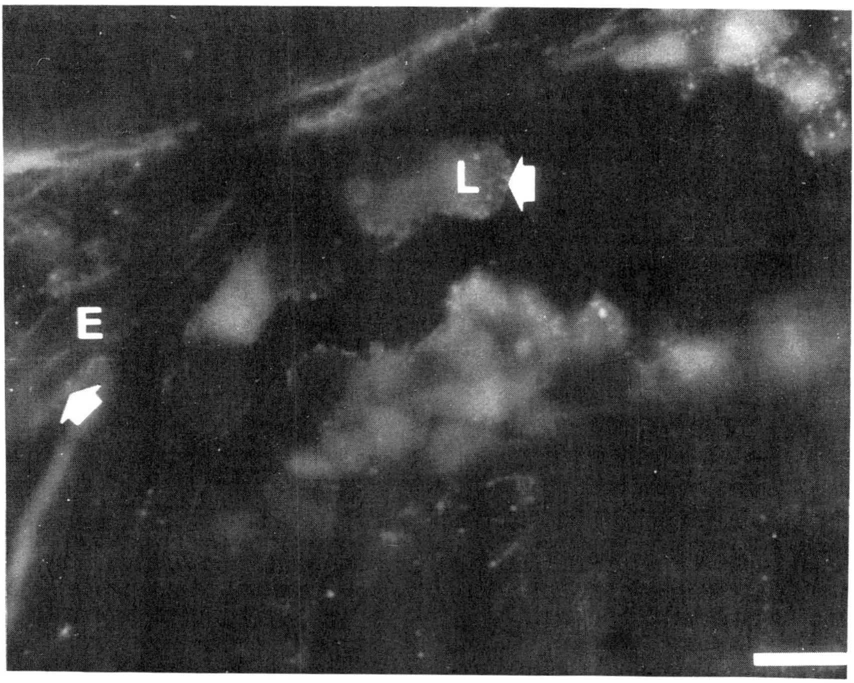

Figure 3. Immunocytochemical detection of Cx43 in adherent leukocytes in the hamster cheek pouch microcirculation. Leukocyte adhesion was induced by topical application of 1 nM PAF during 5 min, followed by 45 min of drug washout. Clusters of adherent leukocytes show prominent Cx43 staining, mostly at cell membrane appositions. Endothelial cells (E) also show surface expression of Cx43. Tissue slices incubated without anti-Cx43 showed background fluorescence (not shown).

noreactivity was assessed by replacing the primary antibody by preimmune serum or blocking solution.

In agreement with previous reports[57], topical PAF application induced a sustained increase in the number of adherent leukocytes, accompanied by a transient drop in the number of rolling cells observed by intravital microscopy (Table 3). Cx43 immunostaining was quite prominent in adherent leukocytes of PAF treated hamsters (Figure 3). Again, labeling was found in membrane appositions, as well as in the cytoplasm. Positive Cx43 reaction was also found in endothelial cells, giving support to the possible formation of leukocyte-endothelium gap junctions. In contrast, control treated animals presented only few sticking cells that were not strongly labeled for Cx43 (not shown).

Adhesion induced by Ischemia-Reperfusion.

Ischemia-reperfusion was induced in the cheek pouch as previously described by Yasuhara et al.[58]. After preparing the tissue for intravital microscopy as mentioned above, two triangular mylar pieces (5 mm base, 3 mm thickness) were placed over the pouch and covered with glass coverslips. Ischemia was induced in the section delimited by one of the triangles by gently applying pressure to the coverslip with a micromanipulator, until blood flow ceased. Flow reperfusion was allowed after 1 h of ischemia by releasing the pressure. Microvascular flow and leukocyte rolling and sticking were monitored on the TV screen. After 1 h reperfusion the animal was killed with an excess of anesthesia, the tissue was washed of circulating blood with intraarterial saline perfusion to prevent fixation of non-adhered leukocytes. Thereafter, the cheek pouch area was fixed and processed for either immunocytochemistry or electron microscopy (EM).

In non-ischemic areas few cells adhered to the venular walls[50]. In contrast, in areas submitted to ischemia-reperfusion many leukocytes were attached to the venular inner wall, frequently forming large clusters (Figure 4). These adherent cells were positive for Cx43 (Fig. 4a). EM observations of areas subjected to ischemia-reperfusion showed leukocytes in close contact with each other and with endothelial cells (Fig. 4b). Higher magnification of membrane appositions revealed small areas of close contact with dimensions similar to those characteristic of gap junctions (Fig. 4c, d). Similar membrane contacts were previously reported between cultured murine macrophages[54].

These findings confirm previous reports in that ischemia-reperfusion induces expression of adhesion molecules responsible for leukocyte attachment[58,59]. Leukocytes became firmly adhered to the endothelium of venules despite the vessels were perfused with saline for 5-10 minutes. We found that synthesis of Cx43 had also occurred within 1 h reperfusion period.

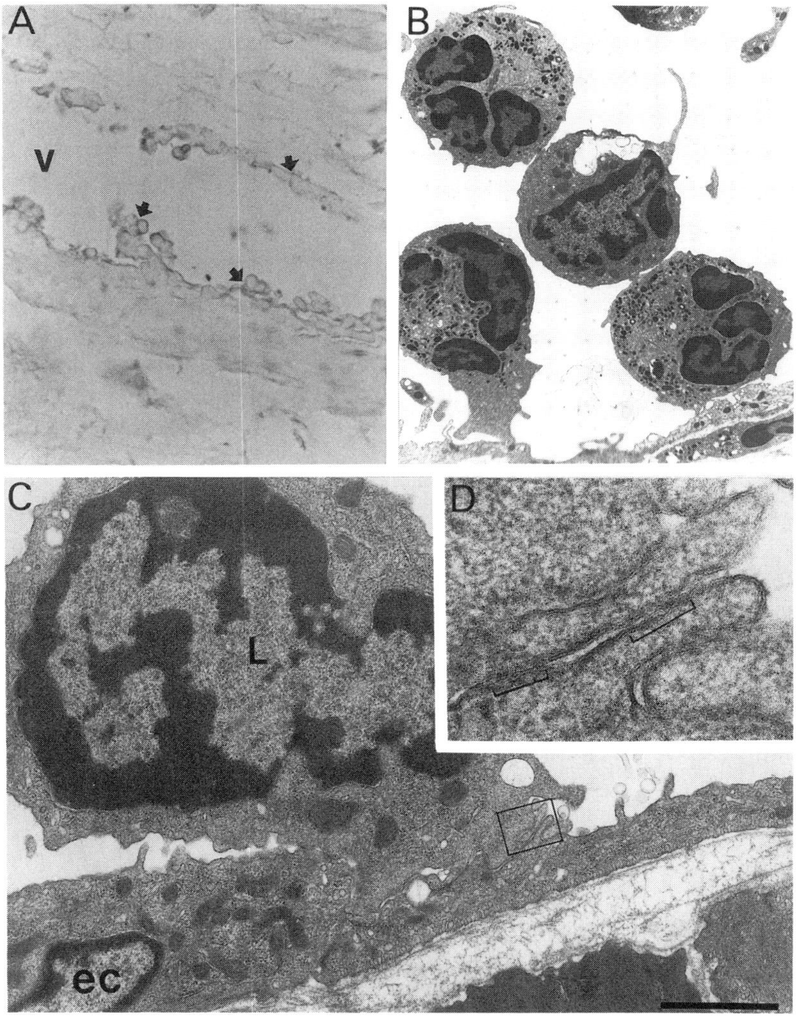

Figure 4. Hamster cheek pouch microcirculation submitted to Ischemia-Reperfusion (I/R). In (A) a small venule (V) within the I/R area is shown with conventional microscopy. Many leukocytes are adhered to the wall and present positive Cx43 staining (arrows), as revealed by the peroxidase-antiperoxidase reaction with anti-Cx43 antibody. Non ischemic areas showed few adherent cells (not shown). In B, C and D, electron micrographs of ischemic areas showing a cluster of neutrophils attached to the endothelium (B). Higher magnification of Leukocyte-Endothelium interaction is shown in C, the enclosed area is enlarged in D. The lines indicate close apposition of plasma membranes, compatible with gap junction formation. Bar represents 60 μm in (A), 5 μm in (B), 1.4 μm in (C), and 0.2 μm in (D).

The induction of Cx43 expression may have been triggered by inflammatory mediators and/or chemoattractants including interleukins, cytokines and PAF released locally by activated blood cells and/or by activated endothelial cells[51,60]. It has been previously reported that the addition of monocytes to human aortic endothelial cells co-cultured with human aortic smooth muscle cells increases the levels of Cx43 mRNA[42].

CONCLUDING REMARKS

Our results obtained *in vivo* in the cheek pouch microcirculation, indicate that leukocyte adhesion induced by either, PAF application or after ischemia-reperfusion, is associated with Cx43 expression. These findings lend support to the possibility that gap junctional communication do occur between leukocytes and endothelium. As we discussed earlier, intimate cell contact mediated by cell adhesion molecules has been shown as a requisite for gap junction formation between other cell types[32-34]. Similar close membrane contacts exist between leukocytes and endothelial cells. In addition, adhesiveness may also be conferred by the interaction between the extracellular loops of connexons of contacting cells. The complexity of connexin regulation may be exemplified by the finding that, contrary to LPS, *in vitro* exposure of hamster or human leukocytes to PAF did not induce a significant Cx43 immunostaining, indicating that this mediator cannot activate leukocytes directly, but other steps, such as endothelium presentation to rolling leukocytes in the presence of selectins is required for activation.

The presence of gap junctions in adherent leukocytes is consistent with *in vitro* reports of dye coupling between blood lymphocytes and endothelial cells[41]. While our data indicate that leukocytes can express Cx43, we do not know whether gap junctions potentially formed during their contact with the endothelium are homotypic or heterotypic. We found Cx43 in venular endothelium in the vicinity of attached leukocytes of the hamster cheek pouch (Fig. 3). In the same tissue, Cx43 and Cx40 have been found in the arteriolar endothelium[47,48]. In addition, Cx43 is expressed by human umbilical endothelial cells in culture[43,46], but we have not explored yet whether other connexin types are also present in venular endothelium or in leukocytes.

Until now there is no direct evidence for *in vivo* cell-to-cell gap junction communication between endothelium and leukocytes, nevertheless, it is appealing to postulate a revised model for leukocyte adhesion and diapedesis in which cytoskeletal and cellular shape changes in leukocytes and endothelium are coordinated by intercellular signaling involving gap junctions (Fig. 1). Leukocyte-endothelium, as well as leukocyte-leukocyte cell contacts may be important in elevating intracellular calcium levels needed to trigger cell shape changes. Intracellular calcium oscillations have been reported in endothelial monolayers incubated with leukocytes[61]. Interestingly, these Ca^{++} oscillations only occurred after a critical number of white cells have established a close contact with the endothelium, suggesting that clusters of leukocytes need to be in contact with the endothelial cell for the activation signal to reach a threshold. Our findings that agglutinated leukocytes show strong Cx43 labeling and present close membrane appositions between them as with the venular wall, supports the possible occurrence of cumulative cell coupling *in vivo*; however, this idea requires experimental demonstration.

REFERENCES

1. G.W. Schmidt-Schoenbein, G.W. Fung, and B.W. Zweifach, Vascular endothelium-leukocyte interaction, sticking shear force in venules, Circ. Res. 36:111 (1975).
2. K. Ley, Lectin-like cell adhesion molecules-1 mediates leukocyte rolling in mesenteric venules in vivo, Blood 77:2553 (1991).
3. K. Ley, M. Cerrito, and K.E. Arfors, Sulfated polysaccharides inhibit leukocyte rolling in rabbit mesentery venules, Am. J. Physiol. 260:H1667 (1991).
4. K. Ley, and P. Gaehtgens, Endothelial, not hemodynamic, differences are responsible for preferential leukocyte rolling in rat mesenteric venules, Circ. Res. 69:1034 (1991).
5. M. Perry, and D.N. Granger, Role of CD11/CD18 in shear rate-dependent leukocyte-endothelial cell interactions in cat mesenteric venules, J. Clin. Invest. 87:1798 (1991).
6. L.A. Lasky, Selectins: interpreters of cell-specific carbohydrate information during inflammation, Science 258:964 (1992).
7. K.E. Arfors, K. Lundberg, L. Lindbom, P.G. Beatty, and J. Harlan, A monoclonal antibody to the membrane glycoprotein complex CD18 inhibits polymorphonuclear leukocyte accumulation and plasma leakage in vivo, Blood 69:338 (1987).
8. D.C.Anderson, The role of β2 integrins and intercellular adhesion molecule type 1 in inflammation. In: Leukocyte Adhesion, D.N. Granger, and G.W. Schmid-Schonbein eds., Oxford university Press, New York. (1995).
9. L.J. Miller, D.F. Bainton, N. Borregaard, and T.A. Springer, Stimulated mobilization of monocyte Mac-1 and p150,95 adhesion proteins from an intracellular vesicular compartment to the cell surface, J. Clin. Invest. 80:535 (1987).
10. T.K. Kishimoto, M. Jutila, E. Berg, and E.C. Butcher, Neutrophil Mac-1 and MEL-14 adhesion proteins inversely regulated by chemotactic factors, Science 245: 1238 (1989).
11. M.S. Diamond, D.E. Stauton, A.R. Defougerolles, S.A. Stacker, J. Garcia-Aguilar, M.L. Hibss, and T.A. Springer, ICAM-1 (CD54): a counter-receptor for Mac-1 (CD11b/CD18), J. Cell Biol. 111:3129 (1990).
12. D.E. Stauton, S.D. Marlin, C. Strratowa, M.L. Dustin, and T.A. Springer, Primary structure of intercellular adhesion molecule 1 (ICAM-1) demonstrates interaction between members of the immunoglobulin and integrin supergene families, Cell 52:925 (1988).
13. U.H. Von Andrian, J.D. Chambers, L.M. McEvoy, R.F. Bargatze, K.E. Arfors, E.C. Butcher, Two-step model of leukocyte-endothelial cell interaction in inflammation: distinct roles for LECAM-1 and the leukocyte b2 integrins in vivo, Proc. Natl. Acad. Sci. USA 88:7538 (1991).

14. M. Lawrence, and T.A. Springer, Leukocytes roll on a selectin at physiologic flow rates: distinction from and prerequisite for adhesion through integrins, Cell 65: 859 (1991).
15. L. W. Argenbright, Inflammatory mediators, cytokines, and leukocyte adhesion in post-capillary venules. In: Leukocyte Adhesion, D.N. Granger, and G.W. Schmid-Schonbein, eds., Oxford university Press, New York. (1995).
16. Y. Tanaka, D.H. Adams, S. Hubscher, H. Hirano, U. Siebenlist, and S. Shaw, T-cell adhesion induced by proteoglycan-immobilized cytokine MIP-1b, Nature 361:79 (1993).
17. J.S Pober, and R.S. Cotran, The role of endothelial cells in inflammation, Transplantation 50:537 (1990).
18. E. Ninio, S Leyvaund., J. Bidault, P. Jurgens, and J. Beveniste, Cell adhesion by membrane-bound paf-acether, Int. Immunol. 3:1157 (1991).
19. D.E. Lorant, K.D. Patel, T.M. McIntyre, R.P. McEver, S.M. Prescott, and G.A. Zimmerman, Coexpression of GMP-140 and PAF by endothelium stimulated by histamine or thrombin: a juxtacrine system for adhesion and activation of neutrophils, J. Cell. Biol. 115:223 (1991).
20. R.E. Lewis, and H.J. Granger, Diapedesis and the permeability of venous microvessels to protein macromolecules: the impact of leukotriene B_4 (LTB_4), Microvasc. Res. 35:27 (1988).
21. R.E. Lewis, and H.J. Granger, Neutrophil-dependent mediation of microvascular permeability, Fed. Proc. 45:109 (1986).
22. R.E. Lewis, R.A. Miller, and H.J. Granger, Acute microvascular effects of the chemotactic peptide N-formyl-methionyl-leucyl-phenylalanine: comparisons with leukotriene B_4. Microvasc. Res. 37:53 (1989).
23. G. Majno, S.M. Shea, and M. Leventhal, Endothelial contraction induced by histamine type mediators. An electron microscopic study, J. Cell. Biol. 42:647 (1969).
24. D. Hülstrom, and E. Svensjo, Intravital and electron microscopy study of bradykinin induced vascular permeability changes using FITC dextran as a tracer, J. Pathol. 129:125 (1979).
25. D.M. Mc Donnald, Endothelial gaps permeability, and leukocytes in venules of rat trachea exposed to inflammatory stimuli, Am. J. Physiol. 266: L61 (1994).
26. M Minnicozzi, W.N. Durán, G.J. Gleich, R.W. Egan, Eosinophil granule proteins increase macromolecular transport in the hamster cheek pouch. J. Immunol. 153: 2664 (1994).
27. D.C. Zawieja, C. Garcia, and H.J. Granger, Oxygen radicals, enzymes, and fluid transport through pericardial interstitium, Am. J. Physiol. 262:H136 (1993).
28. D. Shepro, Endothelial cell, inflammatory edema, and the microvascular barrier: comments by a "free radical", Microvasc. Res. 35:247 (1988).
29. J.S. Alexander, H.B. Hetchman, and D. Shepro, Phalloidin enhances endothelial barrier function and reduces inflammation permeability in vitro, Microvasc. Res. 35:308 (1988).
30. J.Doukas, D. Shepro, and H.B. Hetchman, Vasoactive amines directly modify endothelial cells to affect polymorphonuclear leukocyte diapedesis in vitro, Blood 69:1563 (1987).
31. D.A. Dubose, D. Shepro, and H.B. Hetchman, Correlation among endothelial cell shape, F-actin arrangements and prostacyclin synthesis, Life Sciences 40:447 (1987).
32. R.W. Keane, P.P. Mehta, B. Rose, L.S. Honig, and W. Loewenstein, Neural differentiation, NCAM-mediated adhesion, and gap junctional communication in neuroectoderm. A study in vitro, J. Cell Biol. 106:1307 (1988).
33. R.M. Mege, F. Matsuzaki, W.F. Gallin, J.I Golber, B.A. Cunniggham, and G.M. Edelman, Construction of epitheloid sheets by transfection of mouse sarcoma cells with cDNAs for chicken cell adhesion molecules, Proc. Natl. Acad. Sci. USA 85:7274 (1988).
34. L.S. Musil, B.C. Cunningham, G.M. Edelman, and D.A. Goodenough, Differential phosphorylation of the gap junction protein connexin43 in junctional communication-

competent and deficient cell lines, J. Cell Biol. 111:2077 (1990).
35. W.M.F. Jongen, D.J. Fitzgeral, M. Asamoto, C. Piccoli, T.J. Slaga, D Gros, M. Takeichi, and H. Yamasaki, Regulation of connexin43 mediated gap junctional communication by Ca^{+2} in mouse epidermal cells is controlled by E-Cadherin, J. Cell Biol. 114:545 (1991).
36. M.V.L. Bennett, L.C. Barrio, T.A. Bargiello, E.L. Hertzberg, D.C. Spray, and J.C. Sáez, Gap junctions. New tools, new answers, new questions, Neuron 6:305 (1991).
37. R. Dermietzel, and D.C. Spray, Gap junctions in the brain: Where, what type, how many and why?, TINS 16:186 (1993).
38. D.F. Hulser, and J.H. Peters, Intercellular communication in phytohemagglutinin-induced lymphocyte agglutinates, Eur. J. Immunol. 1:494 (1971).
39. D.F. Hulser, and J.H. Peters, Contact cooperation in stimulated lymphocytes. II electrophysiological investigation on intercellular communication, Tissue Cell Res. 74:319 (1972).
40. G.M Oliveira-Castro, M.A. Barcinski, and I.F. Gaziri, Intercellular communication in lymphocytes. In: Concepts of Membranes in Regulation and Excitation, M. Rocha e Silva, and G. Suarez-Kurtz, eds. Raven Press, New York. (1975).
41. E.C. Guinan, B. Smith, P.F. Davies, and J.S. Pober, Cytoplasmic transfer between endothelium and lymphoytes: quantitation by flow cytometry, Am. J. Pathol. 132:406 (1988).
42. M. Navad, F. Liao, G.P. Hough, L.A. Ross, B.J. Van Lenten, T.B. Rajavashisth, A.J. Lusis, H. Laks, D.C. Drinkwater, and A.M. Fogelman, Interaction of monocytes with co-cultures of human aortic wall cells involves interleukins 1 and 6 with marked increases in connexin43 message, J. Clin. Invest. 87:1763 (1991).
43. E.C. Beyer, and T.H. Steinberg, Evidence that the gap junction protein connexin-43 is the ATP-induced pore in mouse macrophages, J. Biol. Chem. 266:971 (1991).
44. D. Polacek, L. Lal, M.V. Volin, and P.F. Davies, Gap junctional communication between vascular cells. Induction of connexin 43 messenger RNA in macrophage foam cells of atherosclerotic lesions, Am. J. Pathol. 142:593 (1993).
45. R. Bruzzone, J.A. Haegfliger, R. Gimlich, and D. Paul, Connexin 40, a component of gap junctions in vascular endothelium, is restricted in its ability to interact with other connexins, Mol. Biol. Cell 4:7 (1993).
46. K.E. Reed, E.M. Westphale, D.M. Larson, H.Z. Wang, R.D. Veenstra, and E.C. Beyer, Molecular cloning and functional expression of human connexin 37, an endothelial cell gap junction protein, J. Clin. Invest. 91:997 (1993).
47. T.A. Little, E.C. Beyer, and B.R. Duling, Connexin 43 and connexin 40 gap junctional protein are present in arteriolar smooth muscle and endothelium in vivo, Am. J. Physiol. 268: H729 (1995).
48. T.A. Little, J. Xia, and B.R. Duling, Dye tracers define differential endothelial and smooth muscle coupling patterns within the arteriolar wall, Circ. Res. 76:498 (1995b).
49. S.S. Segal, and B.R. Duling, Conduction of vasomotor responses in arterioles: a role for cell-to-cell coupling?, Am. J. Physiol. 256:H838 (1989).
50. P.I. Jara, M.P. Boric, and J.C. Sáez, Leukocytes express connexin 43 after activation with lipopolysaccharide and appear to form gap junctions with endothelial cells after ischemia-reperfusion, Proc. Natl. Acad. Sci. USA. 92:7011 (1995).
51. G.A., Zimmerman, and H.R. Hill, Inflammatory mediators stimulate granulocyte adherence to cultured endothelial cells, Thromb. Res. 35:203 (1984).
52. E.J. Wing, in Hemopoietic Growth Factors and Mononuclear Phagocytes, ed. van Furth, R. (Karger, Basel, Switzerland) pp. 157-167. (1993)
53. V.M. Berthoud, M.B. Rook, O. Traub., E.L. Hertzberg, and J.C. Sáez, On the mechanisms of cell uncoupling induced by a tumor promoter phorbol ester in clone 9 cells, a rat liver epithelial cell line, Eur. J. Cell Biol. 62:384 (1993).

54. T.B. Issekutz, A.C. Issekutz, and H.Z. Movat, The in vivo quantitation and kinetics of monocyte migration into acute inflammatory tissue, Am. J. Pathol. 103:47 (1981)
55. M.P. Boric, M.V. Donoso, A. Fournier, S. St. Pierre, and J.P. Huidobro-Toro, Endothelin reduces microvascular blood flow by acting on arterioles and venules of the hamster cheek pouch, Eur. J. Pharmacol. 190:123 (1990).
56. M.P. Boric, A. Martínez, M.V. Donoso, and J.P. Huidobro-Toro, Neuropeptide Y is a vasoconstrictor and adrenergic modulator in the hamster microcirculation by acting on Y_1 and Y_2 receptors, Eur. J. Pharmacol. 294:391 (1995).
57. P.K. Dillon, M.F. Fitzpatrick, A.B. Ritter, and W.N. Durán, Effect of platelet-activating factor on leukocyte adhesion to microvascular endothelium, Inflammation 12:563 (1988).
58. H. Yasuhara, R.W. Hobson II, P.K. Dillon, and W.N. Durán, A new model for studying ischemia-reperfusion injury in hamster cheek pouch, Am. J. Physiol. 261:H1626 (1991).
59. D.N. Granger, J.N. Benoit, M. Susuki, and M.B. Grisham, Leukocyte adherence to venular endothelium during ischemia-reperfusion, Am. J. Physiol. 257:G683 (1989).
60. W.N. Durán, W.D. Suval, M.P. Boric, and R.W. Hobson II, Microcirculatory disfunction in the pathophysiology of skeletal muscle ischemia. In: Vascular Surgery: Principles and Practice, F.J. Veith, R.W. Hobson II, R.A. Williams, and S.E. Wilson, eds. McGraw-Hill, Inc., New York (1994).
61. S. Pfau, D. Leitberg, H. Rinder, B.R. Smith, R. Pardi, and J.R. Bender, Lymphocyte adhesion-dependent calcium signaling in human endothelial cell, J. Cell Biol. 128:969 (1995).

REGULATION OF GAP JUNCTIONS IN RAT LIVER DURING ACUTE AND CHRONIC CCl_4-INDUCED LIVER INJURY

Claudia G. Sáez[1], Eliseo Eugenin[1], Elliot L. Hertzberg[2] and Juan C. Sáez[1]

[1]Departamento de Ciencias Fisiológicas, Facultad de Ciencias Biológicas, Pontificia Universidad Católica de Chile, Santiago, Chile
[2]Departments of Neuroscience and of Anatomy and Structural Biology, Albert Einstein College of Medicine, Bronx, NY, USA.

INTRODUCTION

Gap junctions are aggregates of intercellular channels that provide a pathway for intercellular exchange of ions and small hydrophilic molecules (< 1.2 kDa)[1]. They are particularly abundant between hepatocytes[2] where they also are permeable to Ca^{2+} and IP_3, allowing synchronized changes in intracellular Ca^{2+} responses induced by hormonal stimulation[3,4]. In rat, receptors for glucogenic hormones, such as glucagon and vasopressin, are mostly expressed by hepatocytes located in zone 3 of the hepatic acini[5,6]. The heterogeneity in receptor density seems to explain vasopressin-induced Ca^{2+} waves that start at zones 3 and propagate in opposite direction to the blood flood through intercellular gap junctional communication[6]. It is likely that such Ca^{2+} waves activate diverse metabolic and mechanic processes along the acini. Therefore, the integrity of gap junctions thought to be crucial for the normal functioning of cell communities.

Each gap junction channel is formed by two hemichannels which are hexamers of homologous proteins, called connexins (Cxs), that are encoded by a gene family[1,7]. Rat hepatocytes located in zone 1 of the hepatic acini form gap junctions that contain Cx26 and Cx32, while hepatocytes of zones 2 and 3 form gap junctions that contain almost exclusively Cx32[5,8,9]. Moreover, nonparenchymal cells, including collangiocytes[10], cells of the Glisson's capsule[5] and perisinusoidal cells[5] form Cx43 gap junctions.

The intercellular exchange of small signal molecules via gap junctions appears to participate in controlling cell growth and differentiation[11,12]. Within 72 h after partial hepatectomy, the regenerating rat liver shows rapid and controlled growth[13-15] and gap junctions undergo a cycle of disappearance and reappearance[16-19]. Hepatocytes that enter the S-phase show a significant reduction of Cx32[18].

CCl_4-induced hepatocellular necrosis is mediated by a highly reactive free radical anion, $CCl\cdot^{-}_3$, generated by the metabolism of CCl_4 by cytochrome P_{450}[20]. CCl_4 intoxication causes necrosis preferentially in zone 3, since hepatocytes located in that zone express the highest levels of cytochrome P_{450} activity along the normal hepatic acini[21]. Thus, acute CCl_4 intoxication might be similar to partial hepatectomy by selectively killing hepatocytes of zone 3. That, a comparable reduction in Cx32 expression that correlates with the degree of cell division occurs during regeneration after CCl_4-induced injury as is seen after partial hepatectomy[22] suggests a mechanistic similarity. While the parallels for Cx32 have been established, the effects of CCl_4-induced liver injury on expression of Cx26 and Cx43 have not been analyzed. Here we report on our analysis of this question as well as whether expression of these three Cxs (26, 32 and 43) are affected during CCl_4-induced fibrosis, a condition similar to that found in cirrhosis[23,24].

MATERIALS AND METHODS

Administration of CCl_4. Female Sprague-Dawley rats (80-100 g of body weight) were used. Acute liver injury was induced by the administration of a single intragastric dose of CCl_4/corn oil (25%/2 ml/100 g body weight). Liver fibrosis was induced in rats that weighed 80-100 g at the beginning of the experiment. Animals were treated with CCl_4 by i.p. injection of 0.2 ml of a mixture of CCl_4 and mineral oil three times a week for 8 weeks. The concentrations of CCl_4 in mineral oil (v/v) were as follows: week 1, 16 %; week 2, 16 %; week 3, 20 % and weeks 4-7, 25 %. All animals were sacrificed three days after the last CCl_4 administration. After 8 weeks of CCl_4 treatment the immunoreactivity to desmin or vimentin was increased indicating that liver fibrosis had occurred[25].

Immunohistochemistry. Frozen liver sections (4 μm thick) were processed for detection of proteins by indirect immunofluorescence as described previously[5]. After incubation with a primary antibody that reacted with the antigen of interest, sections were incubated with the corresponding secondary antibody coupled to FITC or rhodamine for 2 h at room temperature and then mounted in Gelvatol airvol 205 with 100 mg/ml DABCO. The sections were examined under a xenon arc lamp on a Nikon epifluorescent microscope.

Immunoblots and Northern blots. Liver samples were homogenized in 1 mM PMSF with or without 100 mM NaF and 100 mM sodium pyrophosphate, to inhibit endogenous phosphatase activity, using a tissue homogenizer (Brinkman homogenizer, Kinematics, Switzerland) and then sonicated (Microson ultrasonic cell disrupter, Heat systems). Protein was measured in aliquots of the cell lysates by the method of Bradford [Bio-Rad labs, Hercules, CA]. Samples containing 70 or 250 µg of proteins were used to analyze Cx32 and Cx43, respectively. In each case proteins were separated on 12.5 % or 8% SDS-polyacrylamide gels, electrophoretically transferred to nitrocellulose sheets which were then incubated sequentially with blocking solution (5% non fat milk in Tris-buffered saline) affinity purified rabbit polyclonal antibodies prepared against rat liver connexins and with anti-rabbit IgG conjugated to alkaline phosphatase. Antigen-antibody complexes were detected with 5-bromo-4-chloro-3-indolyl phosphate and nitroblue tetrazolium (Sigma Chemical Co.).

Total RNA was isolated as described previously by Panduro et al.[26]. Hybridization analysis of total RNA transferred to a membrane filter was carried out as described[27].

RESULTS

Cell distribution of connexins in liver of rats treated acutely and chronically with CCl_4. Twenty four hours after administration of a single intragastric dose of CCl_4, the immunoreactivity of Cx32 as well as Cx26 was reduced in hepatocytes. This reduction in both Cxs was maximal 48 h after treatment (Fig. 1B). The distribution and intensity of labeling of both Cxs subsequently showed a gradual recovery reaching a pattern close to control by the 5th post-treatment day.

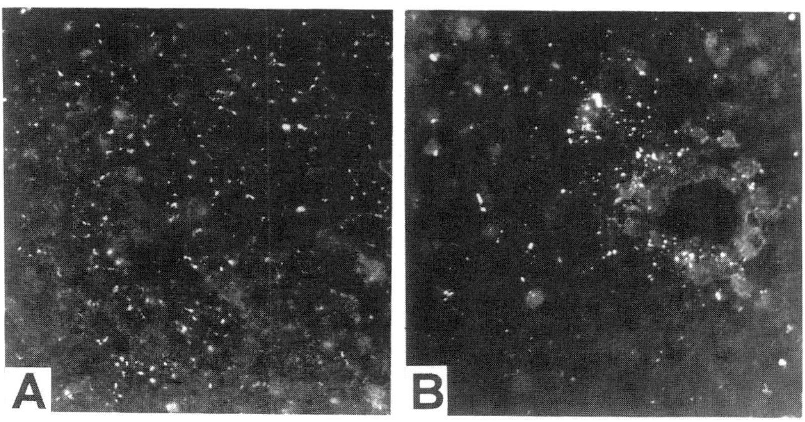

Figure 1. Cx26 immunoreactivity in normal rats and in rats treated acutely with CCl_4. Fluorescent views of frozen sections of a normal rat liver (A), and 48 h after acute CCl_4-treatment (B), immunostained for Cx26 using anti-rabbit FITC-labeled IgG as secondary antibody.

In normal liver, Cx43 was found only in nonparenchymal cells located along the sinusoids. Only a fraction of Cx43 positive cells was recognized as fat storing (Ito) cells by their desmin reactivity (Fig. 2A and B, arrrows) while other perisinusoidal cells were positive to Cx43 but not to desmin (Fig. 2A and B, arrow head), suggesting that other perisinusoidal cells express Cx43.

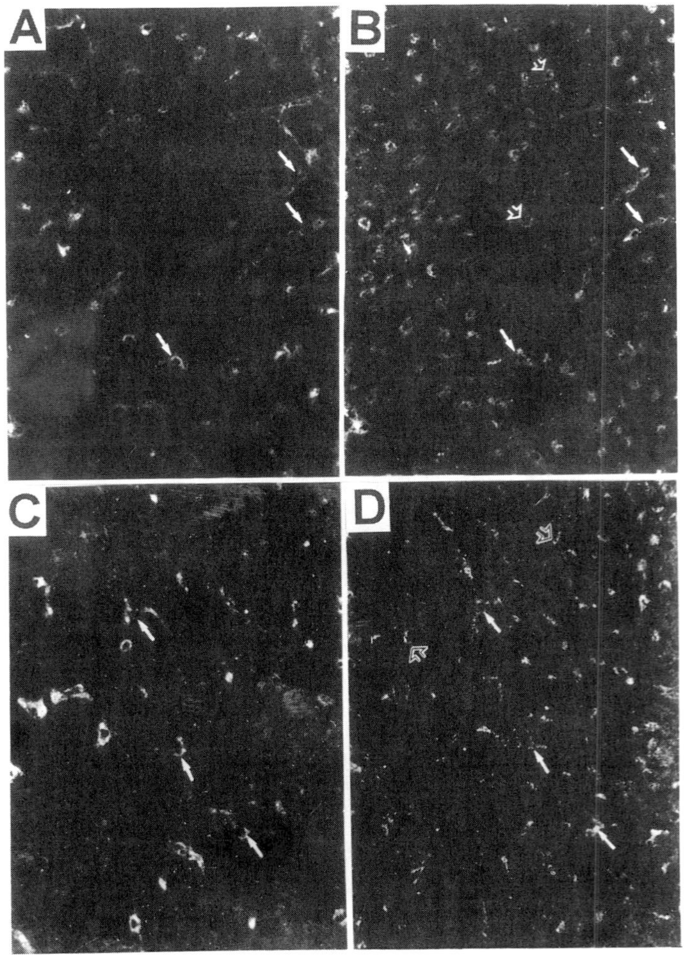

Figure 2. Identification of Cx43 positive cells in normal liver. Frozen sections of normal rat liver were double labeled for Cx43 (A) and desmin (B) or for Cx43 (C) and ED2 (D). Numerous perisinusoidal cells were positive for Cx43 (A and C); a fraction of Cx43 positive cells were also positive to desmin (B, arrows) while others were negative (B, arrowheads). Similarly, many Cx43 positive cells were positive to ED2 (D, arrows), while others were negative (D, arrowheads).

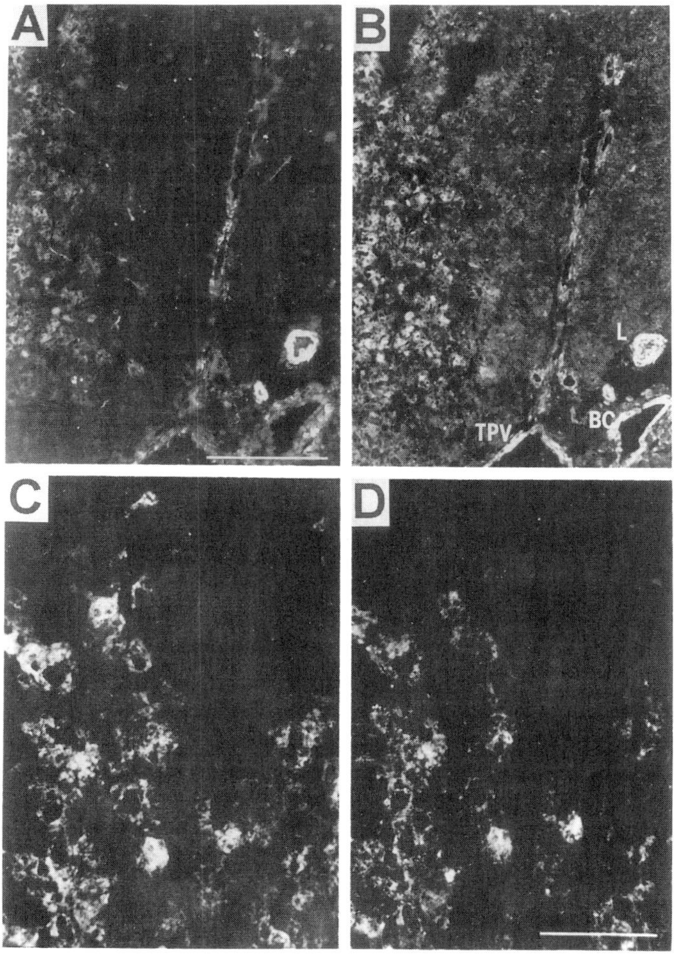

Figure 3. Identification of Cx43 positive cells in liver sections of rats 48 h after acute CCl_4 treatment. Frozen sections of rat liver were obtained 48 h after a single CCl_4 administration and double stained for Cx43 (A) and desmin (B) or Cx43 (C) and ED2 (D). Numerous cells were positive to Cx43 along perisinusoidal areas but were even more abundant in zones 3 or pericentral regions. Although many Cx43 cells were also positive to desmin (B, arrows) or ED2 (D) others were negative to desmin (B, arrowheads) or ED2 (D, arrowheads).

The possibility that Kupffer cells might be another perisinusoidal cell type that express Cx43 was also investigated. In sections of normal liver, a fraction of cells were reactive to both anti-ED2 and Cx43 antibodies (Fig. 2C and D arrows), while other were reactive only to Cx43 (Fig. 2C and D, open arrowhead), consistent with the high number of cells positive to Cx43 and desmin as described above.

Twenty four hours after a single dose of CCl_4, Cx43 immunoreactivity was increased in cells located preferentially in zones 3. The maximal effect was detected 48 h after CCl_4 administration. Many Cx43 positive cells were also reactive to anti-desmin (Fig. 3A and 3B) or to anti-ED2 (Fig. 3C and D) antibodies, suggesting that they were Ito cells and macrophages, respectively. It was difficult to evaluate whether part of Cx43 was located in the plasma membrane, since the intracellular labeling of Cx43 in the cytoplasmic compartments was too abundant. From the 3rd day post-CCl_4 treatment, a progressive recovery in Cx43, desmin and ED2 immunoreactivity occurred attaining their normal pattern of distribution and intensity by the 5th day after CCl_4 administration.

The possibility that CCl_4-induced necrosis is followed by an increase in endothelial cells in the injured area that may be yet another cell type that expresses Cx43 48 h after acute CCl_4 administration was investigated. Many more spindle shaped cells positive to von Willebrand factor were detected 48 h after CCl_4 treatment (Fig. 4B), as compared to normal liver (Fig. 4A). Double labeling of Cx43 and von Willebrand factor was not attempted since both available antibodies were raised in rabbits.

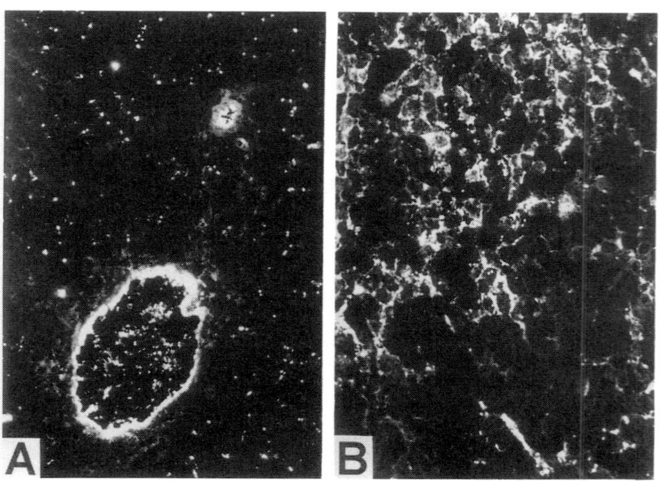

Figure 4. Detection of endothelial cells in normal liver and after acute CCl_4-induce injury. von Willebrand positive cells were stained in frozen section of normal liver (A) or 48 h after CCl_4 administration (B) using indirect immunofluorescence. While few spindle shape cells were positive in normal liver along the sinusoids, many positive cells were detected in zones 3.

Liver sections of rats chronically treated with CCl_4 were analyzed for their content and distribution of Cx26, 32 and 43. Although a high content of extracellular matrix was detected, as described for fibrotic livers, only a slight increase in intracellular labeling of Cx32 was detected in hepatocytes. Expression of Cx26 and Cx43 were not significantly different from that in controls.

Relative levels of Cx32 and its transcript in rat liver after acute or chronic treatment with CCl_4. Twenty four hours after administration of a single CCl_4 dose, the relative levels of Cx32 measured by immunoblot were drastically reduced, and showed a more pronounced reduction 48 h after CCl_4 administration (Fig. 5A). Later on, a gradual recovery was detected reaching essentially normal levels by about the 5th post-treatment day (Fig. 5A). In rats treated chronically with CCl_4 levels of Cx32 were slightly lower than those detected in normal rats of the same age (Fig 5B).

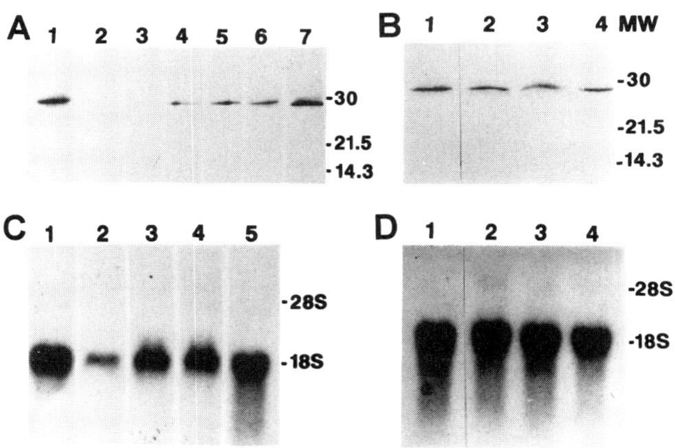

Figure 5. Levels of Cx32 and of its mRNAs after acute and chronic CCl_4 treatment. Total liver homogenates were resolved by 12.5% SDS-PAGE, and subjected to analysis by immunoblotting. (A) Cx32 levels were measured in liver samples of normal rat (lane 1) and 48, 72, 96, 120, 144 and 168 h after acute CCl_4 treatment (lanes 2, 3, 4, 5, 6 and 7, respectively). (B) Cx32 levels in liver homogenate of a normal rat (lane 1) and after 8 weeks of chronic treatment with CCl_4 (lanes 2-4). Numbers on the right of A and B correspond to molecular weight of standard proteins. (C) Northern blot of Cx32 mRNA in control liver (lane 1) and after 48, 72, 96, and 120 h after acute CCl_4 (lanes 2, 3, 4, and 5, respectively). (D) Northern blot of Cx32 mRNA in control liver (lane 1) and after 8 weeks of chronic treatment with CCl_4 (lanes 2-4). On the right side of C and D are indicated the electrophoretic mobilities of ribosomal RNAs 18S and 28S, respectively.

One day after the administration of a single CCl_4 dose, levels of Cx32 mRNA were drastically lower than in controls. They started to recover by the 2nd day and returned to normal by the 3rd post-treatment day (Fig. 5C). In contrast, in rats treated chronically with CCl_4 the levels of Cx32 mRNA were comparable to those found in control animals (Fig. 5D).

Relative levels of Cx43 in liver of rats treated acutely or chronically with CCl_4. In normal rat liver, the unphosphorylated (NP) form of Cx43 predominates (Fig. 6). A rise in Cx43 levels was detected by the 2nd day after acute CCl_4 administration mostly at the expense of an increase in levels of its phosphorylated forms (P2) (Fig. 6A). Levels of Cx43 as well as its state of phosphorylation were as in controls between the 4th and 5th post treatment day. The same pattern of Cx43 forms was detected in liver samples obtained in the presence of 100 mM sodium pyrophosphate and 100 mM NaF (Fig. 6B), suggesting that the absence of other phosphorylated forms of Cx43, such as P3

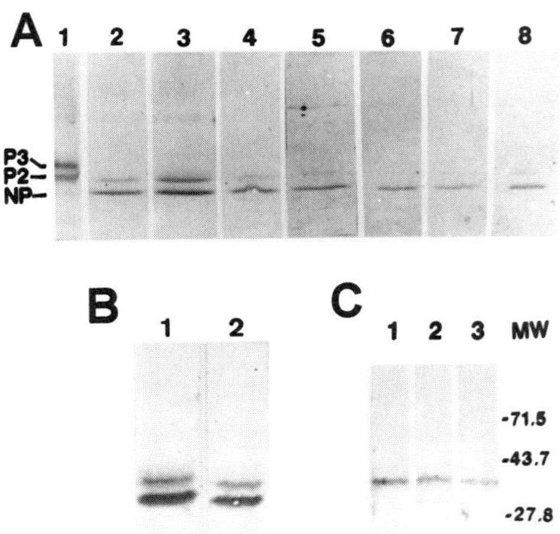

Figure 6. Levels of Cx43 in rat liver after acute or chronic CCl_4 treatment. Aliquots (250 μg of protein) of total liver homogenates were resolved by 8% SDS-PAGE and subjected to immunoblot analysis for Cx43. (A) Cx43 was detected in normal liver (lane 8), and 24, 48, 72, 96, 120, and 144 h after acute treatment with CCl_4 (lanes 2, 3, 4, 5, 6, and 7, respectively). In an aliquot of total rat heart homogenate three Cx43 bands known to correspond to the unphosphorylated form (NP) and two phosphorylated forms (P_2 and P_3) were detected. (B) Immunoblot analysis of Cx43 in liver samples obtained 48 h after acute CCl_4 treatment; samples were homogenized in buffer with (lane 2) or without (lane 1) phosphatase inhibitors. (C) Cx43 in normal liver (lane 1) and in liver of rats treated for 8 weeks with CCl_4 (lanes 2 and 3).

and HP, were not converted to P1 or NP by endogenous phosphatase activities during tissue processing.

Levels and phosphorylation states of Cx43 in liver of chronically CCl_4-treated rats were similar to those observed in controls (Fig. 6C).

DISCUSSION

CCl_4-induced hepatic necrosis promotes a regenerative response similar to that resulting from partial hepatectomy involving removal of two thirds of the liver mass[28,29]. During the regenerative response induced by partial hepatectomy, gap junctions undergo a cycle of disappearance and reappearance[16]. Here we have demonstrated that acute liver injury induced by a single dose of CCl_4 is followed by transient and cell specific changes in the amount of Cx26, Cx32 and Cx43. In contrast, in livers that have become fibrotic because of chronic CCl_4-treatment, the levels and cell distribution of Cx26 and Cx43 did not significantly change and Cx32 showed only a slight reduction both at the protein and mRNA levels along with a small redistribution to cytoplasmic compartments. It has been shown previously that after acute CCl_4-induced liver injury, survivor hepatocytes show a transient reduction in Cx32 similar to that detected after partial hepatectomy[22]. The transient reduction in Cx32 mRNA reported here suggests that changes in Cx32 levels are likely to be the consequence of a reduce protein synthesis. Insofar as acute CCl_4-intoxication induces both a regenerative and an inflammatory response, and the levels of Cx32 mRNA during partial hepatectomy-induced liver regeneration[19] or endotoxin-induced inflammation[30] are primarily controlled by posttranscriptional mechanisms, it is likely that similar mechanisms control Cx32 mRNA levels after acute CCl_4-induced liver injury.

Simultaneous with the Cx32 reduction that occurred after acute CCl_4-treatment was a drastic reduction in Cx26, similar to that observed by others after partial hepatectomy[19]. The genes that encode for these two protein contain some common regulatory elements[31] thus, it is likely that under both conditions of liver regeneration similar mechanisms control the expression of Cx26 and Cx32, leading to functional cell uncoupling.

The reduction in hepatocyte Cx levels might be the consequence of an alteration of humoral factors that control their expression. After acute CCl_4-induced liver injury the number of glucagon receptors shows a transient reduction[32] with a time course similar to that observed here for Cx32. Under these conditions, the number of insulin receptors remains unaltered[32]. As might be anticipated, agents that increase the intracellular cAMP levels, such as glucagon and membrane permeant cAMP derivatives, increase the abundance of gap junctions and the levels of expression of Cx32[27,33,34]. In contrast, insulin down regulates Cx32 expression[34]. Alternatively, CCl_4-induced liver damage might cause down regulation of cell adhesion proteins similar to that which occurs after partial hepatectomy[35]. A deficiency in cell adhesion destabilizes gap junctions and hinders their formation[36-38].

In normal rat liver Cx43 has been detected in perisinusoidal cells[5]. In addition, cell lines derived from fat storing (Ito) cells isolated from chronically CCl_4-injured liver express Cx43[24]. Using double labeling immunohistochemistry we found that two perisinusoidal cell types are positive for Cx43: resident macrophages (Kupffer cells) and fat storing (Ito) cells.

Consistent with a previous study[5] Western blot analysis of Cx43 in rat liver indicates that it is found predominantly as an unphosphorylated form. The presence of diffuse intracellular immunolabeling is consistent with the demonstration that the unphosphorylated form of Cx43 is localized to cytoplasmic compartments rather than assembled in the plasma membrane[39].

The release of chemoattractant factors from liver acutely injured promotes macrophage infiltration into the necrotic tissue[40] which, in CCl_4-induced liver damage, occurs preferentially in zones 3[41]. Moreover, proliferation of Kupffer cells occurs after CCl_4-induced liver damage[42]. These observations are consistent with ours in that a large number of cells labeled with anti-ED2 antibody, and that also showed elevated endogenous peroxidase activity, were localized to zone 3. These cells also contained high Cx43 immunoreactivity, consistent with previous reports demonstrating Cx43 in macrophages[43-45].

We also found that after acute CCl_4 intoxication many cells were immunoreactive to desmin, consistent with a known proliferative response of Ito cells during an inflammatory process[25]. Moreover, double labeling studies revealed that many desmin positive cells also contain Cx43, indicating that Ito cells also contributed to the increase in Cx43 found after acute injury.

It has recently been reported that oval cells in culture express Cx43[46]. During chronic or severe injury to the liver a dramatic expansion of periportally-located oval cells has been observed[47,48]. However, we did not detected a significant increase in Cx43 positive cells in periportal regions, suggesting that oval cells did not significantly contributed to the increase in Cx43 found 48 h after acute CCl_4-treatment.

An increase in von Willebrand positive cells in zones 3 was also detected after acute CCl_4-induced liver damage. This finding might reflect an angiogenic response that occurs during tissue repair. Interestingly, angiogenesis can be promoted by basic fibroblast growth factor which also increases Cx43 expression in microvascular endothelial cells[49]. Thus, the increase in Cx43 levels detected after acute CCl_4-induced liver injury is likely to reflect the increased number of cells expressing Cx43, including macrophages, fat storing (Ito) cells and endothelial cells. Further studies will be required to determine the relative contribution of each cell type to the increase in Cx43.

Unlike the case for acute CCl_4 administration, the expression of all three connexins studied was not significantly affected by 8 weeks treatment with an increasing CCl_4 dose up to the one administrated acutely as a single dose. Although the latter suggests that changes in the expression of gap junction might not be involved with metabolic alterations found in CCl_4-induced cirrhosis, it remains to be determined if post-translational modification of Cxs, such as acylation or phosphorylation that might alter the functional state of gap junctions, occurs.

In conclusion, acute but not chronic intoxication with CCl_4 induces a drastic reduction in intercellular gap junctional communication between parenchymal cells through diverse mechanisms, including early events such as closure of gap junction channels[50], increase in the size the gap junction area that correlates with increase in fluidity of the plasma membrane[51] and late events as discussed above. Although further studies will be required to elucidate the functional implications of each effect, we speculate that early uncoupling may serve to isolate injured from uninjured cells, while subsequent reduced coupling would be the consequence of an increased cell proliferation. Moreover, the increase in Cx43 expression by Kupffer cells and macrophages might be related to their functions, such as phagocytosis and digestion of cell debris at necrotic zones.

Acknowledgment: We would like to thank Ms. Gladys Gacés for her technical assistance. This work was partially supported by Fondo Nacional de Ciencia y Tecnología grants (postdoc-3930011 to C.G.S. and 1960559 to J.C.S.) and by a Fogarty International Research Collaboration Award-National Institute of Health grant (RO3-TW00363 to E.L.H. and J.C.S.).

REFERENCES

1. M.V.L. Bennett, T.A. Bargiello, L.C. Barrio, D.C. Spray, E.L. Hertzberg, and J.C. Sáez, Gap junctions: New Tools, New Answers, New Questions, *Neuron* 6:305 (1991).
2. D.C. Spray, J.C. Sáez, E.L. Hertzberg, and R. Dermietzel, Gap Junctions, in liver: composition, function and regulation, in: The Liver: Biology and Pathobiology, I.M. Arias, W.B. Jakoby, H. Popper, D. Schachter, and D.A. Shafritz, eds., Raven Press Ltd., New York (1994).
3. J.C. Sáez, J.A. Connor, D.C. Spray, M.V.L. Bennett, Hepatocyte gap junctions are permeable to the second messengers inositol 1,4,5-trisphosphate and calcium ions, *Proc. Natl. Acad. Sci. USA* 86:2708 (1989).
4. M.H. Nathanson, J. Boyer, Mechanisms and regulation of bile secretion, *Hepatology* 14:551 (1991).
5. V.M. Berthoud, V. Iwanij, A.M. Garcia, J.C. Sáez, Connexins and glucagon receptors during development of the rat acinus, *Am. J. Physiol.* 263:G650 (1992).
6. M. Nathanson, A.D. Burgstahler, A. Mennone, M.B. Fallon, C. González, and J.C. Sáez, Organization of intercellular Ca^{2+} waves in rat liver, *Am. J. Physiol.* 269:G167 (1995).
7. K. Willecke, H. Henneman, E. Dahl, R. Jungbluth, and R. Heynkes, The diversity of connexin genes encoding gap junctional proteins, *Eur J Cell Biol* 56:1 (1991).
8. M.B. Fallon, M.H. Nathanson, A. Mennone, J.C. Sáez, and J.M. Anderson, Altered expression and function of hepatocyte gap junctions following common bile duct ligation, *Am. J. Physiol.* 268:C1186 (1995).
9. O. Traub, P.M. Druge, and K. Willecke, Degradation and resynthesis of gap junction protein in plasma membranes of regenerating liver after partial

hepatectomy or cholestasis, Proc. Natl. Acad. Sci. USA 80:755 (1983).
10. J.C. Sáez, and N.F. LaRusso, Intrahepatic bile duct epithelial cells are functionally coupled by connexin43-enriched gap junctions, *Gastroenterology* 1993;A-266 (1993).
11. W.R. Loewenstein, Junctional intercellular communication and the control of growth, *Biochem. Biophy. Acta* 560:1 (1979).
12. M.M. Atkinson, and J.D. Sheridan, Reduced junctional permeability in cells transformed by different viral oncogenes, in: Gap junctions, M.V.L. Bennett, D.C. Spray, eds. Cold Spring Harbor Laboratory, Cold Spring Harbor, (1985).
13. N. Bucher, Regeneration of mammalian liver, *Int. Rev. Cytol.* 15:245 (1963).
14. B.A. Haber, K.L. Mohn, R.H. Diamond, and R. Traub, Induction patterns of 70 genes during nine days after hepatectomy define the temporal course of liver regeneration, *J. Clin. Invest.* 91:1319-1326 (1993).
15. J.M. Albrecht, J.S. Hoffman, B.T. Kren, and C.J. Steer, Cuclin and cyclin-dependent kinase 1 mRNA expression in models of regenerating liver and human liver diseases, *Am. J. Physiol.* 265:G857 (1993).
16. A.G. Yee, and J.P. Revel, Loss and reappearance of gap junctions in regenerating liver, *J. Cell Biol.* 78:554 (1978).
17. S.B. Yancey, E. Easter, and J.P. Revel, Cytological changes in gap junctions during liver regeneration, *J. Ultrastruct. Res.* 67:229 (1979).
18. R. Dermietzel, S.B. Yancey, O. Traub, K. Willecke, and J.P. Revel, Major loss of the 28-kD protein of gap junction in proliferating hepatocytes, *J. Cell Biol.* 105:1925 (1987).
19. B.T. Kren, N.M. Kumar, S.Q. Wang, N.B. Gilula, and C.J. Clifford, Differential regulation of multiple gap junction transcripts and proteins during rat liver regeneration, *J. Cell Biol.* 123:707 (1993).
20. W.J. Brattin, E.A.Jr. Glende, and R.O. Recknagel, Pathological mechanisms in carbon tetrachloride hepatotoxicity, *J. Free Radical. Med.* 1:27 (1985).
21. J. Baron, J.A. Redich, and F.P. Gengerich, Immunohistochemical localization of cytochrome P_{450} in rat liver, *Life Sci.* 23:2627 (1978).
22. T. Miyashita, A. Takeda, M. Iwai, and T. Shimazu, Single administration of hepatotoxic chemicals transiently decreases the gap-junction levels of connexin 32 in rat liver, *Eur. J. Biochem.* 96:37 (1991).
3. A. Panduro, F. Shalaby, L. Biempica, D.A. Shafritz, Changes in albumin, -fetoprotein and collagen gene transcription in CCl_4 induced hepatic fibrosis, *Hepatology* 5:64 (1988).
24. P. Greenwel, J. Rubin, M. Schwartz, E.L. Hertzberg, and M. Rojkind, Liver fat-storing cell clones obtained from a CCl_4-cirrhotic rat are heterogeneous with regard to proliferation, expression of extracellular matrix components, interluekin-6, and connexin43, *Lab. Invest.* 69:210-216 (1993).
25. A. Geerts, J.M. Lazou, P. De Bleser, and E. Wisse, Tissue distribution, quantitation and proliferation kinetics of fat-storing cells in carbon tetrachloride-injured rat liver, *Hepatology* 13:1193 (1991).
26. A. Panduro, F. Shalaby, F.R. Weiner, L. Biempica, M.A. Zern, D.A. Shafritz, Transcriptional switch from albumin to alpha-fetoprotein and

changes in transcription of other genes during carbon tetrachloride induced liver regeneration, *Biochemistry* 25:1414 (1986).
27. J.C. Sáez, W.A. Gregory, R. Dermietzel, E.L. Hertzberg, T. Watanabe, L. Reid, M.V.L. Bennett, and D.C. Spray, cAMP delays disappearance of gap junctions between pairs of rat hepatocytes in primary cultures, *Am. J. Physiol.* 257:C1-C11 (1989).
28. R. Carlsson, E. Engvall, A. Freeman, and E. Ruoslahti, Laminin and fibronectin in cell adhesion: enhanced adhesion of cells from regenerating liver to laminin, *Proc. Natl. Acad. Sci. USA* 78:2403 (1981).
29. M. Goyette, C.J. Petropoulus, P.R. Shank, and N. Fausto, Expression of a cellular oncogene during liver regeneration, *Science* 219:510 (1983).
30. C. Gingalewski, K. Wang, M.G. Clemens, and A. De Maio, Posttranscriptional regulation of connexin-32 expression in liver during acute inflammation, *J. Cell Physiol.* 166:461 (1996).
31. H. Henneman, G. Kozek, G. Kozjek, E. Dahl, B. Nicholson, and K. Willecke, Molecular cloning of mouse connexin26 and -32 similar genomic organization but distinct promoter sequences of two gap junctions, *Eur. J. Cell Biol.* 58:81 (1992).
32. M. Mourelle, and B. Rubalcava, Regeneration of the liver after carbon tetrachloride Differences in adenylate cyclase and pancreatic hormone receptors, *J. Biol. Chem.* 256:1656 (1981).
33. O. Traub, L. Look, D. Paul, and K. Willecke, Cyclic adenosine monophosphate stimulates biosynthesis and phosphorylation of the 27 kDa gap junction protein in cultured mouse hepatocytes, *Eur. J. Cell Biol.* 43:48 (1987).
34. J.C. Sáez, M.V.L. Bennett, and D.C. Spray, Hepatocyte gap junctions: metabolic regulation and possible role in liver metabolism, in:Transduction in Biological Systems, C. Hidalgo, J. Bacigalupo, E. Jaimovich, J. Vergara, eds., Plenum. Publ. Co., New York, (1990).
35. P. Odin, and B. Öbrink, The cell-suface expression of the cell adhesion molecule cellCAM 105 in rat fetal tissues and regenerating liver, *Exp. Cell Res.* 179:89 (1988).
36. B.A. Imhof, P.H. Vollmers, S.L. Goodman, and W. Birchmeier, Cell-cell interaction and polarity of epithelial cells: specific perturbation using a monoclonal antibody, *Cell* 35:667 (1983).
37. R.M. Mege, F. Matsuzaki, W.F. Gallin, J.I. Golber, B.A. Cunningham, and G.M. Edelman, Construction of epithelioid sheets by transfection of mouse sarcoma cells with cDNAs for chicken cell adhesion molecules, *Proc. Natl. Acad. Sci. USA* 85:72 (1988).
38. L.S. Musil, B.C. Cunningham, G.M. Edelman, and D.A. Goodenough. Differential phosphorylation of the gap junction protein connexin43 in junctional communication-competent and -deficient cell lines, *J. Cell Biol.*111:2077 (1990).
39. L.S. Musil, and D.A. Goodenough, Biochemical analysis of connexin43 intracellular transport, phosphorylation, and assembly into gap junctional plaques, *J. Cell Biol.* 115:1357 (1991).
40. M.J. Czaja, A. Geerts, J. Xu, P. Schmiedeberg, and Y. Ju, Monocyte chemoattractant protein 1 (MCP-1) expression occurs in toxic rat liver injury and human liver disease, J. Leukocyte Biol. 55:120 (1994).

41. J. Baron, J.A. Redich, and F.P. Gengerich, Immunohistochemical localization of cytochrome P_{450} in rat liver, *Life Sci.* 23:2627 (1978).
42. S.J. Johson, J.E. Hines, and A.D. Burt, Macrophage and perisinusoidal cells kinetics in acute liver injury, J. Pathol. 166:351 (1992).
43. D. Polacek, R. Lal, M.V. Volin, and P.F. Davies, Gap junctional communication between vascular cells. Induction of connexin43 messenger RNA in macrophage foam cells of atherosclerotic lesions, *Am.J. Pathol.* 142:593 (1993).
44. E.C. Beyer, and T.H. Steinberg, Evidence that the gap junction protein connexin-43 is the ATP-induced pore of mouse macrophages. *J Biol Chem* 266:7971 (1991).
45. P.I. Jara, M.P. Boric, and J.C. Sáez, Leukocytes express connexin43 after activation with lipopolysaccharide and appear to form gap junctions with endothelial cells after-ischemia-reperfusion, *Proc. Natl. Acad. Sci. USA* 92:7011 (1995).
46. M. Zhang, and S.S. Thorgeirsson, Modulation of connexins during differentiation of oval cells into hepatocytes, *Exp. Cell Res.* 213:37 (1994).
47. X.X. Fu, C. Su, Y. Lee, R. Hintz, L. Biempica, R. Snyder, and C.E. Rogler, Insulin-like growth factor II expression and oval cell proliferation associated with hepatocarcinogenesis in woodchuck hepatitis virus carriers, *J. Virol.* 62: 3422 (1988).
48. J. Travis, The search for the liver stem cells picks up, *Science* 259:1829 (1993).
49. M.S. Pepper, and P. Meda, Fibroblast growth factor increases junctional communication and connexin 43 expression in microvascular endothelial cells, *J. Cell Physiol.* 153:196 (1992).
50. J.C. Sáez, M.V.L. Bennett, and D.C. Spray, Carbon tetrachloride at hepatotoxic levels blocks reversibly gap junction between rat hepatocytes, *Science* 236:967 (1987).
51. J.L. James, D.S. Friend, J.R. MacDonald, and Smuckler, Alterations in hepatocyte plasma membrane in carbon tetrachloride poisoning. Freeze-fracture analysis of gap junction and electron spin resonance analysis of lipid fluidity, *Lab. Invest.* 54:268 (1986).

CARDIAC MYOCYTES GAP JUNCTIONS: PHOSPHORYLATION OF CX43 THROUGH A PROTEIN KINASE C-DEPENDENT PATHWAY

Juan C. Sáez[1], Angus C. Nairn[2], Andrew J. Czernik[2], Glenn I.Fishman[3], David C. Spray[4], and Elliot L. Hertzberg[4,5]

[1]Departamento de Ciencias Fisiológicas, Facultad de Ciencias Biológicas, Pontificia Universidad Católica de Chile, Santiago, Chile.
[2]Laboratory of Molecular and Cellular Neuroscience, The Rockefeller University, New York, NY 10021-6399 U.S.A.
[3]Department of Medicine, Division of Cardiology, Albert Einstein College of Medicine, Bronx, NY 10461, U.S.A.
[4]Department of Neuroscience, Albert Einstein College of Medicine, Bronx, NY 10461, U.S.A.
[5]Department of Anatomy and Structural Biology, Albert Einstein College of Medicine, Bronx, NY 10461, U.S.A.

INTRODUCTION

Gap junction are aggregates of intercellular channels permeable to ions and small molecules, providing a pathway by which cells may respond synchronously to extracellular signals[1]. In the heart, gap junctions allow myocytes to function as syncytia, providing a means for the propagation of electrical activity between cells and enabling the propagation of contraction waves along the myocardium[2,3].

Connexins (Cxs), the protein subunits of gap junctions, comprise a family of homologous molecules[4]. All Cxs analyzed present four transmembrane domains and show both the amino- and carboxy-termini at the cytoplasmic surface of the plasma membrane. The carboxy-terminus is one of the protein domains that shows the greatest divergence in primary sequence and lenght between Cxs. In several Cxs, this domain presents a number of putative consensus sequences for phosphorylation, suggesting that protein phosphorylation may influence the function of gap junctions in a Cx specific manner[5].

Cx43 is a phosphoprotein[6,7] and in heart is mainly found at intercalated disks[8-11]. In heart, two phosphorylated forms of Cx43 can be detected either by immunoblotting or autoradiographic analysis of ^{32}P-labeled Cx43 separated in SDS-PAGE. Frequently, the unphosphorylated form of Cx43 resides in the cytoplasmic compartment(s)[7,11]. Inhibition of intracellular vesicular trafficking with monensin or brefeldin A leads to the accumulation in cytoplasmic compartments of a intermediate phosphorylated form of Cx43[9,12], indicating that at least some phosphorylation events occur prior to assembly of Cx43 into gap junctions.

On the other hand, several studies have shown that activation of various second messenger pathways can affect intercellular gap junctional communication of adult[13] and neonatal cardiac gap junctions[15,16]. Changes in cardiac myocyte coupling have been associated to phosphorylation of Cx43 mediated by cAMP-, cGMP-dependent protein kinase, or protein kinase C (PKC)[14,15]. In addition, functional changes associated to protein phosphorylation mediated by tyrosine kinase or MAP kinase have been observed in cell lines that express Cx43 gap junctions[6,16-18]. Phosphorylation of Cx43 by the tyrosine kinase src gene product induces inhibition of intercellular communication[6,18]. In a rat liver epithelial cell line (T51B), epidermal growth factor reduces gap junctional communication and Cx43 phosphorylation in a PKC-independent manner, through a mechanism involving MAP kinase activation[16,17]. In several cell types, activation of PKC-dependent pathways leads to a rapid inhibition of gap junctional communication[5]. The PKC-dependent reduction in cellular communication observed in Clone 9 cells was associated with phosphorylation of seryl residues in Cx43 that are distinct from those that are phosphorylated under basal condition, and correlated with internalization of junctional membranes[19]. In Novikoff hepatoma cells, TPA reduced the insertion of Cx43 into the plasma membrane[20]. Moreover, it has been proposed as the mechanism responsible for alterations in the unitary conductance of the channel[21]. On the contrary, in rat Sertoli cells and human smoth muscle cells of the corpus cavernosus, which express Cx43[22,23], activation of PKC-dependent pathway increases cell-cell coupling[22,24].

Here we demonstrate the involvement of a PKC-dependent pathway in the regulation of the state of phosphorylation of Cx43 in neonatal rat cardiocytes. Two-dimensional tryptic phosphopeptide maps of ^{32}P-labeled Cx43 were used to identify several phosphopeptides whose state of phosphorylation was mediated by a PKC-dependent pathway. We also demonstrate that a recombinant Cx43 C-terminal polypeptide, corresponding to residues 243-382 (Cx43[243-382]), was phosphorylated in vitro by PKC at Ser-368 and Ser-372. However, the pattern of phosphopeptide obtained from in vitro phosphorylation of Cx43[243-382] with PKC did not correspond to the TPA-regulated phosphopeptides that were identified in the phosphopeptide maps of Cx43 derived from ^{32}P-prelabeled myocytes. Our data indicate that the phorbol ester-dependent phosphorylation of Cx43 observed in vivo is not mediated directly by PKC.

MATERIAL AND METHODS

Chemicals. Pancreatin, RPMI 1640 medium and fetal bovine serum were from GIBCO. Tosylphenylalanyl chloromethyl ketone-treated trypsin was from Pierce (Rockford, IL, U.S.A.). Formalin-fixed Staphylococcus aureus Cowan cell (Pansorbin) and phenylmethylsulfonyl fluoride (PMSF) were from Calbiochem. Staphylococcus Aureus V8 protease was from Boehringer. [gamma^{32}P]ATP and [^{32}P]- orthophosphoric acid were from New England Nuclear. Acetonitrile was from Burdick-Jackson. XAR-5 X-ray film and cellulose TLC sheets were from Eastman Kodak. Phosphocellulose P-81 paper was from Whatman. ATP, DL-phosphoserine, DL-phosphothreonine, DL-phosphotyrosine, L-glutamine, Triton X-100, dithiothreitol (DTT), Nonidet P-40, diolein, phosphatidylserine, bovine serum albumin, EDTA, Tris, TPA, 4 -phorbol, staurosporine (ST) and all other reagents were obtained from Sigma. PKC was purified from rat brain essentially as described[25]. MAP kinase ($p44^{npk}$), and $p34^{cdc2/cyclinB}$ preparations purified from sea star (Pisaster ochraceus) oocytes were generously provided by Dr Steven Pelech (University of British Columbia). A well-characterized rabbit anti-Cx43 antibody[19,26] was used for immunofluorescence, immunoblotting and immunoprecipitation studies.

Cultures of rat neonatal cardiac myocytes. Neonatal rat cardiac myocytes were prepared as described by Blondel et al.[27].

Bacterial expression of the rat Cx43 carboxy-terminus polypeptide. The carboxy terminal cytoplasmic tail of rat Cx43 was expressed using the T7 RNA polymerase system[28]. A cDNA fragment encoding amino acids 243-382 of rat Cx43, with predicted molecular mass of 15.3 kDa, was derived by the polymerase chain reaction, using the following primer pair:

5'-AAGCATATGAGCGACCTTACCA-3' (sense)
5'-AGGGATCCTAGATCTCCAGGTCATCATCAGG-3' (antisense)

Restriction sites for Nde1 and BamH1 (underlined), were included to facilitate cloning into the pET3c expression vector. The Nde1 site resulted in the substitution of a methionine for the lysine normally found at position 243. The resulting plasmid, pET3C-Cx43 [243-382], was transformed into E. coli strain BL21 (DE3) pLysS. Overnight cultures, grown in Luria Broth medium supplemented with ampicillin and chloramphenicol (25 µg/ml each), were diluted 10-fold into fresh media and antibiotics and grown for 4 h. Isopropyl thiogalactoside to induce the -galactosidase was added (0.4 mM final concentration) and bacteria were grown for one additional hour. Cells were then pelleted, resuspended in 50 mM Tris-HCl, pH 7.5, 10 mM EDTA, 0.1% Triton X-100, 30% sucrose, leupeptin and chymostatin (0.5 mg/ml each). The cells were lysed by repeated cycles of freezing and thawing, followed by brief sonication to shear the nucleic acid. Lysates were clarified by centrifugation at 5,000 x g for 15 min and then applied to an Amicon 30 Centriprep column (W.R. Grace and Co., Beverly, MA, U.S.A.). The filtrate was then applied to an Amicon 10 centriprep column, where the 15.3 kDa protein was concentrated in the retentate. These selective filtration steps resulted in a preparation of approximately 80% purity, as judged by Coomassie blue staining and immunoblot analysis.

Immunoblotting. Protein concentration of cell lysates was measured by the method of Bradford[29]. Aliquots of heart cell homogenates containing equal amount of proteins (50 μg) were analyzed by immunoblot as previously described for other cell types[19].

Radiolabeling of cardiomyocytes and immunoprecipitation of Cx43. Cells were rinsed with phosphate-free minimal essential medium (MEM, Gibco), medium was added containing [^{32}P]-$PO_{4^-}{}^{3}$ (Dupont, NEN, Boston, MA, U.S.A.; 0.25 mCi/2.5 ml/plate), and cells were labeled for 2.5 h. Cells were treated with drugs for different periods of time as indicated and then harvested in 2 mM PMSF in PBS, pH 7.4. Cell suspensions were centrifuged, and lysed in a 1:1 mixture of lysis buffer/buffer A (lysis buffer: 2% SDS, 100 U/ml trasylol, 200 μg/ml soybean trypsin inhibitor, 1 mg/ml benzamidine, 1 mg/ml epsilon-aminocaproic acid, 2 mM PMSF; buffer A: 5% Nonidet P-40, 20 mM EDTA, 250 mM NaCl, 20 mM $Na_2P_2O_7$, 100 mM NaF, 100 mM phosphate buffer, pH 7.4). To reduce non-specific adsorption, the homogenates were precleared by incubation with 15 μl Pansorbin for 15 min, centrifuged and the pellet discarded. The supernatant was incubated with anti-Cx43 antibody for 2 h and then with 50 μl Pansorbin for 30 min. The suspensions were centrifuged, supernatant fractions discarded, the pellets rinsed once in buffer B (190 mM Tris-HCl, 6 mM EDTA, 2.5% Triton X-100, 100 U/ml trasylol, 50 mM Tris-HCl, pH 8.3) and subsequently 3 times with PBS, pH 7.4. All steps of the immunoprecipitation included 2 mM PMSF to reduce proteolysis. The pellets were resuspended in Laemmli buffer[30], proteins were separated by 8% SDS-PAGE, electro-transferred to nitrocellulose and detected by autoradiography.

Two-dimensional tryptic phosphopeptide mapping and phosphoamino acid analysis. Gel pieces containing each of the ^{32}P-labeled forms of Cx43 were cut from the dried gel using the autoradiogram as a guide. Two-dimensional tryptic mapping was performed as described[31].

Aliquots from the tryptic digests were hydrolyzed in 6 N HCl and phosphoamino acid analysis was performed as described[32].

Phosphorylation of Cx43[243-382]. Phosphorylation of peptide Cx43[243-382] by PKC was carried out as described previously for other peptides[33], using aliquots of a fraction enriched in the polypeptide expressed by transformed E. coli. An equivalent fraction of bacteria transformed with vector alone was used to control for ^{32}P-incorporation into unrelated endogenous polypeptides. For phosphorylation by MAP kinase, the reaction mixture contained 25 mM MOPS, pH 7.2, 20 mM Mg acetate, 1 mM EGTA, 1 mM DTT, approximately 25 ng of the purified kinase, and 150 μM [gamma-^{32}P]ATP. Incubations were performed at 30°C for 30-180 min. For phosphorylation with cdc2 kinase, the reaction mixture contained 50 mM Tris-HCl, pH 8.0, 10 mM Mg acetate, 1 mM EGTA, 1 mM DTT, 150 μM [gamma- ^{32}P]ATP and approximately 1-2 μg $p34^{cdc2/cyclinB}$. Incubations were performed at 30°C for 30-180 min.

Purification and microsequencing of phosphopeptides. After phosphorylation by PKC, aliquots of the recombinant polypeptide were separated by SDS-PAGE using 15% acrylamide gels, and then subjected to two-dimensional phosphopeptide mapping as described[31]. Solid TCA was added to the remainder of the PKC reaction mix to yield a final concentration of 10% TCA. After 10 h at 0°C, the protein precipitate was recovered by centrifugation, and the pellet was washed twice with acetone to remove traces of TCA. The polypeptide was then digested with TPCK-treated trypsin (2 μg) in 50 mM NH_4HCO_3, pH 8.0, 1 mM DTT, 1 M urea for 12 h at 37°C. The peptide samples were loaded onto a hydroxyethyl aspartamide column (0.4x25 cm, PolyLC, Inc.) in start buffer (0.1% TFA, 85% CH_3CN, adjusted pH 4.5 with triethylamine). Gradient elution was in the same buffer, using decreasing concentrations of acetonitrile. Phosphoeptides were detected by monitoring absorbance at 214 nm and ^{32}P-content by Cerenkov counting. Protein sequencing was carried using an Applied Biosystems (Foster City, CA, U.S.A.) AB-470 gas phase sequencer. The ^{32}P-labeled phosphopeptides were derivatized with ethanethiol as described[34] prior to automated Edman degradation. Identification of the phenylthiohydantoin amino acid derivatives was accomplished by on-line C_{18} HPLC analysis and was semiquantitative. The appearance of phenylthiohydantoin-(PTH) derivatives of S-ethyl-cysteine in the analysis of each Edman cycle indicated the position of an original phosphoseryl residue.

RESULTS

Regulation of the state of phosphorylation of Cx43. Immunoblot analysis of rat heart ventricle or cultured neonate rat cardiac myocytes showed that Cx43 exist as a major doublet of phosphorylated forms migrating at about 43-45 kDa (P2 and P3), and a minor band of 41 kDa (NP), representing the unphosphorylated form of Cx43 (Fig. 1)[7,9,10]. Cultured neonatal cardiomyocytes were treated with ST or TPA. Both the total amount of Cx43 protein and state of phosphorylation of Cx43 evaluated by immunoblotting were similar under all conditions studied (Fig. 1A).

The steady-state level of ^{32}P-labeling of Cx43 was also evaluated in cardiocytes prelabeled with $^{32}P_i$. Cells were treated for 60 min with 300 nM ST, or with 300 nM ST with the addition of 600 nM TPA during the final 30 min. ST reduced the amount of ^{32}P-incorporation, mainly in the P3 form of Cx43 (Fig. 1B, lane 2). The ST-induced reduction in ^{32}P-incorporation was substantially reversed by the addition of 600 nM TPA during the final 30 min of incubation in the presence of ST (Fig. 1B, lane 3). Treatment with 600 nM 4 -phorbol, an inactive phorbol ester analog, did not change the ^{32}P-incorporation into Cx43 as compared to the level detected in ST-treated cells (not shown). Phosphoamino acid analysis of P2 and P3 froms immunoprecipitated from control cells or cells submitted to any of the drug treatments indicated that Cx43 was phosphorylated almost exclusively at seryl residues (not shown).

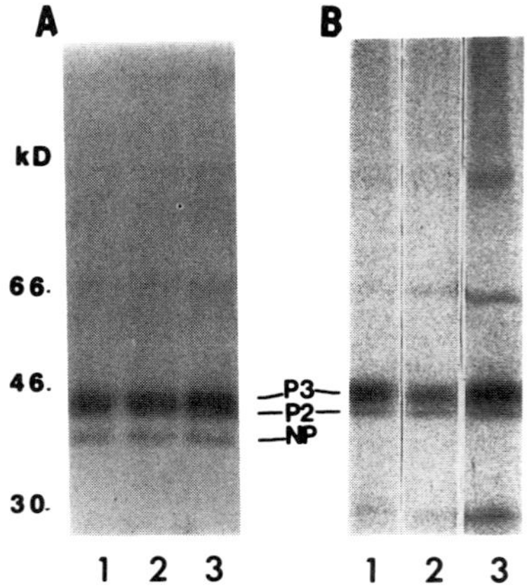

Figure 1. State of phosphorylation of Cx43 in cardiocytes treated with ST and TPA. (A) Proteins from neonatal rat cardiac myocytes were separated in 8% SDS-PAGE, transferred to nitrocellulose, and Cx43 was immunodetected with anti-Cx43 antibody and ^{125}I-protein A. Treatment with 300 nM ST for 60 min (lane 2) or 300 nM ST for 30 min plus 600 nM TPA for additional 30 min (lane 3) did not significantly affect the levels of each Cx43 species as compared to control (lane 1). (B) In ^{32}P prelabeled cells only P2 and P3 incorporated ^{32}P (lane 1) and their state of phosphorylation was evaluated after treatment with 300 nM ST for 60 min (lane 2); or 300 nM ST for 30 min plus 600 nM TPA for additional 30 min (lane 3).

Phosphopeptide maps of Cx43. Two-dimensional tryptic phosphopeptide maps of the two forms (P2 and P3) of Cx43 isolated by immunoprecipitation from ^{32}P-labeled control cardiocytes or cardiocytes treated with ST or ST plus TPA were performed. While the P2 map showed 7 main phosphopeptides (1, 4, 5, 6, 7, 8 and 10) (Fig. 2D), the P3 map contained the same 7 phosphopeptides detected in the P2 map plus 5 distinct additional ones (2, 3, 9, 11 and 12) (Fig. 2A). ST treatment reduced the ^{32}P-incorporation in three phosphopeptides (1, 6 and 8; Fig. 2E) present in the map of P2 of control cells (Fig. 4D); the ST-induced inhibition of peptide 6 phosphorylation was less prominent in P3 (Fig. 2B) than in P2 (Fig. 2E). In P3 (Fig. 2B), a ST-induced reduction in ^{32}P-incorporation was detected in phosphopeptides 2 and 3 in addition to those that were detectably decreased in the P2 phosphopeptides. Treatment with TPA increased the incorporation of ^{32}P in essentially the same phosphopeptides of P2 and P3 (1, 2, 3 and 6) that were reduced by ST (Fig. 2C, F). An additional increase in ^{32}P-incorporation induced by TPA was also detected in phosphopeptides 2 and 3 in P2 (Fig. 4F) and in phosphopeptide 7 in P3 (Fig. 2C).

Phosphorylation of Cx43[243-382] by PKC *in vitro* and identification of the sites of phosphorylation. Cx43[243-382] was phosphorylated with purified PKC for 30 min and the samples analyzed by SDS-PAGE (Fig. 3A, lane 3). A prominent Coomassie stained band at 15 kDa was observed, consistent with the predicted size for this polypeptide. Bands with higher electrophoretic mobility presumably correspond to degradation products of Cx43[243-382], since they were not detected in the lane loaded with a protein fraction obtained from untransformed E. coli (Fig. 3A, lane 2). The 15-kDa band was phosphorylated rapidly by PKC (Fig. 3A, lane 3, and data not shown) and phosphoamino acid analysis revealed only phosphoserine (Fig. 3B).

A tryptic digest of Cx43[243-382] phosphorylated by PKC was subjected to C_{18}-reversed-phase chromatography to isolate the ^{32}P-labeled phosphopeptides. Two major ^{32}P-labeled phosphopeptides were separated. Phosphopeptide 1 eluted in the void volume, whereas phosphopeptide 2 was retained on the column and eluted at approximately 28% acetonitrile. Phosphopeptide 1 was purified futher by chromatography using a hydroxyethyl aspartamide column. Purified phosphopeptides 1 and 2 were derivatized with ethanethiol and then subjected to peptide sequencing analysis. Phosphopeptide 1 yielded the sequence corresponding to residues 367-370 of Cx43, with a phosphorylation site identified at Ser-368. Phosphopeptide 2 was determined to be a mixture of two overlapping phosphopeptides in a ratio of approximately 10:1. The major phosphopeptide corresponded to residues 371-382, with Ser-372 as the phosphorylation site. The minor phosphopeptide corresponded to residues 367-382, with both Ser-368 and Ser 372 as the phosphorylation sites.

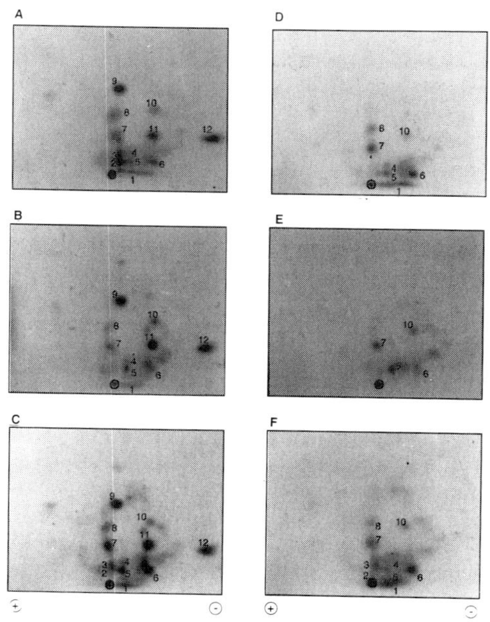

Figure 2. Two-dimensional phosphopeptide maps of Cx43. Cx43 forms were excised form the gels shown in Fig. 1 and were subjected to two-dimensional tryptic mapping. The P3 band (A, B and C) and the P2 band (D, E, F) correspond to control, treatment with ST and treatment with ST plus TPA, respectively. Phosphopeptides are labeled from 1-12. The origin is indicated at bottom middle.

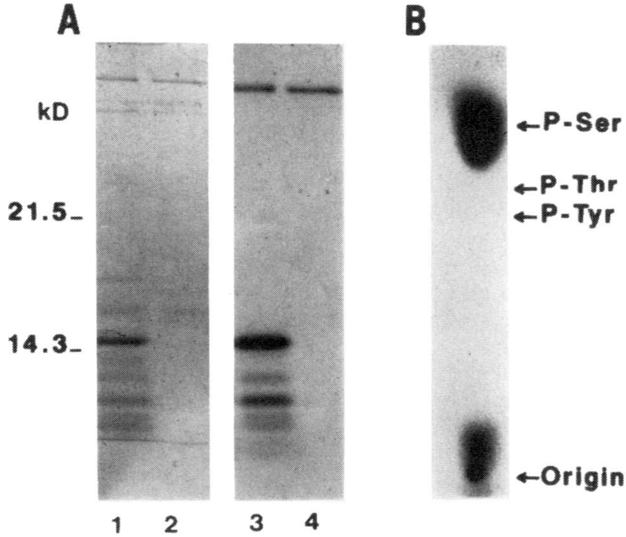

Figure 3. Phosphorylation of Cx43[243-382]. (A) E. coli cell extracts enriched in the 15.3 kDa Cx43[243-382] (lanes 1 and 3) were phosphorylated by PKC and the samples analyzed by SDS-PAGE (15% acrylamide). Extracts from E. coli cells not transfected with plasmid are shown as controls (lanes 2 and 4). Coomassie blue staining (lanes 1 and 2) was followed by autoradiography (lanes 3 and 4). In extracts from non-transformed cells only autophosphorylated PKC was detected (A, lane 4). (B) Phosphoamino acid analysis of the 15 kDa polypeptide revealed only phosphoserine. Radioactivity close to the origin corresponded to partially hydrolyzed polypeptide.

The failure to cleave the longer peptide at Arg-370 resulted from the fact that both Ser-368 and Ser-372 are phosphorylated.

Analysis of two-dimensional tryptic phosphopeptide maps of Cx43[243-382] phosphorylated *in vitro* by PKC. Two-dimensional tryptic phosphopeptide maps of Cx43[243-382] phosphorylated by PKC revealed one major phosphopeptide and two minor ones (Fig. 2D). Mixing experiments revealed that the major phosphopeptide from Cx43[243-382] did not comigrate with any prominant phosphopeptide from Cx43 isolated from ^{32}P-labeled cardiocytes (Fig. 2H). Two-dimensional V-8 protease phosphopeptide maps of Cx43 and Cx43[243-382] also showed a lack of co-migration of the respective phophopeptides (data not shown).

Analysis of two-dimensional tryptic phosphopeptide maps of Cx43[243-382] phosphorylated *in vitro* by MAP kinase and cdc2 kinase. The observation that the ST/TPA-responsive phosphopeptides in Cx43 were distinct from those obtained upon in vitro phosphorylation of Cx43[243-382] by PKC was surprising. Furthermore, this disparity raised questions regarding the identity of the protein kinase(s) that are responsible for phosphorylation of these and other sites in Cx43 in rat cardiocytes. We therefore analyzed whether Cx43[243-382] could serve as a substrate for two other protein kinases, MAP kinase and cdc2 kinase. Both enzymes rapidly phosphorylated

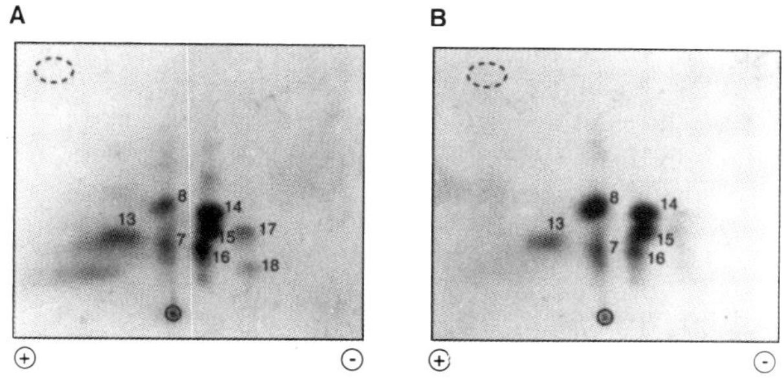

Figure 4. Two-dimensional phosphopeptide maps of Cx43[243-382] phosphorylated by cdc2 and MAP kinase. (A) and (B) represent Cx43[243-382] phosphorylated by cdc2 or MAP kinase, respectively. Circled areas with dotted lines indicate the mobility of phenol red used as indicator. The origin is indicated at bottom middle. Phosphopeptides 7 and 8 showed similar mobilities to those detected in P2 and P3 of Cx43. Phosphopeptides with mobilities different to those detected in P2 and P3 of Cx43 are numbered 13-16.

Cx43[243-382] (data not shown). Two-dimensional phosphopeptide mapping was used to analyze samples of Cx43[243-382] phosphorylated by MAP kinase or cdc2 kinase. Both protein kinases phosphorylated Cx43[243-382] exclusively on seryl residues (data not shown), and the pattern of phosphopeptides revealed by phosphopeptide mapping indicated that both protein kinases appeared to be phosphorylated at the same sites (Fig. 4). However, these phosphopeptides did not appear to co-migrate with any of the ST/TPA-responsive phosphopeptides (compare Fig. 2 and 4).

DISCUSSION

We have previously shown that treatment with ST, an inhibitor of several classes of protein kinases, including PKC, Ca^{2+}/calmodulin-dependent kinase, tyrosine kinases and cyclic nucleotide-dependent kinases[35-37], induced a significant reduction both in dye and electrical coupling of neotal rat cardiomyocytes, without affecting the cellular distribution of Cx43[10]. Consistent we a previous report[10], we found that this effect was associated with a reduction in the steady-state ^{32}P- incorporation into Cx43, supporting the notion that the state of phosphorylation of Cx43 is related to the functional state of Cx43 junctional channels[1,5], instead of changes in the number of channels comprising the functional gap junctions. We also found that the inhibitory effect of ST on intercellular coupling and the state of phosphorylation of Cx43 were reversed toward normal values by acute treatment with TPA, presumably because this treatment overcomes the inhibition of PKC brought about by ST. These results suggest therefore that PKC activity regulates coupling between neonatal rat cardiocytes in a positive fashion.

In other cell types that express Cx43-containing gap junction channels, activation of PKC was found to induce cell uncoupling[5]. In these studies, PKC activation increased the phosphorylation of unphosphorylated form of Cx43

(Cx43-NP), resulting in a reduction of the electrophoretic mobilities of Cx43 to that of the phosphorylated forms (Cx43-P2 and-P3). In clone9 cells, activation of PKC induces further phosphorylation of the phosphorylated forms, Cx43-P2 and Cx43-P3, shifting their electrophoretic mobilities to forms with even lower electrophoretic mobilities, termed Cx43-HP[3]. These changes in the mobility of Cx43 have only been observed in cell types where under basal conditions the majority of Cx43 is the Cx43-NP form. In contrast, in this study in control cardiocytes Cx43 was predominantly phosphorylated and the immunoblot pattern of Cx43 was not affected by ST alone or with ST plus TPA.

The turnover rate of phosphate on Cx43 in neonatal rat heart myocytes under resting conditions is similar to the half-life of the protein, and it has been suggested that the protein remains phosphorylated during its life span[9]. It is possible that such studies might not easily detect rapidly turning over phosphorylation sites in a large population of phosphorylation sites which turnover much more slowly. Here we have demonstrated that a fraction of the ^{32}P-incorporation into Cx43 was reduced by a short exposure to ST and then rapidly increased upon addition of TPA. Rapid turnover of phosphate in Cx43 has been observed in vole fibroblasts[6] and in ovarian follicle cells following stimulation with luteinizing hormone[38].

In other cell types, PKC-dependent phosphorylation of unique sites correlated with aspects of gap junction assembly[20] or disassembly[19], In the present study, we have found that in cardiocytes treated with ST, the TPA-induced increase in ^{32}P- incorporation into Cx43 occurred among seryl residues that were phosphorylated under basal conditions. Thus, phosphopeptides unresponsive to regulation by ST or TPA may represent constitutive phosphorylation sites or sites phosphorylated prior to insertion of Cx43 into the plasma membrane[12].

In this study, we have found that Cx43[243-382] was phosphorylated by purified PKC at Ser-368 and Ser-372. In agreement with the identification of the sites of phosphorylation in the recombinant protein, we had previously reported that a synthetic peptide, Cx43[360-375], was phosphorylated by PKC at Ser-368 and Ser-372[10]. Two-dimensional tryptic maps of Cx43[243-382] phosphorylated by PKC in vitro showed a major phosphopeptide that did not co-migrate with any of the major tryptic phosphopeptides detected in Cx43 isolated from ^{32}P-labeled cells. This observation suggests the strong possibility that the PKC sites identified in vitro are not phosphorylated by PKC in vivo. Alternatively, in vivo phosphorylation of seryl residues distinct from Ser-368 and Ser-372 might limit the ability of trypsin to cleave at normal proteolytic cleavage sites, yielding phosphopeptides from cardiocyte Cx43 that migrate differently then those obtained from Cx43[243-382] phosphorylated by PKC. It has been observed that, the presence of a negative charge next to a trypsin cleavage site inhibits trypsin action[39]. Hence, while our data implicate a protein kinase other than PKC in phosphorylating Cx43 in cells treated with TPA, direct analysis of isolated Cx43 will be necessary to provide a clearer definition of the issue. Notably, the sequence surrounding Ser-368 and Ser-372 conform to consensus sequences for phosphorylation by PKC[40] and these sites are conserved in the sequence of both the rat and the human isoforms[41,42]. Thus, it seems possible that Ser-268 and Ser-372 are physiological targets for PKC under certain conditions, perhaps in cell types where activation of PKC results in uncoupling. Another possible explanation of the present results is that PKC was able to activate a different protein kinase that in turn phosphorylated Cx43. Previous studies have indicated that MAP kinase is activated by PKC[43]. In the present study we found that both MAP kinase and cdc2 kinase phosphorylated Cx43[243-382] at the same site(s). However, these site(s) were also different from the ST/TPA-responsive sites found in Cx43 phosphorylated in intact cells. Thus, these protein kinases are unlikely to be the final mediator of the TPA effect reported here. In support of these conclusion, phosphorylation of Cx43 in seryl residues 255, 279 and 282 by MAP kinase has been associated with cell uncoupling[16,17].

The recent demonstration that Cx43 can be phosphorylated via a cAMP-dependent pathway[38] offers a link between the functional and biochemical changes of Cx43 channels induced by inotropic signals mediated by -adrenoceptors[14]. Other positive inotropic agents function through activation of a_1-adrenoceptors, a mechanism that occurs without an increase in myocardial cAMP levels[44]. Activation of a_1-adrenoceptors stimulates phosphatidylinositol metabolism, generating IP_3 that induces Ca^{2+} release from intracellular stores, and thereby increases force of contraction in cardiac muscle[45]. Simultaneously with IP_3 generation there is release of diacylglycerol that activates PKC[46]. Therefore, the increase in coupling mediated by the PKC-dependent pathway reported here might favor the positive inotropic effect of agents that act through a_1-adrenoceptors.

Acknowledgments: This work was supported in part by a FONDECYT grant 1960559 (to J.C.S.), National Institute of Health grants GM-30667 (to E.L.H.) and TW00363 (FIRCA to E.L.H. and J.C.S.), DK-41368 and HL-38449 (to D.C.S.), and Physician Scientist Award HL-02391 (to G.I.F.). Protein sequence analysis was provided by The Rockefeller University Protein/DNA Technology Center, which is supported in part by NIH shared instrumentation grants and by funds provided by the US Army and Navy for purchase of equipment. We thank Dr. Steven Pelech for providing MAP kinase and cdc2 kinase, Ms. Delia M. Viera for preparation of primary cultures of rat neonatal cardiac myocytes.

REFERENCES

1. D.C. Spray, M.B. Rook, A.P. Moreno, J.C. Sáez, G.J. Christ, A.C. Campos de Carvalho, and G.I. Fishman, Cardiovascular gap junctions: gating properties, function and dysfunction, in: "Ion Channels in the Cardiovascular system: Function and Dysfunction", P.M. Spooner, A.M. Brown, W.A. Catterall, G.J. Kaczorowski, and H.C. Strauss, eds., Futura Publishing, New York, (1994).
2. M.S. Forbes, and N. Sperelakis, Intercalated discs of mammalian heart: a review of structure and function, *Tissue & Cell* 17:605 (1985).

3. M. Oyamada, H. Kimura, Y. Oyamada, A. Miyamoto, H. Ohshika, and M. Mori, The expression, phosphorylation, and location of connexin 43 and gap-junctional intercellular communication during the establishment of a synchronized contraction of cultured neonatal rat cardiac myocytes, *Exp. Cell Res.* 212:351 (1994).
4. M.V.L. Bennett, T.A. Bargiello, L. Barrio, D.C. Spray, E.L. Hertzberg, and J.C. Sáez, Gap junctions: New Tools, New Answers, New Questions, *Neuron* 6:305 (1991).
5. J.C. Sáez, V.M. Berthoud, A.P. Moreno and D.C. Spray, Multiplicity of controls in differentiated and undifferentiated cells and possible functional implications, in: "Advances in Second Messenger and Phosphoprotein Research", S. Shenolikar and A.C Nairn, eds., Raven Press, New York, (1993).
6. D.S. Crow, E.C. Beyer, D.L. Paul, S.S. Kobe, and A.F. Lau, Phosphorylation of connexin43 gap junction protein in uninfected and Rous sarcoma virus-transformed mammalian fibroblasts, *Mol. Cell. Biol.* 10:1754 (1990).
7. L.S. Musil, B.C. Cunningham, G.M. Edelman, and D.A. Goodenough, Differential phosphorylation of the gap junction protein connexin43 in junctional communication-competent and -deficient cell lines, *J. Cell Biol.* 111:2077 (1990).
8. D.W. Laird, and J.-P, Revel. Biochemical and immunochemical analysis of the arrangement of connexin43 in rat heart gap junction membranes, *J. Cell Sci.* 97:109 (1990).
9. K.L. Puraman, D.W. Laird, & J.-P. Revel, Trapping an intermediate form of connexin43 in the Golgi, *Exp. Cell Res.* 206:85 (1993).
10. J.C. Sáez, A.C. Nairn, A.J. Czernik, D.C. Spray, and E.L. Hertzberg, Rat heart connexin43: regulation by phosphorylation in heart. in: "Gap Junctions", Progress in Cell Research, J.E. Hall, G.A. Zampighi, and R.M. Davis, eds., Elsevier Science Publishers,(1993).
11. L.S. Musil, and D.A. Goodenough, Multisubunit assembly of an integral plasma membrane channel protein, gap junction connexin43, occurs after exit from the ER. *Cell* 74:1065 (1993).
12. D.W. Laird, M. Castillo, L. Kasprzak, Gap junction turnover, intracellular trafficking, and phosphorylation of connexin43 in brefeldin A-treated rat mammary tumor cells, *J. Cell Biol.* 131:1193 (1996).
13. W.C. De Mello, Increase in junctional conductance caused by isoproterenol in heart cell pairs is suppressed by cAMP-dependent protein-kinase inhibitor, *Biochem. Biophys. Res. Commun.* 154:509 (1988).
14. J.M. Burt, and D.C Spray, Inotropic agents modulate gap junctional conductance between cardiac myocytes, *Am. J. Physiol.* 254:H1206 (1988).
15. B.R. Takens-Kwak, and H.J. Jongsma, Cardiac gap junctions: three distinct single channel conductances and their modulation by phosphorylating treatments, *Pflügers Arch.* 422:198 (1992).
16. B.J. Warn-Cramer, P.D. Lampe, W.E. Kurata, M.Y. Kanemitsu, L.W. Loo, W. Eckhart, and A,F. Lau, Characterization of the mitogen-activated protein kinase phosphorylation sites on the connexin-43 gap junction protein, *J Biol Chem* 271:3779 (1996).

17. M.Y. Kanemitsu, and A.F. Lau, Epidermal growth factor stimulates the disruption of gap junctional communication and connexin43 phosphorylation independent of 12-O-tetradecanoylphorbol 13 acetate-sensitive protein kinase C: the possible involvement of mitogen-activated protein kinase, *Mol. Biol. Cell* 4:837 (1993).
18. K.I. Swenson, H. Piwnica-Worms, H. McNamee, and D.L. Paul, Tyrosine phosphorylation of the gap junction protein connexin43 is required for the $pp60^{v\text{-}src}$-induced inhibition of communication, *Cell Regulation* 1:989 (1990).
19. V.M. Berthoud, M. Rook, E.L. Hertzberg, and J.C. Sáez, On the mechanism of cell uncoupling induced by a tumor promoter phorbol ester in clone 9 cells, a rat liver epithelial cell line, Eur. *J. Cell Biol.* 62:384 (1993).
20. D. Lampe, Analyzing phorbol ester effects on gap junctional communication: a dramatic inhibition of assembly, *J. Cell Biol.* 127:1895 (1994).
21. A.P. Moreno, J.C. Sáez, G.I. Fishman, and D.C. Spray, Human connexin43 gap junction channels: regulation of unitary conductances by phosphorylation of the channel protein, *Circ.Res.* 74:1050 (1994).
22. A.P. Moreno, A.C. Campos de Carvalho, G. Christ, and D.C. Spray, Gap junctions between corpus cavernosum smooth muscle cells: gating properties and unitary conductance, *Am. J. Physiol.* 264:C80 (1993).
23. M. Risley, I. Tan, C. Roy, and J.C. Sáez, Cell, age, and stage-dependent distribution of connexin43 gap junctions in testes, *J. Cell Sci.* 103:81 (1992).
24. F. Grassi, L. Monaco, G. Fratamico, S. Dolci, E. Ianni, M. Conti, F. Eusebi, and M. Stefanni, Putative second messengers affect cell coupling in the seminiferous tubules. *Cell Biol. Inter. Report.* 10:631 (1986).
25. J.R. Woodget, and T. Hunter, Isolation and characterization of two distinct forms of protein kinase C, *J. Biol. Chem.* 262:4836 (1987).
26. T. Yamamoto, A. Ochalski, E.L. Hertzberg, J.I. Nagy, LM and EM immunolocalization of the gap junctional protein connexin 43 in rat brain, *Brain Res.* 508:313 (1990).
27. R. Blondel, I, Roijen, and J.P. Cheneval, Heart cell in culture: a simple method for increasing the proportion of myoblasts, *Experientia* 27:356 (1971).
28. F.W. Studier, and B.A. Moffatt, Use of bacteriophage T7 RNA polymerase to direct selective high-level expression of cloned genes, *J. Mol. Biol.* 189: 113 (1986).
29. M.M. Bradford, A rapid and sensitive method for the quantitation of microgram quantities of protein utilizing the principle of protein-dye binding, *Anal. Biochem.* 72:248 (1976).
30. U.K. Laemmli, Cleavage of structural proteins during the assembly of the head of bacteriophage T4, Nature 227:680 (1970).
31. M.R. Picciotto, J.A. Cohn, J.A., G. Bertuzzi, P. Greengard, and A.C. Nairn, Phosphorylation of the cystic fibrosis transmembrane conductance regulator, *J. Biol. Chem.* 267:12743 (1992).
32. A.C. Nairn, and P. Greengard, Purification and characterization of Ca^{2+}/calmodulin dependent protein kinase I from bovine brain, *J. Biol. Chem.* 262:7273 (1987).

33. J.C. Sáez, A.C. Nairn, A.J. Czernik, D.C. Spray, E.L. Hertzberg, P. Greengard, and M.V.L. Bennett, Phosphorylation of connexin32, a hepatocyte gap junction protein, by cAMP-dependent protein kinase, protein kinase C and Ca^{2+}/ calmodulin-dependent protein kinase II, *Eur. J. Biochem.* 192:263 (1990).
34. C.F.B. Holmes, A new method for the selective isolation of phosphoserine-containing peptides, *FEBS Lett* 204:21 (1987).
35. D.M. Gadbois, Hamaguchi, J.R., Swank, R.A., and E.M., Bradbury, Staurosporine is a potent inhibitor of p34cdc2 and p34cdc2-like kinases, *Biochem. Biophys. Res. Commun.* 184:80 (1992).
36. S. Yamamoto, H. Jiang, K. Nishikawa, M. Ishihara, J.C. Wang, and R. Kato, Protein kinase C-dependent and independent actions of a potent protein kinase C inhibitor, staurosporine, *Eur. J. Mol. Pharmacol.* 227:113 (1992).
37. N. Yanagihara, E. Tachikawa, F. Izumi, S. Yasugawa, H. Yamamoto, and E. Miyamoto, Staurosporine: an effective inhibitor of Ca^{2+}/calmodulin-dependent protein kinase II, *J. Neurochem.* 56:294 (1991).
38. I. Granot and N. Dekel, Phosphorylation and expression of connexin-43 ovarian gap junction protein are regulated by luteinizing hormone, *J. Biol. Chem.* 269:30502 (1994)
39. W.J. Boyle, van der Geer, P., and T. Hunter, Phosphopeptide mapping and phosphoamino acid analysis by two-dimensional separation on thin-layer cellulose plates, in: Protein Phosphorylation, T. Hunter, and B.M. Sefton, eds., Academic Press, Inc., San Diego. *Methods in Enzymol.* 201:110 (1991).
40. P.J. Kennelly, and E.G. Krebs, Consensus sequences as substrate specificity determinants for protein kinases and protein phosphatases, *J. Biol. Chem.* 266:15555 (1991).
41. E.C. Beyer, D.L. Paul, and D.A. Goodenough, Connexin43: a protein from rat heart homologous to a gap junction protein from liver, *J. Cell Biol.* 105:2621 (1987).
42. G.I. Fishman, D.C. Spray, and L.A. Leinwand, Molecular characterization and functional expression of the human cardiac gap junction channel, *J. Cell Biol.* 111:589 (1990).
43. B. Marquardt, D. Frith, and S. Stabel, Signalling from TPA to MAP kinase requires protein kinase C, raf and MEK: reconstitution of the signalling pathway in vitro, *Oncogene* 9:3213 (1994).
44. H. Scholz, R. Brückner, A. Mügge, and C. Reupcke, Myocardial alpha-adrenoceptors and positive inotropy, *J. Mol. Cell. Cardiol.* 18:79 (1986).
45. T.M. Nosek, M.F. Williams, S.T. Zeigler, and R.E. Godt, Inositol trisphosphate enhances calcium release in skinned cardiac and skeletal muscle, *Am. J. Physiol.* 250:C807 (1986).
46. M.J. Berridge, and R.F, Irvine. Inositol phosphates and cell signalling, *Nature* 341:197 (1989).

REGULATION OF LENS GAP JUNCTIONS

Viviana M. Berthoud, Eileen M. Westphale, and Eric C. Beyer

Department of Pediatrics
Washington University School of Medicine, Box 8116
One Childrens Place
St. Louis, MO 63110

INTRODUCTION

Gap junctions are membrane specializations that contain groups of intercellular channels. These channels are permeable to ions and small molecules up to 1,000 Da. They communicate the cytoplasm of adjacent cells, thereby providing a direct pathway for the passage of ions, nutrients, metabolites and second messengers between coupled cells. Gap junctional channels are made of oligomeric assemblies of members of a family of related proteins called connexins (Cx). These proteins cross the plasma membrane four times leaving the amino and carboxyl termini on the cytoplasmic side (reviewed by Beyer)[1].

The half lives of several connexins have been determined in tissues [2,3] and in cultured cells[4-9]. All determinations yield values of only a few hours. This relatively rapid turnover rate suggests that gap junctions are rather dynamic structures and that their number and size can change in a few hours in response to physiological demands. Thus, the degree or extent of intercellular communication between adjacent cells can be regulated not only through gating of gap junctional channels but also through alterations in connexin turnover.

Regulation of gap junctions can be exerted at many different levels in the life history of a connexin molecule. These include transcription, translation, trafficking, insertion into the plasma membrane, assembly into gap junctional plaques, gating

From Ion Channels to Cell-to-Cell Conversations
Edited by Latorre and Sáez, Plenum Press, New York, 1997

of gap junctional channels, removal from the plasma membrane and degradation. The mechanisms controlling the length of a connexin's life cycle are poorly understood. However, post-translational modification of connexins is likely to be involved. Most connexins are phosphoproteins [8,10-15]. This modification has been implicated in gating of gap junctional channels[16] as well as in connexin trafficking[17], insertion into the plasma membrane[7], assembly into gap junctional plaques[18,19], retrieval of the channels from the plasma membrane[20] and degradation[21,22].

We have been studying regulation of gap junctions in the lens. The lens is an avascular organ composed of an anterior epithelium and lens fiber cells which form the bulk of the organ. During development, epithelial cells in the equatorial region differentiate into fiber cells, a process characterized by loss of intracellular organelles, mitochondria and nuclei. This process continues throughout the lifespan of the organism. Lens gap junctions have been detected by electron microscopy between epithelial cells and between fiber cells (reviewed by Goodenough)[23]. Because of the permeability properties of gap junctions and the lack of a blood supply to the lens, gap junctions may provide a pathway for exchange of nutrients and metabolites between lens cells. Thereby, gap junctional intercellular communication may play a pivotal role for cellular survival in the eye lens[24,25].

Characterization of connexin expression by lens cells has been carried out in a number of different species. In all species studied so far, lens epithelial cells express Cx43[26,27], which is a rather ubiquitous connexin. Fiber cells, however, express other connexins that are abundant only in the lens. These connexins include ovine MP70[28], rat Cx46[12], mouse Cx50[29], chicken Cx56[30] and Cx45.6[31], bovine

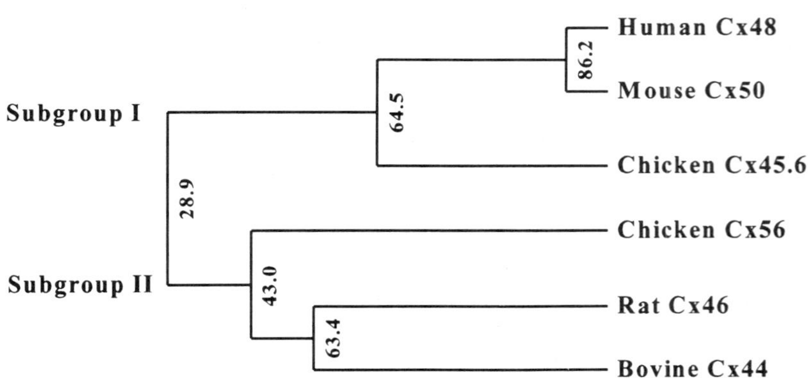

Figure 1. Dendrogram of the relationships of the derived amino acid sequences of the lens fiber connexins human Cx48, mouse Cx50, chicken Cx45.6, chicken Cx56, rat Cx46 and bovine Cx44 constructed using CLUSTAL. The percentage of amino acid sequence identity between subgroups I and II and between subgroup members is indicated by the numbers shown between the sequences.

Cx44[15], and human Cx48[32]. Cloned fiber cell connexins can be grouped into two subgroups according to their amino acid sequence identities (Fig. 1). Human Cx48, mouse Cx50, and chicken Cx45.6, sharing an overall 64.5% identity, belong to subgroup I, and rat Cx46, bovine Cx44, and chicken Cx56, sharing an overall 43% identity, belong to subgroup II. (There is an overall 28.9% identity between the two subgroups).

Lens connexins are modified by phosphorylation[12,14,15,31]. The kinases, the specific amino acid residue(s) involved, and the functional implication(s) of this modification are not well understood. Of the lens connexins, it is known that MP70 but not MP38, its membrane-embedded domain, can be phosphorylated by cAMP-dependent protein kinase[33]. MP70 is also a substrate for a recently described 35 kDa membrane bound protein kinase present in lens tissue which is cAMP- and Ca^{2+}-independent[34]. This kinase phosphorylates MP70 on a serine, but the location of this amino acid within the connexin molecule is unknown[34].

In the following sections, we will summarize the results obtained from studies of Cx56 expression during development of the chicken eye lens, and of the regulation of fiber cell gap junctions in lens cells in culture.

EXPRESSION OF Cx56 IN LENS DURING DEVELOPMENT

Levels and distribution of Cx56 during development of the chicken eye lens have been assessed by Northern blotting, immunoblotting, and immunofluorescence. Levels of Cx56 mRNA increase between embryonic days 12 and 19, then levels decrease in adult chickens to values similar to those observed at embryonic day 12[30]. Low levels of Cx56 are observed by immunoblotting at embryonic day 4, when the first fiber cells are detected[35]. At this age, Cx56 is detected as a single 66 kDa band by one-dimensional electrophoresis (Fig. 2A); this band comigrates with *in vitro* translated Cx56[36]. Levels of Cx56 increase after embryonic day 4; they remain relatively constant between embryonic days 15 and 20; and, then they decrease after hatching. Together with increased levels, the number of Cx56 bands increases; by embryonic day 18, as many as 11 bands are detected (Fig. 2A).

Different phosphorylated forms of several connexins have different electrophoretic mobilities resulting in multiple immunoreactive bands in immunoblots[6,7,15,27,37]. Treatment of homogenates from the 18th embryonic day lens with alkaline phosphatase reduces the pattern of Cx56 bands to a doublet[36]. Thus, the multiple bands result primarily from phosphorylation of Cx56. However, a higher degree of complexity of Cx56 is detected when samples are resolved by two-dimensional electrophoresis. Two isoforms with isoelectric points of 5.6 and 5.9 are detected at embryonic day 4 and several isoforms with isoelectric points ranging from 5.0 to 6.7 are detected at embryonic day 19[38].

Immunoreactivity to anti-Cx56 antibodies localizes at appositional membranes along the length of the newly differentiated fiber cells at embryonic day 4 (Fig. 2B). By embryonic day 20, the first signs of a gradient of immunoreactivity along appositional membranes of fiber cells are observed: immunoreactivity is more intense in the lens cortex than deep in the lens nucleus (Fig. 2C). At later ages, this gradient becomes more pronounced. Thus, the appearance of Cx56 in the chicken eye lens correlates with differentiation of epithelial cells into fiber cells. Moreover, modification of Cx56, an early event during lens development, becomes more complex as development proceeds.

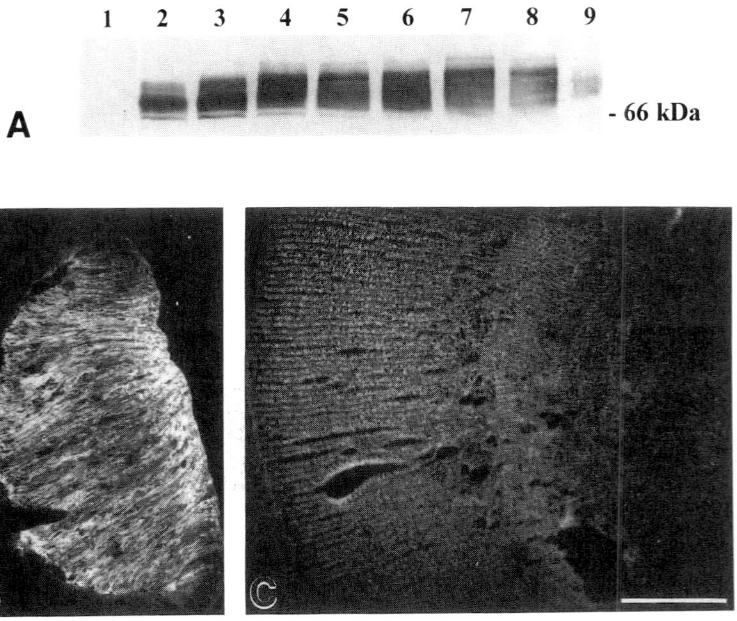

Figure 2. Levels and distribution of Cx56 during development of the chicken eye lens. **A.** Immunoblot of Cx56 in homogenates of chicken lens from 4- (lane 1), 6- (lane 2), 9- (lane 3), 12- (lane 4), 15- (lane 5), 18- (lane 6), and 20-day old embryos (lane 7) and from 1- (lane 8) and 30-day old chickens (lane 9). **B, C**. Localization of immunoreactivity to anti-Cx56 antibodies in a section of chicken lens from the 4th (**B**) or the 20th embryonic day. The sections have been oriented so that anterior is to the left. Bar: **B**, 120 µm; **C**, 185 µm.

LENS CULTURES AS A MODEL OF THE EYE LENS

Embryonic lens cells in culture differentiate to form multicellular structures called lentoids in a process that mimics differentiation of the lens *in vivo*[39]. Similar to lens cells *in vivo*, lentoid cells are communicated by gap junctions[40]. In these cultures the pattern of Cx56 bands is similar to that of a lens homogenate from the 18th embryonic day (Fig. 3A, lane 1). Treatment of homogenates from lentoid-containing cultures with alkaline phosphatase collapsed the multiple Cx56 bands to a doublet (Fig. 3A, lane 2) as seen for lens homogenates[36]. All these characteristics make these cultures an excellent model to study regulation of Cx56.

TURNOVER RATE OF Cx56

Since during differentiation of epithelial cells into fiber cells there is loss of organelles and because differentiated fiber cells must persist for the lifespan of the organism, it might be predicted that the half-life of any fiber cell protein would be extremely long. By performing pulse-chase experiments in chicken lentoid-containing cultures, we have determined the presence of at least two pools of Cx56, one with a short half-life of <2-3 h and one with a long half-life of >2-3 days (Fig. 3B). Thus, the turnover of the pool of Cx56 with a short half-life is similar to that described for non-fiber cell connexins[4-9] while another pool of the protein turns over much more slowly.

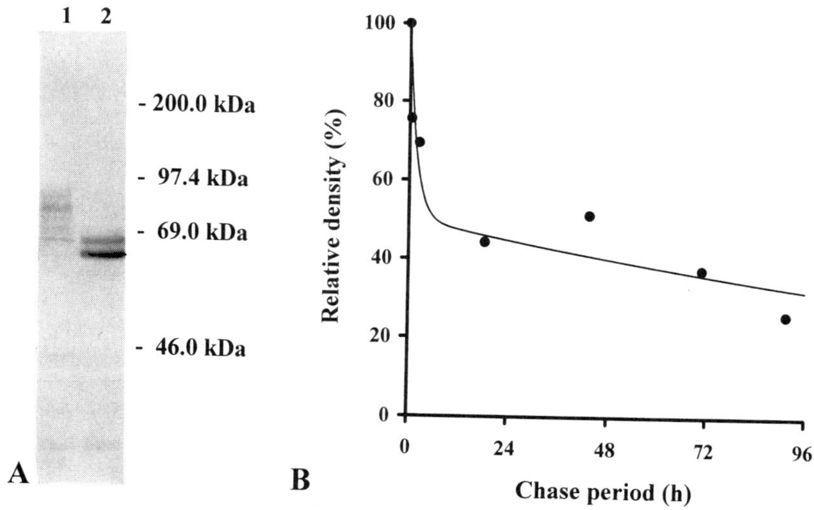

Figure 3. Immunoblot pattern and turnover dynamics of Cx56 in lens cells in culture. **A.** Immunoblot of Cx56 from a homogenate of a lentoid-containing culture after incubation with alkaline phosphatase buffer alone (lane 1) or incubation with alkaline phosphatase (lane 2) prior to resolution of the proteins by SDS-PAGE. **B.** Graph of the curve obtained for the decay of immunoprecipitated ^{35}S-labeled Cx56 from lens cultures labeled with ^{35}S-methionine for 2 h followed by a chase for 0.5, 2.5, 18.5, 43.5, 71 or 91.5 h. (Ordinate: relative densitometric value of the immunoprecipitated Cx56 bands; abscissa: chase period in h).

REGULATION OF Cx56 BY PROTEIN KINASE C

Because Cx56 is a phosphoprotein, we assessed regulation of lentoid gap junctions after activation of protein kinase C (PKC). In lentoid-containing cultures we have identified the presence of several PKC isoforms. Treatment of these cultures with 200 nM 12-O-tetradecanoyl-phorbol-13-acetate (TPA), a known activator of PKC[41,42], leads to activation of PKCγ in lentoid cells, i.e. upon TPA treatment this enzyme translocates from the cytoplasmic compartment to the membrane. In cultures under control conditions extensive intercellular transfer of Lucifer yellow is observed (Fig. 4B). Treatment with 200 nM TPA for 30 min or 3 h induces a drastic reduction in intercellular transfer of Lucifer yellow (Fig. 4D).

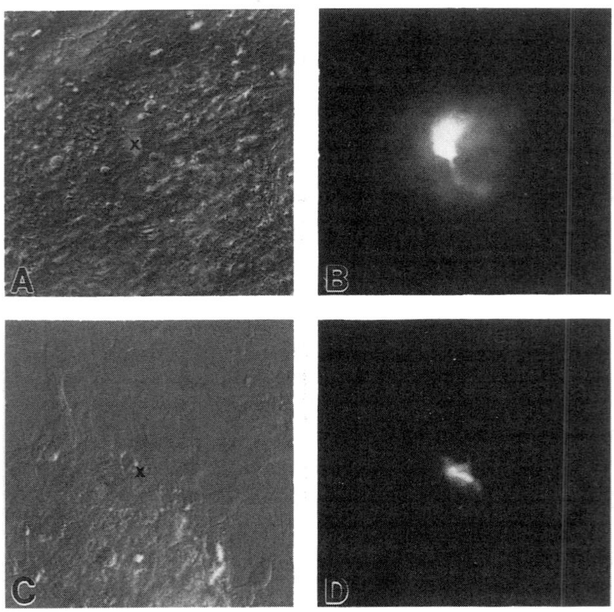

Figure 4. Effect of TPA on transfer of microinjected Lucifer yellow between lentoid cells. Extensive intercellular dye coupling was observed between lentoid cells in cultures under control conditions (B). Treatment with 200 nM TPA induced a drastic decrease in dye passage (D). Phase-contrast micrographs corresponding to B and D are shown in A and C, respectively. The position of the injected cells in panels B and D are indicated by "X".

Immunoblots from samples of control and TPA treated cultures show that the TPA-induced decrease in dye coupling between lentoid cells is associated with a relative increase in the 73-77 and 83 kDa Cx56 bands without substantial changes in the total amount of immunoreactive Cx56. Immunoreactivity to anti-Cx56 antibodies shows a rather smooth distribution along appositional membranes of lentoid cells from control cultures (Fig. 5A). TPA treatment leads to a time-dependent decrease in immunoreactive Cx56 at appositional membranes (Fig. 5B). Similar results have been obtained for chicken Cx45.6.

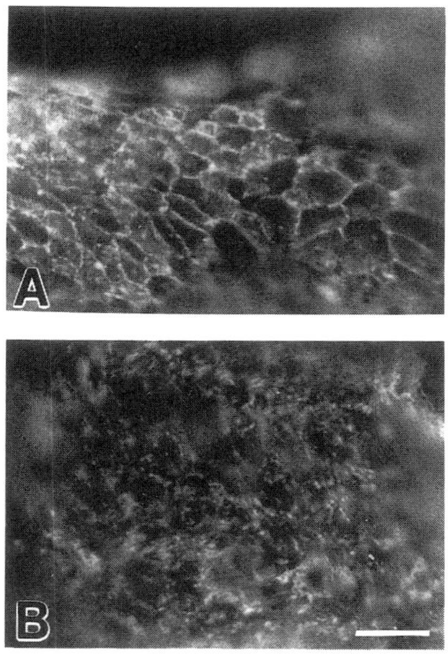

Figure 5. Effects of TPA on the cellular distribution of Cx56. Photomicrographs showing the distribution of immunoreactivity to anti-Cx56 antibodies in lentoid-containing cultures under control conditions (A) or after treatment with 200 nM TPA for 3 h (B). Bar: 34 μm.

These data suggest that TPA treatment affects Cx56-containing channels at appositional membranes. However, since two pools of Cx56 are present in lentoid-containing cultures, it was of interest to assess whether TPA affects primarily one or both Cx56 pools. When TPA is present during the pulse or a chase for 3 h, a decrease in the amount of immunoprecipitable ^{35}S-methionine labeled Cx56 is observed (Fig. 6, compare lanes 1 and 2, and lanes 3 and 4). These changes are associated with changes in the electrophoretic pattern of immunoprecipitated Cx56 (Fig. 6, compare lanes 1 and 2, and lanes 3 and 4) that mimic the changes detected by immunoblotting. When TPA is present for the last 3 h of a 24 h or 48 h chase, no significant changes in the amount of immunoprecipitable ^{35}S-methionine labeled Cx56 are observed. The simplest explanation for these results is that TPA induces accelerated degradation of the short half-life pool of Cx56.

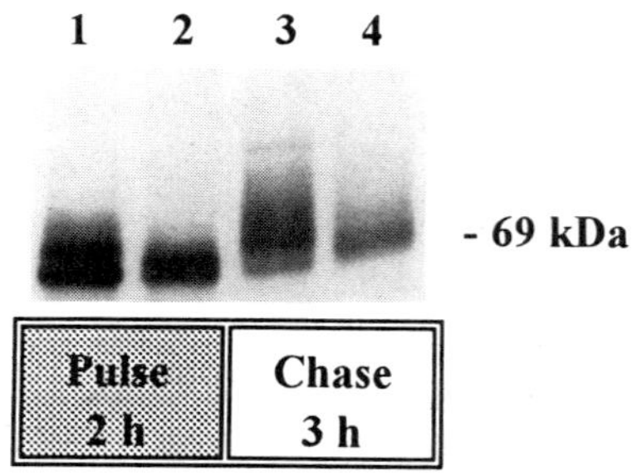

Figure 6. Effect of TPA on Cx56 turnover. ^{35}S-labeled Cx56 was immunoprecipitated from lens cells in culture after a pulse of ^{35}S-methionine for 2 h (lanes 1 and 2) and after a chase with cold methionine for 3 h (lanes 3 and 4). Cultures were left untreated (lanes 1 and 3) or treated throughout the pulse (lane 2) or the 3 h chase (lane 4) with 200 nM TPA.

CONCLUSIONS

The levels, distribution and modification of Cx56 are developmentally regulated in the chicken eye lens. Modification by phosphorylation may be involved in regulation of connexin turnover or gating of gap junctional channels.

The TPA-induced effects on dye coupling between lentoid cells, the immunoblot pattern and cellular distribution of Cx56, and degradation of the short half-life pool of Cx56 might involve direct phosphorylation of Cx56 or they might result as a consequence of other TPA-induced cellular changes. We have shown an increase in phosphorylation of Cx56 at a specific residue after treatment with TPA[43]. Therefore, it is possible that treatment with TPA activates PKCγ which, in turn, phosphorylates Cx56 in a site-specific manner, and this phosphorylation acts as a signal for degradation of Cx56 with or without previous closure of gap junctional channels.

In several other systems, TPA has been shown to induce phosphorylation and subsequent degradation of specific proteins including human IκB-α[44], the type B interleukin-8 receptor[45], and receptor tyrosine kinases[46]. The data presented here suggest that this could also be the case for Cx56 and that site-specific phosphorylation could be a generalized mechanism targeting connexins for degradation.

ACKNOWLEDGMENTS

We thank Drs. Jean Jiang and Daniel Goodenough from Harvard University for kindly supplying us with anti-Cx45.6 antibodies. These studies were supported by NIH grant EY08368.

REFERENCES

1. E.C. Beyer, Gap junctions, *Int. Rev. Cytol.* 137C:1 (1993).
2. R.F. Fallon and D.A. Goodenough, Five-hour half-life of mouse liver gap-junction protein, *J. Cell Biol.* 90:521 (1981).
3. S.B. Yancey, B.J. Nicholson, and J.-P. Revel, The dynamic state of liver gap junctions, *J. Supramol. Struct. Cell Biochem.* 16:521 (1981).
4. O. Traub, J. Look, D. Paul, and K. Willecke, Cyclic adenosine monophosphate stimulates biosynthesis and phosphorylation of the 26 kDa gap junction protein in cultured mouse hepatocytes, *Eur. J. Cell Biol.* 43:48 (1987).
5. O. Traub, J. Look, R. Dermietzel, F. Brümmer, D. Hülser, and K. Willecke, Comparative characterization of the 21-kD and 26-kD gap junction proteins in murine liver and cultured hepatocytes, *J. Cell Biol.* 108:1039 (1989).
6. D.S. Crow, E.C. Beyer, D.L. Paul, S.S. Kobe, and A.F. Lau, Phosphorylation of connexin43 gap junction protein in uninfected and Rous sarcoma virus-transformed mammalian fibroblasts, *Mol. Cell Biol.* 10:1754 (1990).
7. L.S. Musil, B.A. Cunningham, G.M. Edelman, and D.A. Goodenough, Differential phosphorylation of the gap junction protein connexin43 in junctional communication-competent and -deficient cell lines, *J. Cell Biol.* 111:2077 (1990).
8. D.W. Laird, K.L. Puranam, and J.P. Revel, Turnover and phosphorylation dynamics of connexin43 gap junction protein in cultured cardiac myocytes, *Biochem. J.* 273:67 (1991).
9. B.J. Darrow, J.G. Laing, P.D. Lampe, J.E. Saffitz, and E.C. Beyer, Expression of multiple connexins in cultured neonatal rat ventricular myocytes, *Circulation Research* 76:381 (1995).
10. J.C. Saez, D.C. Spray, A.C. Nairn, E. Hertzberg, P. Greengard, and M.V.L. Bennett, cAMP increases junctional conductance and stimulates phosphorylation of the 27-kDa principal gap junction polypeptide, *Proc. Natl. Acad. Sci. U.S.A.* 83:2473 (1986).
11. A.F. Lau, V. Hatch-Pigott, and D.S. Crow, Evidence that heart connexin43 is a phosphoprotein, *J. Mol. Cell Cardiol.* 23:659 (1991).
12. D.L. Paul, L. Ebihara, L.J. Takemoto, K.I. Swenson, and D.A. Goodenough, Connexin46, a novel lens gap junction protein, induces voltage-gated currents in nonjunctional plasma membrane of Xenopus oocytes, *J. Cell Biol.* 115:1077 (1991).
13. J.G. Laing, E.M. Westphale, G.L. Engelmann, and E.C. Beyer, Characterization of the gap junction protein, connexin45, *J. Membr. Biol.* 139:31 (1994).
14. E. TenBroek, M. Arneson, L. Jarvis, and C. Louis, The distribution of the fiber cell intrinsic membrane proteins MP20 and connexin46 in the bovine lens, *J. Cell Sci.* 103:245 (1992).
15. V.K. Gupta, V.M. Berthoud, N. Atal, J.A. Jarillo, L.C. Barrio, and E.C. Beyer, Bovine connexin44, a lens gap junction protein: molecular cloning,

immunological characterization, and functional expression, *Invest. Ophthalmol. Vis. Sci.* 35:3747 (1994).

16. A.P. Moreno, G.I. Fishman, and D.C. Spray, Phosphorylation shifts unitary conductance and modifies voltage dependent kinetics of human connexin43 gap junction channels, *Biophys. J.* 62:51 (1992).
17. K.L. Puranam, D.W. Laird, and J.P. Revel, Trapping an intermediate form of connexin43 in the Golgi, *Exp. Cell Res.* 206:85 (1993).
18. L.S. Musil and D.A. Goodenough, Biochemical analysis of connexin43 intracellular transport, phosphorylation, and assembly into gap junctional plaques, *J. Cell Biol.* 115:1357 (1991).
19. P.D. Lampe, Analyzing phorbol ester effects on gap junctional communication: a dramatic inhibition of assembly, *J. Cell Biol.* 127:1895 (1994).
20. V.M. Berthoud, M.B. Rook, O.Traub, E.L. Hertzberg, and J.C. Saez, On the mechanisms of cell uncoupling induced by a tumor promoter phorbol ester in clone 9 cells, a rat liver epithelial cell, *Eur. J. Cell Biol.* 62:384 (1993).
21. S.Y. Oh, C.G. Grupen, and A.W. Murray, Phorbol ester induces phosphorylation and down-regulation of connexin 43 in WB cells, *Biochim. Biophys. Acta* 1094:243 (1991).
22. M. Elvira, J.A. Diez, K.K. Wang, and A. Villalobo, Phosphorylation of connexin-32 by protein kinase C prevents its proteolysis by μ-calpain and m-calpain, *J. Biol. Chem.* 268:14294 (1993).
23. D.A. Goodenough, The crystalline lens. A system networked by gap junctional intercellular communication, *Semin. Cell Biol.* 3:49 (1992).
24. D.A. Goodenough, Lens gap junctions: a structural hypothesis for nonregulated low-resistance intercellular pathways, *Invest. Ophthalmol. Vis. Sci.* 18:1104 (1979).
25. Mathias, R.T. and J.L. Rae, Cell to cell communication in lens, in: "Cell Interactions and Gap Junctions," N. Sperelakis and W.C. Cole, eds., CRC Press, Inc., Boca Raton (1989).
26. E.C. Beyer, J. Kistler, D.L. Paul, and D.A. Goodenough, Antisera directed against connexin43 peptides react with a 43-kD protein localized to gap junctions in myocardium and other tissues, *J. Cell Biol.* 108:595 (1989).
27. L.S. Musil, E.C. Beyer, and D.A. Goodenough, Expression of the gap junction protein connexin43 in embryonic chick lens: molecular cloning, ultrastructural localization, and post-translational phosphorylation, *J. Membr. Biol.* 116:163 (1990).
28. J. Kistler, B. Kirkland, and S. Bullivant, Identification of a 70,000-D protein in lens membrane junctional domains. *J. Cell Biol.* 101:28 (1985).
29. T.W. White, R. Bruzzone, D.A. Goodenough, and D.L. Paul, Mouse Cx50, a functional member of the connexin family of gap junction proteins, is the lens fiber protein MP70, *Mol. Biol. Cell* 3:711 (1992).
30. D.M. Rup, R.D. Veenstra, H.Z. Wang, P.R. Brink, and E.C. Beyer, Chick connexin56, a novel lens gap junction protein. *J. Biol. Chem.* 268:706

(1993).
31. J.X. Jiang, T.W. White, D.A. Goodenough, and D.L. Paul, Molecular cloning and functional characterization of chick lens fiber connexin45.6, *Mol. Biol. Cell* 5: 363 (1994).
32. R.L. Church, J.H. Wang, and E. Steele, The human lens intrinsic membrane protein mp70 (cx50) gene -clonal analysis and chromosome mapping, *Current. Eye Research.* 14:215 (1995).
33. C.E. Voorter and J. Kistler, cAMP-dependent protein kinase phosphorylates gap junction protein in lens cortex but not in lens nucleus, *Biochim. Biophys. Acta* 986:8 (1989).
34. M.L. Arneson, H.-L. Cheng, and C.F. Louis, Characterization of the ovine-lens plasma-membrane protein-kinase substrates, *Eur. J. Biochem.* 234:670 (1995).
35. T.M. Miller and D.A. Goodenough, Evidence for two physiologically distinct gap junctions expressed by the chick lens epithelial cell, *J. Cell Biol.* 102:194 (1986).
36. V.M. Berthoud, A.J. Cook, and E.C. Beyer, Characterization of the gap junction protein connexin56 in the chicken lens by immunofluorescence and immunoblotting, *Invest. Ophthalmol. Vis. Sci.* 35:4109 (1994).
37. J.X. Jiang, D.L. Paul, and D.A. Goodenough, Posttranslational phosphorylation of lens fiber connexin46: a slow occurrence, *Invest. Ophthalmol. Vis. Sci.* 34:3558 (1993).
38. V.M. Berthoud, E.M. Westphale, A.J. Cook, M. Katar, H. Maisel, and E.C. Beyer, Expression and regulation of Cx56 in chicken lens cultures, *Mol. Biol. Cell* 5:1902a (1994).
39. A.S. Menko, K.A. Klukas, and R.G. Johnson, Chicken embryo lens cultures mimic differentiation in the lens, *Dev. Biol.* 103:129 (1984).
40. A.S. Menko, K.A. Klukas, T.F. Liu, B. Quade, D.F. Sas, D.M. Preus, and R.G. Johnson, Junctions between lens cells in differentiating cultures: structure, formation, intercellular permeability, and junctional protein expression, *Dev. Biol.* 123:307 (1987).
41. M. Castagna, Y. Takai, K. Kaibuchi, K. Sano, U. Kikkawa, and Y. Nishizuka, Direct activation of calcium-activated, phospholipid-dependent protein kinase by tumor-promoting phorbol esters. *J. Biol. Chem.* 257:7847 (1982).
42. J.E. Niedel, L.J. Kuhn, and G.R. Vanderbark, Phorbol diester receptor copurifies with protein kinase C. *Proc. Natl. Acad. Sci. USA* 80:36 (1983).
43. V.M. Berthoud, P.D. Lampe, W.E. Kurata, A.F. Lau, and E.C. Beyer, Phosphorylation of connexin56 by protein kinase C and protein kinase A, *Mol. Biol. Cell* 6:189a (1995).
44. E.B.-M. Traenckner, H.L. Pahl, T. Henkel, K.N. Schmidt, S. Wilk, and P.A. Baeuerle, Phosphorylation of human IKB-α on serines 32 and 36 controls IKB-α proteolysis and NF-KB activation in response to diverse stimuli, *EMBO J.* 14:2876 (1995).

45. S.G. Mueller, W.P. Schraw, and A. Richmond, Activation of protein kinase C enhances the phosphorylation of the type B interleukin-8 receptor and stimulates its degradation in non-hematopoetic cells, *J. Biol. Chem.* 270:10439 (1995).
46. K. Seedorf, M. Shearman, and A. Ullrich, Rapid and long term effects of protein kinase C on receptor tyrosine kinase phosphorylation and degradation, *J. Biol. Chem.* 270:18953 (1995).

MOLECULAR PHYSIOLOGY OF GAP JUNCTION CHANNELS FORMED BY CONNEXIN43

David C. Spray[1,2,4], Renato Rozental[1,4], Yang Gao[1], Monique Vink[1], and Glenn I. Fishman[2,3]

[1]Department of Neuroscience, Albert Einstein College of Medicine, Bronx, NY, USA
[2]Department of Medicine, Albert Einstein College of Medicine, Bronx, NY, USA
[3]Department of Molecular Genetics, Albert Einstein College of Medicine, Bronx, NY, USA
[4]Department of Internal Medicine, Federal University of Goias, Goiania- Goias, Brazil

INTRODUCTION

Multicellular animals have evolved tissues composed of cell types that are specialized for individual physiological functions. With few exceptions, cells of the same type are connected to one another through gap junction channels, thereby forming communication compartments within the tissues. In addition, compartments of different cell types are in many cases coupled to one another, providing intercellular communication among these compartments. The channels that serve as pathways for direct intercellular communication are generally organized into plaques at points or regions of cell contact; in early electron microscopic studies of liver junctional membranes, these plaques were termed "gap junctions" to emphasize that they contained membrane-spanning proteins as distinct from cytoplasmic continuities due to fusions of lipid bilayers in these regions (see Goodenough and Revel, 1970).

The proteins that form gap junction channels are encoded by the connexin multi-gene family, which has at least thirteen members in rodents (see White et al, 1995; Bennett, et al, 1994). Connexins are named according to the molecular weight (in kD) of the amino acid sequences predicted from the cDNAs (Beyer et al, 1990). Connexin expression is cell-type specific within most tissues, although there is considerable redundancy; thus, different cell types can express the same gap junction protein, and most cells express multiple connexins.

Connexin43 (Cx43) appears to be the most ubiquitously expressed gap junction protein in various types of mammalian cells; as is most relevant to data presented in this chapter, it is the most abundant gap junction protein in

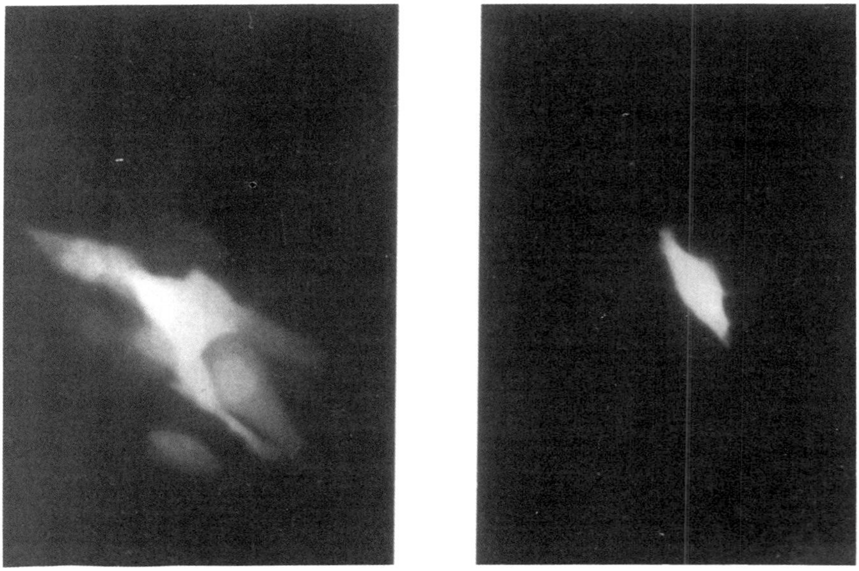

Figure 1. Exogenous expression of human connexin43 in a human hepatoma cell line, SKHep1. *Left).* Parental SKHep1 cell line is communication-deficient, as demonstrated by lack of Lucifer Yellow diffusion. *(Right).* Human Cx43 transfected colonies surviving antibiotic selection were injected with Lucifer Yellow and those exhibiting robust dye transfer were subcloned for further analysis. Modified from Fishman et al., 1990.

astrocytes within the nervous system (Yamamoto et al, 1990; Dermietzel et al, 1989), in smooth and striated muscle of the cardiovascular system (Beyer et al, 1990), and in epithelial cells of the lens (see Donaldson et al, 1994).

A central issue in the field of general physiology is to determine the role of ion channels in normal tissue function, and to discover whether and by what mechanisms the disruption of channel expression and function may result in specific pathological conditions. Such studies have been greatly aided in the last decade by the application of molecular biological approaches, which have provided novel experimental preparations for physiological analysis. The goal of this chapter is to summarize our progress along such lines for gap junction channels formed by connexin43, where biophysical properties and physiological roles have been revealed by expression of this protein in otherwise communication-deficient cell lines and have most recently been evaluated in tissues and cells from transgenic mice in which the gene encoding this gap junction protein has been deleted by homologous recombination.

PROPERTIES OF GAP JUNCTION CHANNELS FORMED OF Cx43

Gap junction channels formed of Cx43 have been studied in cell pairs obtained from a variety of mammalian tissues, including cardiac myocytes (Burt and Spray, 1988; Rook et al, 1988), astrocytes (Dermietzel, et al, 1991; Giaume et al, 1991), and lens epithelial cells (Donaldson et al. 1994). In addition, rodent Cx43 has been expressed exogenously in pairs of *Xenopus* oocytes (e.g., Werner et al, 1985; Dunham et al, 1992), and both rat and human Cx43 in several mammalian cell lines in which endogenous gap junction expression is low or altogether absent (Fishman et al, 1990; Zhu et al, 1991; Elfgang et al, 1995). We have favored the use of mammalian expression systems, which allow single channel recordings from cell pairs and may provide a cellular background that is more similar to that in which these channels are endogenously expressed. In initial experiments

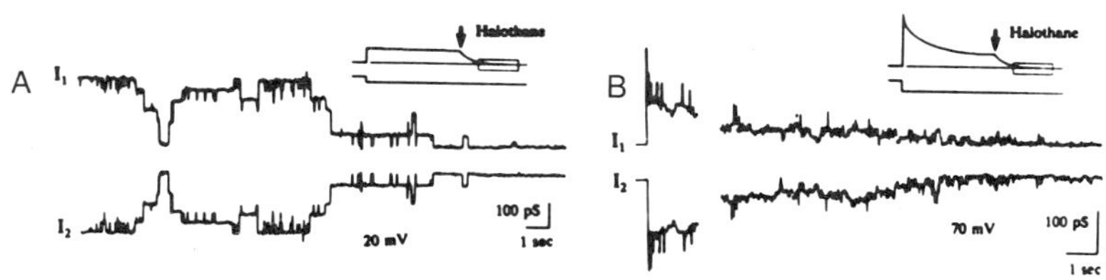

Figure 2. Unitary conductances of hCx43 channels expressed in SKHep1 cells, revealed by halothane treatment of cell pairs during sustained low or high V_j (insets). A. Channels are of two sizes (about 60-70 and 90-100 pS) when driving force is low. Largest channel conductances are favored under dephosphorylating conditions. B. Channels are 30 pS when driving force is high. Modified from Moreno et al, 1994a.

using human Cx43, we stably transfected SKHep1 cells (a highly metastatic human hepatoma cell line) using a vector in which human Cx43 cDNA was driven by the HSV promoter and cotransfected with a plasmid conferring resistance to the antibiotic neomycin. Colonies were initially selected for antibiotic resistance, and then were screened for robust Cx43 expression by injection of the highly fluorescent gap junction permeant molecule Lucifer Yellow (Fig. 1). Coupled colonies expressed mRNA of a size corresponding to that of the coding region of Cx43, and cell pairs demonstrated junction-like plaques of Cx43 immunoreactive material at appositional membranes. Subsequent studies by our group and others have successfully utilized a variety of mammalian vectors to stably express connexins in a number of mammalian cell types (see Waltzmann and Spray, 1995), including the recent use of a tetracycline repressible system in which connexin expression of the target gene can be experimentally manipulated (Fishman et al, 1995).

Unitary Conductance

Current fluctuations corresponding to openings and closings of gap junction channels can be recorded directly by applying the dual whole cell recording technique to pairs of coupled cells (Neyton and Trautmann, 1985; White et al, 1985). In this method, high resistance seals are made to the cell membrane, after which the membrane beneath the pipette is ruptured, allowing low resistance access to the cell interior; each cell is voltage clamped to a common holding potential and junctional currents are recorded in one cell in response to voltage commands imposed in the other cell. When many channels are open, macroscopic behavior of junctional conductance in response to various stimuli can be evaluated; under conditions where few junctional channels are active, elemental currents corresponding to openings and closures of single gap junction channels can be observed and quantified. Single gap junction channel currents have been evaluated in poorly coupled cell pairs in which only a few channels are stably present, in uncoupled cell pairs during the process of junction formation, and in well coupled cell pairs treated with lipophilic agents such as halothane (Fig. 2), heptanol, octanol and arachidonic acids, all which reversibly reduce the number of open channels. Single channel conductances measured in weakly coupled pairs not exposed to pharmacological uncoupling agents are not detectably different from those obtained in the presence of these drugs (see Burt and Spray, 1989).

Depending on the experimental conditions, Cx43 gap junction channels can display three distinct unitary conductances ($_j$) of about 30, 70 and 110 pS, when recorded using an internal solution in which CsCl is the major electrolyte (Moreno et al, 1994 a,b). The smallest of these conductances is favored at higher driving forces (Fig. 2B) and corresponds to a channel substate, as is discussed in the section on voltage sensitivity below. The 70 and 110 pS channels (Fig. 2A) are related to phosphorylation state of the Cx43 protein, such that the higher conductance events appears to correspond to the dephosphorylated protein and the lower $_j$ values to phosphorylated Cx43. Evidence that this modification changes unitary conductance is primarily pharmacological, in that the largest channel size is obtained in cell pairs that are internally dialyzed with phosphatase and in cells in which phosphorylation is inhibited by staurosporin, whereas the 70 pS events predominate after treatment of cells with the potent phosphatase inhibitor okadeic acid and after exposure of cells to tumor promoting phorbol esters.
Phosphate incorporation studies performed under similar conditions indicate consistent changes in Cx43 phosphorylation, and studies using synthetic peptides indicate that a serine-rich region in the carboxyl-terminus of Cx43 is the region involved (see chapter by Saez et al, this volume). Additional support for this association of phosphorylation state with unitary conductance has been provided recently by site-directed mutagenesis of the suspected serine residues; substitution by charged residues leads to a single population of smaller cconductances, whereas substitution of serines with neutral residues results in channels with larger unitary conductances (G.I. Fishman, L.K. Moore and D.C. Spray, manuscript in preparation).

It has been reported that mutations in this serine-rich region of Cx43 are found in a small number of patients exhibiting the profound left-right axis disorder, viscero-atrial heterotaxia (Britz-Cunningham et al, 1995), and these mutations result in less efficient Cx43 phosphorylation when the corresponding junctional proteins are analyzed as substrates in mammalian cell lines. Although genetic screens of additional VAH patients by others have not added to the examples of such Cx43 mutations (Casey and Ballabio, 1995; Splitt et al, 1995), the implied role of Cx43 in developmental axial symmetry is consistent with developmental defects in amphibia in which gap junctions are blocked by antibodies to Cx43 (Davids et al, 1997) and with results from experiments in which over- or underexpression of Cx43 in transgenic mice has been associated with axial asymmetry (C.W. Lo, personal communication).

Additional phosphorylation sites may also influence function of Cx43 channels (Warn-Cramer et al, 1996). For example, studies on cells transfected with human and rat Cx43 indicate that the rodent channels are responsive to cGMP, whereas the human channels are not (Kwak et al, 1995). It is intriguing that a potential cGMP-sensitive serine residue is among the six sequence differences between human and rat Cx43 (Fishman et al, 1990).

Voltage Sensitivity

For connexins expressed in mammalian systems, gap junction channel open probability (P_o) is maximal when no voltage is imposed across the junctional membrane, and P_o is gradually reduced with increasing transjunctional voltages (V_j) of either sign. However, even at very high V_j values, a substantial voltage insensitive component of g_j remains, which we termed g_{min} in our initial studies on endogenous junctional channels in amphibian embryos (Spray et al, 1979). For either polarity of V_j pulses, steady state junctional conductance (g_j) is fit well by a form of the Boltzmann equation (Spray et al, 1981):

$$g_j = (g_{max} - g_{min})/\{1 + \exp[A(V-V_0)]\} + g_{min}$$

Each connexin forms a gap junction channel with a distinctive voltage sensitivity, as illustrated for the three cardiac gap junction proteins (Cx40, Cx43 and Cx45) in Figure 3. For human and rat Cx43 gap junctions, $g_{min}/g_{max} = 0.4$ and V_0 (the voltage at which G_j is reduced by 50%) is about 60 mV. The equivalent number of gating charges, calculated from the slope A in the above equation, is about 2.5 to 3. The steady-state voltage sensitivity of Cx43 gap junctional conductance is not appreciably affected by agents that alter channel phosphorylation; however, the kinetics of channel closure by V_j are more rapid under dephosphorylating conditions (Moreno et al, 1993). We initially attributed the residual voltage insensitive component of junctional conductance (g_{min}) to either a population of channels with low voltage sensitivity, alteration of voltage sensitivity induced by the voltage field, or to a channel substate induced by the transjunctional voltage (Spray et al, 1981). The last of these possibilities now seems to be the most likely, based on studies using hCx43 transfectants which revealed that the larger channels were prevalent at low V_j and the smallest

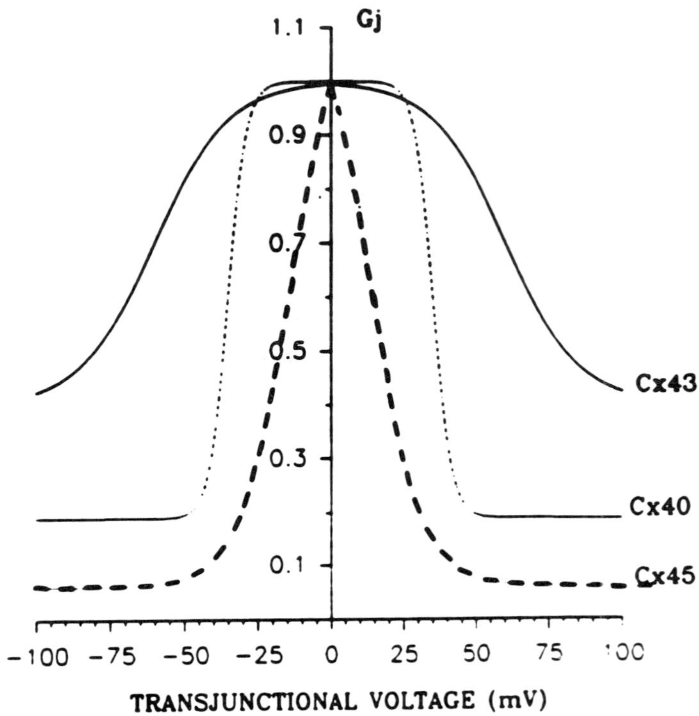

Figure 3. Voltage sensitivity of channels formed of cardiac connexins (Cx43, Cx40, and Cx45). Lines are best fits to Boltzmann equations describing decrease in junctional conductance (G_j) with transjunctional voltages (V_j) of either sign.

channels appeared as V_j was increased (Moreno et al, 1994a). Although the Cx43 channel substate is induced at high V_j values, its open probability is little reduced by further increases in V_j, and it is in this sense voltage-insensitive. Such substates have not been observed in response to lipophilic uncoupling agents, suggesting that distinct mechanisms might underlie channel closure by voltage and chemical stimuli.

DIFFERENCES IN CELL FUNCTIONS ACQUIRED AFTER Cx43 TRANSFECTION

The availability of otherwise communication-deficient mammalian cells stably transfected with individual connexins has made it possible to evaluate the impact of coupling on cell functions. In such studies, specific behaviors of the stable transfectants are compared to those of the parental line or to "mock transfectants", where communication-deficient colonies are clonally selected after transfection with vectors lacking the connexin cDNA.

Cell growth and bystander cell killing

The first mammalian cells in which gap junctions were introduced by stable transfection were our studies using a highly metastatic human hepatoma line (SKHep1) and those by C. Naus' group on the rat C6 glioma cell line (Eghbali et al, 1990; Fishman et al, 1990; Zhu et al, 1991). Because of the longstanding hypothesis that gap junctions might play a role in coordinating cell growth, both laboratories measured growth rates of transfected cells *in vitro* and of tumors resulting from implantation of cells into immunosuppressed mice. Our studies of growth rates of SKHep1 cells transfected with either rat Cx32 or human Cx43 showed no difference between the communication deficient parental line and the well-coupled transfectants (Eghbali et al, 1991; Elshami et al, 1996; Fig. 4A). Similarly, C6 cells transfected with Cx32 showed no slowing of growth *in vitro* (Bond et al, 1994); interestingly, the most highly expressing C6-Cx43 transfectants grew more slowly, an effect that was mimicked by exposure of parental cells to medium in which the transfectants were grown, suggesting that the inhibitory signal was secreted, rather than diffusing through an intercellular pathway (Zhu et al, 1991; Naus et al, 1992). By contrast to the lack of effect of gap junction communication on growth control that has been observed by most authors *in vitro* (but see Mehta et al, 1991), experiments on animals have shown that growth of tumors composed of cells transfected with Cx26, Cx32 or Cx43 is markedly slower than of tumors of parental cells or poorly coupled transfectants (Eghbali et al, 1991; Naus et al, 1992; Mesnil et al, 1994). However, all connexins may not be equally effective: Cx37 transfected N2A cells formed tumors that grew at same rates as the tumors of parental cells (F.D. Vrionis, J.K. Win, P. Qi, M. Waltzmann, V. Cherington and D.C. Spray, submitted).

These studies have not yet addressed the fundamental question of how it is that connexin expression modifies tumor growth *in vivo*. One possibility is that the coupling of Cx43 transfectants to those surrounding the tumor allows the transfer of growth-slowing molecules from normal cells to those of the tumor, so that the functional compartment represented by the tumor would no longer be separated from host tissue. This concept has received support from studies by H. Yamasaki and colleagues, who have shown that coupling within hepatomas may be maintained in the absence of coupling to surrounding liver cells (see Yamasaki and Naus, 1996). It may also explain why certain connexins would be more or less effective (and some, such as Cx26, might even function as tumor suppressors (Lee et al, 1991). In this regard, the recent unpublished finding of increased incidence of hepatomas in Cx32 knockout mice (K. Willecke, personal communication) is especially significant, though it remains to be determined whether the underlying cause is deletion of Cx32 or the reduced Cx26 expression that is also seen in livers of these animals (Nelles et al, 1996).

Abundant evidence has been provided in the decades since the growth control hypothesis was advanced that most tumor cells are coupled and that it is not the loss of coupling *per se* that distinguishes a tumorigenic from a normal cell. In fact, the coupling present among tumor cells has recently been used as a mechanism to target cell death to these cells. The phenomenon that is used therapeutically is termed "the bystander effect", in which cells infected with the thymidine kinase from *Herpes simplex* transfer their sensitivity to the anticancer drug gancyclivor to coupled neighbors (Bi et al, 1993). The substance transferred appears to be a highly phosphorylated gancyclivor derivative, whose

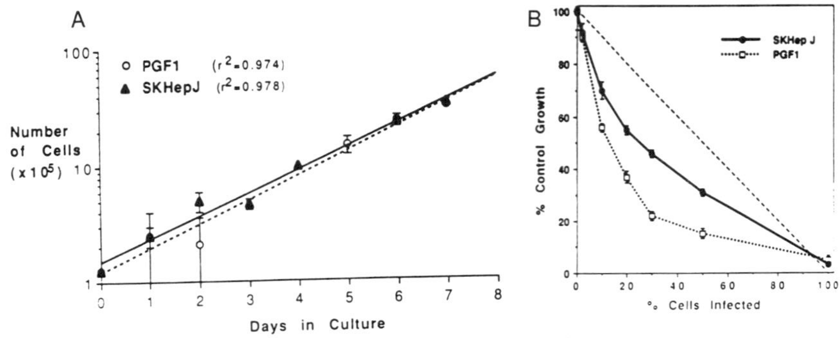

Figure 4. Bystander cell killing in hCx43-transfected SKHep1 cells. A. Growth rates of parental SKHep1 cells (SkHepJ) and hCx43, transfectants (PGF1) are not significantly different in culture. B. Bystander cell killing is significantly enhanced in hCx43 transfected cells; dashed line represents expected curve if only uncoupled cells expressing thymidine kinase were killed. Modified from Elshami et al, 1996.

molecular weight (about 300 Da) is well within the size range of junction permeant molecules. The bystander effect is not seen in cells that are not coupled to one another but is conferred on cells after constitutive expression of Cx43 (Fig. 4B) and by regulated expression of Cx32 (Elshami et al, 1996). That gap junctions play a central role in the bystander-effect *in vivo*, as well as *in vitro*, is indicated in recent experiments comparing growth of tumors from parental and Cx37 transfected N2A neuroblastoma cells (Vrionis et al, *op cit*).

In conclusion, tumor cells appear to be coupled to one another, although whether reduced strength of homotypic coupling or uncoupling from surrounding tissue contribute to tumorigenesis remain possibilities for further study. Whether some gap junction proteins may be more or less tumor suppressive or more or less effective in bystander cell killing remains to be completely resolved. Nevertheless, the gap junction mediated intercellular communication that is responsible for the phenomena termed "kiss of death" by Dickerman and Tischfield (1978) and "metabolic cooperation" by Subak-Sharpe et al (1969) appears to offer the therapeutic opportunity to optimize destruction of tumor cells by targeted gap junction expression.

Neuronal maturation

Neurons in adult mammals are in most cases not coupled electrially to one another and, when coupling is present, it generally is much weaker than that seen in other cell types such as cardiac myocytes, liver hepatocytes or even glial cells of the nervous system. Nevertheless, the cells from which neurons derive during development (early embryonic cells and neuronal precursor cells) are strongly coupled to one another by way of gap junction channels formed of Cx43. We have recently found that the progressive transition of neuroblasts from multipotent stem cells to neuronal or glial lineages is accompanied by a profound decrease in coupling (see Rozental and Spray, 1996).

In order to explore whether uncoupling was a prerequisite for differentiation, we have manipulated connexin expression and asked whether these manipulations affected the acquisition of neuronal phenotypic markers. For one type of experiment, we have used one immortalized hippocampal neuroblast cell line, which differentiates when exposed to retinoic acid (Rozental et al, 1996). For a second type of experiment, we have stably introduced connexin genes into the neuron-like pc12 cell line, which differentiate along the neuronal pathway when treated with nerve growth factor (Andrade-Rozental, et al, 1996).

In the case of these immortalized neuroblasts, cells were treated with the uncoupling agents heptanol and glyccerhetinic acid together with retinoic acid in order to test whether decreased intercellular coupling affected the differentiating response. In addition, Cx43 was transiently expressed in these cells to determine whether high levels of constitutively expressed Cx43 would affect the outcome; green fluorescent protein (GFP) was cotransfected in order to directly visualize the transfected cells. The results of these experiments were remarkably clear: in cases where gap junctional conductance was reduced, neuroblasts responded more rapidly and extensively to retinoic acid, extending long and complex neurites, expressing voltage sensitive ion channels, and demonstrating elevated levels of cytoplasmic neurofilament proteins. By contrast, transfectants constitutively expressing Cx43 were markedly less responsive, remaining flat, inexcitable and with low levels of expression of neuronal markers.

Studies on stably transfected pc12 cells have given equally unequivocal results (illustrated in Fig. 5): clones of parental cells with a low level of Cx37 expression are less responsive to NGF than communication-deficient clones, whereas well-coupled Cx43 transfectants are quite refractory to actions of NGF.

These experiments indicate that Cx43 expression is inversely and intimately tied to the process of neuronal development in these cell lines, such that uncoupling accelerates differentiation, whereas constitutive coupling retards it. The signal transduction mechanism linking intercellular communication with neuronal differentiation is as yet undefined, although our working hypothesis is that the uncoupling either terminates the delivery of an immaturity factor from proliferating cells to those destined to differentiate or allows accumulation of endogenously generated differentiating signals.

Whether all connexins are equally efficient in limiting neuronal differentiation is an issue that remains to be explored. Connexins 37 and 43 appear to have similar action in pc12 cells, but connexins with steeper voltage sensitivity might be expected to contribute an additional functional mechanism to this process. The gap junction channels expressed during early amphibian development are adequately voltage sensitive that uncoupling can be regenerative under conditions when a modest steady transjunctional voltage is generated (Harris et al, 1983). Such a situation would be predicted to occur as cells acquire various complements of ionic pumps and ligand-activated channels during development and might contribute to changes in coupling seen during neural tube induction (Blackshaw and Warner, 1976).

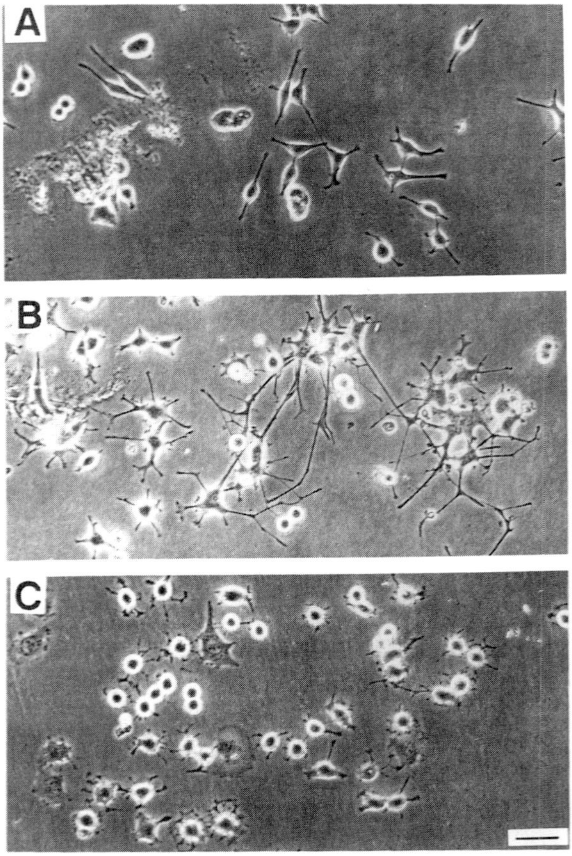

Figure 5. Strength of intercellular communication influences neuronal differentiation. A. pc12 cell line expressing Cx37. B. pc12 cell line uncoupled and fully differentiated. C. pc12 cell line stably transfected with Cx43.

TISSUE AND CELL ABNORMALITIES IN Cx43 KNOCKOUT MICE

Mice in which Cx43 was deleted by homologous recombination (Cx43 KO mice) were generated by a collaboration between Janet Roussant and Jerry Kidder's laboratories in 1995 (Reaume et al, 1995) and are now commercially available from the Jackson Laboratory (Bar Harbor, ME). These animals die at birth, due to hyperplasia of right ventricle and stenosis of the pulmonary artery, leading to reduced or total loss of circulation to the lung. However, cells derived from these animals can be maintained in culture, allowing comparison of their properties with those obtained from wildtype littermates.

Cardiovascular function

Cardiac action potential propagation depends on active currents in the nonjunctional membrane and also on electrical coupling between the cells provided by gap junction channels. Multiple connexins (primarily Cx40, Cx43 and Cx45) are expressed in the rodent heart: Cx40 is primarily found in the conduction system, Cx45 is primarily found in atrial and nodal regions, and Cx43 is the most prevalent gap junction protein in ventricular muscle, although it is also abundantly co-expressed with the other connexins in other regions (for reviews, see Fishman et al, 1996, Saffitz, et al, 1995).

In cardiac myocytes obtained from Cx43 KO mice and wildtype littermates, we have compared junctional conductance, single junctional channel properties, dye coupling, as well as synchrony of spontaneous contractions (Fishman et al, 1996). Junctional conductance in pairs of isolated myocytes primarily from ventricular tissue was reduced by half (4.6 ± 0.9 vs 10.7 ± 1.6, nS; $p<0.05$). However, transfer of intracellularly injected Lucifer Yellow was reduced to barely detectable levels in Cx43 KO myocytes, as contrasted with strong dye coupling to contiguous and second order mycoytes from wildtype littermates. Consistent with the reduced junctional conductance and Lucifer Yellow transfer, propagation of Ca^{2+} waves between myocytes under Ca^{2+} overload conditions was four times slower and markedly less extensive in Cx43 KO than in wildtype mycoytes. When contraction synchrony was compared, interbeat intervals of Cx43 KO myocytes were twice as long as for wildtype cells and showed much greater dispersion, reflected in a five fold greater variability in inter-beat intervals (Fig. 6).

Single channel recordings reveal that the channels mediating coupling in Cx43 KO myocytes are different than in wildtype animals. Unitary conductances of wildtype myocytes were primarily of sizes associated with phosphorylated Cx43 (60-90 pS), whereas in KO myocytes channels were of exclusively different sizes (57 ± 0.7 and 160 ± 1.2 pS).

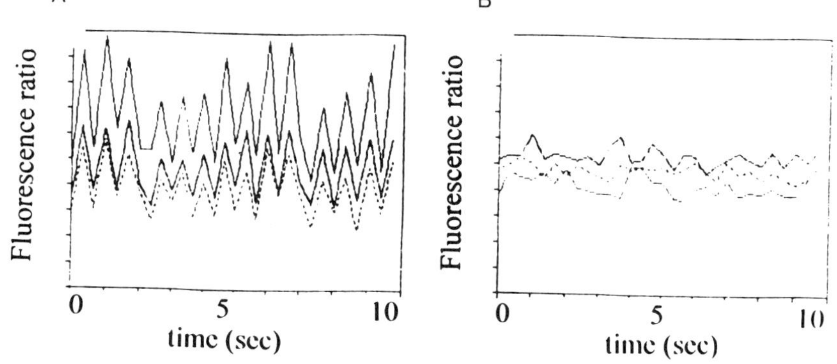

Figure 6. Comparison of synchrony of contractions in cultured cardiac myocytes from Cx43 KO wildtype (A) and Cx43 KO (B) mice. Note decreased rhythmicity in Cx43 KO myocytes.

We conclude that cardiac myocytes of Cx43 KO mice are deficient in intercellular coupling. Dye coupling is more profoundly reduced than electrical coupling, presumably reflecting the decreased anion permeability of the other cardiac gap junction proteins (Cx40 and Cx45) compared to Cx43. Conceivably, the reduced anion permeability of gap junctions in Cx43 KO ventricular myocytes may contribute to the developmental abnormality, by limiting diffusion of anionic morphogens.

The nervous system

Cx43 is the predominant gap junction protein expressed between astrocytes in the central nervous system, and Cx43 channels are believed to provide a pathway connecting astrocytes into a vast reservoir for buffering K^+ released by neurons as well as pathways for trophic and Ca^{2+} signalling throughout the brain and between neuronal and glial cells. We have compared the strength of electrical and dye coupling as well as the properties of gap junction channels connecting Cx43 KO astrocytes with those of wildtype littermates. Brain size, morphology and cortical lamination were grossly normal in Cx43 KO mice, and the limited behavioral repertoire of newborn mice was remarkably unimpaired. However, junctional conductance, Lucifer Yellow diffusion and rate and extent of Ca^{2+} wave spread (Fig. 7) were markedly depressed in the Cx43 KO astrocytes, although the degree of depression of these indices of intercellular coupling were quite different. Junctional conductance in Cx43 KO astrocytes averaged 0.22 ± 0.06 nS, about 1% of that of normal astrocytes (14.9 ± 3 nS).

The channels present most predominantly in astrocytes from wildtype brain were typically relatively voltage insensitive, with unitary conductances of 60 to 90 pS. In Cx43 KO astrocytes, channel size was much smaller (~30-40 pS) and conductance was much more voltage sensitive, consistent with the presence of Cx43 in normal brain and detectable levels of Cx45 and Cx46 in Cx40 KO astrocytes. Although it was formally possible that these additional connexins were compensatorily expressed in the Cx43 KO brain, Northern blot analyses revealed expression at comparable levels, suggesting that the presence of these additional connexins is normally masked by the abundant Cx43 expression in wildtype astrocytes.

Lens

The lens consists of a single layer of epithelial cells overlying lens fiber cells which differentiate from the epithelial cells at the margins or bow regions of the lens. Lens epithelial cells are interconnected by gap junctions formed of Cx43, and are believed to form junctions with underlying lens fiber cells by pairing Cx43 hemichannels contributed by the epithelial cells with Cx46 and/or Cx50 hemichannels contributed by the lens fiber cells. It has been postulated for many years that epithelial gap junctions might contribute either to fiber cell differentiation during development or to maintenance of lens transparency, by providing a pathway for nutrient exchange between extraocular fluid and the vascular lens fiber cell compartment (see Goodenough, 1992).

The availability of Cx43 KO animals has provided a test of these hypotheses. Comparing lens morphology and expression of the lens fiber protein

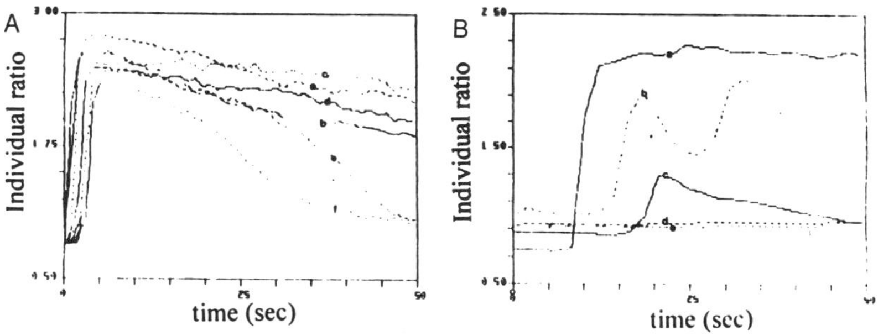

Figure 7. Ca^{2+} wave transfer in confluent dishes of wildtype and Cx43 KO astrocytes. Recordings were obtained from Indo1AM-loaded cells using a Nikon RCM 8000 scanning confocal microscope. Traces in A are from six wildtype astrocytes chosen to fall at progressively larger distances from the stimulated cell and were acquired at 1 sec intervals. In response to a brief mechanical stimulus, Ca^{2+} signals spread from one cell to the next with a velocity of about 1 cell diameter/sec. Traces in B obtained from the field in which we observed the strongest coupling of all Cx43 KO astrocyte cultures examined (four litters). Images here showed less rapid spread, with many cells totally refractory to the Ca^{2+} wave entry. Moreover, the plot shows that the delay between Ca^{2+} wave spread in contiguous cells was on the order of 5 sec, much longer than the latency between invasion of neighboring wildtype astrocytes. Note also the presence of secondary Ca^{2+} spread into one of the regions, resembling the phenomenon of electrical re-entry seen in cardiac tissues under pathological conditions.

MIP in wildtype and Cx43 KO lenses, we have found no gross differences in differentiation of fiber cells in the bow region. However, between epithelial cells, between epithelial and outer cortical lens fiber cells (Fig. 8A,C), and between deep cortical lens fibers (Fig. 8B,D), extracellular space is severely dilated and deep cortical fiber cells are packed with vacuoles in Cx43 KO lenses. Such separation of appositional membranes occurs in early stages of cataractous conditions generated by a variety of other treatments, leading to loss of lens transparency. Although the separation of lens fiber cells from epithelial cells suggests that a cause of the abnormality may indeed involve compromise of nutrient exchange between these compartments, it remains possible that dilation of extracellular space results from alterations of extracellular osmolarity, which might result from reduced coupling among the cells of the eye that are specialized for fluid secretion. Whether metabolite delivery from epithelial to lens fiber cells is actually altered in the Cx43 KO animal is currently under investigation.

SUMMARY AND CONCLUSIONS

As illustrated in this chapter, molecular genetic approaches have already begun to provide novel preparations in which gap junction properties and cell functions can be evaluated. For connexin43, the major gap junction channel

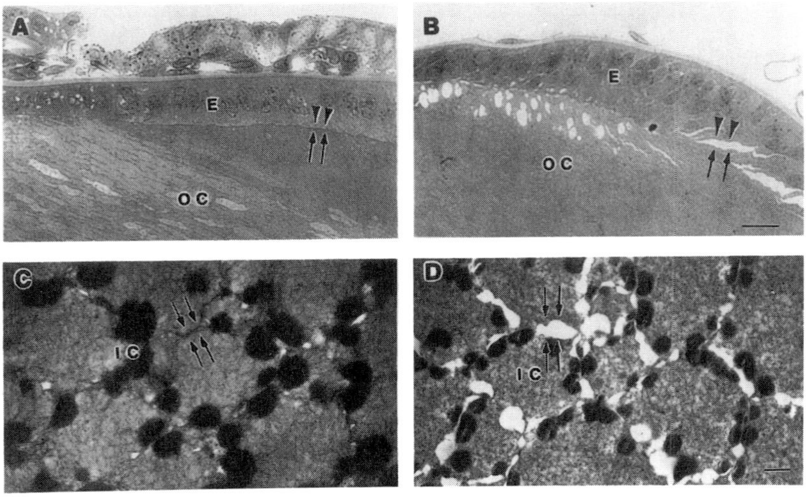

Figure 8. Morphological differences between lenses of Cx43 KO (B,D) and wildtype (A,C) mice. Equatorial thin sections (A,B) stained with Toluidine Blue viewed using a light microscope. The field of view includes epithelial cells (E) and fiber cells of outer cortex (OC). In lens of wildtype mouse (A) epithelial cells are tightly apposed to one another and to fiber cells (arrowheads and arrows). The lens of Cx43 KO mouse (B) exhibits a much looser appositional area between epithelial cells and large spaces are present between epithelial and fiber cells (arrowheads and arrows). Scale bar = 20μm. In the inner cortex (IC) viewed by transmission electron microscopy (C,D), fiber cells of wildtype lens (C) maintain appositional membrane contract with each other (arrows), but those of Cx43 KO mice (D) are characterized by dilation of extracellular space or large vacuolar spaces between fiber cells (arrows). Scale bar = 1μm.

protein expressed in cardiac myocytes, glial cells of the nervous system, as well as in many other tissues, stable expression in the otherwise communication-deficient mammalian cell line has revealed details of channel gating, including the existence of alternative conductance states influenced by phosphorylation state of the protein and by the voltage imposed across the junctional membrane. Differences in functional behaviors of cells and cell populations expressing Cx43 as compared to parental and mock transfected cells support a role of these gap junction channels in slowing tumor growth. Moreover, a role for gap junctions in bystander cell killing offers a therapeutic strategy for extending targeted cell death beyond individually targetted cells within a tissue. Experimental manipulation of coupling via Cx43 in neuron-like cells indicates an intimate relationship between disappearance of cell coupling and neuronal differentiation, and resolution of the underlying signal transduction will hopefully reveal processes responsible for aberrant neural circuit excitability.

Connexin43 knockout mice exhibit an unpredicted and still puzzling developmental defect in the heart, yet the heart still contracts. Junctional conductance between cultured myocyte pairs is only reduced by 50%, yet rhythmicity, dye coupling, and second messenger diffusion are profoundly affected. Thus, although junctional channels contributed by additional cardiac connexins compensate in part for the lack of Cx43 in these animals, their abundance and/or properties appear inadequate to provide normal conduction and distribution of signalling molecules throughout this tissue. Long range signalling among glial cells by Ca^{2+} waves and by electrical currents is severely attenuated in astrocytes from Cx43 KO animals, yet brain development and motor activities are not grossly impaired. Presumably, deficits in nervous system function will be demonstrable when older animals are generated using strategies of tissue-specific or regulated Cx43 ablation. The marked alterations in lens morphology of Cx43 KO mice are consistent with a role for Cx43 in providing nutrient exchange from epithelial to fiber cells, although the mechanisms responsible for generating these changes remain to be fully resolved.

The strategy outlined in this chapter, where transfected cells used to evaluate connexin-specific properties and roles in cell functions are complemented by studies involving cells isolated from connexin-ablated mice, is now being applied to other gap junction proteins. For example, mice lacking Cx32 have been recently generated by Klaus Willecke's group (Nelles et al, 1996). This gap junction protein is normally found in hepatocytes in the liver and in oligodendroglia and Schwann cells of the nervous system, and coding region mutations are responsible for the X-linked form of Charcot-Marie-Tooth disease, a progressively developing peripheral neuropathy (Bergoffen et al, 1993; Deschenes et al, 1996). Channels formed of Cx32 in transfected cells exhibit biophysical properties distinct from Cx43 (Moreno et al, 1991a,b), yet the coupling provided by these channels is also effective in showing tumor growth and in mediating bystander cell killing (Bond et al, 1994; Elshami et al, 1996). Cx32 KO animals exhibit a deficit in glucose mobilization in response to sympathetic nerve stimulation, reflecting reduced metabolic coupling in the liver acinus (Nelles et al, 1996). In addition, preliminary studies indicate impaired nerve conduction in older animals (mentioned in Nelles et al, 1996); which would be consistent with the human genetic CMTX disease, and a profoundly greater incidence of both spontaneous and chemically reduced hepatomas (K. Willecke, personal communication).

REFERENCES

Andrade-Rozental, A.F., Rozental, R., Urban, M., El-Sabban, M., and Spray, D.C. ,1996,. Connexin43 transfection delays neuronal differentiation. *Soc. Neurosci. Abst.* 22:1463.

Bennett, M.V., Zheng, X., and Sogin, M.L. ,1994,. The connexins and their family tree. *Soc. Gen. Physiol. Series* 49:223.

Bergoffen, J., Scherer, S.S., Wang, S. , et al. ,1993,. Connexin mutations in X-linked Charcot-Marie-Tooth disease. *Science* 262:2039.

Beyer, E.C., Paul, D.L., and Goodenough, D.A. ,1990,. Connexin family of gap junction proteins. *J. Membr. Biol.* 116:187

Bi, W.L., Parysek, L.M., Warnick, R., et al. ,1993,. In vitro evidence that metabolic cooperation is responsible for the bystander effect observed with HSV tk retroviral gene therapy. *Human Gene Therapy* 4:725.

Blackshaw, S.E. and Warner, A.E. ,1976, Alterations in resting membrane properties during neural plate stages of development of the nervous system. *J. Physiol.* 255:231.

Bond, S.L., Bechberger, J.F., Khoo, N.K.S., et al. ,1994,. Transfection of C6 glioma cells with connexin32: the effects of expression of a nonendogenous gap junction protein. *Cell Growth & Diff* 5:179.

Britz-Cunningham, S.H., Shah, M.M., Zuppan, C.W., and Fletcher, W.H. ,1995, Mutations of the connexin43 gap-junction gene in patients with heart malformations and defects of laterality. *New Engl. J. Med.* 332:1323.

Burt, J.M. and Spray, D.C. ,1988,. Single channel events and gating behavior of the cardiac gap junction channel. *Proc. Natl. Acad. Sci. USA* 85:3431

Burt, J.M. and Spray, D.C. ,1989, Volatile anesthetics reversibly reduce gap junctional conductance between cardiac myocytes.*Circ. Research* 65:829.

Casey, B. and Ballabio, A. ,1995,. Letter to editor. *New Engl. J. Med.* 333:941.

Cornell-Bell, A.H., Finkbeiner, S.M., Cooper, M.S., et al. ,1990,. Glutamate induces calcium waves in cultured astrocytes: Long-range glial signaling. *Science* 247:470.

Davids, M., Heydrich, U., Hofer, A., Kremer, M., Spray, D.C. and Dermietzel, R. ,1997,. Microinjection of antibody to connexin43 into cleavage stage Xenopus embryos results in specific developmental defects. *Anatomy and Development,* in press.

Dermietzel, R., Hertzberg, E.L., Kessler, J.A., and Spray, D.C. ,1991,. Gap junctions between cultured astrocytes: Immunocytochemical, molecular, and electrophysiological analysis. *J. Neurosci.* 11:1421.

Dermietzel, R., Traub, O., Hwang, T.K., Beyer, E., Bennett, M.v., Spray, D.C. and Willecke, K. ,1989,. Differential expression of three gap junction proteins in developing and mature brain tissues. *Proc. Nat. Acad. Sci. of USA* 86:10148.

Deschenes, S., Bone, L.J., Fishbeck, K.H. and Scherer, S. ,1996,. Connexin32 and X-Linked-Charcot-Marie-Tooth Disease, in: "Gap Junctions in the Nervous System", D.C. Spray and R. Dermietzel, eds., R.G. Landes Company, Georgetown, Texas.

Dickerman, L.H. and Tischfield, J.A. ,1978, Comparative effects of adenine analogs upon metabolic cooperation between Chinese hamster cells with different levels of adenine phosphoribosyltransferase activity. *Mutation Research* 49:83.

Donaldson, P.J., Roos, M., Evans, C., Beyer, E., and Kistler, J. ,1994,. Electrical properties of mammalian lens epithelial gap junction channels. *Invest. Opth. & Visual Sci.* 35:3422.

Dunham, B., Liu, S., Taffet, S., Trabka-Janik, E., Delmar, M., Petryshyn, R., Zheng, S., Perzova, R. and Vallano, M.L. ,1992,. Immunolocalization and expression of functional and nonfunctional cell-to-cell channels from wild-type and mutant rat heart connexin43 cDNA. *Circ. Res.* 70:1233.

Eghbali, B., Kessler, J.A. and Spray, D.C. ,1990,. Expression of gap junction channels in a communication incompetent cell line after transfection with connexin32 cDNA.*Proc. Natl. Acad. Sci. USA* 87:1328.

Eghbali, B., Kessler, J.A., Reid, L.M., Roy, C., and Spray, D.C. ,1991,. Involvement of gap junctions in tumorigenesis: Transfection of tumor cells with connexin32 cDNA retards growth *in vivo*. *Proc. Natl. Acad. Sci. USA* 88:10801.

Elfgang, C., Eckert, R., Lichtemberg-Frate, H., Butterweck, A., Traub, O., Klein, R.A., Hulser, D.F., and Willecke, K. ,1995,. Specific permeability and selective formation of gap junction channels inconnexin-transfected HeLa cells. *J. Cell. Biol.* 129:805.

Elshami, A.M., Saavedra, A., Shang, H., Kucharczuk, J.C., Spray, D.C., Fishman, G.I., Kaiser, L.R. and Albelda, S.M. ,1996,. Gap junctions play a role in the "bystander effect" of the herpes simplex virus thymidine kinase/ganciclovir system in vitro. *Gene Therapy* 3:85.

Fishman, G.I., Gao, Y., Hertzberg, E.L. and Spray, D.C. ,1995,. Reversible intercellular coupling by regulated expression of a gap junction channel gene. *Cell Adhesion and Communication* 3:353.

Fishman, G., Moreno, A.P., Spray, D.C. and Leinwand, L.A. ,1991,. Functional analysis of human cardiac gap junction channel mutants. *Proc. Natl. Acad./ Sci. USA* 88:3525.

Fishman, G., Spray, D.C. and Leinwand, L.A. ,1990,. Molecular characterization and functional expression of the human cardiac gap junction channel. *J. Cell. Biol.* 111:589.

Fishman, G.I., Vink, M.J., and Spray, D.C. ,1996,. Can we learn about conduction using genetic approaches?, in: "Discontinuities in Cardiac Conduction", American Heart Association, P. Spooner, J. Jalife, and R. Joyner, eds., Dallas, Tx, in press.

Giaume, C., Fromaget, C., Aoumari, A.E., et al. ,1991,. Gap junctions in cultured astrocytes: Single-channel currents and characterization of channel-forming protein. *Neuron* 6:133.

Goodenough, D.A. ,1992,. The crystalline lens. A system networked by gap junctional intercellular communication. *Sem. in Cell. Biol.* 3:49.

Goodenough, D.A. and Revel, J.P. ,1970,. A fine strutural analysis of intercellular junctions in the mouse liver. *J. Cell. Biol.* 45:272.

Harris, A.L., Spray, D.C., and Bennett, M.V.L. ,1983,. Control of intercellular communication by voltage dependence of gap junctional conductance. *J. Neurosci.* 3:79.

Kwak, B.R., Saez, J.C., Wilders, R., Chanson, M., Fishman, G.I., Hertzberg, E.L., Spray, D.C. and Jongsma, H.J ,1995,. Effects of c-GMP-dependent phosphorylation on rant and human connexin43 gap junction channels. *Pflug. Archiv. Eur. J. Physiol.* 430:770.

Lee, S.W., Tomasetto, C., and Sager, R. ,1991,. Positive selection of candidate tumor-suppressor genes by subtractive hybridization. *Proc. Natl. Acad. Sci. USA* 88:2825.

Loewenstein, W.R. ,1981,. Junctional intercellular communication: the cell-cell membrane channel. *Physiol. Rev.* 61:829.

Mehta, P.P., Hotz-Wagenblatt, A., Rose, B., Shalloway, D. and Loewenstein, W.R. ,1991,. Incorporation of the gene for a cell-cell channel protein into transformed cells leads to normalization of growth. *J. Membr. Biol.* 124:207.

Mesnil, M., Krutovskikh, V., Piccoli, C., Elfgang, C., Traub, O., Willecke, K. and Yamasaki, H. ,1995,. Negative growth control of HeLa cells by connexin genes: connexin species specificity. *Cancer Res.* 55:629.

Moreno, A.P., Eghbali, B., and Spray, D.C. ,1991a,. Connexin32 gap junction channels in stably transfected cells: unitary conductance. *Biophys. J.* 60:1254.

Moreno, A.P., Eghbali, B., and Spray, D.C. ,1991b,. Connexin32 gap junction channels in stably transfected cells. Equilibrium and kinetic properties. *Biophys. J.* 60:1267.

Moreno, A.P., Fishman, G.I. and Spray, D.C. ,1992,. Phosphorylation shifts unitary conductance and modifies voltage dependent kinetics of human connexin43 gap junction channels. *Biophys. J.* 62:51.

Moreno, A.P., Fishman, G.I., Eghbali, B. and Spray, D.C. ,1993,. Unmasking electrophysiological properties of connexins32 and 43: transfection of communication-deficient cells with wild type and mutant connexins. *Prog. Cell Res.* 3:127.

Moreno, A.P., Rook, M.B., Fishman, G.I., and Spray, D.C. ,1994a,. Gap junction channels: Distinct voltage-sensitive and insensitive conductance states. *Biophys. J.* 67:113-119.

Moreno, A.P., Saez, J.C., Fishman, G.I. and Spray, D.C. ,1994b,. Human connexin43 gap junction channels: Regulation of unitary conductances by phosphorylation. *Circ. Research* 74:1050.

Naus, C.C.G., Elisevich, K., Zhu, D., et al. ,1992,. In vivo growth of C6 glioma cells transfected with connexin43 cDNA. *Cancer Res.* 52:4208.

Nelles, E., Butzler, C., Jung, D., Temme, A., Gabriel, H.-D., Dahl, U., Traub, O., Stumpel, F., Jungermann, K., Zielasek, J., Toyka, K.V., Dermietzel, R. and Willecke, K. Defective propagation of signals generated by sympathetic nerve stimulation in the liver of connexin32-deficient mice. *Proc. Natl. Acad. Sci. USA* 93:9565.

Neyton, J. and Trautmann, A. ,1985,. Single-channel currents of an intercellular junction. *Nature* 317:331.

Reaume, A.G., deSousa, P.A., Kulkarni, S., et al. ,1995,. Cardiac malformation in neonatal mice lacking connexin43. *Science* 267:1831.

Rook, M.B., Jongsma, H.J., and van Ginneken, A.C. ,1988,. Properties of single gap junctional channels between isolated neonatal rat heart cells. *Am. J. Physiol.* 255:H770.

Rozental, R. and Spray, D.C. ,1996,. Temporal expresison of gap junctions during neuronal ontogeny. in: "Gap Junctions in the Nervous System", D.C. Spray and R. Dermietzel, eds., R.G. Landes Company, Georgetown, Texas.

Rozental, R., Urban, M., Andrade-Rozental, A.F., Fishman, G.I., Eves, E.M., Rosner, M. and Spray, D.C. ,1996,. Strength of intercellular communication has a pronounced influence on neuronal differentiation. *Soc. Neurosci. Abst.* 22:1205.

Saffitz, J.E., Davis, L.M., Darrow, B.J., Kanter, H.L., Laing, J.G., and Beyer, E.C. ,1995,. The molecular basis of anisotropy: role of gap junctions. *J. Card. Electro.* 6:498.

Splitt, M.P., Burn, J. and Goodship, J. ,1995,. Letter to editor. *New Engl. J. Med.* 333:941.

Spray, D.C., Harris, A.L., and Bennett, M.V.L. ,1979,. Voltage dependence of junctional conductance in early amphibian embryos. *Science* 204:432.

Spray, D.C., Harris, A.L. and Bennett, M.V. ,1981,. Equilibrium properties of a voltage-dependent junctional conductance. *J. Gen. Physiol.* 77:77.

Subak-Sharpe, H., Burk, R.R., and Pitts, J.D. ,1969,. Metabolic co-operation between biochemically marked mammalian cells in tissue culture. *J. Cell. Science.* 4:353.

Waltzmann, M.N. and Spray, D.C. ,1995,. Exogenous expression of connexins for physiological characterization of channel properties: Comparison of methods and results, in: "Intercellular Communication Through Gap Junctions", Progr. Cell Research, Y. Kanno, K. Kataoka, Y. Shiba, Y. Shibata, and T. Shimazu, eds.,

Warn-Cramer, B.J., Lampe, P.D., Kurata, W.E., Kanemitsu, M.Y., Loo, L.W., Eckhart, W., and Lau, A.F. ,1996,. Characterization of the mitogen-activated protein kinase phosphoryaltion sites on the connexin-43 gap junction protein. *J. Bio. Chem.* 271:3779.

Werner, R., Miller, T., Azarnia, R., et al. ,1985,. Translation and functional expression of cell-cell channel mRNA in Xenopus oocytes. *J. Membr. Biol.* 87:253.

White, R.L., Spray, D.C., Campos de Carvalho, A.C., Wittenberg, B.A. and Bennett.M.V. ,1985,. Some electrical and pharmacological properties of gap junctions between adult ventricular myocytes. *Am. J. Physiol.* 249:C447.

White, T.W., Bruzzone, R., and Paul, D.L. ,1995,. The connexin family of intercellular channel forming proteins, *Kid. Int.* 48:1148.

Yamamoto, T., Ochalski, A., Hertzberg, E.L., and Nagy J.L. ,1990,. On the organization of astrocytic gap junctions in rat brain as suggested by LM and EM immunohistochemistry of connexin43 expression. *J. Comp. Neurol.* 302:853.

Yamasaki, H. and Naus, C.C.G. ,1996,. Role of connexin genes in growth control. *Carcinogenesis* 17:1199.

Zhu, D., Caveney, S., Kidder, G.M., et al. ,1991,. Transfection of C6 glioma cells with connexin43 cDNA: analysis of expression, intercellular coupling, and cell proliferation. *Proc. Natl. Acad. Sci. USA* 88:1883.

A POSSIBLE PHYSIOLOGICAL ROLE OF THE MAJOR INTRINSIC PROTEIN OF THE OCULAR LENS

Grischa Chandy[1], Guido Zampighi[2] and James E. Hall[1]

[1]Department of Physiology and Biophysics
UC Irvine, Irvine, CA 92717

[2]Department of Anatomy and Cell Biology
UCLA, Los Angeles, CA 90024

INTRODUCTION

Although there has been a long standing interest in determining the physiological role of the Major Intrinsic Protein of lens (MIP), this goal has remained elusive. Because of the double membrane structures in which it was found, its molecular weight, and its biochemical properties, MIP was first thought to be a gap junction protein, but after it was cloned [1] this view has become more and more untenable. Several studies showed that MIP could form ion channels in planar lipid bilayers [2-4], but channels with comparable properties could not be found in lens fiber cells. As a result of the availability of the DNA sequence coding for MIP, similar proteins were cloned at an increasing pace, and there are now at least 24 members of the MIP family. A crucial finding was made by Agre et al [5-7] who demonstrated that a related protein, CHIP (now called Aquaporin 1 [5]) greatly increased the water permeability of *Xenopus* oocytes. MIP expressed in oocytes is so much less effective as a water channel than CHIP that experimenters at first had some difficulty in demonstrating its water permeability. But, as we will show in this paper, at least some of the reasons for this difficulty have been resolved.

LENS ANATOMY AND PHYSIOLOGY

Lens Structure

The lens is an avascular tissue formed of two cell types: the anterior epithelial layer which faces the aqueous humor and the fiber cells which form the bulk of the lens [8]. Only cells in the epithelial layer are mitotic. As the epithelial

From Ion Channels to Cell-to-Cell Conversations
Edited by Latorre and Sáez, Plenum Press, New York, 1997

cells in the center of the epithelial layer divide, they push each other towards the equator of the lens, and as the cells near the equator divide, they push each other toward the interior of the lens and begin to differentiate into fiber cells. The closer to the center of the lens, the more advanced the differentiation becomes. Lens fiber cells are long (some 10 mm in length) thin cells of hexagonal cross-section with two long sides and four short sides. Fiber cells are packed very regularly with minimal extracellular space between them. They lack most cellular organelles and contain large concentrations of soluble protein called crystallines. The high soluble-protein concentration increases the index of refraction of the lens and aides in image formation. The lack of extracellular space is crucial in preventing light scattering and consequent opacity.

Lens Physiology

The function of the lens is to focus light from both near and distant objects onto the retina. Thus a functional lens must remain transparent, maintain a high index of refraction and be deformable. Since the lens is an avascular tissue, all nutrients and metabolites must enter and exit the lens by diffusion or by specialized transport mechanisms of the lens itself. Lens extra-cellular space is so severely restricted, and the cells of the lens are so well-coupled electrically that the bulk of the lens can be thought of as a huge membrane-divided synctium[9-12]. The low density of organelles, the high concentrations of crystallines, and the limited extracellular space help maintain the clarity and flexibility of the lens. Because the lens is relatively large and has a very restricted energy budget for transporting fluid, ions and metabolites to and from the interior, any disruption of the lens's normal function is likely to lead to opacity or cataract.

MIP IN THE LENS

Thus far, lens fiber cells are the only cell type found to endogenously express MIP, which is up-regulated during fiber cell differentiation[13]. Eventually, MIP forms 50-80% of lens membrane protein[14-16]. As mentioned previously, the lens is extremely well-coupled, this correlates with the fact that 60 to 70% of lens fiber-cell membrane is junctional in nature[17,18]. For a number of years it was believed that MIP was primarily responsible for the cell-to-cell coupling in the lens. This idea was subsequently refuted (reviewed by Ehring[19]) following the observation that MIP is present in the lens in both non-junctional and junctional structures; but in the latter, MIP is not found opposing itself in both leaflets of the bilayer[20-23]. Furthermore, MIP has no morphological or sequence similarity to the connexins, proteins demonstrated to form gap junctions[19,24,25].

MIP STRUCTURE & BIOCHEMISTRY

The cDNA of MIP was originally cloned from a bovine lens library. Analysis of the deduced amino-acid sequence revealed five hydrophobic

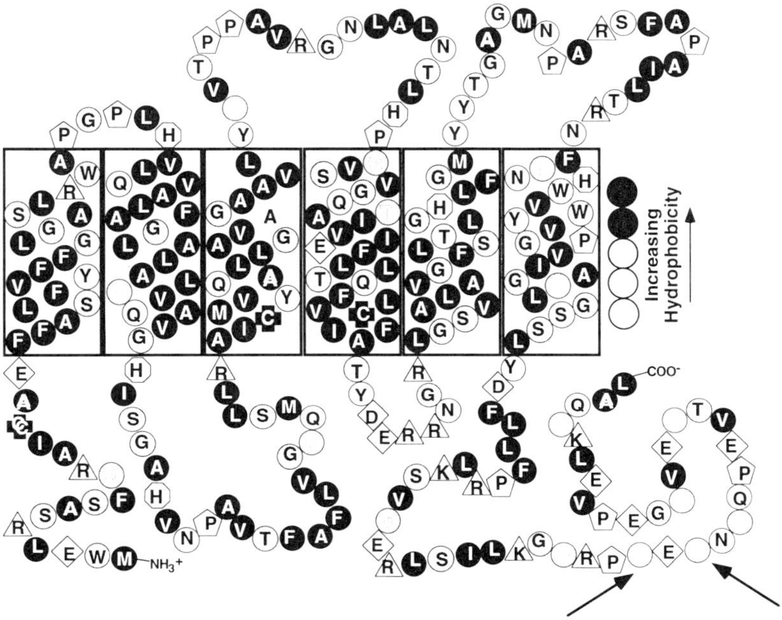

Figure 1. Predicted Membrane Topology of MIP. Residue hydrophobicity increases with increasing gray level. Triangles indicate positive residues and diamonds indicate negative residues. The identified *in vivo* phosphorylation sites[16] are marked by the phosphates. The boxed phosphate is the PKA regulated site. A glycosylation consensus site is indicated with a solid arrow and the site at which cleavage occurs to produce MIP-22 is indicated with an open arrow

domains and one amphipathic domain. All six of these domains are hypothesized to traverse the lipid bilayer with the amino- (N) and carboxyl- (C) termini on the cytoplasmic side[20,22,26-29]. This putative secondary structure of MIP, based on hydropathy analysis[30], is shown in Figure 1. Since the first half of MIP has significant (31%) identity to the second half, it has been proposed that the ancestor of MIP arose from a tandem, intragenic, duplication event[31]. The human MIP gene has been characterized[32] and assigned to chromosome 12q13.2 to q13.3[33].

MIP can be biochemically modified in many ways: it is palmitoylated[34], phosphorylated by protein kinase A (PKA)[16,35,36], and protein kinase C (PKC)[16,37] its N- and C-termini are proteolytically cleaved[38], it interacts with calmodulin[39] and can be glycated after it has been inserted into the plasma membrane.[40,41] MIP is tetrameric,[42] naturally forming crystalline-like arrays in

lens fiber-cell membranes and when reconstituted into vesicles at high protein to lipid ratios [22,43].
Biochemical and structural work on one of MIP's homologues, Aquaporin 1 (AQP1) supports the structural data available for MIP. The N- and C-termini of AQP1 are cytoplasmic [44-46]. AQP1 spans the endoplasmic reticulum membrane only four times [47,48], while the mature plasma membrane version has six membrane-spanning domains [49,50]. Other structural studies (biochemical, freeze fracture and high resolution electron microscopy of two-dimensional AQP1 crystals) support the notion that AQP1 assembles as a tetramer in the plasma membrane with multiple membrane spanning domains [45,46,51-55]. By inference, these structural studies on AQP1 support the data available for MIP.

FUNCTIONAL STUDIES OF MIP

Although MIP is found in great abundance in the lens, it has been extremely difficult to study its physiology *in vivo* because of technical difficulties in maintaining fiber-cells outside of the lens. Fortunately it has been relatively easy to purify protein for *in vitro* functional studies. MIP has been reconstituted into both vesicles and lipid bilayers (for review see Ehring [19]). Recently functional studies have been made on *Xenopus* oocytes exogenously expressing MIP.

Vesicles

The earliest functional work on MIP was carried out by reconstituting fragments of membranes rich in lens junctional complexes into proteoliposomes. These proteoliposomes were permeable to molecules up to 1.1 kD in size [56]. Further work carried out on similar systems using purified protein showed that MIP is permeable to molecules up to 1.5 kD [57-60]. The increases in permeability were modulated by calmodulin in the presence of Ca^{2+} [61,62], pH [58] and enzymatic proteolysis [61,62]. That the MIP-induced permeability was a specific effect of MIP was called into question by the fact that glycophorin, a non-channel membrane protein, also increased the permeability of vesicles to the same degree as MIP [63]. Moreover, van Hoek *et al* [64] found that reconstituted MIP did not increase vesicle permeability to urea but did seem to increase membrane water permeability (P_f).

Bilayer

HPLC purified MIP has been extensively characterized in planar lipid bilayers and forms a large-conductance (360 pS) ion channel in 100 mM KCl. These channels have weak anion selectivity, are symmetrically voltage dependent and have fast opening and closing rates [43]. Ca^{2+} and H^+ ion concentration have no effect on the channel gating properties [43], but phosphorylation by PKA enhances voltage dependence of MIP channels, while dephosphorylated MIP forms almost voltage-independent channels [65]. Trypsin cleavage of the C-terminus mimics the effect of dephosphorylation [66].

Xenopus oocytes

While MIP forms ion channels in the bilayer, no new ionic conductances are detected when it is expressed in *Xenopus* oocytes. Since homologues of MIP function as water channels (See below.), investigators have attempted, with conflicting results, to determine if MIP increases water permeability of oocyte membranes [47,66,67]. Recently Kushmerick *et al.* [68] have shown that MIP may have multiple functions in oocytes including fluxing glycerol.

MIP-RELATED PROTEINS (MRPS)

MIP is a member of a super-family of genes. All MRPs have two conserved asparagine-proline-alanine (NPA) boxes. The first is found in the first putative endofacial loop and the second is found in the third putative exofacial loop (Fig. 1). MRPs are found in diverse species amongst several phyla. Table 1 summarizes the presently known MRPs, their species of origin, and their functions, if known. Especially in vertebrates multiple MRPs have been found in a diverse collection of tissues ranging from lens to kidney.
Since MIP belongs to this super-family and has a degree of structural similarity to all of these proteins, understanding the function of MRPs may illuminate MIP's possible physiological roles. Most of the identified MRPs increase the water permeability of biological membranes. Other MRPs with an identified function appear to flux small neutral solutes such as urea and/or glycerol. We briefly describe some of the key proteins known to increase water permeability.

Aquaporins (Aqp)

Aquaporin-1. Aquaporin-1 , the protein formerly known as CHIP [6], was the first isolated and cloned protein shown to function as a water channel [69-71]. AQP1 was isolated in a complex procedure combining N-terminal-sequencing of purified protein from erythrocytes, polymerase chain reaction and traditional cloning from a human fetal liver cDNA library and a human bone marrow cDNA library [72]. The AQP1 gene has been identified and assigned to human chromosome 7p14 [73,74]. AQP1 is the protein responsible for the high water permeability of red blood cells (RBC) and the constitutively high water permeability of the proximal [71,75,76]. It may also account for the RBCs high glycerol permeability [77]. A single channel water permeability (p_f) has been estimated for AQP1 (~4 to 12 x10^{-13} cm^3/sec) by measuring P_f of proteoliposomes reconstituted with AQP1 at known protein to lipid ratios [46,54,78]. A p_f (1x10^{-14} cm^3/sec) for AQP1 has also been estimated in oocytes, combining P_f measurements and quantitative immunoblotting to determine AQP1 expression [69]. Some of the variability has been attributed to the different osmolytes used in the various measurements [79].
AQP1 has been identified in many reabsorptive and secretory epithelia and capillary endothelia [75,80-82]. Because AQP1 has been identified in all of these cells types, it is thought to play an important role in water reabsorption and secretion.

Table 1. Known MIP-related proteins.

Protein	Abbrev.	Source	Function/%	Reference
26 KD Major Intrinsic Protein of the lens	**MIP-26**	*Bos taurus* *Rattus norvegius* *Gallus gallus* ° *Homo sapiens* *Xenopus laevis*	Water Channel, Ion Channel/>94	(1) (128) (129) (32) (130)
Channel-forming Integral Protein	**CHIP-28**	*Homo sapiens* *Rattus norvegicus*	Water channel/45	(72) (131,132)
	CHIP-28k	*Rattus norvegicus*		(71)
	CHIP-29	*Bos taurus*		(133)
Delayed Early Response gene-2	**DER-2**	*Mus musculus*		(134)
(Aquaporin 1)	**(AQP-1)**	Rana esculenta		(135)
Water Channel Collecting Duct	**WCh-CD**	*Rattus rattus*	Water Channel* 59	(86,93)
(Aquaporin 2)	**(AQP-2)**	*Homo sapiens*		(23,91)
Aquaporin-3	**AQP-3**	Rat	Water Channel, Urea and Glycerol Facilitator/32	(112,136)
Glycerol Intrisic Protein	**GLIP**			(137)
		Homo sapiens		(138,139)
Aquaporin-4	**AQP-4**	Rat	Water Channel/42	(111)
Mercury Insensitive Water Channel	**MIWC**	*Rattus norvegicus*		(110)
Aquaporin-5	**AQP-5**	Rat	Water Channel/49	(140)
Water channel upregulated with dehydration	**WCH3**	*Rattus norvegicus*	ND/49	(141)
Big brain	**Bib**	*Drosophila Melanogaster*	?/ND	(142)
Cicadella intrinsic protein	**CIC**	*Cicadella Viridis*	?/35	(143)
Tonoplast Intrisic Protein	**αTIP**	*Phaseoulus valgaris* *Arabidopsis thaliana*	Water Channel^/30	(144) (145)
Tonoplast Intrinsic Protein	**βTIP**	*Arabidopsis thaliana*	?/ND	(145)

Table 1. Continued.

Tonoplast Intrinsic Protein (Aquaporin *Arab. thal.*)	**γTIP (AQP At1)**	*Arabidopsis thaliana*	Water Channel/30	(145)
		Oryza Sativa		(146)
Soybean Putative Channel Proteins	**SPCP1 & 2**	*Glycine max*		(147)
Tobacco root membrane protein	**TobRB7**	*Nicotiana tabacum*	?/30	(148,149)
Dark Intrinsic Protein	**DiP RtTIP**	*Antirrhinum majus*		(150)
Pea shoot turgor responsive gene	**7a/TUR**	*Pissum Sativa*	Water Channel/30	(151)
Responsive to Dessication-28	**RD-28**	*Arabidopsis thaliana*		(152)
Plasma membrane intrinsic protein-2	**PIP2a, b**	*Arabidopsis thaliana*		(99)
(Water stress induced TIP)	**p8-2 (WTIP)**	*Atriplex canescens*		(153)
Plasma membrane intrinsic protein-1	**PIP1a,b,c**	*Arabidopsis thaliana*	Water Channel/ND	(99,154)
Turgor-Responsive cDNA	TMP-A TMP-B TMP-C	*Arabidopsis thaliana*	?/ND	(155-157)
Tomato Ripening-associated Membrane Protein	TRAMP	*Lysopersicon esculentum*	?/30	(158)
Plant-encoded Nodulin-26	Nod-26	*Glycine max*	Ion Channel/35	(100,159)
Rice MIP	rMIP	*Oryza sativa*	?/ND	(160)
Fdp1 suppressor	FPS1	*Saccharomycces cerveisiae*	Glycerol Facilitator/ ND	(161)
Glycerol Facilitator	GlpF	*Escheria Coli*	Glycerol Facilitator/ 20	(162)
		Bacillus subtilis		(163)
		Streptomyces coelicolot °		(164)
Protein 5' to the X-prolyl dipeptidyl aminopeptidase-encoding gene	ORF	*Lactococcus lactis*	?/ND	(165)
Synechococcus membrane protein	SmpX	*Synechococcus* species PCC7942	?/ND	(166)
E. Coli membrane protein	AQPz	*E. Coli*	Water Channel	(167)

Orthologous sequences have been clustered together. ° (partial sequence), ? (unknown function), # (regulated by anti-diuretic hormone), ^ (protein kinase A-phosphorylation dependent), % (One protein from each orthologous group was compared to rat MIP to obtain a percent identity. These are approximate because they varied a few percentage points depending on which sequence was used and which portion of the sequences were compared). If multiple names for a protein have been published the preferred name has been bolded. A function has been included only if there is reasonably good evidence for that function). Citations refer to the published sequence.

However, humans who have AQP1-gene mutations resulting in non-functional AQP1 are phenotypically normal except for a reduced RBC P_f [83]. Possibly other water channels can substitute for AQP1 in the epithelial and endothelial cells.

Aquaporin-2. It has been possible to characterize novel water-channel cDNA sequences using AQP1 derived probes (Table 1). One of these is WCH-CD renamed AQP2 [5]. AQP2 is found only in the kidney collecting duct and is one of MIPs closest relatives (58% identity and 80% similar). Most of the dissimilarity is in the cytoplasmic tail of the C-terminus (Fig. 1). AQP2 is the protein responsible for the antidiuretic hormone (ADH)-induced increase in water permeability of the kidney collecting duct which allows mammals to make a concentrated urine. Within the collecting duct, ADH regulates membrane permeability by modulating a cAMP dependent signaling pathway that controls the excocytosis and endocytosis of AQP2-containing vesicles [75,84-86]. And in Xenopus oocytes expressing AQP2, increased cAMP increases water permeability without increasing AQP2 membrane expression [87]. Mutations in the AQP2 gene (assigned to human chromosome 12q13.1) are responsible for a rare recessive form of nephrogenic diabetes insipidus [23,33,88-92]. WT-AQP2 increases plasma membrane P_f whereas the mutants do not [86,93]. From experiments in oocytes it was determined that these mutant AQP2s are most likely non-functional because they are ineffectively transported to the plasma membrane [94].

Plant Aquaporins. Besides providing a better grasp of kidney physiology and pathology, the discovery of the water channel family has had a profound influence on our understanding of plant physiology [95]. Most plant cells contain large storage vacuoles. The tonoplast separates plant cytoplasm from vacuolar contents. Certain tonoplast proteins are presumed to be essential in maintaining vacuole-cytoplasm concentration gradients and osmotic equilibrium. A number of tonoplast-intrinsic proteins (TIP) are members of the MIP family. gTIP has been expressed in *Xenopus* oocytes and shown to increase water permeability [96]. Another plant MRP, aTIP, induced a small increase in water permeability which was enhanced by PKA activators [97]. Two other plant MRPs (See Table 1) localized to the plant-cell membrane, RD28-PIP and PIP1, are water channels [98,99].

Nodulin-26 (NOD26) Legumes and rhizobia bacteria maintain a nitrogen-fixing symbiosis. A bacterial "infection" causes the formation of root nodules whose infected cells contain a novel organelle, the symbiosome. The symbiosome membrane (SM) isolates the bacteria from the plant's cytoplasm and protects the bacteria from the plant's defense mechanisms. But the symbiosome membrane also adds a barrier through which nutrients and metabolites must pass between bacterium and plant. This barrier is overcome by nodule-specific membrane proteins which promote transport between the bacteroid space and the plant cytoplasm. NOD26 is a plant-encoded SM membrane [100,101] possibly involved in the transport of malate, an important metabolic energy source for the bacteria. *In vivo* phosphorylation of NOD26 correlates with an increased rate of malate uptake across the peribacteriod membrane[102].

NOD26 has been reconstituted into planar lipid bilayers and, like MIP, forms a large (1.2 nS in 0.2 mM KCl), slightly anion selective channel[103]. Phosphorylation by calmodulin-like domain protein kinase (CDPK) at serine 262 alters the channel properties of the protein just as phosphorylation of MIP does. Unphosphorylated NOD26 channels remain completely open at high membrane potentials. However, phosphorylated NOD26 shows increased voltage-dependent gating and closure to lower subconductance states at high membrane potentials[104,105]. The similarity to MIP holds when Nodulin-26 is expressed in Xenopus oocytes where it also produces water channels[106] (also Chandy, Roberts and Hall, work in progress).

WATER CHANNELS.

A number of the MRPs are known or thought to be water channels. Classical water channels are characterized by five physical properties, reviewed in[107,108].

a. A large P_f. Water flux is proportional to P_f times the osmotic gradient.

$$\Phi \,\alpha\, P_f \,\Delta \mathrm{Osm}$$

b. $P_f / P_d > 3$. P_d is a coefficient that describes the rate of water diffusion across a membrane in the absence of osmotic or pressure gradients. The flux of tracer water is proportional to P_d times the gradient of tracer.

$$\overset{*}{\Phi} \,\alpha\, P_d \Delta \overset{*}{H_2O}$$

c. The large p_f is blocked by mercurial sulfhydryls.

d. The mercurial block is reversed by reducing agents.

e. Water flux through water channels has a very low activation energy, $E_a < 5$ Kcal/mol.

Historically these properties have been used to define water channels. But as we have learned more about the molecular nature of water channels, these criteria have been refined. Recently it has been shown that a cysteine (C189) in the third exofacial loop is responsible for the mercurial sensitivity in AQP-1 and probably AQP-2[71,109]. At least two water channels,[110,111] and RD28-PIP[98], have been identified that are *not* blocked by mercury and do *not* contain a cysteine at the position homologous to AQP1's C189. However, AQP3 and gTIP are sensitive to mercury but do not contain a cysteine at the position homologues to AQP1's C189[96,112]. Thus mercurial block is not a universal characteristic of water channels.

The criterion that water flux through water channels must have an activation energy less than 5 kcal/mol is somewhat arbitrary and was derived from the observation that water flux across lipid membranes has a high activation energy (10-25 Kcal/mol). Water fluxes through membranes containing water channels have previously been found to have extremely low activation energies[107], but as we shall see some water channels have water fluxes with higher activation energies than others. Perhaps some mutant water channels will be made with even higher activation energies, but with substantially the same transport mechanisms as their more effective relatives. Thus this criterion too may have to be relaxed or generalized.

Although there have been major advances in the cloning and structure of MIP and MRPs, a well-defined functional role for MIP in the lens still remains elusive. Although many of the MRP's are water channels, the evidence that MIP is a water channel is contradictory. To partly resolve this issue, we *quantitatively* compared the water-fluxing properties of MIP to those of known water channels as described below.

METHODS

Oocytes

Stage V and VI oocytes were harvested from *Xenopus laevis* (purchased from NASCO, Wisconsin) and treated with 2 mg/ml collagenase (Boerhinger Mannheim, Indiana) in OR2 (in mM 82.5 NaCl, 2.5 KCl, 1 $MgCl_2$, 5 HEPES, pH 7.6) to remove the follicular layer. Complementary RNA (cRNA) was injected into oocytes using a positive-displacement micropipette (Nanoject, Drummond; Broomall, PA) 24 hours after harvest. Control oocytes were injected with 50 nl of water. Oocytes were maintained in ND96 (in mM; 96 NaCl, 2 KCl, 1 $MgCl_2$, 1.8 $CaCl_2$, 5 HEPES pH 7.6) supplemented with 10 mg/ml aminobenzylpenicillin/streptomycin, and 2.5 mM NaPyruvate, at 18°C. ND96 was changed daily. Oocytes were incubated 48-96 hours before measurements were performed.

Expression Constructs and RNA preparation

The *Xenopus* expression constructs for MIP and AQP1 (gifts of Dr. Peter Agre) were used for cRNA synthesis. MIP-stop was generated from the MIP clone using a 4-primer, 2-stage PCR strategy modified from Ho *et al.*, (1989) [113]. A stop codon was inserted at lysine 228, the site at which *in vivo* proteolytic cleavage occurs producing MIP-22 (Gorin et al., 1984; Lampe and Johnson, 1990) [1,16]. The mutation was confirmed by dideoxy sequencing (Sanger et al., 1977) [114].

Swelling Assay

The method for measuring water flux in oocytes has changed little from that used 65 years ago on sea urchin eggs by Lucké and McCutcheon [115]. Volume changes due to water flow driven by an osmotic gradient are estimated by measuring diameter or cross-sectional area of the osmotically-challenged eggs or oocytes as a function of time. Lucké and McCutcheon made visual observations every few minutes. Modern improvements include a video camera, a frame grabber and a digital computer to capture an image, calculate the area and plot the results. The modern method for volume flux measurements on *Xenopus* oocytes has been described [116]. Our method differed slightly in details. Water permeability was calculated from the rate of volume increase which was estimated from automated measurements of the increase in cross-sectional area of the oocyte in response to a dilution of ND96 Unless indicated ND96 was diluted 70% (7 ml of distilled water for 3 ml of ND96). Full-strength ND96 has an osmolality of about 185 to 195 mOsm. Seventy percent diluted ND96 has an osmolality of about 60 mOsm). Precise osmolalities for each experiment were determined using a Wescor-5500 osmometer (459 South Main Street, Logan, Utah 84321). Water permeability measurements were made at temperatures covering the range between 4¡C and 40¡C. Temperature was electrically controlled using a Model 5000 KT stage controller (20/20 Technology) and the bath temperature was measured using a micro thermistor probe (TH-1 Thermistor Probe, N.B. Daytner, P.O. Box 81062, Wellesley Hills, MA 02181) placed within 2 mm of the oocyte. Temperature was read out using a YSI Tele-Thermometer (Box 279, Yellow Springs, Ohio 45387).

Capacitance Measurements.

Capacitance of oocytes was measured using the two-electrode voltage-clamp technique (Soreq and Seidman, 1992). Data were collected using a Warner Instruments (Hamden, CT) oocyte clamp driven by P-clamp (version 5.5.1, Axon Instruments, Burlingame, CA) and data were analyzed using AxoGraph software (version 2.0, Axon Instruments). Whole oocytes were held at -50 mV and pulsed from -100 mV to 0 mV in 20 mV increments. The area under the elicited current spike was integrated to give the charge moved by a given voltage step. Charge was plotted against voltage, and the slope of the fitted line was used as a measure of the membrane capacitance.

Electron Microscopy

Fixation. After measuring the appropriate functional property, oocytes were fixed by sandwiching them between two glass slides separated by 200 mm thick spacers. This procedure flattened the oocyte from a sphere into an oblate, a geometry that greatly simplified the collection of extensive areas of fracture-faces needed for particle density quantification. Oocytes were fixed in 3-3.5% glutaraldehyde in 0.2 M Na-cacodylate, pH 7.4. After 15 minutes, the oocytes were removed from the slides and immersed in the same fixative solution for 1 hour at room temperature.

Particle Measurements

Freeze-Fracture. The oblate oocytes were infiltrated with 25% glycerol in 0.2 M Na-cacodylate pH 7.4. The cells were cut into smaller pieces (4-6), placed on Balzers specimen holders and frozen by immersion in liquid propane cooled in a liquid nitrogen bath. Oocyte pieces were oriented on the holders with their external surfaces upward to improve the chances of exposing P faces rather than E fracture faces. The frozen oocytes were transferred into a Balzers 400K freeze-fracture-etch apparatus, fractured at either -150¡C or -120¡C and at 1 x 10^{-7} mbar of partial pressure. The fractured surfaces were immediately coated with platinum at an 80° angle and carbon at 90°. Shadowing at 80° showed particles with shorter shadows and greatly improved the accuracy of the quantification of the particle density. Cleaning of the replicas was facilitated by coating the replicated specimen with 0.5% collodion in amyl acetate followed by mechanical agitation in fresh bleach solution [117].

Sampling and Quantification. For each oocyte, we obtained a minimum of 3 replicas to assure that the density of intramembrane particles (IMP) from different regions of the oocyte were included in the quantification. The oolemma was identified by the presence of microvilli. The E face is characterized by the presence of large intramembrane particles (14 nm) at a density of 800 particles/mm^2 and the P face by the presence of particles with a diameter of 7-nm at a density of about 200 particles/mm^2. Heterologous proteins appear as particles on the P face. We imaged all P faces in each replica at 25,000X.

From this pool of negatives we selected images to be quantified using NIH image 1.57 running on a Power Macintosh 8100/100. Each selected image contained 1 to 2 mm^2 of uninterrupted P fracture face. The density of intra membrane particles in the microvilli was quantified separately to assure that it was similar to that measured in the plasma membrane. The selected negatives were enlarged to 75000X and scanned at 200 dpi on a Hewlett Packard ScanJet Iicx flatbed scanner (Cupertino California). From the digitized images we measured the total number of particles on the P face and the area in which the particles were contained using NIH image 1.57. For each oocyte we found empirically that after counting 10 mm^2 of area, the particle density did not change. In high-expressing oocytes (4000-5000 particles/mm^2) it was only necessary to measure 2 mm^2.

Biochemistry

Crude Membrane Isolation. Crude membranes of 30 oocytes were isolated by placing them in a cold buffered hypotonic solution (7.5 mM Na_2HPO_4, pH 7.4) containing protease inhibitors (10 mg/ml PMSF, 1 mg/ml Aprotinin, 1 mg/ml Leupeptin, 1 mg/ml Pepstatin A and 3 mM EDTA [49,109,111]). Yolk protein was removed by spinning at 500g for 5 minutes at 4°C. The supernatant was centrifuged at 16,000g for 30 minutes at 4°C. The pellet containing the membrane fraction was washed with hypotonic buffer. Lens membranes were purified as previously described [43]. SDS-PAGE analysis was done according to the method of Laemmli [118].

Glycosidase treatment of MIP-stop and MIP. Membranes were solubilized and sonicated in phosphate buffer A (35 mM sodium phosphate pH 7.0, 5 mM EDTA, 0.5% w/v SDS, 50 mM b-mercaptoethanol, and protease inhibitors (see above)). Reactions were performed after adding an equal volume of phosphate buffer B (5% NP-40, 35 mM sodium phosphate, 5 mM EDTA and protease inhibitors). Each sample was split in two: one set served as controls and the other received 1 unit of N-glycosidase F (Boerhinger Mannheim). All samples were incubated at 32°C to 35°C for ~20 hours.

Western Blots. Five oocyte equivalents were run in all oocyte lanes. Two initial probes were used; a monoclonal IgM, BBC3 (a gift from Drs. D.L. Paul and D.A. Goodenough) that recognizes MIP's cytoplasmic C-terminus and a polyclonal antibody , E80, generated against amino acids 180 to 195 of MIP (a gift from Drs. Horowitz and Revel). Supernatant from monoclonals secreting BBC3 was diluted 1:10, and E80 serum was diluted 1:500. BBC3 and E80 antibody-staining were visualized with a peroxidase-linked goat anti-mouse antibody and a peroxidase-linked goat anti-rabbit antibody (both purchased from Sigma). Chemiluminescence (ECL, Amersham, England) was used to visualize the antibodies.

RESULTS

Mip- and Aqp1-Injected Oocytes Swell Faster than Mip-Stop- or Water-Injected Oocytes

Volume changes due to water flow driven by an osmotic gradient were estimated by measuring the cross-sectional area of the osmotically challenged oocytes as a function of time. Oocytes injected with AQP1 and MIP cRNAs swelled faster than control oocytes exposed to the same osmotic challenge (Fig. 2a,b). Swelling rates depend on the water permeability of the membrane, osmotic gradient and oocyte size. Because permeability is an intrinsic property of the membrane, swelling rates were used to calculate membrane permeability, P_f, for the oolemma according to the formula

$$P_f = [d(V/V_o)/dt][V_o/S_o]/[\Delta_{osm} V_w (S_a/S_e)].$$

V is the volume as a function of time, V_o is the volume at time = 0, S_o is the estimated geometric surface area of the oocyte at time = 0, Δ_{osm} is the osmotic gradient, V_w is the partial molar volume of water, S_a is the actual area of the oolemma after correcting for the presence of folds and microvilli and S_e is the estimated area calculated by assuming a sphere. Figure 2c summarizes P_f measurements from two batches of oocytes injected and measured on separate occasions. MIP-expressing oocytes ($P_f = 7.9 \pm 2.1 \times 10^{-4}$ cm/s, n=15) consistently have higher permeability than control oocytes ($P_f = 0.5 \pm 0.11 \times 10^{-4}$ cm/s, n=7) as do AQP1 ($P_f = 20 \pm 3.6 \times 10^{-4}$ cm/s, n=8) expressing oocytes. MIP-induced water permeability was not inhibited by $HgCl_2$ (data not shown). MIP-stop cRNA injection does not induce an increased oocyte P_f (Fig. 2). Nodulin-26 cRNA injection also induces an increase in oocyte water permeability that is $HgCl_2$ sensitive (data not shown).

We measured the water permeabilities of oocytes injected with different amounts of cRNA. Permeability was nearly proportional to the amount of cRNA injected for both AQP1and MIP (data not shown). Coinjection of cRNA encoding the non-functional MIP-stop along with wild-type MIP-cRNA did not alter wild-type function .

Biochemistry

Western blotting showed that oocytes injected with either MIP or MIP-stop actually made the corresponding protein. The monoclonal antibody,

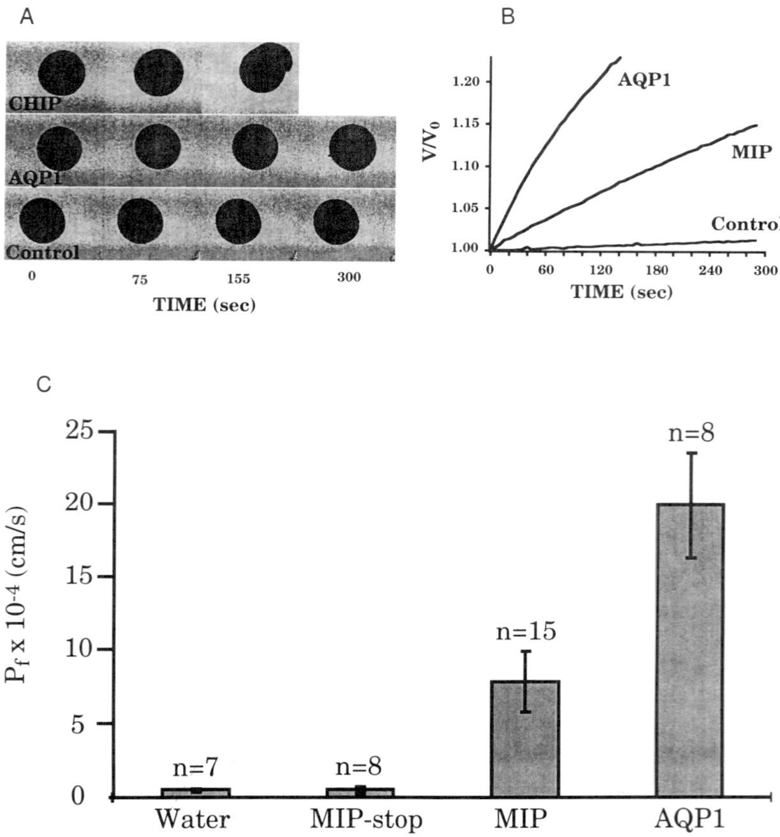

Figure 2. Swelling Assay. Oocytes were injected with Water, MIP or AQP1 cRNA. Panel A. Black circles are images taken of the oocytes swelling over time. Panel B. The same oocytes in panel A are represented by the relative volume changes over time. Panel C.Summary of Pf from two representative experiments. Data are Mean ±SD.

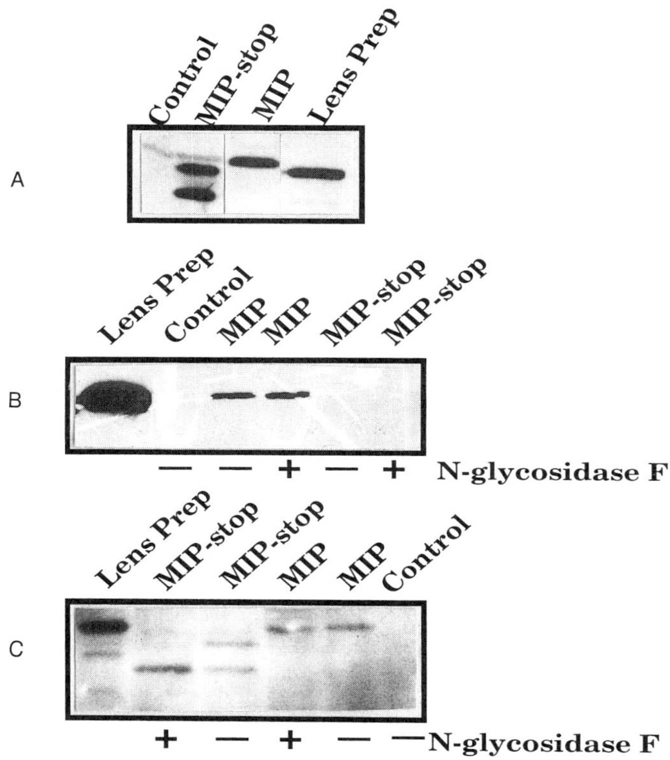

Figure 3. Western Blots of Oocyte Membrane Isolations. Each lane was loaded with ~5 oocyte equivalents of sample. 10 to 20 ng of lens prep. protein was loaded per lane. **Panel A.** Primary Ab was E80 Lane 1 is water injected, Lane 2 is MIP-stop injected, Lane 3 is MIP injected, Lane 4 is lens membrane preparation. **Panel B.** Oocytes plus and minus treatment with 1 unit of N-glycosidase F. Primary Ab was BBC3. Lane 1 is lens membrane preparation, Lane 2 is water injected, Lane 3 & 4 are MIP injected. Lane 5 & 6 are MIP-stop injected. **Panel C.** Primary Ab was E80. Lane 1 is lens membrane preparation, Lane 2 & 3 are MIP-stop injected, Lane 4 & 5 are MIP injected and lane 6 is water injected.

BBC3, detects protein with MW near 26 kD in the MIP-injected oocytes and in lens membrane preparations (Figure 3b). BBC3 detected no protein in water-injected or MIP-stop injected oocytes. However, the polyclonal anti-MIP (E80) detected a 26 kD protein in MIP-injected oocytes and two proteins of 22 and 25 kD in MIP-stop-injected oocytes (Figure 3a). E80 only detected the 22 kD protein band in N-glycosidase-treated membranes from oocytes expressing MIP-stop while continuing to detect the 26 kD band from MIP-expressing oocytes (Figure 3c). Thus both MIP and MIP-stop are synthesized, but unlike wild type MIP, a fraction of MIP-stop is glycosylated.

Estimation of single-molecule water permeability

By combining the results of the above experiments with measurement of surface protein expression it is possible to obtain an accurate estimate of the single-molecule water permeability for MIP and AQP1. On the basis of measurements made on oocytes expressing high levels of MIP and AQP1, we previously reported a single-molecule water permeability (p_f) of 2.1 x 10^{-16} cm^3/sec and 1.5 x 10^{-14} cm^3/sec respectively [119]. Here we report a more precise value based on a larger data set, certain corrections to the membrane area, and improved analysis.

We measured swelling rate, capacitance and IMP densities on each individual oocyte expressing MIP, AQP1, and controls. We used the capacitance measurements to obtain an estimate of the ratio of the actual surface area to the geometric surface area at a given level of protein expression. We used this ratio and the measured geometric area for each oocyte to calculate the effective surface area available for water permeation. Using this surface area and the measured swelling rate, we calculated the values of P_f.

Intra membrane particle (IMP) density increases with increased expression levels. We plotted the IMP density as a function of P_f in order to estimate the single-molecule permeability (Figure 4). The slope of the lines

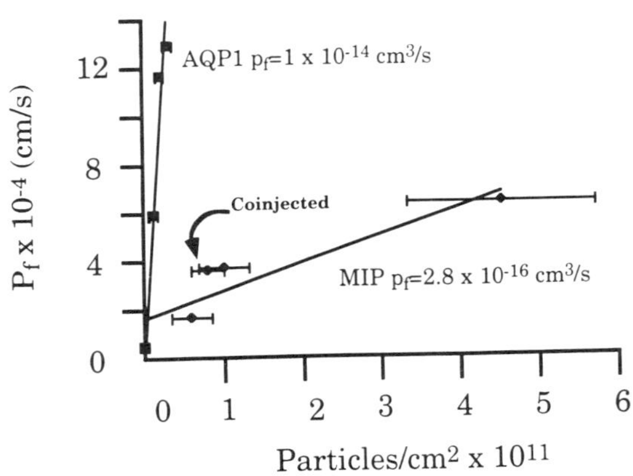

Figure 4. Single protein water conductance of MIP and AQP1. Varying density of protein expressed in the oolemma vs. the water permeability of that oocyte. The slope of the fitted line predicts the p_f for MIP (2.8 x 10^{-16} cm^3/sec) and AQP1 is (1.0 x 10^{-16} cm^3/sec).

then allows us to calculate the p_fs for AQP1 (1.0×10^{-14} cm^3/s) and MIP (2.8×10^{-16} cm^3/s) assuming each particle is a tetramer[22,43,45,46,51-55,120]. MIP-stop, which did not increase membrane P_f or the IMP density, provides an excellent negative control for the method. Though Western blot analysis showed that MIP-stop was in fact synthesized by the oocytes at levels comparable to MIP (Figure 3) there was no increase in permeability induced by MIP-stop, nor was there any increase in IMP density (data not shown).

Activation energy of MIP- and AQP1- induced permeability is less than that of water permeability through a lipid bilayer

A principal characteristic for classical water channels is a lowered activation energy for water flux across the lipid bilayer. To test if this was true of MIP-induced water flux, we measured the water permeability induced by both MIP and AQP1 as a function of temperature. Control oocyte permeability was subtracted from the total permeability of oocytes expressing heterologous protein at each temperature leaving only the permeability induced by the protein. This correction had little effect on the value of the activation energy of water flux induced by AQP1, but did lower the activation energy estimated for the water flux induced by MIP. Activation energies of water fluxes attributable to MIP (6.9 ± 2.5 Kcal/mole), AQP1 (2.4 ± 1.1 Kcal/mole) and Nodulin-26 are much lower than that of water flux through a bare lipid bilayer (20.6 ± 4.9 Kcal/mole).

DISCUSSION

Properties of Mip Water Transport in Oocytes Resemble Those of Aqp1

We have shown that the water transport properties through MIP and AQP1 differ from the properties of water transport through the bare oolemma. (Since water transport through the oolemma closely resembles water transport through lipid bilayers, we assume that water transport in the bare oolemma is through the lipid bilayer pathway.) Both MIP and AQP1 increase the water permeability of the oolemma in proportion to the surface density of molecules expressed in the oolemma. Both reduce the activation energy from about 20 Kcal/mole to 3-7 Kcal per mole[69]. (The activation energy for the self diffusion of water in water is about 4 Kcal/mole[107]). These conclusions add to and confirm the results of Mulders *et al.*[67] Our values for activation energy of MIP water transport differs slightly from theirs, and we provide a measure of the number of molecules responsible for a given water flux. The essential difference between MIP and AQP1 is thus *quantitative*, not *qualitative*.

Single-Molecule Permeability of Mip is Low, but is Sufficient to Produce a Significant Water Permeability in Fiber Cells

We find that a single molecule of MIP is approximately 36 times less effective in promoting water flux at 10°C than a single molecule of AQP1. The water transport induced by both MIP and AQP1 is proportional to the number of particles induced in the oolemma. Thus the permeabilities induced by MIP and AQP1 are due to the presence of the respective molecules and not to an artifact of oocyte manipulation. A possible reason that previous investigators failed to see a robust increase in water permeability induced by MIP was a failure to achieve sufficient levels of expression in the oocyte system. Although expression level is not always strictly proportional to amount of cRNA injected, we did have to inject much larger amounts of MIP cRNA than AQP1 cRNA to attain comparable levels of water permeability. It thus seems incontrovertible that MIP increases water permeability of oocyte membranes and that it very likely does so by the same mechanism as does AQP1.

Properties of MIP in Artificial Systems

MIP appears to increase membrane permeability to molecules up to 1.5 kD, when reconstituted into proteoliposomes[58-60,121]. In planar lipid bilayers MIP clearly forms a large conductance, voltage-dependent channel modulated by phosphorylation[43,65]. These properties are reproducible in the artificial systems in which they are measured, but they have not been observed in *vivo* expression systems such as the *Xenopus* oocyte or in lens fiber cells themselves. It thus seems likely that these properties of MIP appear rarely, if at all, in *in vivo* systems.

MIP-Stop

MIP-stop was made to mimic MIP-22, the natural degradation product of MIP in the lens. There is a correlation between increased levels of MIP-22 with cataract[122-124]. We wished to examine any functional differences there might be between MIP and MIP-22. However, removing the c-terminal 35 amino acids prevents MIP-stop from being properly transported to the oocyte plasma membrane. Interestingly a normally unused glycosylation site, predicted by hydropathy analysis to be in or very near the hydrocarbon bilayer, becomes accessible for glycosylation (See Figure 1). Whether failure of transport to the oolemma and glycosylation are related cannot be ascertained at this time, but there are known mutations which alter protein folding, resulting in failure of the protein to reach the plasma membrane. Of these the most widely known mutation is DF508 of CFTR[125]. AQP2 mutations which result in diabetes insipidus prevent its transport to the plasma membrane. These mutant AQP2s are also abnormally glycosylated in the oocyte[94]. Two mouse mutants *Lop* and Cat^{Fr} get cataracts resulting from defects in their MIP. *Lop*-mouse MIP fails to properly fold and does not reach the plasma membrane but instead remains localized to the endoplasmic reticulum[126].

We also tested for interaction of MIP and MIP-stop by co-injecting their cRNAs into oocytes. Possible results from this experiment are: MIP and MIP-stop associate and are transported to the plasma membrane where the heteromer is either (1) functional or (2) non-functional, (3) MIP-stop prevents MIP transport to the oolemma, and (4) they do not associate MIP-stop remains intracellular while MIP would be transported to the plasma membrane normally. Since coinjection of MIP-stop cRNA did not alter MIP function, we can conclude that the a functional heteromer is not formed and that MIP-stop does not alter MIP transport. Thus either MIP-stop and MIP do not associate or they do so in a manner that does not affect MIP function or transport. Without the addition of freeze fracture measurements we could not have distinguished between these possibilities. Because the function/particle remains constant with or without the expression of MIP-stop we conclude that MIP-stop remains intracellular while WT-MIP is transported normally to the oolemma.

CONCLUSIONS

As the most abundant lens membrane protein [14-16], MIP surely plays a role in lens physiology and thus its pathology. The central question is, "What role (or roles) does MIP play in lens physiology?" Our results only partially answer this question, but they do clearly demonstrate and quantify the per-molecule water permeability of MIP.

Knowing the p_f of MIP allows us to calculate the permeability of any membrane with a known density of MIP monomers. The center-to-center spacing of the unit cell in MIP tetragonal arrays is 6.7 nm [25]. From this number the density of MIP monomers in plaques would be 9.2×10^{12} monomers/cm^2. Thus if MIP plaques occupy about 50% of fiber cell surface area, the permeability of a lens fiber cell would be $9.2 \times 10^{-12} \times 0.5 \times 2 \times 10^{-16} = 9 \times 10^{-4}$ cm/sec. Lipid bilayers composed of phosphatidyl choline have a permeability of about $20\text{-}50 \times 10^{-4}$ cm/sec, but lipid bilayers containing mostly sphingomyelin and cholesterol, the principal lipids in the lens, have permeabilities which are about an order of magnitude lower [127]. Thus MIP would likely increase the water permeability of lens fiber cells substantially.

Our results are consistent with results previously published on MIP-induced water permeability, even the results of investigators who concluded that MIP is not a water channel. Van Hoek *et al.* [64] reconstituted MIP into vesicles and found that the difference in water permeability between lipid vesicles and MIP-containing vesicles was about 10×10^{-4} cm/sec. They estimated the lipid to protein ratio in these vesicles was 15.3, weight to weight.

Assuming 100 square angstroms as the area of a lipid molecule, this would give a density of MIP molecules of 4×10^{12} /cm^2. Using our value of the single-molecule water permeability of MIP gives a permeability of 8×10^{-4} cm/sec, in very good agreement with the permeability determined for MIP-containing vesicles. Subsequent data on the qualitative similarity of the permeabilities induced by MIP and AQP1 would seem to argue that these earlier results support rather than mitigate against the notion that MIP induces an increased water permeability.

Our results indicate that increasing water permeability is a *plausible* function for MIP in the ocular lens. However, this may not be its only function.

Acknowledgments

This work was supported by grants from the National Eye Institute EY 05661 to JEH and EY 4110

REFERENCES

1. M.B. Gorin, S.B.Yancey, J.Cline, J.P. Revel, and J. Horwitz, The major intrinsic protein (MIP) of the bovine lens fiber membrane: characterization and structure based on cDNA cloning, *Cell.* 39:49 (1984).
2. P. Donaldson, and J. Kistler, Reconstitutions of channels from preparations enriched in lens gap junction protein MP70, J. Membr. Biol. 129:155 (1992).
3. G.A. Zampighi, J.E. Hall, and M., Kreman, Purified lens junctional protein forms channels in planar lipid films, *Proc. Natl. Acad. Sci. USA.* 82:8468 (1985).
4. G.R. Ehring, G.A. Zampighi, J. Horwitz, D. Bok, and J.E. Hall, Properties of channels reconstituted from the major intrinsic protein of lens fiber membranes, *J. Gen. Physiol.* 96:631 (1990).
5. P. Agre, S. Sasaki, and M.J. Chrispeels, Aquaporins - A Family Of Water Channel Proteins, *Am. J. Physiol.* 265:F461 (1993).
6. P. Agre, G.M. Preston, B.L. Smith, J.S., Jung, S. Raina, C. Moon, W.B. Guggino, and S. Nielsen, Aquaporin CHIP: the archetypal molecular water channel, *Am J Physiol.* 265:F463 (1993).
7. P. Agre, and G.M. Preston, Isolation of the cDNA for erythrocyte integral mebrane protein of 28 kilodaltons: Member of an ancient family, *Proc. Natl. Acad. Sci. USA.* 88:11110 (1992).
8. H. Maisel, "The Ocular lens : structure, function, and pathology," Dekker, New York (1985).
9. R.T. Mathias, and J.L. Rae, Transport properties of the lens, *Am. J. Physiol.* 249:C181 (1985).
10. R.T. Mathias, J.L. Rae, and R.S. Eisenberg, The lens as a nonuniform spherical syncytium, *Biophys. J.* 34:61 (1981).
11. J.L. Rae, and R.T. Mathias, The Physiology of the Lens, in: "The Ocular Lens. Structure, Function, and Pathology" .H. Maisel, ed., Marcel Dekker, Inc. New York (1985).
12. R.T. Mathias, J.L. Rae, and G.J. Baldo, Physiological Properties of the Normal Lens, *Am J Physiol* [in press] (1996).
13. R.G. Johnson, K.A. Klukas, L.T. Hong, and D.C. Spray, Antibodies to MP28 are localized to lens junctions, alter intercellular permeability and demonstrate increased expression during development, in: "Gap Junctions" E.L. Hertzberg and R. Johnson, eds., Alan R. Liss, New York.
14. H. Bloemendal, A. Zweers, F. Vermorken, I. Dunia, and E.L. Benedetti, The plasma membranes of the eye lens fibres. Biochemical and structural characterization, *Cell Diff.* 1:91 1972).
15. J. Alcala, N. Lieska, N., and H. Maisel, Protein composition of bovine lens cortical fiber cell membranes, *Exp. Eye Res.* 21:591 (1975).
16. P.D. Lampe, and R.G. Johnson, Amino acid sequence of the in vivo phosphorylation sites in the main intrinsic protein (MIP) of lens membrane, Eur. *J. Biochem.* 194:541 (1990).
17. J.R. Kuszak, H. Maisel, C.V. Harding, Gap junctions of chick lens fiber cells, *Exp. Eye Res.* 27:495 (1978).

18. H. Maisel, C.V. Harding, J.R.Alcala, J.R., Kuszak, R. Bradley. The morphology of the lens, in: "Molecular and Cellular Biology of the Eye Lens" .H. Bloemendal, ed., John Wiley and Sons, New York.
19. G.R. Ehring, G.A. Zampighi, and J.E. Hall, Does MIP play a role in cell-cell communication? in: "Gap Junctions" J.E. Hall, G.A. Zampighi, and R.M. Davis, eds., Elsevier, Amsterdam.
20. D.L. Paul, and D.A. Goodenough, Preparation, characterization, and localization of antisera against bovine MP26, an integral protein from lens fiber plasma membrane, *J. Cell Biol.* 96:625 (1983).
21. P.G. Fitzgerald, D. Bok, and J. Horwitz, The distribution of the main intrinsic membrane polypeptide in ocular lens, *Curr. Eye Res.* 4:1203 (1985).
22. G.A. Zampighi, J.E. Hall, G.R. Ehring, and S.A. Simon, The structural organization and protein composition of lens fiber junctions, *J. Cell Biol.* 108:2255 (1989).
23. S. Sasaki, K. Fushimi, H. Saito, F. Saito, S. Uchida, K. Ishibashi, M. Kuwahara, T. Ikeuchi, K. Inui, and K. Nakajima, et al. Cloning, characterization, and chromosomal mapping of human aquaporin of collecting duct, *J. Clin. Invest.* 93:1250 (1994).
24. G.A. Zampighi, S.A. Simon. and J.E. Hall, The specialized junctions of the lens, in press.(1997).
25. G. Zampighi, S.A. Simon, J.D. Robertson, T.J. McIntosh, M.J. Costello, On the structural organization of isolated bovine lens fiber junctions, *J. Cell Biol.* 93:175 (1982).
26. D. Bok, J. Dockstader, and J. Horwitz, Immunocytochemical localization of the lens main intrinsic polypeptide (MIP26) in communicating junctions, *J. Cell Biol.* 92:213 (1982).
27. P. Keeling, K. Johnson, D. Sas, K. Klukas, P. Donahue, and R. Johnson, Arrangement of MP26 in lens junctional membranes: analysis with proteases and antibodies, *J. Membr. Biol.* 74:217 (1983).
28. D.L. Paul, D.A. Goodenough, In vitro synthesis and membrane insertion of bovine MP26,an integral protein from lens fiber plasma membrane, *J. Cell Biol.* 96:636 (1983).
29. D.F. Sas, M.J. Sas, and R.G. Johnson, Lens junctional protein: analyzing MP26 with monoclonal andtibodies, *Curr. Eye Res.* 11:1171 (1985).
30. J. Kyte, and R.F. Doolittle, A simple method for displaying the hydropathic character of a protein, *J. Mol. Biol.* 157:105 (1982).
31. J. Reizer, A. Reizer, and M.H.Jr. Saier, The MIP family of intergral membrane proteins: Sequence comparisons, evolutionary relationships, reconstructed pathway of evolution and proposed functional differentiation of the two repeated halves of the proteins, *Crit. Rev. Biochem. Mol. Biol.* 28:235 (1993).
32. M.M. Pisano, and A.B. Chepelinsky, Genomic Cloning, Complete Nucleotide Sequence, And Structure Of The Human Gene Encoding The Major Intrinsic Protein (Mip) Of The Lens, *Genomics.* 11:981 (1991).
33. F. Saito, S. Sasaki, A.B. Chepelinsky, K. Fushimi, F. Marumo, T. Ikeuchi, Human AQP2 and MIP genes, two members of the MIP family, map within chromosome band 12q13 on the basis of two-color FISH, *Cytogenet. Cell Genet.* 68:45 (1995).

34. S. Manenti, I. Dunia, and E.L. Benedetti, Fatty Acid Acylation of Lens Fiber Plasma Membrane Proteins Mp26 and Alpha-Crystallin are Palmitoylated, *FEBS Lett.* 262:356 (1990).
35. K.R. Johnson, S.S. Panter, and R.G. Johnson, Phosphorylation of lens membranes with a cyclic AMP-dependent protein kinase purified from the bovine lens, *Biochim. Biophys. Acta.* 844:367 (1985).
36. C.F. Louis, R.G. Johnson, K.R. Johnson, and J. Turnquist, Characterization of the bovine lens plasma membrane substrates for cAMP-dependent protein kinase, Eur. J. Biochem. 150:279 (1985).
37. P.D. Lampe, and R.G. Johnson, Phosphorylation of MP26, a lens junction protein, is enhanced by activators of protein kinase C, *J. Membr. Biol.* 107:145 (1989).
38. L.J. Takemoto, and M. Takehana, Covalent change of major intrinsic polypeptide (MIP26K) of lens membrane during human senile cataractogenesis, *Biochem. Biophys. Res. Commun.* 135:965 (1986).
39. C.F. Louis, P. Hogan, L. Visco, and G. Strasburg, Identity Of The Calmodulin-Binding Proteins In Bovine Lens Plasma Membranes, *Exp. Eye Res.* 50:495 (1990).
40. M. Prabhakaram, and B.J. Ortwerth, Glycation Of Mp26 And Mp22 In Bovine Lens Membranes, Biochem. Biophys. Res. Commun. 185:496 (1992).
41. M.S. Swamy, and E.C. Abraham, Glycation Of Lens Membrane Intrinsic Proteins, *Curr. Eye Res.* 11:833 (1992).
42. T. Aerts, J. Xia, H. Slegers, J. de Block, J. Clauwaert, Hydrodynamic characterization of the major intrinsic protein from the bovine lens fiber membranes, *J. Biol. Chem.* 265:8675 (1990).
43. G.R. Ehring, G. Zampighi, J. Horwitz, D. Bok, and J.E. Hall, Properties of channels reconstituted from the major intrinsic protein of lens fiber membranes, *J. Gen. Physiol.* 96:631 (1990).
44. J. Farinas, A.N. Van Hoek, L.B. Shi, C. Erickson, and A.S. Verkman,. Nonpolar environment of tryptophans in erythrocyte water channel CHIP28 determined by fluorescence quenching, *Biochemistry.* 32:11857 (1993).
45. B.L. Smith, and P. Agre, Erythrocyte Mr 28,000 transmembrane protein exists as a multisubunit oligomer similar to channel proteins, *J. Biol. Chem.* 266:6407 (1991).
46. M.L. Zeidel, S. Nielsen, B.L. Smith, S.V. Ambudkar, A.B. Maunsbach, Agre, P. Ultrastructure, pharmacologic inhibition, and transport selectivity of aquaporin channel-forming integral protein in proteoliposomes, *Biochemistry.* 33:1606 (1994).
47. L.B. Shi, W.R. Skach, T. Ma, and A.S. Verkman, Distinct biogenesis mechanisms for the water channels MIWC and CHIP28 at the endoplasmic reticulum, *Biochemistry.* 34:8250 (1995).
48. W.R. Skach, L. Shi, C.M. Calayag, A. Frigeri, V.R. Lingappa, and A.S. Verkaman, Biogenesis and Transmembrane Topology of the CHIP28 Water Channel at the Endoplasmic Reticulum, *J. Cell Biol.* 125:803 (1994).
49. G.M. Preston, J.S. Jung, W.B. Guggino, and P. Agre, Membrane topology of aquaporin CHIP. Analysis of functional epitope-scanning mutants by vectorial proteolysis, *J. Biol. Chem.* 269:1668 (1994).

50. W. Skach, and A.S. Verkman, *Biophys. J.* 68:A344 (1995).
51. J.M. Verbavatz, D. Brown, I. Sabolic, G. Valenti, D.A. Ausiello, A.N. Van Hoek, T. Ma, and A.S. Verkman, Tetrameric assembly of CHIP28 water channels in liposomes and cell membranes: a freeze-fracture study, *J. Cell Biol.* 123:605 (1993).
52. A.K. Mitra, M. Yeager, A.N. Van Hoek, M.C. Wiener, A.S. Verkman, Projection structure of the CHIP28 water channel in lipid bilayer membranes at 12-A resolution, *Biochemistry* 33:12735 (1994).
53. T. Walz, B.L. Smith, P. Agre, and A. Engel, The three-dimensional structure of human erythrocyte aquaporin CHIP, Embo.J. 13:2985 (1994).
54. T. Walz, B.L. Smith, M.L. Zeidel, A. Engel, and P. Agre, Biologically active two-dimensional crystals of aquaporin CHIP, *J. Biol. Chem.* 269:1583 (1994).
55. B.K. Jap, and H. Li, H. Structure of the osmo-regulated H2O-channel, AQP-CHIP, in projection at 3.5 A resolution, *J. Mol. Biol.* 251:413 (1995).
56. H. Nikaido, and E.Y. Rosenberg, Functional reconstitution of lens gap junction proteins into proteoliposomes, *J. Membr. Biol.* 85:87 (1985).
57. C. Peracchia, and S.F. Girsch, Permeability and gating of lens gap junction channels incorporated into liposomes, *Curr. Eye Res.* 4:431 (1985).
58. M. Gooden, D. Rintoul, M. Takehana, and L. Takemoto, Major Inrinsic Polypeptide (MIP26K) from Lens Membrane: Reconstitution into Vesicle and Inhibition of Channel Forming Activity by Peptide Antiserum, *Biochem. Biophys. Res. Comm.* 128:993 (1985).
59. B.A. Scaglione, and D.A. Rintoul, A Fluorescence-Quenching Assay for Measuring Permeability of Reconstituted Lens MIP26, Invest. Ophtal. & Visual Science 30:961 (1989).
60. L. Shen, P. Shrager, S.J. Girsch, P.J., Donaldson, and C. Peracchia, Channel Reconstitution In Liposomes And Planar Bilayers With Hplc-Purified Mip26 Of Bovine Lens, *J. Membr. Biol.* 124:21 1991).
61. S.J. Girsch, and C. Peracchia, Lens cell-to-cell channel protein: II. Conformational change in the presence of calmodulin, *J. Membr. Biol.* 83:227 1985).
62. C. Peracchia, and S.J. Girsch, Functional modulation of cell coupling: evidence for a calmodulin-driven channel gate, *Am. J. Physiol.* 248:H765 (1985).
63. L.J. Jarvis, and C.F. Louis, The Permeability Of Reconstituted Liposomes Containing The Purified Lens Fiber Cell Integral Membrane Protein-Mp20, Protein-Mp26 And Protein-Mp70, *J. Membr. Biol.* 130:251 (1992).
64. A.N.Van Hoek, M. Wiener, S. Bicknese, L. Miercke, J. Biwersi, and A.S. Verkman, Secondary structure analysis of purified functional CHIP28 water channels by CD and FTIR spectroscopy, *Biochemistry.* 32:11847 (1993).
65. G.R. Ehring, N. Lagos, G.A. Zampighi, and J.E., Hall, Phosphorylation modulates the voltage dependence of channels reconstituted from the major intrinsic protein of lens fiber membranes. *J. Membr. Biol.* 126:75 (1991).

66. G.R. Ehring, P. Lang, and J.E. Hall, Proteolytic Digestion with Trypsin Reduces the Voltage-dependent Closing of MIP Channels Incorporated into Planar Lipid Bilayers (BLM), *Biophys. J.* 59:Abstract (1991).
67. S.M. Mulders, G. M. Preston, P. M. Deen, W.B. Guggino, C.H. Van Os, P. Agre, Water channel properties of major intrinsic protein of lens, *J. Biol. Chem.* 270:9010 (1995).
68. C. Kushmerick, S.J. Rice, G.J. Baldo, and H.C. Haspel, R.T. Mathias, Cloning, Expression, and Functional Studies of Frog Lens MIP, *Biophys. J.* (Abstracts). 66: abstract (1984).
69. G.M. Preston, T.P. Carroll, W.B. Guggino, and P. Agre, Appearance of water channels in Xenopus oocytes expressing red cell CHIP28 protein, *Science* 256:385 (1992).
70. T.H. Ma, A. Frigeri, S.T. Tsai, J.M. Verbavatz, and A.S. Verkman, Localization And Functional Analysis Of Chip28K Water Channels In Stably Transfected Chinese Hamster Ovary Cells, *J. Biol. Chem.* 268:22756 (1993).
71. R.B. Zhang, W. Skach, H. Hasegawa, A.N. Vanhoek, and A.S. Verkman, A.S. Cloning, Functional Analysis And Cell Localization Of A Kidney Proximal Tubule Water Transporter Homologous To Chip28, *J. Cell Biol.* 120:359 (1993).
72. G.M. Preston, and P. Agre, Isolation Of The cDNA For Erythrocyte Integral Membrane Protein Of 28-Kilodaltons - Member Of An Ancient Channel Family, *Proc. Natl. Acad. Sci. USA.* 88:11110 (1991).
73. P.M. Deen, D.O. Weghuis, A. Geurs van Kessel, B. Wieringa, C.H. Van Os, The human gene for water channel aquaporin 1 (AQP1) is localized on chromosome 7p15-->p14, *Cytogenet. Cell Genet.* 65:243 (1994).
74. C. Moon, G.M. Preston, C.A. Griffin, E.W. Jabs, and P. Agre, The Human Aquaporin-Chip Gene - Structure, Organization, And Chromosomal Localization, *J. Biol. Chem.* 268:15772 (1993).
75. S. Nielsen, B.L. Smith, E.I. Christensen, and P. Agre, Distribution of the aquaporin CHIP in secretory and resorptive epithelia and capillary endothelia, *Proc Natl Acad Sci U S A.* 90:7275 (1993).
76. B.L. Smith, R. Baumgarten, S. Nielsen, D. Raben, M.L. Zeidel, and P. Agre, P. Concurrent expression of erythroid and renal aquaporin CHIP and appearance of water channel activity in perinatal rats [see comments], *J Clin Invest.* 92:2035 (1993).
77. L. Abrami, F. Tacnet, and P. Ripoche, Evidence For a Glycerol Pathway Through Aquaporin 1 (Chip28) Channels, *Pflügers Archiv-Eur. J. Physiol.* 430:447 (1995).
78. A.N. Van Hoek, and A.S. Verkman, Functional reconstitution of the isolated erythrocyte water channel CHIP28, *J. Biol. Chem.* 267:18267 (1992).
79. T. Zeuthen, Molecular mechanisms for passive and active transport of water, in: "International Review of Cytology" Anonymous Academic Press,
80. C. Bondy, E. Chin, B.L. Smith, G.M. Preston, and P. Agre, Developmental Gene Expression And Tissue Distribution Of The Chip28 Water-Channel Protein, *Proc. Natl. Acad. Sci. USA.* 90:4500 (1993).

81. H. Hasegawa, R. Zhang, A. Dohrman, and A.S. Verkman, Tissue-Specific Expression Of Messenger RNA Encoding Rat Kidney Water Channel Chip28K By Insitu Hybridization, *Am. J. Physiol.* 264:C237 (1993).
82. H. Hasegawa, S. Lian, W.E. Finkbeiner, and A.S. Verkman, Extrarenal Tissue Distribution of CHIP28 Water Channels by in situ Hybridization and Antibody Staining, *Am. J. Physiol.* 266:C893 1994).
83. G.M. Preston, B.L. Smith, M.L. Zeidel, J.J. Moulds, and P. Agre, Mutations in aquaporin-1 in phenotypically normal humans without functional CHIP water channels, *Science* 265:1585 (1994).
84. S. Nielsen, T. Pallone, B.L. Smith, E.I. Christensen, P. Agre, and A.B. Maunsbach, Aquaporin-1 water channels in short and long loop descending thin limbs and in descending vasa recta in rat kidney, *Am J Physiol.* 268:F1023 (1995).
85. K. Fushimi, S. Sasaki, T. Yamamoto, M. Hayashi, T. Furukawa, S. Uchida, M. Kuwahara, K. Ishibashi, M. Kawasaki, I. Kihara, et al. Functional characterization and cell immunolocalization of AQP-CD water channel in kidney collecting duct, *Am J Physiol.* 267:F573 (1994).
86. T. Ma, H. Hasegawa, W. R. Skach, A. Frigeri, and A.S. Verkman, Expression, functional analysis, and in situ hybridization of a cloned rat kidney collecting duct water channel, *Am J Physiol.* 266:C189 (1994).
87. M. Kuwahara, K. Fushimi, Y. Terada, L.Q. Bai, F. Marumo, S. Sasaki, Camp-Dependent Phosphorylation Stimulates Water Permeability Of Aquaporin-Collecting Duct Water Channel Protein Expressed In Xenopus Oocytes, *J. Biol. Chem.* 270:10384 (1995).
88. T.M. Fujiwara, K. Morgan, and D.G. Bichet, Molecular biology of diabetes insipidus, *Ann. Rev Med.* 46:331 (1995).
89. N. Knoers, A. van den Ouweland, J. Dreesen, M. Verdijk, L.A. Monnens, and B.A. van Oost, B.A. Nephrogenic diabetes insipidus: identification of the genetic defect, *Pediatr Nephrol.* 7:685 (1993).
90. P.M. Deen, M.A. Verdijk, N.V. Knoers, B. Wieringa, L.A. Monnens, C.H. Van Os, and B.A. van Oost, Requirement of human renal water channel aquaporin-2 for vasopressin-dependent concentration of urine, *Science* 264:92 1994).
91. S. Uchida, S. Sasaki, K. Fushimi, and F. Marumo, Isolation of human aquaporin-CD gene, *J. Biol. Chem.* 269:23451 (1994).
92. A.F. van Lieburg, M.A. Verdijk, V.V. Knoers, A.J. van Essen, W. Proesmans, R. Mallmann, L.A. Monnens, B.A. van Oost, C.H. Van Os, and P.M. Deen, Patients with autosomal nephrogenic diabetes insipidus homozygous for mutations in the aquaporin 2 water-channel gene, *Am. J. Hum. Genet.* 55:648 (1994).
93. K. Fushimi, S. Uchida, Y. Hara, Y. Hirata, F. Marumo, and S. Sasaki, Cloning and expression of apical membrane water channel of rat kidney collecting tubule, *Nature* 361:549 (1993).
94. P.M. Deen, H. Croes, R.A. van Aubel, L.A., Ginsel, and C.H. Van Os, Water channels encoded by mutant aquaporin-2 genes in nephrogenic diabetes insipidus are impaired in their cellular routing, *J.Clin.Invest.* 95:2291 (1995).

95. M.J. Chrispeels, and C. Maurel, C. Aquaporins: The Molecular Basis of Facilitated Water Movement Through Living Plan Cells? *Plant. Physiol.* 105:9 (1994).
96. C. Maurel, J. Reizer, J.I. Schroeder, and M.J. Chrispeels, the vacuolar membrane protein gamma-tip creates water specific channels in xenopus-oocyte, *EMBO J.* 12:2241 (1993).
97. C. Maurel, R.T. Kado, J.Guern, and M.J.Chrispeels, Phosphorylation regulates the water channel activity of the seed-specific aquaporin alpha-tip camp-dependent phosphorylation stimulates water permeability of aquaporin-collecting duct water channel protein expressed in xenopus oocytes, *EMBO. J.* 14:3028 (1995).
98. M.J. Daniels, T.E. Mirkov, M.J. Chrispeels, The plasma membrane of arabidopsis thaliana contains a mercury-insensitive aquaporin that is a homolog of the tonoplast water channel protein TIP, *Plant Physiol.*106:1325 (1994).
99. W. Kammerloher, U. Fisher, G.P. Piechottka, A.R. Scheffner, Water channels in the plant plasma membrane cloned by immunoselection from a mammalian epression system, *Plant J.* 6:187 (1994).
100. M.G. Fortin, N.A. Morrison,D.P.Verma, Nodulin-26 a peribacteroid membrane nodulin is expressed independently of the development of the peribacteroid membrane nodulinis expressed independently of the development of the peribacteroid compartment, *Nucleic Acids Res.*15:813(1987).
101. N.N. Sandal, K.A. Marcker, Soubean nodulin 26 is homologuos to the major intrinsic protein of the bovine lens fiber membrane, *Nucleic Acids Res.*16:9347(1988).
102. L.J. Ouyang, J. Whelan, C.D. Weaver, D.M.Roberts, D.A. Day, The protein phosphorylation stimulates the rate of malate uptake across the peribacteroid membrane of soybean nodules, Febs Lett.293:188(1991).
103. C.D. Weaver, N.H. Shomer, C.F. Louis and D.M. Roberts, Nodulin 26, a nodule-specific symbiosome membrane protein from soybean, is an ion channel, *J. Biol. Chem.* 269:17858(1994).
104. C.D.Weaver and D.M. Roberts, Determination of the site of phosphorylation of nodulin 26 by the calcium-dependent protein kinase from soybean nodules, *Biochem.*31:8954(1992).
105. J.W. Lee, Y.Zhang, C.D. Weaver, N.H.Shomer, C.F. Louis and D.M. Roberts, Phosphorylation of nodulin 26 on serine 262 effects its voltage-sensitive channel activity in planar lipid bilayers, *J. Biol. Chem.* 270:27051(1995).
106. G., Chandy, D.M.,Roberts, and J.E. Hall, Water channel activity of wild type and mutant Nodulin-26 expressed in *Xenopus Laevis* oocytes, *Biophys J.* 70:A196 (1996).
107. A., Finkelstein, "Water Movement Through Lipid Bilayers, Pores, and Plasma Membranes. Theory and Reality,"John Wiley and Sons;New York (1987).
108. A.S. Verkman, "Water Channels", R.G. Landes Company; Austin (1993).
109. G.M.,Preston, J.S., Jung, W.B. Guggino, and P. Agre, The mercury-sensitive residue at cysteine 189 in the CHIP28 water channel, *J. Biol. Chem.* 268:17 (1993).

110. H., Hasegawa, T. Ma, W. Skach, M.A. Matthay, and A.S. Verkman, Molecular cloning of a mercurial-insensitive water channel expressed in selected water-transporting tissues, *J. Biol. Chem.* 269:5497 (1994).
111. J.S. Jung, R.V. Bhat, G.M. Preston, W.B. Guggino, J.M. Baraban, and P. Agre, Molecular characterization of an aquaporin cDNA from brain: candidate osmoreceptor and regulator of water balance, *Proc. Natl. Acad. Sci. USA.* 91:13052 (1994).
112. K. Ishibashi, S. Sasaki, K. Fushimi, S. Uchida, M. Kuwahara, H. Saito, T. Furukawa, K. Nakajima, Y. Yamaguchi, and T. Gojobori, Molecular cloning and expression of a member of the aquaporin family with permeability to glycerol and urea in addition to water expressed at the basolateral membrane of kidney collecting duct cells, *Proc. Natl. Acad. Sci. USA.* 91:6269 (1994).
113. S.N. Ho, H.D.Hunt, R.M.Horton, J.K. Pullen, and L.R., Site-directed mutagenesis by overlap extension using the polymerase chain reaction. *Gene* 77:51 (1989).
114. F. Sanger, S. Nicklen, and A.R. Coulson, DNA sequencing with chain-terminating inhibitors, *Proc .Natl .Acad. Sci .USA.* 74:5463 (1977).
115. B. Lucka, and M. McCutcheon, The living cell as an osmotic system and its permeability to water, *Physiol. Rev.* 12:68 (1932).
116. R.B. Zhang, and A.S. Verkman, Water and urea permeability properties of Xenopus oocytes: expression of mRNA from toad urinary bladder, *Am J Physiol.* 260:C26 (1991).
117. G. Zampighi, M. Kreman, F. Ramon, A.L. Moreno, and S.A. Simon, Structural characteristics of gap junctions. I. Channel number in coupled and uncoupled conditions, *J. Cell Biol.* 106:1667-1678(1988).
118. U.K. Laemmli, Cleavage of structural proteins during the assembly of the head of bacteriophage T4, *Nature.* 227:680 (1970).
119. G.Zampighi, M. Kreman, K.J. Boorer, D.D.F. Loo, F. Bezanilla, G. Chandy, J.E. Hall, and E.M. Wright, A method for determining the unitary functional capacity of cloned channels and transporters expressed in *Xenopus laevis* oocytes, *J. Membr. Biol.* 148:65 (1995).
120. T. Aerts, J.Z. Xia, H. Slegers, and J.C. de Block, Hydrodynamic characterization of the Major Intrisic Protein from the Bovine Lens Fiber Membranes. Extraction in N-octyl-beta-D-glucopyranoside and Evidence for a Tetrameric Structure, *J. Biol. Chem.* 265:8675 (1990).
121. S.J. Girsch, and C. Peracchia, Lens cell-to-cell channel protein: I. Self-assembly into liposomes and permeability regulation by calmodulin, *J. Membr. Biol.* 83:217 (1985).
122. J. Alcala, R.J. Cenedella, and M. Katar, Limited proteolysis of MP26 in lens fiber plasma membranes of the U18666A-induced cataract in rats, *Curr. Eye Res.* 4:1001 (1985).
123. J. Alcala, N.J. Unakar, M. Katar, J.Y. Tsui, Limited proteolysis of MP26 in lens fiber plasma membranes of the galactose-induced cataract in the rat, *Curr. Eye Res.* 5:697 (1986).
124. L.L. David, L.J. Takemoto, R.S. Anderson, T.R. Shearer, Proteolytic changes in main intrinsic polypeptide (MIP26) from membranes in selenite cataract, *Curr. Eye Res.* 7:411 (1988).

125. S.H. Cheng, R.J. Gregory, J. Marshall, S. Paul, D.W. Souza, G.A. White, C.R. O'Riordan, and A.E. Smith, Defective intracellular transport and processing of CFTR is the molecular basis of most cystic fibrosis, *Cell.* 63:827 (1990).
126. A. Shiels, and S. Bassnett, Mutations in the founder of the MIP gene family underlie cataract development in the mouse, Nature Genetics. 12:212-215(1996).
127. R. Fettiplace, and D.A. Haydon, Water permeability of lipid membranes, *Physiol. Rev.* 60:510 (1980).
128. N.A. Kent, and A. Shiels, Nucleotide and derived amino-acid sequence of the major intrinsic protein, *Nucleic Acids Res.* 18:4256(1990).
129. R. Kodama, K. Agata, M. Mochii, and G. Eguchi, Partial amino acid sequence of the major intrinsic protein (MIP) of the chicken lens deduced from the nucleotide sequence of a cDNA clone, *Exp. Eye Res.* 50:737 (1990).
130. L.A.R. Austin, S.J. Rice, G.J. Baldo, A.J. Lange, H.C. Haspel, and R.T. Mathias, The cDNA sequence encoding the major intrinsic protein of frog lens, *Gene* 124:303 (1993).
131. J. Li, S. Nielsen, Y. Dai, K.W. Lazowski, E.I. Christensen, L.A. Tabak, and B.J. Baum, Examination of rat salivary glands for the presence of the aquaporin CHIP, *Pflügers Arch.* 428:455 (1994).
132. P.M. Deen, J.A. Dempster, B. Wieringa, and C.H. Van Os, Isolation of a cDNA for rat CHIP28 water channel: high mRNA expression in kidney cortex and inner medulla, *Biochem. Biophys. Res. Commun.* 188:1267 (1992).
133. R.V. Patil, X. Yang, I. Saito, M. Coca-Prados, and M.B.Wax, Cloning of a novel cDNA homologous to CHIP28 water channel from ocular ciliary epithelium, *Biochem Biophys Res Commun.* 204:861 (1994).
134. A. Lanahan, J.B. Williams, L.K. Sanders, and D. Nathans, Growth factor-induced delayed early response genes, Mol. Cell. Biol. 12:3919 (1992).
135. L. Abrami, M. Simon, G. Rousselet, P. Ripoche, and J. Buhler, Nucleotide and Deduced Amino Acid Sequence of a Frog CHIP Aquaporin, Unpublished. Genbank: (1993).
136. M. Echevarria, E.E. Windhager, S.S. Tate, and G. Frindt, Cloning and expression of AQP3, a water channel from the medullary collecting duct of rat kidney, *Proc. Natl. Acad. Sci. USA.* 91:10997 (1994).
137. T. Ma, A. Frigeri, H. Hasegawa, and A.S. Verkman, Cloning of a water channel homolog expressed in brain meningeal cells and kidney collecting duct that functions as a stilbene-sensitive glycerol transporter, *J. Biol. Chem.* 269:21845 (1994).
138. N. Inase, K. Fushimi, K. Ishibashi, S. Uchida, M. Ichioka, S. Sasaki, and F. Marumo, Isolation of human aquaporin 3 gene, *J. Biol. Chem.* 270:17913 (1995).
139. K. Ishibashi, S. Sasaki, F. Saito, T. Ikeuchi, and F. Marumo, Structure and Chromosomal Localization Of a Human Water Channel (Aqp3) Gene, *Genomics* 27:352 (1995).
140. S. Raina, G.M. Preston, W.B. Guggino, and P. Agre, Molecular cloning and characterization of an aquaporin cDNA from salivary, lacrimal, and respiratory tissues, *J. Biol. Chem.* 270:1908 (1995).

141. T. Ma, A. Frigeri, W. Skach, and A.S. Verkman, Cloning of a novel rat kidney cDNA homologous to CHIP28 and WCH-CD water channels, *Biochem. Biophys. Res. Commun.* 197:654 (1993).
142. Y Rao, Y., L.Y. Yan, and Y.N. Jan, Similarity of the product of Drosophila neurogenic gene big brain to transmembrane channel protein, *Nature* 345:163 (1990).
143. F. Beuron, F. Le Caherec, M.T. Guillam, A. Cavalier, A. Garret, J.P. Tassan, C. Delamarche, P. Schultz, V. Mallouh, J.P. Rolland, et al. Structural analysis of a MIP family protein from the digestive tract of Cicadella viridis, *J. Biol. Chem.* 270:17414 (1995).
144. K.D. Johnson, H. Hofte, H., and M.J. Chrispeels, An intrinsic tonoplast protein of protein storage vacuoles in seeds is structurally related to a bacterial solute transporter (GlpF), *Plant Cell.* 2:525 (1990).
145. H. Hofte, L. Hubbard, J. Reizer, D. Ludevid, E.M. Herman, and M.J. Chrispeels, Vegetative And Seed-Specific Forms Of Tonoplast Intrinsic Protein In The Vacuolar Membrane Of Arabidopsis thaliana, *Plant Physiol.* 99:561 (1992).
146. Q. Liu, M. Umeda, and H. Uchimiya, Isolation and Characterization of a Rice cDNA (RTIP) similar to the Gene Encoding Tonoplast Intrinsic Protein (gamma TIP) from Arabidopsis, Genbank. Accession #D25534:(1994).
147. G.H. Miao, Z. Hong, and D.P. Verma, Topology and phosphorylation of soybean nodulin-26, an intrinsic protein of the peribacteroid membrane, *J. Cell Biol.* 118:481 (1992).
148. Y.T. Yamamoto, C.L. Cheng, and M.A. Conkling, Root-specific genes from tobacco and Arabidopsis homologous to an evolutionarily conserved gene family of membrane channel proteins, *Nucleic Acids Res.* 18:7449 (1990).
149. Y.T. Yamamoto, C.G. Taylor, G.N. Acedo, C.L. Cheng, and M.A. Conkling, Characterization of cis-acting sequences regulating root-specific gene expression in tobacco, Plant Cell. 3:371 (1991).
150. F.A. Culianez-Macia, and C. Martin, DIP - A Member of the MIP Family of Membrane Proteins that is Expressed in Mature Seeds and Dark-Grown Seedlings of Antirrhinum majus, *Plant J.* 4:717 (1993).
151. F.D. Guerrero, J.T. Jones, and J.E. Mullet, Turgor-Responsive Gene Transcription And RNA Levels Increase Rapidly When Pea Shoots Are Wilted - Sequence And Expression Of 3-Inducible Genes, *Plant Mol. Biol.* 15:11 (1990).
152. K. Yamaguchi-Shinozaki, M. Koizumi, S. Urao, and K. Shinozaki, Molecular Cloning And Characterization Of 9 cDNAs For Genes That Are Responsive To Desiccation In Arabidopsis thaliana - Sequence Analysis Of One cDNA Clone That Encodes A Putative Transmembrane Channel Protein, *Plant & Cell Physiol.* 33:217 (1992).
153. J. Cairney, R.J., Newton, E.A. Funkhouser, S. Chang, and D. Hayes, Nucleotide sequence of a cDNA for a water channel protein (aquaporin) homolog from Atriplex canescens (Pursh.) Nutt, *Plant Physiol.* 108:1291 (1995).

154. R. Kaldenhoff, A.Kolling, and G.A. Richter, A Novel Blue Light-Inducible and Abscisic Acid-Inducible Gene Of Arabidopsis-Thaliana Encoding an Intrinsic Membrane Protein, *Plant Mol. Biol.* 23:1187 (1993).
155. T. Shagan, D. Meraro, and D., Bar-Zvi, Nucleotide sequence of an Arabidopsis thaliana turgor-responsive TMP-B cDNA clone encoding transmembrane protein with a major intrinsic protein motif, *Plant Physiol.* 102:689 (1993).
156. T. Shagan, and D. Bar-Zvi, Nucleotide sequence of an Arabidopsis thaliana turgor-responsive cDNA clone encoding TMP-A, a transmembrane protein containing the major intrinsic protein motif, *Plant Physiol.* 101:1397 (1993).
157. T. Kinoshita, I. Hara-Nishimura, H. Siraishi, K. Okada, Y. Shimura, and M. Nishimura. Nucleotide sequence of a transmembrane protein (TMP-C) cDNA in Arabidopsis thaliana, *Plant Physiol.* 105:1441 (1994).
158. R.G. Fray, A. Wallace, D. Grierson, and G.W. Lycett, Nucleotide sequence and expression of a ripening and water stress-related cDNA from tomato with homology to the MIP class of membrane channel proteins, *Plant Mol. Biol.* 24:539 (1994).
159. N.N. Sandal, and K.A. Marcker, Soybean nodulin 26 is homologous to the major intrinsic protein of the bovine lens fiber membrane, *Nucleic Acids Res.* 16:9347 (1988).
160. Q. Liu, M. Umeda, and H. Uchimiya, Isolation and expression analysis of two rice genes encoding the major intrinsic protein, *Plant Mol Biol.* 26:2003 (1994).
161. L. Van Aelst, S. Hohmann, F.K. Zimmermann, A.W.H. Jans, and J.M. Thevelein, A Yeast Homologue Of The Bovine Lens Fibre Mip Gene Family Complements The Growth Defect Of A Saccharomyces-Cerevisiae Mutant On Fermentable Sugars But Not Its Defect In Glucose-Induced Ras-Mediated Camp Signalling, *EMBO J.* 10:2095 (1991).
162. S. Muramatsu, and T. Mizuno, Nucleotide sequence of the region encompassing the glpKF operon and its upstream region containing a bent DNA sequence of Escherichia coli, *Nucleic Acids Res.* 17:4378 (1989).
163. C. Holmberg, L. Beijer, B. Rutberg, and L. Rutberg, Glycerol catabolism in Bacillus subtilis: nucleotide sequence of the genes encoding glycerol kinase (glpK) and glycerol-3-phosphate dehydrogenase (glpD), *J. Gen. Microbiol.* 136:2367 (1990).
164. C.P. Smith, and K.F. Chater, Structure and regulation of controlling sequences for the Streptomyces coelicolor glycerol operon, *J. Mol. Biol.* 204:569 (1988).
165. B. Mayo, J. Kok, K. Venema, W. Bockelmann, M. Teuber, H. Reinke, and V. Gerard, Molecular Cloning and Sequence Analysis of the X-Prolyl Dipeptidyl Aminopeptidase Gene from Lactococcus lactis subsp. cremoris, *Appl. & Environm. Microbiol.* 57:38 1991).
166. S. Kashiwagi, K. Kanamuru, and T.A. Mizuno, Synechococcus gene encoding a putative pore-forming intrinsic membrane protein, *Biochim Biophys Acta.* 1237:189 (1995).

167. G. Calamita, W.R. Bishai, G.M. Preston, W.B. Guggino, and Agre, Molecular cloning and characterization of AqpZ, a water channel from Escherichia coli, *J. Biol. Chem.* 270:29063 (1995).

REGULATION OF SMOOTH MUSCLE ACTIVITY AND GAP JUNCTIONS BY SEXUAL HORMONES IN THE OVIDUCT

Manuel Villalón,[1] Marcela Hermoso,[1] Mateo Budinich,[1]
Julio Aguilera,[1] and Juan C. Sáez[2]

[1]Unidad de Reproducción y Desarrollo
[2]Departamento de Ciencias Fisiológicas
Facultad de Ciencias Biológicas
Pontificia Universidad Católica de Chile
Casilla 114-D
Santiago, Chile.

INTRODUCTION

In the mammalian oviduct, gamete transport results from a complex and still poorly understood interaction of muscle contractions, ciliary activity, and secretory function. Detailed measurements of transport of supravitally stained eggs in cumulus inside the oviduct recorded by microscope-cinematography have revealed that phasic muscle contractions of the tube are responsible for the typical random pendular motions of eggs which have been observed in the oviduct of a variety of species, including primates (Halbert et al. 1976; Verdugo et al., 1976; Villalón et al., 1991). These pendular motions are produced by contractions of the circular smooth muscle of the oviductal wall and strongly suggest that myoelectric activity and the mechanical performance of the myosalpinks have a strong random component (Portnow et al., 1977). Now it is well accepted that tubal transport should be regarded as a stochastic process (Verdugo and Villalón, 1993; Harper, 1994).

Coordination of smooth muscle activity is often provided by intercellular communication mediated by gap junctions, which are intercellular channel aggregates formed by protein subunits termed connexins (Cxs) (Bruzzone et al., 1996). For example, in the uterus, frequent and coordinate myometrial contractions that occur during labor are preceded by upregulation of gap junctions (Blennerhassett and Garfield, 1990; Tabb et al. 1992; Winterhager et al., 1991). In the oviduct, gap junctions between cells of the muscularis might also coordinate contraction waves. In the other oviductal cell strata, the epithelium, gap junctions might allow for synchronization of the ciliary beat as in the respiratory tract epithelium (Sanderson et al., 1990). Both, coordination of smooth muscle and synchronization of ciliary activity would favor the oocyte and embryo transport toward the uterus.

It is now well established that levels of sexual hormone can drastically and reproducibly modify tubal transport rates in the mammalian oviduct (Croxatto et al., 1991). For example, in pregnant rats estradiol consistently accelerates the transport of eggs (Ortiz et al., 1979), but this effect can be blocked by simultaneous administration of progesterone (Fuentealba et al., 1988). It is not clear however, how steroids can influence the tubal mechanical effectors that generate the forces that promote movement of gametes and zygotes in the oviduct. In this study, we determine the effect of estradiol and progesterone upon oviductal transport velocities of eggs surrogate and measure the content of Cxs in the oviduct.

MATERIAL AND METHODS

Animals. Sprague-Dawley rats were kept from 6:00 A.M. to 9:00 P.M. under artificial illumination and controlled temperature (21-24°C). Water and pelleted food were supplied *ad libitum*. Vaginal smears were taken daily and each female found in proestrus was caged overnight with a male for mating. The following morning the presence of spermatozoa and cornified cells in the vaginal smear was taken as evidence that insemination and ovulation had occurred. If both events had occurred that day, it was designated day 1 (d1) of pregnancy.

Rats on d1 of pregnancy were divided in four groups. In group I, each female received a s.c. injection of 1 µg 17-ß-estradiol dissolved in 0.1 ml olive oil. Group II received a s.c. injection of 5 mg of progesterone and group III received both estradiol (1 µg) and progesterone (5 mg). Control animals were injected with the same volume of vehicle.

Oviductal Microesphere Insertion. Dextran blue microespheres of 120-140 µm in diameter (Density Marker Beads, Pharmacia, Sweden) were transferred to the rat oviductal lumen at 10:00 A.M. on d1 of pregnancy, according to the technique previously described by Moore and Croxatto (1988). Briefly, rats were anesthetized and 4 to 6 microespheres were injected into each oviduct, using a microsyringe, through flank incisions to exposed the genital tract. Transfers were performed under a surgical microscope. At 6:00 P.M. of d1 a total of 3 rats per group were injected with hormones according to the protocol described above.

Assessment of Microspheres Transport. At 3 P.M. of d2 of pregnancy (12 h after treatment) movement of dextran blue microespheres was observed and recorded *in situ*, under ether anesthesia. For this purpose, oviducts (isthmus segment) were exposed through the same incision used to transfer the microespheres. During the observation period, temperature and moisture of the organs exposed were maintained approximately constant with the aid of the lamp and frequent irrigation with warm saline solution. Recording of the microsphere movements was done using a digital video camera Sony Hi 8 mm (Model CCD-V5000). Single video image frame were transferred to a computer

Leading Edge (model 386x) equipped with an image frame grabber (Data Translation 2856) and the software Imagepro. Digital image analysis of the videotapes (recorded at 30 frames per second) were used to measure the distance traveled by each microsphere and instant transport velocities were calculated.

Antisera. Cx43 immunoreactivity was studied using a previously characterized rabbit antiserum against a synthetic peptide corresponding to Cx43 aminoacids 346-360 (Yamamoto et al., 1990; Berthoud et al., 1993), as deduced from the cDNA sequence for Cx43 mRNA (Beyer et al., 1987). An affinity-purified rabbit antiserum against a synthetic peptide corresponding to Cx26 aminoacids 104-120, as deduced from the cDNA sequence for Cx26 mRNA (Zhang and Nicholson, 1992) was used. This antibody recognizes a 26 kDa band in Western blot analysis of hepatocyte membranes and binds to gap junctions of hepatocytes that surround terminal portal venules (Fallon et al., 1995). Immunodetection of Cx43 and Cx26 antisera binding was accomplished using affinity-purified, goat anti-rabbit IgG conjugated to fluorescein isothiocyanate (FITC) or horseradish peroxidase (Sigma).

Tissue Preparation. Immunoblotting studies were performed in the smooth muscle layer of the oviduct after mechanically removing with forceps, the epithelial layer. Tissue was obtained 15 h post-treatment since the onset of ovum transport acceleration begins at about 11 h post-estradiol administration (Ortiz et al., 1979; Croxatto et al., 1991).

Immunoblotting. For immunoblot analysis, the smooth muscle layer from the oviduct was placed in a minimal volume of lysis buffer (containing: 150 mM NaCl, 10% glycerol, 1% Triton X-100, 1.5 mM $MgCl_2$, 1 mM EGTA, 10 μg/ml leupeptin, 10 μg/ml trasylol, 1 mM PMSF, 100 mM sodium pyrophosphate, 100 mM NaF, 200 μM okadaic acid and 50 mM Hepes, pH 7.5) and homogenized using a Polytron (Kinematica GimbH CH 6005 Luzern, Steinhofhalde, Switzerland) followed by 30 sec sonication (Ultrasonic Cell Disrupter, Microson, Heat Systems Inc., Medisonic Inc., Farmingdall, NY/USA). Fifty micrograms of protein, determined by Bradford's technique [Bradford et al., 1976] were dissolved in sample buffer (Laemmli, 1970) and subjected to electrophoresis in parallel to prestained low range molecular weight standards (Sigma) on 0.75 mm 8% SDS-PAGE mini gels. Gels were then blotted to nitrocellulose and proteins were electrically transferred (Towbin et al., 1979). Gels were then stained with Coomassie blue R-250 to assess the uniformity of transfer.

Nitrocellulose blots were blocked by incubation for 2 h at room temperature (or overnight at 4°C) in blotto (containing: 0.05 M Tris buffered saline (TBS), pH 7.2, 5% dried non-fat milk and 0.1% sodium azide). Blots were then incubated 2 h in blotto containing 10 mg/ml bovine serum albumin (BSA), essentially free of immunoglobulin and 1:2.000 dilution of anti-Cx43 antiserum or 1:1.000, 1:500, 1:200 dilutions of mouse monoclonal anti-Cx32 antibody. Blots were rinsed 5 times for 5 min each in TBS and then incubated for 1 h in blotto containing 1:2.000 dilutions of goat anti-rabbit IgG conjugated to alkaline phosphatase. The alkaline phosphatase activity was detected by color development during incubation of the blots in Tris-HCl buffer (pH 9.5) containing a BCIP/NBT tablet (1 tablet in 10 ml).

RESULTS

Microspheres Transport

Transport of dextran blue microespheres in the isthmus of the rat showed a typical pendular motion towards the ovarian or uterine direction, with periods of rest or undetectable movement in control and experimental groups. Calculated instant average velocities (Table 1) showed no statistical differences between groups, with mean velocity values ranging between 0.12 and 4.35 μm/sec. Maximum instant velocities of 923 μm/sec and 384 μm/sec were measured in estradiol treated and control animals, respectively, indicating a 2.5 fold increase in estradiol treated animals compared to the control group. Progesterone treatment alone produced a significant reduction in the range of maximum instant velocity (95-106 μm/sec). However, this reduction in instant velocity was even more noticeable in the group that received progesterone and estradiol, showing the lowest mean (0.12 μm/sec) and range of instant velocity (65-78 μm/sec) of all groups.

Table 1. Summary of the statistic analysis of instant velocity measurements of microesphere displacement in the oviduct in control (C), estradiol (E), progesterone (P) or estradiol and progesterone (E+P) treated animals. Mean and standard deviation (SD) of instant velocities are expressed in μm/sec.

	TREATMENT			
	C	E	P	E + P
Mean	0.77[1]	4.36[1]	0.49[1]	-0.12[1]
SD	30.9	82.6	17.8	9.6
n	1118	1178	1166	1170
Range	-383 384	-868 923	-95 106	-78 65
Mode	0	0	0	0
Skewness	0.54	0.96[2]	0.11	-0.28

[1] No difference statistically significant p>0.10

[2] Significantly different p< 0.05

Frequency distribution of calculated instant velocities for each group are shown in figure 1. We observe, that all groups showed a normal distribution, centered at zero velocity (mode = 0) indicating the random direction of

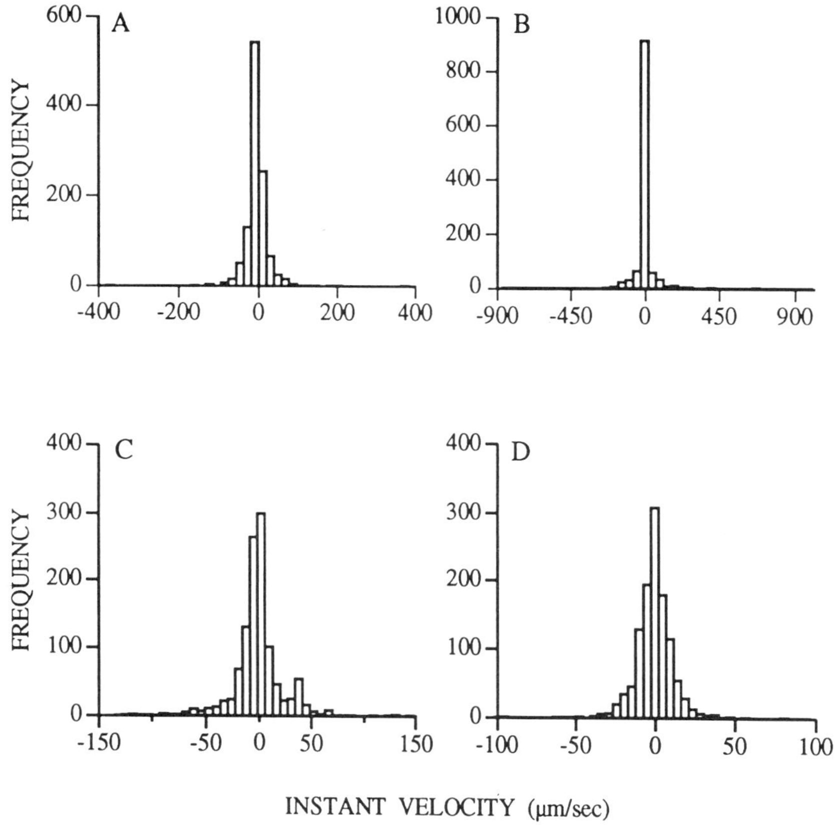

Figure 1. Frequency distribution of calculated instant velocities of microesphere displacements in the oviduct in control (A), estradiol (B), progesterone (C) or estradiol and progesterone (D) treated animals. Observe that all groups show a normal distribution, centered at zero velocity (mode = 0) indicating the random direction of movement towards the uterus or the ovary.

movement towards the uterus or the ovary. Only in the estradiol treated group we observed a statistically significant skewing ($p<0.05$) to the right of the distribution ($s = 0.95$) of the velocities toward the uterus, indicating a higher probability of displacement at higher velocities of the microespheres toward the uterus in this group.

Oviductal Cx43 is Upregulated by Estradiol

Western blot analysis of d1 pregnant rat oviducts revealed three Cx43 bands, known to correspond to phosphorylated (P2 and P3) and unphosphorylated (NP) forms of the protein (Musil et al., 1990).

Treatment of adult pregnant rats on d1 with 1 μg 17-ß-estradiol induced an increase in the relative levels of the different forms of Cx43 as compared

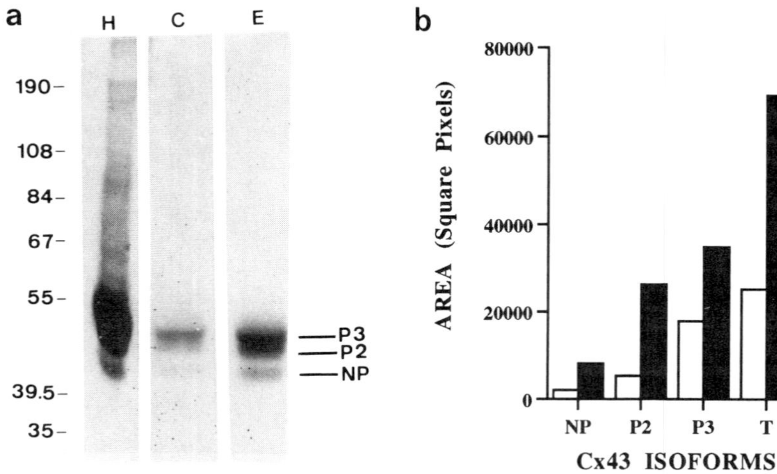

Figure 2. Estradiol regulation of relative levels and state of phosphorylation of Cx43 in oviductal smooth muscle of adult rats. **a.** Immunoblot of Cx43 in total homogenates (50 μg of protein) of rat heart (lane H) and oviduct of rats treated with 1 μg of estradiol (lane E) or its vehicle (lane C) on d1 of pregnancy. Positions of the non-phosphorylated form (NP), and phosphorylated forms (P2 and P3) of Cx43 are indicated. Levels of all Cx43 forms increased in rats treated with 17-ß-estradiol. **b.** Densitometric measurements of each Cx43 band revealed about three fold increase in total Cx43 content after estradiol treatment (black bars). The increase was mostly due to an augmentation of the phosphorylated P2 and P3 forms. Levels of Cx43 content are expressed in square pixels.

with the control group, which received only the vehicle. The increase in Cx43 isoforms in the muscle fraction homogenate of each experimental group is shown in Figure 2a. Smooth muscle oviductal tissue from estradiol treated animals contained higher levels of total Cx43. Densitometric measurements of each Cx43 band revealed that the increase in total Cx43 content induced by estradiol treatment (Figure 2b) occurred mostly as an augmentation of the phosphorylated forms P2 and P3. In the oviduct muscularis of rats treated or untreated with estradiol on d1 of pregnancy, Cx32 was not detected by immunoblotting (not shown).

DISCUSSION

The statistic analysis of instantaneous velocity of microespheres obtained in this study showed that the probability of prouterine versus proovarian motion follows the typical features of a Gaussian distribution, similar to the motion pattern described for other species (Villalón et al., 1992; Bourdage and Halbert, 1984; Halbert et al., 1976).

Previous studies have shown that egg surrogate movements in the ampullar segment of the monkey oviduct show a Gaussian distribution, shifted approximately 34 µm/sec in a prouterine direction, suggesting that ovum transport is probably a process where random (muscle activity) and nonrandom (ciliary activity) components interact with each other to produce the typical pendular movement of the egg in the oviduct (Verdugo and Villalón, 1993). Since we did not observed a shift in the distribution of instant velocities for any of the groups, it is probably that the contribution of ciliary activity to egg transport through the isthmus is minimal or non-existent in the rat oviduct. This is further supported by the observation that there is a reduction in the number of ciliated cells present in the luminal surface, coincident with an increase in the smooth muscle layer thickness in the isthmic region of the mammalian oviduct (Woodruff and Pauerstein, 1969).

In this study, we found that estradiol induced a significant rise in Cx43 content of the smooth muscle layer of the rat oviduct. A similar effect in myometrial gap junctions has been detected in ovariectomized immature or pregnant rats after treatment with estradiol (Anderson et al., 1993; Burghardt et al., 1987; Hendrix, 1992; Mackenzie and Garfield, 1985; Wu et al., 1994) . This is consistent with the existence of estrogen response elements in the Cx43 gene (Wu et al., 1994). After estradiol treatment, Cx43 in the oviductal smooth muscle was detected mostly as the phosphorylated P2 and P3 forms. Thus, it is likely that the smooth muscle cells were better coupled after estradiol treatment since phosphorylated forms of Cx43 are known to be abundant in the plasma membrane (Musil et al., 1990) forming functional channels (Sáez et al., 1993). The observation that higher velocities, were measured in animals treated with estradiol, also supports the idea that synchronized muscle activity in the oviduct will be improved by the presence of gap junctions between smooth muscle cells, allowing for propagation of contraction waves for longer distances and with increased force. Experiments are under way to determine the effect of progesterone on the content of Cxs in the oviduct, in order to explain the change in motion pattern of microespheres observed in this group.

To our knowledge, this is the first report on changes in Cx43 phosphorylation state *in vivo* that are related to changes in the physiological state of the oviduct. In the myometrium while estradiol causes an increase, progesterone induces a reduction in Cx43 levels (Petrocelli and Lye, 1993). Therefore, in the female genital tract intercellular gap junctional communication mediated by Cx43 can be regulated at various levels. Further studies will be required to determine the underlying mechanisms involved in the regulation of gap junction functionality by sexual hormones in the oviduct.

Acknowledgments

We kindly thank Dr. E.L. Hertzberg of the Albert Einstein College of Medicine for providing antibodies against Cx26 and Cx43. M.A. Hermoso is a recipient of a scholarship from the Pontificia Universidad Católica de Chile. We also thank Ms. J. González and Ms. G. Garcés for technical assistance. The work has been supported by FONDECYT grants 1940691 (to M.V.), 2950062 (to M.H.) and 1960559 (to J.C.S.).

REFERENCES

Andersen, J., Grine, E., Eng, C.L.Y., Zhao, K., Barbieri, R.L., Chumas, J.C., and Brink, P.R., 1993, Expression of connexin-43 in human myometrium and leiomyoma, *Am. J. Obstet. Gynecol.* 169:1266.

Berthoud, V.M., Rook, M., Traub, O., Hertzberg, E.L., and Sáez, J.C., 1993, On the mechanisms of cell uncoupling induced by a tumor promoter phorbol ester in clone 9 cells, a rat liver epithelial cell line, *Eur. J. Cell Biol.* 62:384.

Beyer, E.C., Paul, D.L., and Goodenough, D.A., 1987, Connexin43: a protein from rat heart homologous to a gap junction protein from liver, *Cell Biol.* 105:2621.

Blennerhassett, M.G., and Garfield, R.E., 1990, Effect of gap junction number and permeability on intercellular coupling in rat myometrium, *Am. J. Physiol.* 261:C1011.

Bourdage, R.J., and Halbert, S.A., 1984, Quantification of intraluminal motion of surrogate ova in the rabbit oviductal isthmus, *Biol. Reprod.* 30:1124.

Bradford, M.M., 1976, A rapid and sensitive method for the quantitation of microgram quantities of protein utilizing the principle of protein-dye binding, *Anal. Biochem.* 72:248.

Bruzzone, R., White, T.W., and Paul, D., 1996, Connections with connexins: The molecular basis of direct signaling, *Eur. J. Biochem.* 238:1.

Burghardt, R.C., Gady-Kurten, D., Burghardt, R.L., Kurten, R.C., and Mitchell, P.A., 1987, Gap junction modulation in rat uterus III.Structure-activity relationships of estrogen receptor-binding ligands on myometrial and serosal cells, *Biol. Reprod.* 36:741.

Croxatto, H.B., Ortiz, M.E., Forcelledo, M.L., Fuentealba, B., Noé, G., Moore, G., Morán, F., and Cárdenas, H., 1991, Hormonal control of ovum transport through the rat oviduct, *Arch. Biol. Med. Exp.* 24:403.

Fallon, M.B., Nathanson, M.H., Mennone, A., Sáez, J.C., and Anderson, J.M., 1995, Altered expression and function of hepatocyte gap junctions following common bile duct ligation, *Am. J. Physiol.* 268:C1186.

Fuentealba, B., Nieto, M., and Croxatto, H.B., 1988, Progesterone abbreviates the nuclear retention of estrogen receptor in the rat oviduct and counteracts estrogen action on egg transport, *Biol. Reprod.* 38:63.

Halbert, S.A., Tam, P.Y., and Blandau, R.J., 1976, Egg transport in the rabbit oviduct. The roles of cilia and muscle. *Science* 191:1052.

Harper, M.J.K., 1994, Gamete and zygote transport, *in:* "The Physiology of Reproduction", Vol. 1, E. Knobil, J.D. Neill, eds, Raven Press, New York.

Hendrix, E.M., Mao, S.T.J., Everson, W., and Larsen, W.J., 1992, Myometrial connexin 43 trafficking and gap junction assembly at term and in preterm labor, *Mol. Reprod. Develop.* 33:27.

Laemmli, U. K., 1970, Cleavage of structural proteins during the assembly of the head of bacteriophage T4, *Nature* 227: 680.

Mackenzie, L.W., and Garfield, R.E., 1985, Effects of 17 ß-Estradiol on myometrial gap junction and pregnancy in the rat, *Can. J. Physiol. Pharmacol.* 64:462.

Moore, G.D., and Croxatto, H.B., 1988, Effects of delayed transfer and treatment with oestrogen on the transport of microspheres by the rat oviduct, *J. Reprod. Fert.* 83:795.

Musil, L.S., Cunningham, B.A., Edelman, G.M., and Goodenough, D.A., 1990, Differential phosphorylation of the gap juction protein connexin43 in junctional communication-competent and -deficient cell lines, *J. Cell. Biol.* 111:2077.

Ortiz M.E., Villalón, M., and Croxatto, H.B., 1979, Ovum transport and fertility following postovulatory treatment with estradiol in rats, *Biol. Reprod.* 21:1163.

Petrocelli, T., and Lye, S.J., 1993, Regulation of transcripts encoding the myometrial gap junction protein, connexin-43, by estrogen and progesterone, *Endocrinology* 133:284.

Portnow, J., Talo, A., and Hodgson, B.J., 1977, A random wall model of ovum transport, *Bull. Math. Biol.* 39:349.

Sáez, J.C., Berthoud, V. M., Moreno, A. P., and Spray, D. C., 1993, Multiplicity of controls in differentiated and undifferentiated cells and possible functional implications, *in*: "Advances in Second Messenger and Phosphoprotein Research". S. Shenolikar, and A.C. Nairn, eds, Raven Press, New York.

Sanderson, M. J., Charles, A. C., and Dirksen, E. R., 1990, Mechanical stimulation and intercellular communication increases intracellular Ca^{2+} in epithelial cells, *Cell Regulation* 1:585.

Tabb T., Thilander, G., Grover, A., Hertzberg, E., and Garfield R., 1992, An immunochemical and immunocytologic study of the increase in myometrial gap junctions (and connexin 43) in rats and humans during pregnancy, *Am. J. Obtet. Gynecol.* 167:559.

Towbin, H., Stachelin, T., and Gordon, J., 1979, Electrophoretic transfer of proteins from polyacrylamide gels to nitrocellulose sheets: Procedure and some applications. *Proc. Nat. Acad. Sci. USA.* 76:4350.

Verdugo, P., Blandau, R.J., Tam P.Y., and Halbert, S.A., 1976, Stochastic elements in the development of deterministic models of egg transport, *in:* "Ovum Transport and Fertility Regulation", M.J.K. Harper, C.J. Pauerstein, C.E. Adams, E. Coutinho, H.B. Croxatto, D.M. Paton, eds, Scriptor, Copenhagen.

Verdugo, P.M., and Villalón, M.J., 1993, Functional anatomy of the Fallopian tube, *in:* "Infertility: Male and Female". V. Insler, B. Lunenfeld, eds, Churchil livingstone, England.

Villalón, M., Verdugo, P., Boling, J.L., and Blandau, R.J., 1991, The transport of cumulus egg masses through the ampullae of the oviducts in the pigtailed, *Macaca nemestrina*, *Arch. Biol. Med. Exp.* 24:385.

Winterhager, E., Stütenkemper, R., Traub, O., Beyer, E., and Willecke, K., 1991, Expression of different connexin genes in rat uterus during decidualization and at term, *Eur. J. Cell Biol.* 55:133.

Woodruff, J.D., and Pauerstein, C.J., 1969, "The Fallopian Tube", The Williams & Wilkins Co., Baltimore.

Wu, Y., Dahl, G., and Werner, R., 1994. The connexin43 gene is responsive to oestrogen, *Proc. R. Soc. Lond. B.* 255:125.

Yamamoto, T., Ochalski, A., Hertzberg, E. L., and Nagy, J. Y., 1990, LM and EM immunolocalization of the gap junctional protein connexin43 in rat brain, *Brain Res.* 508:313.

Zhang, J.T., and Nicholson, B., 1992, Sequence and tissue distribution of a second protein of hepatic gap junctions, Cx26, as deduced from its cDNA, *J. Cell Biol.* 109:3391.

INTERCELLULAR CALCIUM SIGNALING IN LIVER

Michael H. Nathanson[1], Michael B. Fallon[1], Angela D. Burgstahler[1], Albert Mennone[1], Stephan F. Schlosser[1], Carlos B. Gonzalez[2], and Juan C. Sáez[3]

[1]Liver Study Unit, Yale Univ. School of Med., New Haven, CT USA
[2]Dept. of Physiology, Universidad Austral de Chile, Valdivia, Chile
[3]Dept. of Physiological Sciences, Pontifica Universidad Catolica de Chile, Santiago, Chile

INTRODUCTION

Cytosolic Ca^{2+} (Ca_i^{2+}) is an important second messenger in hepatocytes, where it regulates diverse cell functions ranging from bile secretion[1] to glucose metabolism[2] to apoptosis[3]. In excitable cell types such as neurons[4,5] and myocytes[6], Ca_i^{2+} signals spread among cells via gap junctions, and this intercellular signaling is thought to permit these tissues to display an integrated, organ-level response to hormonal stimulation. Intercellular Ca_i^{2+} signaling also can occur in hepatocytes, as was first suggested by the observation that both Ca^{2+} and inositol 1,4,5-trisphosphate (IP_3) can cross gap junctions in isolated rat hepatocyte couplets[7]. Here we review subsequent work by us which demonstrates that intercellular Ca_i^{2+} signaling occurs under physiological circumstances in hepatocytes. First, we describe gap junction-mediated synchronization of hormone-induced Ca_i^{2+} signals in isolated rat hepatocyte couplets[8], an isolated cell system in which polarity[9] and gap junction function[10] are preserved. We then describe whether this organization is altered in couplets isolated from livers following bile duct obstruction[11], an experimental model of impaired bile secretion in which gap junction expression may be altered. We also discuss whether intercellular Ca_i^{2+} signaling occurs *in vivo*, based on our studies examining hormone-induced Ca_i^{2+} signaling in the isolated perfused rat

From Ion Channels to Cell-to-Cell Conversations
Edited by Latorre and Sáez, Plenum Press, New York, 1997

liver[12]. Finally, since hepatocytes contain ATP receptors[13] and there is ATP in bile[14], we examined and review our preliminary evidence that isolated hepatocytes are capable of paracrine, non-gap junction-mediated Ca_i^{2+} signaling via ATP secretion[15].

METHODS

Male Sprague-Dawley rats (180-250 g; Camm Research Lab Animals, Wayne, NJ) maintained on Purina rodent chow under a constant light cycle were used for all experiments. Common bile duct ligation was performed in selected rats as described previously[11]. Arg^8-vasopressin, phenylephrine, ATP and octanol were obtained from Sigma Chemical Company (St. Louis, MO), Fluorescent Ca^{2+} dyes (rhod-2/AM and fluo-3/AM) were from Molecular Probes (Pitchford, OR), and all other chemicals were of the highest quality commercially available.

Isolated rat hepatocytes and hepatocyte couplets were prepared in the Hepatocyte Isolation Core of the Yale Liver Center as described previously[9,16-18]. Briefly, rat livers were perfused with medium containing 0.05% collagenase (Boehringer Mannheim Biochemicals, Indianapolis, IN) and 0.8 U trypsin inhibitor (Sigma Chemical, St. Louis, MO)/U tryptic activity. Livers were then excised, minced, passed through serial nylon mesh filters, and the resultant cells washed. These cells were suspended in Liebovitz L-15 medium (GIBCO, Grand Island, NY) containing 50 U penicillin and 50 ug streptomycin per ml, and plated onto glass coverslips or coverslip fragments. Cells were incubated at 37^0C and used 2-6 hr after plating. Cell viability by trypan blue exclusion was measured 2 hr after plating and generally exceeded 95%.

Ca_i^{2+} signals were observed in isolated rat hepatocytes and hepatocyte couplets using confocal microscopy[8,19,20]. Couplets were prepared as described above, then loaded with fluo-3/AM (6 mM)[21,22] for 20 min at 37^0C in medium containing 10% fetal calf serum. Coverslips containing the couplets were transferred to a chamber on the stage of a Zeiss Axiovert microscope, perifused at 37^0C with a HEPES-buffered solution and observed using a BioRad MRC-600 confocal imaging system. A krypton-argon laser was used to excite the dye at 488 nm and emission signals above 515 nm were collected. Optical sections between 0.5-1.0 mm in thickness were obtained; typically, these cells are spherical and have a diameter of 20-25 mm. Neither autofluorescence nor other background signals was detectable at the machine settings that were used. There was also no change in size, shape or location of cells during the experiments.

Hepatocyte couplets isolated from normal livers and from livers 1 day after common bile duct ligation were observed in response to stimulation with vasopressin (10 nM), with or without pretreatment with octanol (250 mM), or in response to phenylephrine (0.1-10 mM). Octanol-treated couplets were perifused for 15 sec with buffer containing the octanol, then switched to octanol-free buffer containing vasopressin. Synchronization of vasopressin- and phenylephrine-induced Ca^{2+} signals in all couplets was assessed using confocal line scanning microscopy[8,19,20]; with this approach, confocal fluorescence is determined at each point along a single line across the couplet, rather than at each point across the entire couplet. This approach permits increased temporal resolution of Ca^{2+} signals, without significant photobleaching or loss of spatial resolution along the line that is scanned. Fluo-3 fluorescence in response to hormonal stimulation was recorded in this way and the resulting line scan was low-pass filtered for noise reduction as described previously[19]. Line scans were displayed as images consisting of 512x512 pixels, with a spatial resolution of either 0.26 or 1.5 mm/pixel (in the *x* direction) and a temporal resolution ranging from 6-500 msec/pixel (in the *y* direction). For each couplet, the degree of synchronization between its two cells was calculated as the difference between the time of the initial increase in fluorescence in each of the cells. Absolute Ca_i^{2+} concentrations were not estimated because fluo-3 can not be ratio imaged[21,22], but since the time difference between Ca^{2+} signals is determined independent of Ca_i^{2+} concentration, such measurements were not limited by use of fluo-3.

In other experiments, Ca_i^{2+} signals were visualized in individual cells using confocal video microscopy[23]. Cells were isolated, loaded with fluo-3/AM, then observed using a BioRad MRC-600 confocal imaging system as described above. Images were obtained at a rate of 1 sec^{-1} using a 63x objective and recorded using a Panasonic TQ3031F optical memory disk recorder[23]. Increases in Ca_i^{2+} in cells were detected by measuring increases in fluorescence intensity. Data were analyzed from the recorded images using standard image processing techniques[23]. In these experiments, individual cells were mechanically stimulated with a microinjection pipette[15]. For some of these studies, cells were (mechanically) stimulated in the presence of the nuclear stain propidium iodide. Here, fields were simultaneously excited with the 488 nm and 568 nm lines of the confocal microscope's krypton-argon laser, and separate images were collected at 522$\pm$32 nm and at >585 nm. In this way, distinct fluorescence images for fluo-3 and propidium iodide could be detected[15].

To examine Ca_i^{2+} signaling *in vivo*, rat livers were isolated as described previously[24] and perfused using a HEPES-buffered solution in a non-recirculating fashion. Livers were then loaded with the Ca^{2+}-sensitive fluorescent dye rhod-2[22] by switching to a recirculating system containing 1.5 mM rhod-2/AM in L-15 medium with 10% fetal calf serum. For imaging[12], livers were switched back to a single-pass perfusion system maintained at a constant flow rate (30 ml/min) and moved to a humidified 37^0C perfusion chamber. The chamber was then placed on the stage of a Zeiss Axiovert microscope coupled to a BioRad MRC-600 confocal imaging system. A krypton-argon laser was used to excite the dye at 568 nm, and emission signals above 585 nm were collected. Virtually no fluorescence was detectable in non-dye-loaded livers at the machine settings used for these studies. In each liver preparation, baseline images were obtained, then continuous infusion of vasopression was begun via a syringe pump to obtain a concentration of 20 pM-200 nM in the perfusate. As was done for single cells, these images were obtained at a rate of 1 sec^{-1} and recorded using an optical memory disk recorder. Data were analyzed from the recorded images using an image processor[23].

Immunochemistry was performed to localize the vasopressin V_{1a} receptor in normal rat livers[12] and connexins 26 and 32 in normal and common bile duct-ligated livers[11]. For immunochemistry of the V_{1a} receptor, liver sections were fixed and embedded in paraffin[12], then 10 mm-sections were blocked with phosphate-buffered saline (PBS) solution containing 1% bovine serum albumin and 0.1 M glycine, then incubated for 2 hr with an affinity-purified polyclonal IgG that reacts with the V_{1a} receptor isolated from rat liver membranes[25]. Immunoblots using this antibody recognize a single 55 kD band in liver and kidney, and immunostaining in kidney colocalizes with the V_1 but not the V_2 receptor[25]. Labeled specimens were washed and exposed to horseradish peroxidase-conjugated goat anti-rabbit 2^0 antibodies, then developed via the 3,3'-diaminobenzidine reaction[16], mildly counterstained with hematoxylin, and examined by light microscopy. To localize the two connexin isoforms, frozen sections of liver were prepared and stained with either a polyclonal rabbit anti-connexin 32 antibody or an affinity-purified polyclonal rabbit antibody directed against the peptide sequence for one of the cytosolic loop regions of connexin 26[11], then counterstained with fluorescein-conjugated goat anti-rabbit antibody and examined by epifluorescence microscopy. Western blots were also performed to examine connexin 26 and 32 expression in normal and common bile duct-ligated livers, as described previously[11].

For statistical analyses, outliers were identified *a priori* using an F test based on the Mahalanobis distance[26]. Comparisons between groups were then made using a two-sample t test. Results were expressed as mean±SD unless stated otherwise.

INTERCELLULAR Ca^{2+} SIGNALING IN HEPATOCYTE COUPLETS

To examine intercellular Ca_i^{2+} signaling in hepatocytes, we compared vasopressin (10 nM)-induced increases in Ca_i^{2+} in isolated rat hepatocytes and hepatocyte couplets[8]. Ca_i^{2+} increases began within 172±77 msec (range, 0-330 msec) in pairs of cells comprising couplets. In cells of couplets pretreated with the gap junction blocker octanol (250 mM), Ca_i^{2+} increases instead began within 1.37±0.56 sec ($p<0.005$). The comparable measure of intercellular variation in isolated single cells is the standard deviation of their latency period[8], which was 1.32±1.07 sec for hepatocytes examined here. This was not significantly different from the intercellular variation seen in either isolated hepatocytes (1.21±0.61 sec) or couplets pretreated with octanol. These results are summarized in Figure 1. Together, these findings demonstrate that vasopressin-induced Ca_i^{2+} signals are much more highly coordinated among cells in hepatocyte couplets than among single hepatocytes, and that this coordinated signaling is mediated by gap junction conductance.

As an alternative measure of intercellular Ca_i^{2+} signaling, phenylephrine (0.1-10 mm)-induced Ca_i^{2+} oscillations were observed in isolated hepatocytes (n=124) and couplets (n=24). The frequency of oscillations was dose-dependent in both single hepatocytes and couplets[8], and at each concentration of phenylephrine the frequency did not differ significantly between single cells and couplets. The oscillation frequency did vary among individual hepatocytes and from couplet to couplet, though. However, Ca_i^{2+} oscillations always were synchronized in the hepatocytes within each individual couplet. These findings are further evidence that Ca_i^{2+} signaling induced by endocrine hormones is highly coordinated in hepatocyte couplets. Since both IP_3 and Ca_i^{2+} can cross gap junctions in hepatocyte couplets[7], and since octanol desynchronizes Ca_i^{2+} signaling in the couplets[8], gap junctions are likely responsible for this intercellular coordination of hormone-induced Ca_i^{2+} signaling.

Ca^{2+} SIGNALING FOLLOWING BILE DUCT OBSTRUCTION

To explore the pathologic consequences of impaired intercellular communication, both expression and function of gap junctions were examined in normal liver and in livers following common bile duct ligation[11].

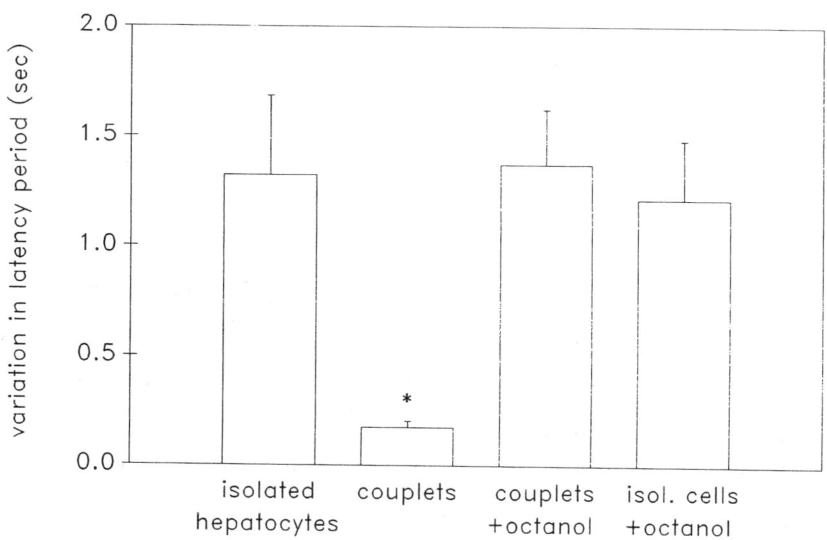

Figure 1. Synchronization of Ca_i^{2+} signals in isolated hepatocytes and hepatocyte couplets stimulated with vasopressin (10 nM) ± octanol (250 mM). The variation in latency period is shown. For isolated hepatocytes, this is the standard deviation of the latency period, while for hepatocyte couplets this is the mean difference in time of onset of Ca_i^{2+} signals in adjacent cells. Values shown are mean±SEM ($^*p<0.01$). Adapted from[8] by copyright permission of the American Society for Cell Biology.

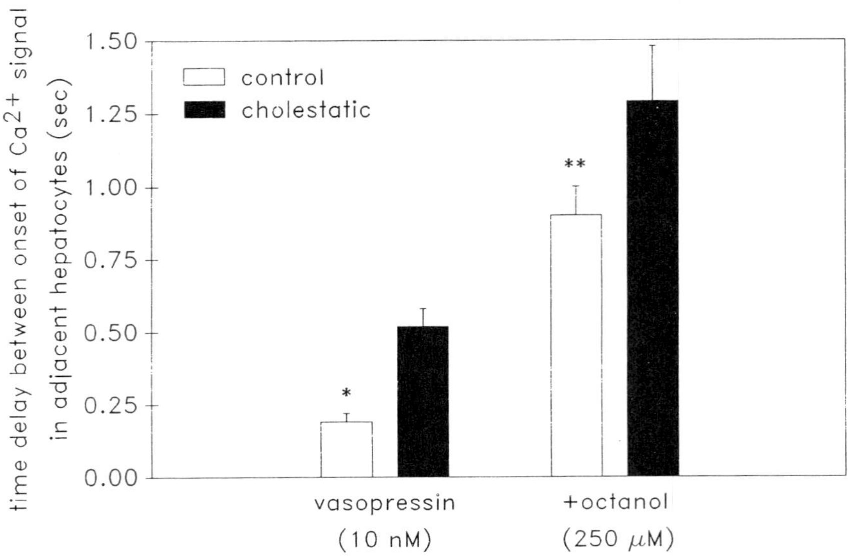

Figure 2. Synchronization of Ca_i^{2+} signals in hepatocyte couplets isolated from normal and common bile duct-ligated (cholestatic) livers, each stimulated with vasopressin (10 nM) ± octanol (250 mM). Signaling is less synchronized in cholestatic couplets than in controls, but octanol further desynchronizes couplets from both groups. There is no significant difference between the two octanol-treated groups. Values shown are mean±SEM ($^*p<0.001$). Reproduced from[11] by copyright permission of the American Physiological Society.

Immunofluorescence revealed connexin 26 labeling throughout periportal hepatocytes under control conditions, but labeling was virtually absent in liver sections 1-7 days after bile duct ligation. Immunoblotting confirmed loss of connexin 26 expression up to 7 days after ligation, although some return of protein expression was evident 9 days after bile duct ligation by both immunofluorescence and immunoblotting[11]. Unlike connexin 26, connexin 32 was distributed throughout the hepatic lobule under control conditions[11]. Like connexin 26, though, connexin 32 could not be detected after bile duct ligation, measured within 1 day by immunofluorescence and within 2 days by immunoblotting[11]. Furthermore, no return of connexin 32 expression was seen within 14 days of ligation. These findings demonstrate that common bile duct ligation causes a rapid and complete loss of gap junction expression in rat liver, and suggest that this loss should be associated with impaired intercellular communication in this tissue.

To determine directly the functional significance of this decrease in gap junction expression, vasopressin-induced Ca_i^{2+} signals were measured in 32 hepatocyte couplets from normal livers (n=19 with and n=13 without octanol) and 39 couplets from cholestatic livers (n=20 with and n=19 without octanol)[11].

In couplets from normal livers, Ca_i^{2+} signals in adjacent hepatocytes began within 0.19±0.10 sec (range 0.1-0.4 sec), which is similar to the value of 0.18 sec we had previously reported[8]. This time interval was increased almost five-fold, to 0.90±0.45 sec (range 0.1-1.5 sec), in those couplets pre-treated with octanol ($p<0.001$), as we had also reported[8]. In couplets from cholestatic livers, Ca_i^{2+} signals in adjacent hepatocytes began within 0.52±0.25 sec (range 0.1-1.0 sec), a time interval almost three times longer than in normal controls ($p<0.001$). This time interval also increased further, to 1.29±0.86 sec (range 0.1-3.1 sec), in cholestatic couplets pre-treated with octanol ($p<0.001$ relative to cholestatic controls). The mean time interval in octanol-treated normal and cholestatic couplets was not significantly different, and neither of these differed significantly from the value of 1.37 sec we previously reported for octanol-treated normal couplets[8]. These results are summarized in Figure 2. Thus, we found that couplets from cholestatic livers respond in a less synchronized fashion than couplets from normal livers, but in a more synchronized fashion than octanol-treated couplets. This suggests that intercellular communication is impaired but not completely abolished in the cholestatic couplets, perhaps in part because gap junction expression is not completely abolished 24 hr after bile duct ligation, the time at which the cholestatic couplets were isolated. To test this, it would be necessary to examine couplets from livers >24 hr after bile duct ligation, but isolation of such cells is technically difficult. It is unlikely that the delay seen in cholestatic couplets in part reflects impaired Ca_i^{2+} signaling mechanisms in individual cells, since vasopressin induced a similar net increase in Ca_i^{2+} in normal and cholestatic single cells[11], and since octanol-treated couplets behaved similarly whether isolated from normal or cholestatic livers. These studies thus provide a measure of gap junction function in normal and cholestatic hepatocytes, and suggest that intercellular communication is impaired but not blocked in hepatocytes 24 hr after common bile duct ligation.

Ca^{2+} SIGNALING IN THE INTACT LIVER

To determine whether intercellular Ca_i^{2+} signaling actually occurs *in vivo*, we examined Ca_i^{2+} signaling in intact perfused livers[12]. Increases in Ca_i^{2+} were detected in 21 experiments using 8 separate isolated perfused rat liver preparations. Each Ca_i^{2+} increase was manifested as a Ca_i^{2+} wave, 15 (71%) of which began in the pericentral region then propagated to the periportal region (Figure 3). Of the remaining 6 Ca_i^{2+} waves that were detected, 2 spread into the periportal region from an adjacent hepatic acinus, rather than originating *de novo* in that periportal region. Waves spread across each hepatic acinus at a constant rate, with no measureable delay at cell boundaries. This pattern of spread is similar to that seen in freshly isolated rat hepatocyte couplets, as described above. Ca_i^{2+} waves were detected in livers stimulated with as little as 25 pM vasopressin, which is similar to the threshold dose of vasopressin (10 pM) required to increase Ca_i^{2+} in isolated hepatocytes[27]. Ca_i^{2+} wave speed increased with increasing concentrations of vasopressin, ranging from a minimum of 23±4 mm/sec (mean±SD) in response to 25 pM vasopressin, to a maximum of 119±24 mm/sec in response to maximal stimulation (20-200 nM vasopressin). This dependence of Ca_i^{2+} wave speed on the dose of agonist also occurs in other epithelial cell types[19], and has been described in *Xenopus* oocytes expressing the vasopressin V_{1a} receptor cloned from rat liver as well[23]. In this work, we detected wave speeds ranging from 90-156 mm/sec in response to nanomolar amounts of vasopressin[12], which corresponds closely to an average wave speed of 109 mm/sec that we had reported in isolated rat hepatocyte couplets[8]. Thus, most aspects of Ca_i^{2+} signaling detected in isolated rat hepatocyte couplet preparations appear to reflect Ca_i^{2+} signaling patterns *in vivo*.

To investigate why vasopressin-induced Ca_i^{2+} waves begin in the pericentral region, we examined the distribution of the vasopressin V_{1a} receptor in rat liver[12]. Immunochemistry using a previously characterized antibody directed against the rat liver V_{1a} receptor[25] demonstrated that the receptor is

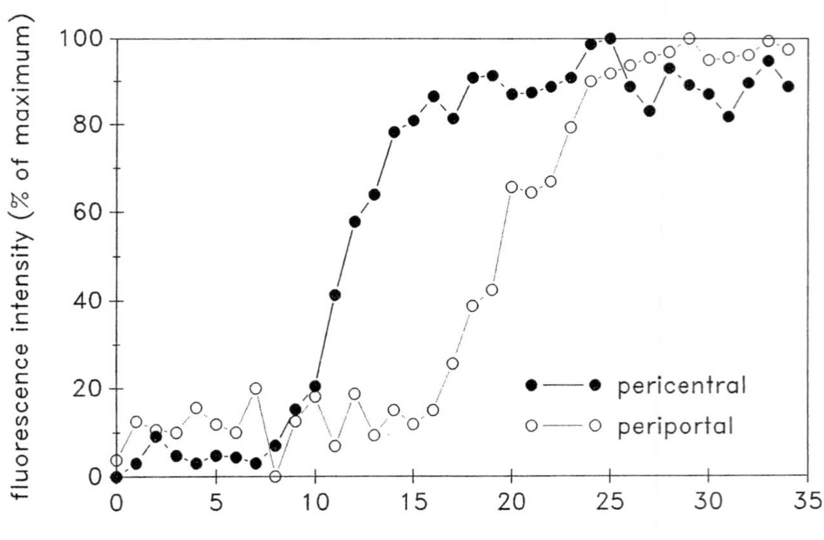

Figure 3. Lobular organization of vasopressin (300 pM)-induced Ca_i^{2+} signals in an isolated perfused rat liver. Curves represent increases in fluorescence intensity in pericentral (closed circles) and periportal (open circles) hepatocytes within a single lobule. The periportal Ca_i^{2+} increase begins 8 sec after the pericentral increase, and these groups of hepatocytes are 576 mm apart, so that this Ca_i^{2+} wave spreads across the lobule at a speed of 72 mm/sec. Reproduced from [12] by copyright permission of the American Physiological Society.

concentrated in hepatocytes near central venules, with little staining in the remaining portions of the hepatic acinus[12]. This observation is consistent with *in situ* hybridization histochemistry showing that mRNA for the V_{1a} receptor also is concentrated in the pericentral regions of rat liver[28]. These findings suggest that vasopressin-induced Ca_i^{2+} waves begin in the pericentral region of the hepatic acinus because the V_{1a} receptor is most concentrated there. Once initiated, Ca_i^{2+} waves likely spread to the periportal region using gap junctions to cross from hepatocyte to hepatocyte[7,8] and by Ca^{2+}-induced Ca^{2+} release within individual cells[29]. There is marked variability in the response of individual isolated hepatocytes to vasopressin[8,30,31], and the varied expression of the V_{1a} receptor among hepatocytes reported here may in part explain that finding as well. The receptor for glucagon also is linked to IP_3-mediated Ca_i^{2+} signaling in hepatocytes[32], and also is concentrated in the pericentral region in rat liver[33]. A number of other proteins are heterogeneously expressed along the hepatic acinus[34], although the functional significance of such lobular gradients had been unclear. It remains to be determined whether other hormone receptors that link to Ca_i^{2+} mobilization in hepatocytes, including the a_{1B}-adrenergic and angiotensin II receptors, are similarly distributed according to a lobular pattern in liver.

NON-GAP JUNCTION-MEDIATED Ca^{2+} SIGNALING

Intercellular communication among hepatocytes can occur via gap junctions[8], as discussed in the preceding sections. Intercellular communication also can occur among cells that lack gap junctions, via ATP secretion which leads to stimulation of nucleotide receptors on target cells[35]. We therefore examined and review here our preliminary evidence that hepatocytes can communicate via nucleotide secretion as well[15]. The effects on Ca_i^{2+} of mechanical stimulation, including microinjection, were examined in isolated rat hepatocytes and isolated bile duct units using confocal fluorescence video microscopy[15]. Stimulation of a single hepatocyte evoked an increase in Ca_i^{2+} in that cell, plus a Ca_i^{2+} rise in neighboring isolated hepatocytes. Perifusion with ATP prior to mechanical stimulation supressed the Ca_i^{2+} increase, and the ATP/ADPase apyrase[36] also (reversibly) inhibited the Ca_i^{2+} rise induced by mechanical stimulation, although apyrase did not block vasopressin-induced Ca_i^{2+} signals. Mechanical stimulation of hepatocytes also induced a Ca_i^{2+} increase in co-cultured isolated bile duct units, which, like hepatocytes[13], are known to express ATP receptors[37]. Finally, this form of Ca_i^{2+} signaling could be elicited in the presence of propidium iodide without nuclear labeling by that dye, demonstrating that this phenomenon does not depend on disruption of the stimulated cell. Thus, mechanical stimulation of isolated hepatocytes can evoke Ca_i^{2+} signals in the stimulated cell as well as in neighboring hepatocytes and bile duct epithelia[15]. This signaling is mediated by release of ATP into the extracellular space. This suggests a novel, alternative form of intercellular communication in liver whereby hepatocytes can locally modulate Ca_i^{2+} signals previously thought to be controlled exclusively by endocrine hormones.

CONCLUSIONS

Here we reviewed evidence that hormone-induced Ca_i^{2+} signals are highly coordinated among hepatocytes, and that this is mediated by gap junctional communication[8]. We also showed that intercellular Ca_i^{2+} signals are further organized *in vivo*, typically occurring in a pericentral-to-periportal direction[12]. What is the functional significance of this coordinated, organized intercellular Ca_i^{2+} signaling pattern[12]? Increased Ca_i^{2+} induces contraction of the actin-myosin network that lines the canalicular membranes of hepatocytes[16,38,39], and these contractions are responsible for peristaltic flow of bile[40]. Furthermore, peristalsis occurs in a pericentral-to-periportal direction[40]. Thus, pericentral-to-periportal Ca_i^{2+} waves may direct canalicular peristalsis. It is also well established that increased Ca_i^{2+} induces glycogenolysis and gluconeogenesis in hepatocytes[2], and the spread of Ca_i^{2+} waves across the entire hepatic lobule may permit vasopressin to induce glucose release from all hepatocytes, even though the V_{1a} receptor is not evenly distributed among them. We also provided preliminary evidence that hepatocytes can secrete ATP[15], which is a Ca_i^{2+} agonist in this cell type[13]. If hepatocytes secrete the ATP into bile (which is known to contain ATP[14]), this would potentially provide an alternative, paracrine mechanism to generate pericentral-to-periportal Ca_i^{2+} waves, since that is the direction of bile flow. The spatial distribution of Ca_i^{2+} waves in intact tissue thus may provide a mechanism for the differential regulation of cells comprising that tissue, which in turn may play an important role in the integrative response that leads to organ-level function. We also found that both gap junction expression and intercellular communication were impaired following common bile duct ligation[11], which is a widely used experimental model of cholestasis. Further studies will be needed to investigate whether this impairment in intercellular communication interferes with canalicular peristalsis and thus plays a role in the pathogenesis of cholestasis.

ACKNOWLEDGEMENTS

This work was supported by grants from the American Heart Association, the American Liver Foundation, the American Gastroenterological Association, and the National Institutes of Health (DK45710), FONDECYT (1960559), and by the Morphology, Cell Isolation, and Liver Perfusion Core Facilities of the Yale Liver Center (NIH DK34989).

REFERENCES

1. M.H. Nathanson, Cellular and subcellular calcium signaling in gastrointestinal epithelium, *Gastroenterology* 106:1349 (1994).
2. J. Exton, Role of Phosphoinositides in the regulation of liver function, *Hepatology* 8:152, 1988.
3. D.E. Clapham, Calcium signaling, *Cell* 80:259, (1995).
4. L. Venance, D. Piomelli, J. Glowinski, and C.Giaume, C. Inhibition by anandamide of gap junctions and intercellular calcium signalling in striatal astrocytes, *Nature* 376:590 (1995).
5. K. Kandler, and L.C. Katz, Neuronal coupling and uncoupling in the developing nervous system, *Curr. Opin. Neurobiol.* 5:98 (1995).
6. H. Kimura, Y. Oyamada, H. Ohshika, M. Mori, and N.Oyamada, Reversible inhibition of gap junctional intercellular communication, synchronous contraction, and synchronism of intracellular Ca^{2+} fluctuation in cultured neonatal rat cardiac myocytes by heptanol, *Exp. Cell Res.* 220:348 (1995).
7. J.C. Sáez, J.A. Connor, D.C. Spray, and M.V.L. Bennett, Hepatocyte gap junctions are permeable to the second messenger, inositol 1,4,5-triphosphate, and to calcium ions, *Proc. Natl. Acad. Sci. USA* 86:2708 (1989).
8. M.H. Nathanson, and A.D. Burgstahler, Coordination of hormone-induced calcium signals in isolated rat hepatocyte couplets: Demonstration with confocal microscopy, *Mol. Biol. Cell* 3:113 (1992).
9. J. Graf, A. Gautam, and J.L. Boyer, Isolated rat hepatocyte couplets: A primary secretory unit for electrophysiologic studies of bile secretory function, *Proc. Natl. Acad. Sci. USA* 81:6516, (1984).
10. D.C. Spray, R.D. Ginzberg, E.A. Morales, Z.Gatmaitan, and I.M. Arias, Electrophysiological properties of gap junctions between dissociated pairs of rat hepatocytes, *J. Cell Biol.* 103:135 (1986).
11. M.B. Fallon, M.H. Nathanson, A. Mennone, J.C. Sáez, A.D. Burgstahler, and J.M. Anderson, Altered expression and function of hepatocyte gap junctions after common bile duct ligation in the rat. *Am. J. Physiol.* 268:C1186 (1995).
12. M.H. Nathanson, A.D. Burgstahler, A. Mennone, M.B. Fallon, C.B. Gonzalez, and J.C.Sáez, Ca^{2+} waves are organized among hepatocytes in the intact organ, *Am. J. Physiol.* 269:G167 (1995).
13. G.R. Dubyak, and C. El-Moatassim, Signal transduction via P_2-purinergic receptors for extracellular ATP and other nucleotides, *Am. J. Physiol.* 265:C577 (1993).
14. R.S. Chari, S.M. Schutz, J.E. Haebig, G.H.Shimokura, P.B. Cotton, J.G. Fitz, and W.C. Meyers, Adenosine nucleotides in bile, *Am. J. Physiol.* 270:G246 (1996).
15. S.F. Schlosser, A.D. Burgstahler, and M.H.Nathanson, Isolated rat hepatocytes can signal to other hepatocytes and bile duct cells by release of nucleotides, *Proc. Natl. Acad. Sci. (USA)*, in press.
16. M.H. Nathanson, A. Gautam, O.C. Ng, R. Bruck, and J.L. Boyer, Hormonal regulation of paracellular permeability in isolated rat hepatocyte couplets, *Am. J. Physiol.* 262:G1079 (1992).

17. A. Gautam, O-C. Ng, and J.L. Boyer, Isolated rat hepatocyte couplets in short-term culture: Structural characteristics and plasma membrane reorganization, *Hepatology* 7:216 (1987).
18. J. Graf, and J.L. Boyer, The use of isolated rat hepatocyte couplets in hepatobiliary physiology, *J. Hepatol.* 10:387 (1990).
19. M.H. Nathanson, P.J. Padfield, A.J. O'Sullivan, A.D. Burgstahler, and J.D. Jamieson, Mechanism of Ca^{2+} wave propagation in pancreatic acinar cells, *J. Biol. Chem.* 267:18118 (1992).
20. M.H. Nathanson, and A.D. Burgstahler, Subcellular distribution of cytosolic Ca^{2+} in isolated rat hepatocyte couplets: Evaluation using confocal microscopy, *Cell Calcium* 13:89 (1992).
21. J.P.Y. Kao, A.T. Harootunian, and R.Y. Tsien, Photochemically generated cytosolic calcium pulses and their detection by Fluo-3, *J. Biol. Chem.* 264:8179 (1989).
22. A. Minta, J.P.Y. Kao, and R.Y. Tsien, Fluorescent indicators for cytosolic calcium based on rhodamine and fluorescein chromophores, *J. Biol. Chem.* 264:8171 (1989).
23. M.H. Nathanson, M.S., Moyer, A.D. Burgstahler, A. -M. O'Carroll, M.J. Brownstein, and S.J. Lolait, Mechanisms of subcellular cytosolic Ca^{2+} signaling evoked by stimulation of the vasopressin V_{1a} receptor, *J. Biol. Chem.* 267:23282 (1992).
24. M.H. Nathanson, A. Gautam, R. Bruck, C.M. Isales, and J.L. Boyer, Effects of Ca^{2+} agonists on cytosolic Ca^{2+} in isolated hepatocytes and on bile secretion in the isolated perfused rat liver, *Hepatology* 15:107 (1992).
25. C.B. Gonzalez, C.E., Caorsi, C.E., Reyes, S. Troncoso, and V. Barra, Identification and purification of the vasopressin receptor from rat liver membranes, *Ann. NY Acad. Sci.* 689:526 (1993).
26. A.A. Afifi, and S.P. Azen, *Statistical Analysis: A Computer Oriented Approach*, 2nd ed., NY: Academic Press (1979).
27. A.D. Burgstahler, and M.H. Nathanson, NO modulates the apicolateral cytoskeleton of isolated hepatocytes by a PKC-dependent, cGMP-independent mechanism, *Am. J. Physiol.* 269:G789 (1995).
28. N.L. Ostrowski, S.J. Lolait, D.J., Bradley, A. -M. O'Carroll, M.J. Brownstein, and W.S. Young, III Distribution of V1a and V2 vasopressin receptor messenger ribonucleic acids in rat liver, kidney, pituitary and brain, *Endocrinology* 131:533 (1992).
29. M.H. Nathanson, A.D. Burgstahler, and M.B. Fallon, Multi-step mechanism of polarized Ca^{2+} wave patterns in hepatocytes, *Am. J. Physiol.* 267:G338 (1994).
30. T.A. Rooney, E.J. Sass, and A.P. Thomas, Characterization of cytosolic calcium oscillations induced by phenylephrine and vasopressin in single Fura-2 loaded hepatocytes, *J. Biol. Chem.* 264:17131 (1989).
31. T. Kawanishi, L.M. Blank, A.T. Harootunian, M.T. Smith, and R.Y. Tsien, Ca^{2+} oscillations induced by hormonal stimulation of individual fura-2-loaded hepatocytes, *J. Biol. Chem.* 264:12859 (1989).
32. M.J.O. Wakelam, G.J. Murphy, V.J. Hruby, and M.D. Houslay, Activation of two signal-transduction systems in hepatocytes by glucagon, *Nature* 323:68 (1986).

33. V.M. Berthoud, V. Iwanij, A.M. Garcia, and J.C. Sáez, Connexins and glucagon receptors during development of rat hepatic acinus, *Am. J. Physiol.* 263:G650 (1992).
34. K. Jungermann, and R.G. Thurman, Hepatocyte heterogeneity in the metabolism of carbohydrates, *Enzyme* 46:33 (1992).
35. Y. Osipchuk, and M. Cahalan, Cell-to-cell spread of calcium signals mediated by ATP receptors in mast cells, *Nature* 359:241 (1992).
36. J.P. Grierson, and J. Meldolesi, Shear stress-induced $[Ca^{2+}]_i$ transients and oscillations in mouse fibroblasts are mediated by endogenously released ATP, *J. Biol. Chem.* 270:4451 (1995).
37. J. McGill, S. Basavappa, A.W. Mangel, G.H. Shimokura, J.P. Middleton, and J.G. Fitz, Adenosine triphosphate activates ion permeabilities in biliary epithelial cells, *Gastroenterology* 107:236 (1994).
38. S. Watanabe, C.R. Smith, and M.J. Phillips, Coordination of the contractile activity of bile canaliculi: Evidence from calcium microinjection of triplet hepatocytes, *Lab. Invest.* 53:275 (1985).
39. Y. Yamaguchi, E. Dalle Molle, and W.G.M. Hardison, Vasopressin and A23187 stimulate phosphorylation of myosin light chain-1 in isolated rat hepatocytes, *Am. J. Physiol.* 261:G312 (1991).
40. N. Watanabe, N. Tsukada, C.R. Smith, and M.J. Phillips, Motility of bile canaliculi in the living animal: Implications for bile flow, *J. Cell Biol.* 113:1069 (1991).

CURRENT ELECTROPHYSIOLOGICAL TECHNIQUES TO STUDY EXOCYTOSIS

Andrés F. Oberhauser

Department of Biology
Faculty of Sciences
University of Chile
and Centro de Estudios Científicos de Santiago
Santiago, Chile

ABSTRACT

Stimulus-secretion coupling has been extensively studied in a wide variety of cells. It is clear that there are many complex interactions between different messenger systems. Although many proteins have been proposed as critical components of the exocytotic machinery in secretory cells, little is known about their function or how they interact with intracellular signals. This is mainly because of the lack of techniques to directly study the activation pathways directing secretory vesicle exocytosis. Over the last years, the patch-clamp and amperometric techniques have created a revolution in the study of exocytosis. Using these techniques one can measure exocytosis at high temporal resolution, while simultaneously controlling the intracellular environment with the patch-pipette solutions. Here, I describe the use of these techniques in the study of exocytosis.

INTRODUCTION

Exocytosis is the term used to describe the fusion of specialized storage organelles, secretory vesicles, with the plasma membrane of a cell, a process

From Ion Channels to Cell-to-Cell Conversations
Edited by Latorre and Sáez, Plenum Press, New York, 1997

that results in the release of vesicular contents into the extracellular milieu. Exocytosis is the final step of the secretory pathway and represents the culmination of a series of intracellular transport events involving the specific targeting, docking, and fusion of vesicles with the plasma membrane. These events are likely to involve protein components localized to the plasma membrane, the transported vesicle, and also cytosolic factors. One of the most critical events in this process remains a mystery: how do the membrane of the secretory vesicle fuses with the cell membrane? Many of the proteins involved in the release of hormones and neurotransmitter have been identified by recent molecular and biochemical studies, but the detailed nature of the vesicle fusion process remains unclear (Söllner et al., 1993; Südhof, 1995; Calakos and Scheller, 1996). At present, it is impossible to assign specific functions to the various proteins that have been characterized.

Recently it has become possible to follow with exquisite detail the exocytosis of single secretory granules by using the patch-clamp and amperometry techniques (reviewed by Monck et al., 1994). By controlling the composition of the intracellular solution with the patch pipette ("pipette biochemistry") it is possible to study more distal parts of the stimulus-secretion coupling pathway that are involved in different steps in exocytosis. For instance, this technique has been used to study the role of several molecules (like Ca^{2+} and guanine nucleotides) in the regulation of the exocytosis in single secretory cells with minimal interference with receptor mediated signaling events or changes in the plasma membrane conductance (e.g. Fernandez et al., 1984; Oberhauser et al., 1992). The patch-clamp technique also allows the measurement of the changes conductance of single exocytotic fusion pores (Breckenridge and Almers, 1987; Alvarez de Toledo and Fernandez, 1988). With amperometry it is possible to monitor the release of oxidizable secretory products by means of a carbon fiber microelectrode in close proximity to a single isolated cell (Leszczyszyn et al., 1990). This technique offers a highly sensitive means of monitoring the release of hormones and neurotransmitters from single cells or even through a single fusion pore (Chow et al., 1992; Alvarez de Toledo et al., 1993; Bruns and Jahn, 1995). In this paper, I will review the properties of the fusion pore and discuss the molecular nature of early fusion pore in the light of novel electrophysiological findings.

METHODS

Preparation of mast cells. Mast cells from normal rats were obtained by peritoneal lavage. Mast cells were extracted after injecting 20-50 ml of RPMI medium (supplemented with 40 mM $NaHCO_3$, 10% fetal bovine serum and 100U/ml penicillin plus 1μg/ml streptomycin) into the peritoneal cavity of the rats. After some gentle massage of the rat's abdomen, the peritoneal cavity was surgically exposed, and 10-30 ml of fluid was recovered. The peritoneal fluid contained large numbers of cells, a small fraction (about 10%) of which were mast cells. The cell suspension was then centrifuged at low speed (200 x g, 7 min) and the cell pellet resuspended in 3 ml of RPMI medium (supplemented with 20 mM HEPES and no bicarbonate or antibiotics). Aliquots of this cell suspension were plated on glass-bottom experimental chambers and stored at room temperature until use.

Solutions. The standard extracellular solution used was (in mM): 130 NaCl, 2 $MgCl_2$, 2 $CaCl_2$, 10 Hepes at pH 7.2, and 10 glucose (310 mmol/kg). The internal solution was (in mM): 140 K-glutamate, 7 $MgCl_2$, 0.2 ATP-Mg, 10 Hepes (pH 7.2), 10 EGTA and 1 $CaCl_2$ and between 1-10 μM GTPγS.

Measurement of exocytosis with patch-clamp. Exocytosis was monitored by measuring the cell membrane capacitance using the whole cell mode of the patch clamp technique (Hamill et al., 1981) in conjunction with a digital phase detector (Joshi and Fernandez, 1988) implemented on system comprising an AT-MIO16x interface (National Instruments) and an Axopatch 1C patch clamp amplifier (Axon, Rapid City, CA, USA). Fire-polished pipettes were coated with silicone elastomer resin (Sylgard) and had resistances of 2-5 MOhm when immersed in the external solution. The computer generated sine wave (833Hz, 54 mV peak-to-peak) was filtered at a corner frequency of 830 Hz with a fixed frequency low pass filter (DT L8L, Frequency Devices Inc.) and then fed into the stimulus input of the patch-clamp amplifier. The current was filtered with an antialiasing filter at a corner frequency of 2 KHz (902, Frequency Devices Inc.) and sampled with a 16-bit A/D converter. To resolve small capacitance changes, the initial values of the cell membrane capacitance and access resistance were canceled with the C-slow and G-series potentiometers of the patch-clamp amplifier. After compensation, a calibration was obtained by unbalancing the compensation circuitry by 100 fF and 1 MOhm. The data acquisition and analysis programs were written in Labview (v4.0, National Instruments).

Measurement of exocytosis with amperometry. Amperometric detection of secretory products was monitored with a carbon fiber electrode with a tip diameter of ~14 μm (Kawagoe et al., 1993). Electrodes were fabricated by aspirating a carbon fiber into a glass capillary pipette which was then pulled in a pipette puller. The carbon fiber was sealed to the glass pipette by briefly immersing the tip in hot (~80°C) epoxy resin (Epon 828 (Shell) mixed with 14% w/w n-phenylenediamine (Sigma)). The excess of epoxy was removed by washing the pipette tip with acetone. The pipettes were cured overnight at 30°C, then at 100°C for 2h, and finally at 150°C for 2h. Before use, the tip was polished, at 35°, using a homemade pipette beveler. The holding voltage applied to the electrode was +650 mV and the redox current was monitored using a Dagan 8900 patch clamp amplifier. The carbon fiber was placed 1 - 2 μm over the cell in order to maximize the detection of secretory events occurring around the entire cell surface.

RESULTS

Membrane Capacitance Measurements with the Patch-clamp Technique

The patch-clamp technique provides a unique tool for studying events at the level of the plasma membrane like the opening and closing of ion channels in the plasma membrane. The patch-clamp technique can also be used to measure capacitive currents. The charge carried by the capacitive current is directly proportional to the cell surface area because membranes have a fairly constant specific capacitance of 1 μF/cm^2. Thus, the membrane capacitance can be used to measure the increase in cell surface area that occur during the exocytotic fusion of secretory granules with the plasma membrane. However, measurements of capacitive currents are not as simple as measurements of ion channel currents. The current that passes through a capacitor is given by:

$$I_C = C\, dV(t)/dt \qquad (1)$$

Therefore, in order to measure capacitance the voltage must be constantly changed. The simplest method is to apply a sinusoidal voltage and measure the membrane current. If we consider a simple equivalent circuit for a patch-clamped cell, as shown in figure 1, with a membrane resistance (due to ion channels) in parallel with a capacitor (lipid bilayer), then the current across these circuit elements will be given by:

$$i_R(t) = V(t)/R = (1/R)\, V_o \cos(\omega t) \qquad (2)$$

$$i_C(t) = C\, (dV(t)/dt) = -\, \omega C V_o \sin(\omega t) = -\, \omega C V_o \cos(\omega t + 90^o) \qquad (3)$$

From this we can see that, one component of the resulting current will be in phase with the stimulus (i_R) and the capacitive current component will be 90° out of phase with respect to the stimulus (V(t)). Neher and Marty (1982) demonstrated that it is possible to follow changes in membrane capacitance by measuring the admittance of a patch-clamped secretory cell. As with ionic currents, the admittance (the alternating current equivalent of conductance) is determined from the measured current using Ohm's law:

$$Y(\omega) = I(\omega)/V(\omega) \qquad (4)$$

where $Y(\omega)$ is the admittance, $\omega = 2\pi f$ and f is the frequency of the sine wave.

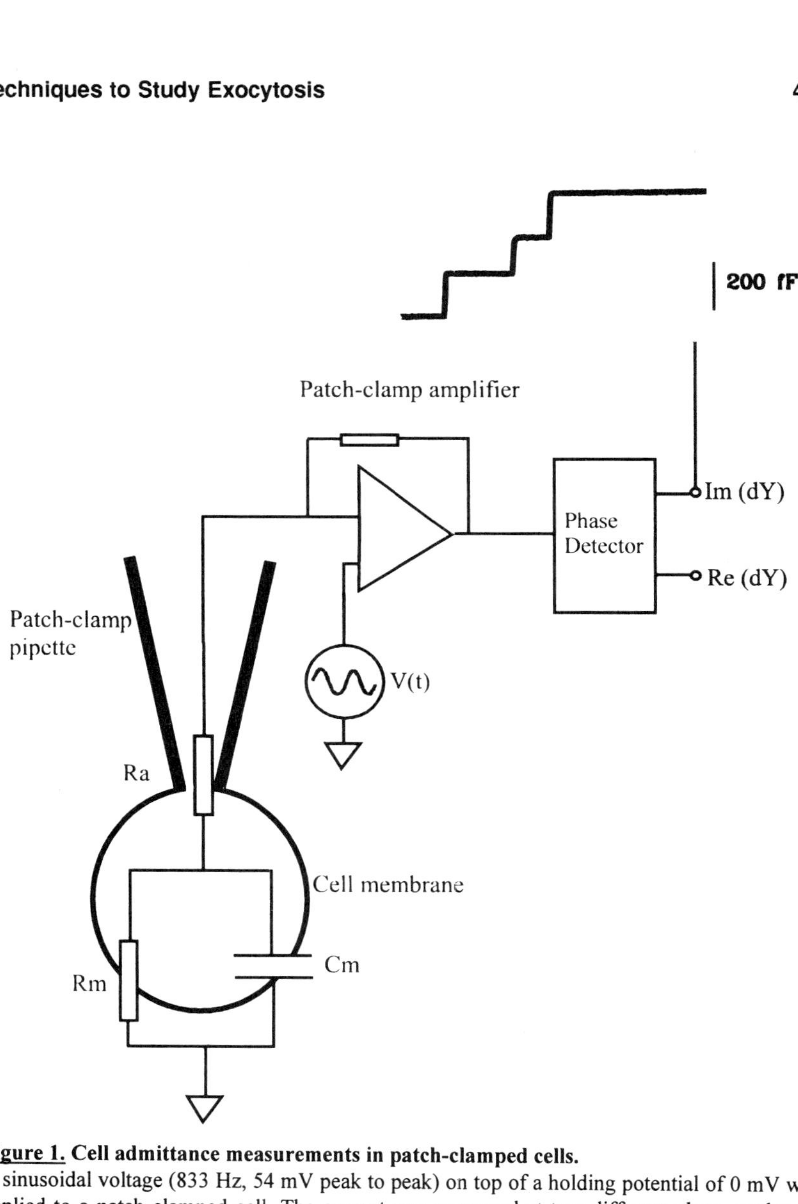

Figure 1. Cell admittance measurements in patch-clamped cells.
A sinusoidal voltage (833 Hz, 54 mV peak to peak) on top of a holding potential of 0 mV was applied to a patch clamped cell. The current was measured at two different phase angles, Φ and Φ+90°, relative to the stimulus. The phase was periodically adjusted using the phase tracking technique (Fidler and Fernandez, 1989). With this method, two outputs are obtained: one that reflects changes in the real part of the cell admittance (Re [ΔY], which is proportional to changes of Ra), and a second output that reflects changes in the imaginary part of the admittance, (Im [ΔY], which is proportional to changes in Cm). These two components of the cell admittance were captured with a temporal resolution of 9.6 ms per point, using a software based phase detector.

From equations (2) and (3) it is clear that the admittance of a patch-clamped cell has two components, one that is in phase with the membrane voltage and is proportional to changes in resistance (Re [ΔY]), and one that is 90° out of phase with the membrane voltage (Im [ΔY]). From this latter component it is possible to determine the membrane capacitance.

Continuous measurements of cell admittance from patch-clamped secretory cells have been made using a phase sensitive detector. This detector

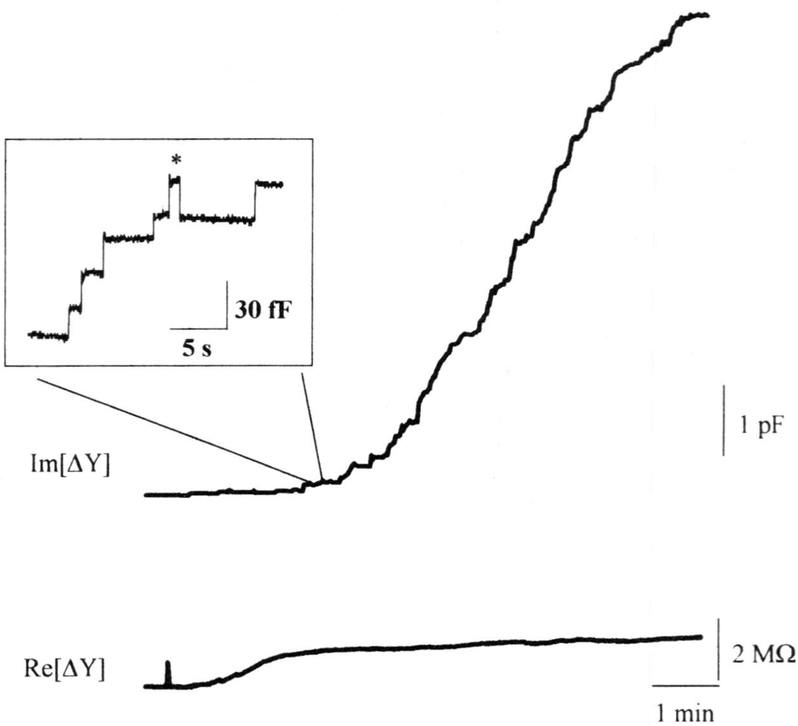

Figure 2. Membrane capacitance measurements in single mast cells.
Time course of the cell membrane capacitance (represented as Im [ΔY]) increase measured during a complete degranulation in a rat peritoneal mast cell. The membrane capacitance increased from 3.7 to 9.9 pF. The access resistance, Re [ΔY], varied only slightly (from 7 MΩ to 10 MΩ). The deflection at the beginning of the trace represents a 1 MΩ calibration (by switching the phase-tracking resistor on and off). The inset shows an expansion of the membrane capacitance recording at the beginning of the degranulation. Both reversible (marked by an asterisk) and irreversible fusion events are shown in this trace. Secretion was stimulated by including 10 μM GTPγS in the pipette solution.

can be either a lock-in amplifier (Neher and Marty, 1982; Lindau and Neher, 1988) or a digital phase detector (Joshi and Fernandez, 1988; Fidler and Fernandez, 1989). The digital phase detector has the advantage of a high sensitivity over a large dynamic range. Also this method avoids the need of extra instrumentation (like an expensive lockin-amplifier), since the phase detection is done by software. Other techniques of measuring cell membrane capacitance with the patch-clamp technique have been described in detail by Lindau and Neher (1988) and Gillis (1995).

Figure 2 shows the time course of the changes in membrane admittance measured in a patch-clamped mast cell. This measurement was done with a digital phase detector. The recording starts shortly after gaining access to the cytosol by disrupting the patch of membrane under the pipette tip by gentle suction. In this "whole-cell" recording mode the cytosol is dialyzed with the solution in the pipette, allowing control of the intracellular environment. In this example the pipette solution included GTPγS. In mast cells, GTPγS bypasses some of the normal stimulus-secretion coupling reactions and stimulates secretion more directly (Fernandez et al., 1984). As the GTPγS took effect, the cell began to secrete and the capacitance increased in a sigmoidal fashion (shown as Im [ΔY]). The membrane capacitance increased from 3.7 pF to a stable value of 9.9 pF. If this trace is expanded, at any point in time, it can be seen that it is made of step increases in capacitance (inset in figure 2). It has been demonstrated that each step in membrane capacitance represent the exocytosis of one secretory vesicle (Fernandez et al., 1984). Therefore, the thick trace represents the exocytotic fusion of hundreds (about 250 in this cell) of secretory granules each one contributing about 25 fF of membrane capacitance (or 2.5 μm^2 of membrane area).

Measurements of the Fusion Pore Conductance

Ultrastructural techniques have shown that an early event in the exocytotic fusion of a mast cell secretory granule is the formation of a narrow, water-filled pore spanning both the granule and plasma membranes and connecting the lumen of the secretory granule to the extracellular environment (Chandler and Heuser, 1980). Subsequent patch-clamp measurements, again in mast cells, revealed that the fusion pore is first apparent as a small 1-2 nm diameter pore (Breckenridge and Almers, 1987; Alvarez de Toledo and Fernandez, 1988; Spruce et al., 1990). This pore opens abruptly and undergoes rapid fluctuations in size, usually culminating in the irreversible expansion and release of the secretory granule contents.

Some of the properties of single fusion pores measured in mast cells can be summarized as follows. The existence of the fusion pore can be divided into several distinct phases. The fusion pore opens abruptly (<100 μs) to about 300 pS (range: 80 - 1000 pS); then it expands to about 5 nS within 20 ms at about 300 pS/ms. This expansion phase is characterized by large, rapid fluctuations in pore conductance ("flicker"). At this point, the fusion pore can either: (i) open to a very large structure (> 20 nm) in less than 100ms, or (ii) it can close completely. This latter observation is important because it indicates that the formation of a fusion pore is not an irreversible process. Although it is not yet clear the physiological relevance of transient fusion, it has been observed in all secretory cells studied to date (Fernandez et al., 1984; Breckenridge and Almers, 1987; Lindau et al., 1993). It has been recently suggested that transient fusion events might be the main form of exocytosis at the neuronal synapse (Bruns and Jahn, 1995).

Recently, similar observations for the properties of the fusion pore have been made during the exocytosis of small vesicles, as small as synaptic vesicles, in human neutrophils (Lollike et al., 1995). These observations suggest that properties of the fusion pore seen in mast cells are conserved in different secretory systems. It is therefore likely that the fusion pore that forms during the exocytosis of a vesicle in a presynaptic terminal shares some of the properties of the fusion pore measured in mast cells.

The Structure of the Exocytotic Fusion Pore in Mast Cells

While it is clear that in the latter stages of expansion, when the size of the pore is several times bigger than any other known macromolecule, the fusion pore must be made of lipids, the structure of the smaller precursors is not known. Several models have been proposed to explain the fusion of the vesicular membrane and the plasma membrane. One of the first model to have been proposed suggested that the fusion pore was formed by a multimeric protein unit (akin to the structure of an ion channel) that underwent a conformational change following stimulation, leading to the opening of a pore in the center of the multimeric protein complex (Almers, 1990). Expansion of the pore would then result from dispersal of the protein subunits and the incorporation of lipids between them. A second model has been described in which it is envisaged that proteins form a "hydrophobic bridge" between the plasma membrane and the secretory granule membrane. According to this model membrane lipids could then cross this bridge with their hydrophobic tails remaining in contact with the hydrophobic surfaces of the "bridge" of proteins (Pollard et al., 1990; Zimmerberg et al., 1991).

Patch-clamp measurements of the activity of individual fusion pores in mast cells have shown that the fusion pore has some unusual and unexpected properties: (i) there is a large flux of lipid through the pore (Monck et al., 1990); (ii) the rate of pore closure has a discontinuous temperature dependency (Oberhauser et al., 1992), and (iii) comparisons of experimental data with theoretical fusion pores (Nanavati et al., 1992) and with breakdown pores (Oberhauser and Fernandez, 1993) supports the view that the fusion pore is initially a pore through a single bilayer, as would be expected for membrane fusion proceeding through a hemifusion mechanism. Based on these observations Monck and Fernandez (1992) presented a model where the fusion pore is initially a pore through a single bilayer. It is envisioned that the formation of a fusion pore is regulated by a macromolecular scaffold of proteins that is responsible for bringing the plasma membrane into a highly curved dimple very close to a tense secretory granule membrane, creating the architecture where the strongly attractive hydrophobic force causes the membranes to form a 'hemifusion' intermediate. Membrane fusion is

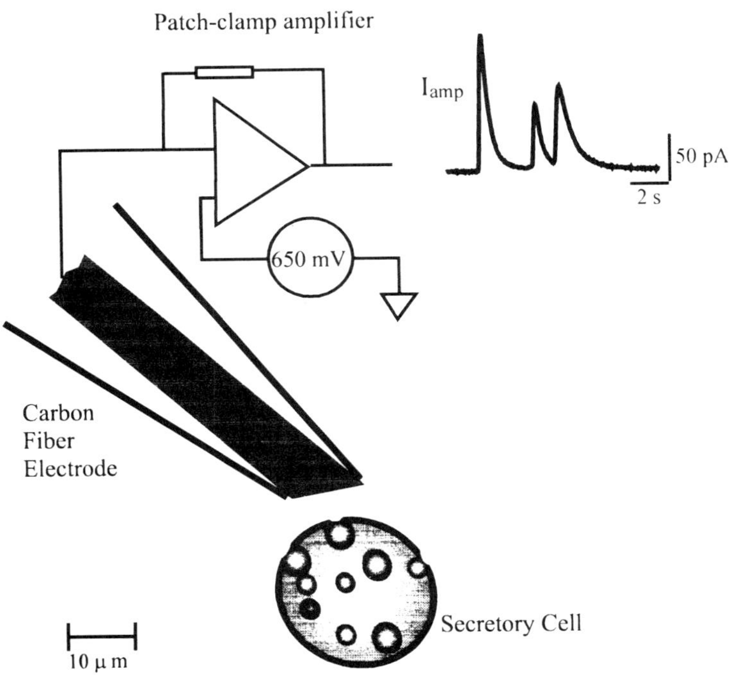

Figure 3. Measurements of neurotransmitter release from single secretory vesicles.
By using a carbon fiber electrode it is possible to monitor the release of secretory products during single exocytotic events. Application of a constant potential through a carbon fiber electrode can oxidize (or reduce) some compounds (e.g. transmitters like serotonin or catecholamines) that are released from the secretory granule, resulting in the generation of a small electric currents (I_{amp}). Each spike in the current trace represents the release of a single granule's contents.

completed by the formation of an aqueous pore after rupture of the shared bilayer. We also proposed that the microenvironment of the interface when the pore first opens, dominated by the charged groups on the secretory vesicle matrix and phospholipids, will greatly influence the release of secretory products (Nanavati and Fernandez, 1993; Monck et al., 1995). Therefore, the amount of the neurotransmitter released might be finely regulated by the kinetics of opening and closure of the fusion pore.

Amperometric Measurements of Exocytosis

An alternative method of measuring secretion consist in monitoring the release of oxidizable secretory products by placing a carbon fiber microelectrode (held at a constant potential) in close proximity to a single isolated cell (Leszczyszyn et al., 1990; Chow and von Rüden, 1995), as shown in figure 3. Release of secretory granule contents results in the generation of brief current spikes as secretory products are oxidized or reduced on the surface of the carbon fiber.

One advantage of the amperometric method of following secretion is that it only monitors the release of secretory components into the extracellular milieu and as such is not complicated by some of the factors affecting capacitance measurements (like membrane retrieval by endocytosis). It is also very sensitive and can detect the release from single vesicles that are too small to be detected as individual events by capacitance measurements using the whole-cell configuration (Chow et al., 1992; Ureña et al., 1994). The development of amperometry to detect the release of secretory products from single cells has yielded exciting new information regarding the mechanisms by which the neurotransmitters are released to the extracellular medium.

Figure 4A shows an amperometric recording obtained from a single degranulating rat mast cell. The cell was stimulated by adding 10 μg/ml of the compound 48/80 to the extracellular solution. The exocytotic release of serotonin from the secretory causes an increase in the recorded amperometric current (Alvarez de Toledo et al., 1993). The characteristic feature of this recording is the large number of overlapping spikes. Some individual spikes are shown in an expanded time scale in figure 4B and 4C. Combined amperometric and capacitance measurements have shown that the detection of a single amperometric spike corresponds with the fusion of a single granule with the plasma membrane, observed as a stepwise increase in the membrane capacitance (Alvarez de Toledo et al., 1993; Oberhauser et al., 1996). Therefore, each spike in figure 4 represents the release of a single granule content.

The time integral of the amperometric current trace is shown as a dotted line in figure 4A. This trace is a measure of the total amount of secretory

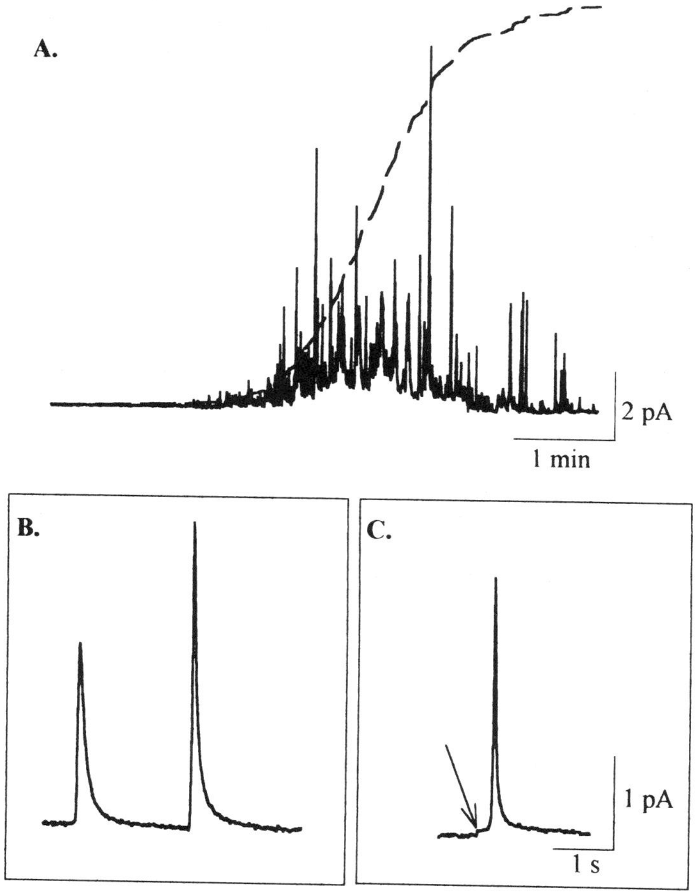

Figure 4. Amperometric measurements of exocytosis in a single rat mast cell.
A. Secretory response recorded from a single rat mast cell with a carbon fiber in the amperometric mode. The current trace shows the exocytotic release of the serotonin contained in the secretory granules. The recording starts 1 min after the addition of 10 μg/ml of 48/80 compound to the extracellular solution. The time integral of the amperometric current trace is shown as a dashed line. B, C. Examples of single amperometric spikes obtained from (A). The spike shown in (C) displays a 'foot' (pointed by the arrow).

products that were released by a cell following its stimulation. Notice that the time integral of the amperometric current trace follows a similar time course to the membrane capacitance recording shown in figure 2. Therefore, time integral of the amperometric current allows a direct comparison of the events measured using the amperometric technique with those measured by capacitance (von Rüden and Neher, 1993; Oberhauser et al., 1996). The integrated amperometric signal can also be used to derive the rate of fusion of vesicles with the plasma membrane assuming a knowledge of the average spike size.

As shown in figure 4C some of the amperometric spikes are preceded by a small 'foot' (pointed by the arrow). This small amount of release before the main phase of release represents the leakage of serotonin through the nascent fusion pore. A 'foot' followed by a 'spike' is seen as a fusion pore that expands slowly and then becomes fully opened. Similar observations have been made in wide variety of secretory cells including chromaffin cells (Chow et al., 1992), glomus cells from the carotid body (Ureña et al., 1994), and Retzius neurons from the leech (Bruns and Jahn, 1995). Therefore, the amperometric technique provides a alternative method of studying the properties of single fusion pores in preparations where the patch-clamp technique does not have enough resolution to resolve the activity of single fusion events.

The amperometry technique has been recently used to demonstrate that the release of secretory products can occur during transient fusion events (Alvarez de Toledo et al., 1993), which leads one to question the current models of membrane recycling.

CONCLUSION

In summary, much progress in understanding the molecular mechanisms that govern the formation and subsequent closure of the exocytotic fusion pore has been made over the past few years. The identification the molecular machinery that participate during the process of exocytotic fusion will require the use of the electrophysiological techniques described here in combination with molecular biological and biochemical techniques.

ACKNOWLEDGMENTS

This work was supported by a Grant No. 1950437 from the Fondo Nacional de Investigaciones (Fondecyt), by a Grant from the Fundación Andes (C-12999/8) and by Grant No. 95-201 RG/BIO/LA from TWAS. The institutional support of a group of Chilean private companies (Empresas CMPC, CGE, COPEC, Minera Escondida, NOVAGAS, Business Design Ass.,

ENERSIS, XEROX Chile) is also recognized. I am also grateful to the support from the "Cátedra Presidencial para la Ciencia" (to Ramón Latorre).

REFERENCES

Almers, W., 1990, Exocytosis. *Annu. Rev. Physiol.* 52: 607-624.

Alvarez de Toledo, G. and Fernandez, J.M., 1988, The events leading to secretory granule fusion, *in*: "Cell Physiology of Blood", R.B. Gumm and J.C. Parker, eds., the Rockefeller University Press. Soc. Gen. Physiol. series 43: 333-344.

Alvarez de Toledo, G., Fernandez-Chacon, R., and Fernandez, J.M., 1993, Release of secretory products during transient vesicle fusion. *Nature*, 363: 554-557.

Breckenridge, L.J., and Almers, W., 1987, Currents through the fusion pore that forms during exocytosis of a secretory vesicle. *Nature* 328: 814-817.

Bruns, D., and Jahn, R., 1995, Real-time measurements of transmitter release from single synaptic vesicles. *Nature*, 377: 62-5.

Calakos, N., and Scheller, R., 1996) Synaptic vesicle biogenesis, docking and fusion: a molecular description. *Physiol. Rev.* 76: 1-29.

Chandler, D. E., and Heuser, J. E., 1980, Arrest of membrane fusion events in mast cells by quick-freezing. *J. Cell Biol.* 86: 666-674.

Chow, R.H., von Rüden, L., and Neher, E. , 1992, Delay in vesicle fusion revealed by electrochemical monitoring of single secretory events in adrenal chromaffin cells. *Nature* 356: 60-63.

Chow, R.H., von Rüden, L., 1995, Electrochemical detection of secretion from single cells. *in*: "Single-channel recording", second edition. B. Sakmann and E. Neher. Plenum Press, New York. pp. 245-275.

Fernandez, J.M., Neher, E. and Gomperts, B.D., 1984, Capacitance measurements reveal stepwise fusion events in degranulating mast cells. *Nature* 312: 453-455.

Fidler, N. and Fernandez, J.M., 1989, Phase tracking: an improved phase detection technique for cell membrane capacitance measurements. *Biophys. J.* 56: 1153-1162.

Gillis, K. D., 1995, Techniques for membrane capacitance measurements, *in*: "Single-channel recording", second edition. B. Sakmann and E. Neher. Plenum Press, New York. pp. 155-198.

Hamill, O.P., Marty, A., Neher, E., Sakmann, B. and Sigworth, F.J. , 1981, Improved patch-clamp techniques for high-resolution current recording from cells and cell-free membrane patches. *Pflügers Arch.* 391: 85-100.

Joshi, C. and Fernandez, J., 1988, Capacitance measurements. An analysis of the phase detector technique used to study exocytosis and endocytosis. *Biophys. J.* 53:885-892.

Kawagoe, K.T., Zimmerman, J.B., and Wightman, R.M. , 1993, Principles of voltammetry and microelectrode surface states. *J. Neurosci. Methods* 48: 225-240.

Leszczyszyn, D.J., Jankowski, J.A., Viveros, H., Diliberto, E., Near, J.A., and Wightman, R.M., 1990, Nicotinic receptor-mediated catecholamine secretion from individual chromaffin cells. *J. Biol. Chem.* 265: 14736-37.

Lindau, M., and Neher, E., 1988, Patch-clamp techniques for time resolved capacitance measurements in single cells. *Plügers Arch.* 411: 137-146.

Lindau, M., and Almers, W., 1995, Structure and function of fusion pores in exocytosis and ectoplasmic membrane fusion. *Curr. Opin. in Cell Biol.* 7: 509-517.

Lollike K., Borregaard, N., and Lindau, M., 1995, The exocytotic fusion pore of small granules has a conductance similar to an ion channel. *J. Cell. Biol.* 99: 99-104.

Monck, J.R., Alvarez de Toledo, G. and Fernandez, J.M. , 1990, Tension in secretory granule membranes causes extensive membrane transfer through the exocytotic fusion pore. *Proc. Natl. Acad. Sci.* 87: 7804-7808.

Monck, J., and Fernandez, J., 1992, The exocytotic fusion pore. *J.Cell. Biol.* 119: 1395-1404.

Monck, J.R., Oberhauser, A.F., Tatham, P.E.R., and Fernandez, J.M., 1994, Structure and regulation of the Exocytotic Fusion Pore, *in*: "Membrane Channels - Molecular and Cellular Physiology", ed., C. Perrachia, pp 557-580.

Monck, J.R., Oberhauser, A.F., and Fernandez, J.M. , 1995, Neurotransmitter Release through the Exocytotic Fusion Pore. *Molec. Memb. Biol.* 12: 151-156.

Nanavati, C., Markin, V., Oberhauser, A.F. and Fernandez, J.M., 1992, The exocytotic fusion pore modeled as a lipidic structure. *Biophys. J.*, 63: 1118-32.

Nanavati, C., and Fernandez, J. M., 1993, The secretory granule matrix: a fast acting smart polymer. *Science* 259: 963-965.

Neher, E., and Marty, A., 1982, Discrete changes of cell membrane capacitance observed under conditions of enhanced secretion in bovine adrenal chromaffin cells. *Proc. Natl. Acad. Sci. USA* 79: 6712-6716.

Oberhauser, A., Monck, J.R., and Fernandez, J.M., 1992, The events leading to the opening and closing of the exocytotic fusion pore have markedly different temperature dependencies. *Biophys. J.*, 61: 800-809.

Oberhauser, A., Monck, J.R., Balch, W.E. and Fernandez, J.M., 1992, Exocytotic fusion is activated by Rab3 peptides. *Nature*, 360: 270-273

Oberhauser, A. and Fernandez, J.M., 1993, Patch-clamp studies of intact secretory granules. *Biophys. J.*, 65:1844-1852.

Oberhauser, A.F., Robinson, I., and Fernandez, J.M., 1996, Simultaneous capacitance and amperometric measurements: a comparison, *Biophys. J.* 71: 1131-1139.

Pollard, H. B., Rojas, E., Pastor, R. W., Rojas, E. M., Guy, R. H., Burns, A. L., 1990, Synexin: molecular mechanism of calcium-dependent membrane fusion and voltage-dependent calcium-channel activity. Evidence in support of the 'hydrophobic bridge hypothesis' for exocytotic membrane fusion. *Ann. NY Acad. Sci.* 635: 328-351.

Söllner,T., Whiteheart, S.W., Brunner, M., Erdjument-Bromage, H., Geromanos, S., Tempst, P., and Rothman, J.E., 1993, SNAP receptors implicated in vesicle targeting and fusion. *Nature* 362: 318-324.

Spruce, A., Breckenridge, L., Lee, A., and Almers, W., 1990, Properties of the fusion pore that forms during exocytosis of mast cell secretory vesicle. *Neuron* 4:643-654.

Südhof, T.C., 1995, The synaptic vesicle cycle: a cascade of protein-protein interactions. *Nature* 375: 645-653.

Ureña, J., Fernandez-Chacon, R., Benot, A. R., Alvarez de Toledo, G., and Lopez-Barneo, J., 1994, Hypoxia induces voltage-dependent Ca^{2+} entry and quantal dopamine secretion in carotid body glomus cells. *Proc. Natl. Acad. Sci. USA* 91: 10208-11.

von Rüden, L., and Neher, E., 1993, A Ca^{2+}-dependent early step in the release of catecholamines from adrenal chromaffin cells. *Science* 262: 1061-1065.

Zimmerberg, J., Curran, M., and Cohen, F. S., 1991, A lipid/protein complex hypothesis for exocytotic fusion pore formation. *Ann. N.Y. Acad. Sci.* 635: 307-317.

INDEX